Lecture Notes in Mathematics

For information about Vols. 1–669, please contact your bookseller or Springer-Verlag.

Vol. 670: Fonctions de Plusieurs Variables Complexes III, Proceedings, 1977. Edité par F. Norguet. XII, 394 pages. 1978.

Vol. 671: R. T. Smythe and J. C. Wierman, First-Passage Perculation on the Square Lattice. VIII, 196 pages. 1978.

Vol. 672: R. L. Taylor, Stochastic Convergence of Weighted Sums of Random Elements in Linear Spaces. VII, 216 pages. 1978.

Vol. 673: Algebraic Topology, Proceedings 1977. Edited by P. Hoffman, R. Piccinini and D. Sjerve. VI, 278 pages. 1978.

Vol. 674: Z. Fiedorowicz and S. Priddy, Homology of Classical Groups Over Finite Fields and Their Associated Infinite Loop Spaces. VI, 434 pages. 1978.

Vol. 675: J. Galambos and S. Kotz, Characterizations of Probability Distributions. VIII, 169 pages. 1978.

Vol. 676: Differential Geometrical Methods in Mathematical Physics II, Proceedings, 1977. Edited by K. Bleuler, H. R. Petry and A. Reetz. VI, 626 pages. 1978.

Vol. 677: Séminaire Bourbaki, vol. 1976/77, Exposés 489–506. IV, 264 pages. 1978.

Vol. 678: D. Dacunha-Castelle, H. Heyer et B. Roynette. Ecole d'Eté de Probabilités de Saint-Flour. VII-1977. Edité par P. L. Hennequin. IX, 379 pages. 1978.

Vol. 679: Numerical Treatment of Differential Equations in Applications, Proceedings, 1977. Edited by R. Ansorge and W. Törnig. IX, 163 pages. 1978.

Vol. 680: Mathematical Control Theory, Proceedings, 1977. Edited by W. A. Coppel. IX, 257 pages. 1978.

Vol. 681: Séminaire de Théorie du Potentiel Paris, No. 3, Directeurs: M. Brelot, G. Choquet et J. Deny. Rédacteurs: F. Hirsch et G. Mokobodzki. VII, 294 pages. 1978.

Vol. 682: G. D. James, The Representation Theory of the Symmetric Groups. V, 156 pages. 1978.

Vol. 683: Variétés Analytiques Compactes, Proceedings, 1977. Edité par Y. Hervier et A. Hirschowitz. V, 248 pages. 1978.

Vol. 684: E. E. Rosinger, Distributions and Nonlinear Partial Differential Equations. XI, 146 pages. 1978.

Vol. 685: Knot Theory, Proceedings, 1977. Edited by J. C. Hausmann. VII, 311 pages. 1978.

Vol. 686: Combinatorial Mathematics, Proceedings, 1977. Edited by D. A. Holton and J. Seberry. IX, 353 pages. 1978.

Vol. 687: Algebraic Geometry, Proceedings, 1977. Edited by L. D. Olson. V, 244 pages. 1978.

Vol. 688: J. Dydak and J. Segal, Shape Theory. VI, 150 pages. 1978.

Vol. 689: Cabal Seminar 76–77, Proceedings, 1976–77. Edited by A.S. Kechris and Y. N. Moschovakis. V, 282 pages. 1978.

Vol. 690: W. J. J. Rey, Robust Statistical Methods. VI, 128 pages. 1978.

Vol. 691: G. Viennot, Algèbres de Lie Libres et Monoïdes Libres. III, 124 pages. 1978.

Vol. 692: T. Husain and S. M. Khaleelulla, Barrelledness in Topological and Ordered Vector Spaces. IX, 258 pages. 1978.

Vol. 693: Hilbert Space Operators, Proceedings, 1977. Edited by J. M. Bachar Jr. and D. W. Hadwin. VIII, 184 pages. 1978.

Vol. 694: Séminaire Pierre Lelong – Henri Skoda (Analyse) Année 1976/77. VII, 334 pages. 1978.

Vol. 695: Measure Theory Applications to Stochastic Analysis, Proceedings, 1977. Edited by G. Kallianpur and D. Kölzow. XII, 261 pages. 1978.

Vol. 696: P. J. Feinsilver, Special Functions, Probability Semigroups, and Hamiltonian Flows. VI, 112 pages. 1978.

Vol. 697: Topics in Algebra, Proceedings, 1978. Edited by M. F. Newman. XI, 229 pages. 1978.

Vol. 698: E. Grosswald, Bessel Polynomials. XIV, 182 pages. 1978.

Vol. 699: R. E. Greene and H.-H. Wu, Function Theory on Manifolds Which Possess a Pole. III, 215 pages. 1979.

Vol. 700: Module Theory, Proceedings, 1977. Edited by C. Faith and S. Wiegand. X, 239 pages. 1979.

Vol. 701: Functional Analysis Methods in Numerical Analysis, Proceedings, 1977. Edited by M. Zuhair Nashed. VII, 333 pages. 1979.

Vol. 702: Yuri N. Bibikov, Local Theory of Nonlinear Analytic Ordinary Differential Equations. IX, 147 pages. 1979.

Vol. 703: Equadiff IV, Proceedings, 1977. Edited by J. Fábera. XIX, 441 pages. 1979.

Vol. 704: Computing Methods in Applied Sciences and Engineering, 1977, I. Proceedings, 1977. Edited by R. Glowinski and J. L. Lions. VI, 391 pages. 1979.

Vol. 705: O. Forster und K. Knorr, Konstruktion verseller Familien kompakter komplexer Räume. VII, 141 Seiten. 1979.

Vol. 706: Probability Measures on Groups, Proceedings, 1978. Edited by H. Heyer. XIII, 348 pages. 1979.

Vol. 707: R. Zielke, Discontinuous Čebyšev Systems. VI, 111 pages. 1979.

Vol. 708: J. P. Jouanolou, Equations de Pfaff algébriques. V, 255 pages. 1979.

Vol. 709: Probability in Banach Spaces II. Proceedings, 1978. Edited by A. Beck. V, 205 pages. 1979.

Vol. 710: Séminaire Bourbaki vol. 1977/78, Exposés 507–524. IV, 328 pages. 1979.

Vol. 711: Asymptotic Analysis. Edited by F. Verhulst. V, 240 pages. 1979.

Vol. 712: Equations Différentielles et Systèmes de Pfaff dans le Champ Complexe. Edité par R. Gérard et J.-P. Ramis. V, 364 pages. 1979.

Vol. 713: Séminaire de Théorie du Potentiel, Paris No. 4. Edité par F. Hirsch et G. Mokobodzki. VII, 281 pages. 1979.

Vol. 714: J. Jacod, Calcul Stochastique et Problèmes de Martingales. X, 539 pages. 1979.

Vol. 715: Inder Bir S. Passi, Group Rings and Their Augmentation Ideals. VI, 137 pages. 1979.

Vol. 716: M. A. Scheunert, The Theory of Lie Superalgebras. X, 271 pages. 1979.

Vol. 717: Grosser, Bidualräume und Vervollständigungen von Banachmoduln. III, 209 pages. 1979.

Vol. 718: J. Ferrante and C. W. Rackoff, The Computational Complexity of Logical Theories. X, 243 pages. 1979.

Vol. 719: Categorial Topology, Proceedings, 1978. Edited by H. Herrlich and G. Preuß. XII, 420 pages. 1979.

Vol. 720: E. Dubinsky, The Structure of Nuclear Fréchet Spaces. V, 187 pages. 1979.

Vol. 721: Séminaire de Probabilités XIII. Proceedings, Strasbourg, 1977/78. Edité par C. Dellacherie, P. A. Meyer et M. Weil. VII, 647 pages. 1979.

Vol. 722: Topology of Low-Dimensional Manifolds. Proceedings, 1977. Edited by R. Fenn. VI, 154 pages. 1979.

Vol. 723: W. Brandal, Commutative Rings whose Finitely Generated Modules Decompose. II, 116 pages. 1979.

Vol. 724: D. Griffeath, Additive and Cancellative Interacting Particle Systems. V, 108 pages. 1979.

Vol. 725: Algèbres d'Opérateurs. Proceedings, 1978. Edité par P. de la Harpe. VII, 309 pages. 1979.

Vol. 726: Y.-C. Wong, Schwartz Spaces, Nuclear Spaces and Tensor Products. VI, 418 pages. 1979.

Vol. 727: Y. Saito, Spectral Representations for Schrödinger Operators With Long-Range Potentials. V, 149 pages. 1979.

continuation on page 409

Lecture Notes in
Mathematics

Edited by A. Dold and B. Eckmann

887

Freddy M. J. van Oystaeyen
Alain H. M. J. Verschoren

Non-commutative Algebraic Geometry

An Introduction

Springer-Verlag
Berlin Heidelberg New York 1981

Authors
Freddy M. J. van Oystaeyen
Alain H. M. J. Verschoren
University of Antwerp. U.I.A., Department of Mathematics
Universiteitsplein 1, 2610 Wilrijk, Belgium

AMS Subject Classifications (1980): 14 A 99, 14 H 99, 16 A 38, 16 A 66

ISBN 3-540-11153-0 Springer-Verlag Berlin Heidelberg New York
ISBN 0-387-11153-0 Springer-Verlag New York Heidelberg Berlin

2141/3140-543210

ACKNOWLEDGEMENT

Both authors thank the National Foundation for Scientific Research, N.F.W.O., for the continuous support they have received; A. Verschoren is supported by NFWO-grant A2/5. We are grateful towards the University of Antwerp U.I.A. for support, hospitality and facilities extended to us. In particular the people at the department of mathematics should be thanked. All of them had to put up with a couple of nervous algebraists, but let us particularly mention J. Van Geel, whose office served as coffee-room, redaction room, typist's room and last hide-out on many occasions.

Talking to M. Artin, when he occasionally visited U.I.A. helped shape the contents of this book. We thank L. Small for some accurate suggestions which affected not only A. Verschoren's thesis but also some chapters in this book.

We also thank J. Murre for the interest he took in our work, for the many conversations as well as his stimulating talks on Algebraic Geometry that kept us interested in this field, even if it was commutative...

Finally we gladly express our gratitude towards Ludwig Callaerts, who actually succeeded in typing ninety-nine percent of the manuscript, before breaking down. He really did an excellent job.

F. VAN OYSTAEYEN　　　　A. VERSCHOREN

CONTENTS

Introduction.

It took us some time before we decided that this book is about "Non-commutative Algebraic Geometry". We realize that any non-commutative generalization of a commutative theory should be rooted in a thorough understanding of the heart of the commutative matter and, whereas the understanding of Algebraic Geometry is itself a non-trivial task, it becomes harder still if one sticks to the point of view that the obtained generalization should shed some new light upon the commutative theory.
Whether this book does or does not satisfy these requirements is subjected to the reader 's attitude towards Geometry. Subjectivity may be very unmathematical but it is perhaps allowable in the foregoing statement because of the extent of the topic as well as the fuzziness of its limits. Our point of view is that Algebraic Geometry describes the structure of geometric objects immersed in affine or projective space, utilizing ring- and sheaf-theoretic methods in exploiting the basic duality between these geometric objects and their morphisms and rings and ring homomorphisms. Therefore our first aim will be to obtain a non-commutative analogue of the Spec functor and the corresponding sheaf theory. However, functoriality of Spec with respect to ring homomorphisms is incompatible with the property that each ring should be recovered as the ring of global sections of the structure sheaf on Spec (R), unless one restricts to commutative rings. On the other hand, utilizing suitable localization in defining structure sheaves, it is not difficult to obtain a sheaf on Spec (R) such that R is the ring of global sections and such that Spec is functorial with respect to extensions of rings, i.e. ring homomorphisms $f:R\to S$, such that $f(R)$ and $Z_R(S) = \{s \in S;\ sr = rs \text{ for all } r \in f(R)\}$ generate the ring S. Then the fact that the stalks of this sheaf need not be local rings, presents another minor problem which can be solved by considering varietal spaces, which are ringed spaces endowed with three (!) interrelated structure sheaves. Now if, as in the commutative case, one wants non-commutative algebraic varieties to be determined by the set of closed points, if one hopes to have an analogue for

Hilbert's Nullstellensatz, if one is hankering after satisfactory theory of products of varieties and subvarieties, etc..., then, step by step, one is led to consider rings satisfying a polynomial identity (P.I. rings) affine over an algebraically closed field.

Now, the theory of rings satisfying polynomial identities has been flourishing the past decade and the prime ideal structure of these rings has been studied extensively, a.o. by C. Procesi, [136], [130], M. Artin [18] and M. Artin, W. Schelter [20], [21]. The only sheaves used in the so-called "geometry of P.I. rings" are sheaves of Azumaya algebras on certain open sets of the spectrum. Roughly speaking an Azumaya algebra can only appear as a stalk, when this stalk is strongly related to the center, hence precisely the non-Azumaya stalks, reflect the non-commutativity of the considered variety. Although a sheaf of Azumaya algebras on a particular dense set in the spectrum may already contain a lot of information, we feel that incorporating non-Azumaya stalks in our structure sheaves on varieties is an essential extra ingredient of the theory expounded here. In fact, an algebraic variety may usually be viewed as a covering of the underlying central variety, and the splitting of a point of the central variety into several points of the variety is reflected in the non-Azumaya-ness of certain stalks, i.e. the defect of certain localization being non-central ! The first part of the book, Chapter I to V, presents the necessary non-commutative algebra as well as the sheaf theoretical technicalities which will become the foundation for the geometrical theory.

We have kept the book as selfcontained as possible, although we have not strived for full generality everywhere. Well-known facts either in commutative Algebraic Geometry or in Ring Theory have been included as propositions without proofs, the exhaustive list of references makes it possible to trace any result used in this book. Note that we have included some results on P.I. rings, usually of a distinguished geometrical flavor, which have not been included in recent books. In the second part of the book, Chapters VI to X, we develop the "language" of algebraic k-varieties over an algebraically closed field. In particular, starting from affine k-varieties we construct the cellular k-varieties, these are algebraic k-varieties such that each point has an affine neighbourhood and these neighbourhoods have affine intersections. As it turns out, cellular varieties are likely to be the most fitting analogues of commutative algebraic varieties. In order to obtain non-commutative analogues of affine and projective space we have to introduce quasivarieties. These are very much the same as varieties but here we do not assume any Noetherian hypothesis. This generalization is being forced upon us because the rings for affine spaces turn out to be non-Noetherian in general. Utilizing the theory of graded rings established in Chapter III, we introduce projective space over an affine k-algebra. This allows to provide a more general framework and perhaps a more solid foundation for certain similar constructions carried out in some special cases by M. Artin in [20]. At this point it should be mentioned that the projective geometry has not been presented in full depth here, for example, the chapter on coherent sheaves over an algebraic k-variety is still rather unfinished, e.g. the properties of coherent sheaves over Proj (R) have scarcely been hinted at.

The definition of closed subvarieties presents no problem and most of their desired properties may be derived without much difficulties. On the other hand, it is impossible to define the categorical product of algebraic k-varieties, and therefore we have introduced the notion of a geometric product. Anyhow, this geometrical product suits our aims well enough and it turns out to be particularly effective in the study of cellular varieties. One should note that, the fact that "non-commutative algebraic groups" are Azumaya varieties, i.e. homeomorphic to the underlying central algebraic group, follows from some homogeneity argument and it has not been forced by the special nature of the geometric product used in the definition.

Chapter XI plays a special role in our set up, as a matter of fact it was only after the Riemann-Roch theorem for curves was established that the title of the book presented no more moral problems to us. The first section, that could have been a separate chapter in his own right, deals with the study of k_o-rational points of a variety in case k_o is not necessarily algebraically closed. The second section uses birationality of varieties in reducing the study of curves "up to birationality" to the study of abstract curves which are given by a variety of maximal orders in a certain central simple algebra i.e. the function algebra of the variety. The latter function algebra represents, in the case of curves, an element of the Brauer group of a function field in one variable. Conversely elements of that Brauer group may be viewed as function algebras of certain non-commutative curves defined over the central curve (everything up to birationality) given by the function field in one variable. The version of the Riemann-Roch theorem we have included is based upon the theory of primes in central simple algebras, cf. J. Van Geel [170] , F. Van Oystaeyen [178] , which yields a suitable generalization of valuation theory of fields. Thus we define the genera of the elements of the Brauer group of an algebraic function field in one variable and express the classical relations between genus and dimensions of certain k-vector spaces of divisors. We obtain the canonical class by looking at the different of the central simple algebra and this is compatible with the results of E. Witt, [202] , but here we do obtain more information about the ring of constants.

In the final Chapter XII, we point out some other results, like M. Artin's version of the "Zariski main theorem", his use of proper and geometric morphisms and the relations between these concepts and integrality of extensions of P.I. rings. The topics in this chapter have not been worked out extensively, since we feel they have not yet reached a (semi-) final form, e.g. regularity of varieties seems to be linked to the theory of hereditary orders over regular rings but the latter is poorly developed as yet; we aim to come back to these problems later. The particular role of Chapter XI is once more highlighted in the problem of tracing back the relations between the set of genera of the elements of the Brauer group of a function field in one variable and the geometry of the commutative curve associated to it. With this we hope to have indicated

how the circle may be closed, now that some applications of the non-commutative theory in the study of commutative varieties seem to evolve.
One thing about notation, since the authors were having daily quarrels concerning the spelling of the term non-commutative (noncommutative) it will sometimes appear as noncommutative (non-commutative).

LEITFADEN

1.

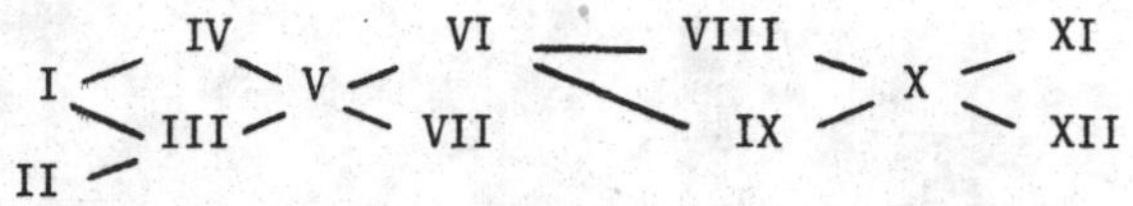

2. The Shortest Way to the Riemann-Roch Theorem.

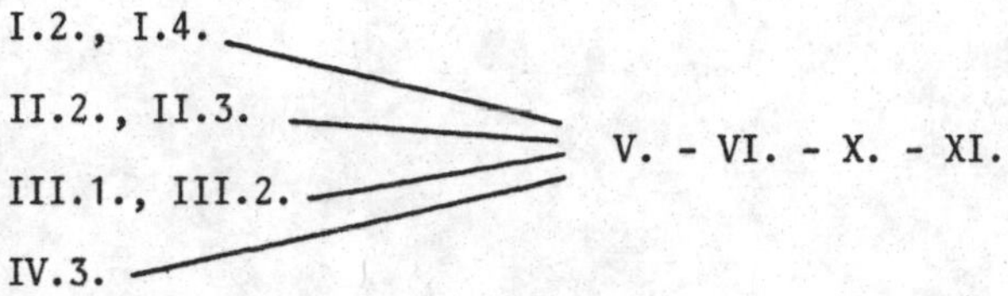

(You will have to take something for granted.)

I. GENERALITIES.

I.1. Simple Artinian Rings.

Throughout, R is an associative ring with unit. A left R-module M is said to be simple if M is non-zero and M has no submodules other than M and o. A direct sum of simple modules is called a semisimple module. It is well-known that $M \in R$-mod is semisimple if and only if every submodule of M is a direct summand in M. Clearly submodules and epimorphic images of semisimple modules are semisimple. A semisimple module is isotypic if it is a direct sum of isomorphic simple modules, the type of a simple module is given by its isomorphism class.
The socle of an arbitrary $M \in R$-mod is the sum of all simple submodules of M.

If $M \in R$-mod is semisimple then M is a right $End_R(M)$-module and M is again semisimple as such. Note also that the R-endomorphism ring of a semisimple $M \in R$-mod is the product of the $End_R(M_\alpha)$, where $M = \bigoplus_\alpha M_\alpha$ is the decomposition of M into isotypic components.

The ring R is semisimple if it is semisimple when considered as a left R-module.

I.1.1. Theorem. The following statements are equivalent :

a. R is semisimple.

b. Every $M \in R$-mod is semisimple.

c. Every $M \in R$-mod is projective.

d. Every $M \in R$-mod is injective.

Proof : Easy. For more detail cf. P. Cohn [39], or P. Ribenboim [145].

I.1.2. Schur's Lemma. If $M \in R$-mod is simple then $End_R(M)$ is a skew field.

This celebrated lemma applies to a celebrated theorem :

I.1.3. Wedderburn's Theorem. A semisimple ring is a direct sum of full matrix rings over skewfields and vice versa. The skew fields appearing are determined up to isomorphism by the structure of the ring and so is the size of the matrices.

A non-zero ring is said to be simple if it is semisimple and if it possesses no non-trivial ideals. By I.1.3. it is clear that a simple ring is Artinian and that every simple ring is a full matrix ring over a skew field and conversely Furthermore, the semisimple rings are just the finite direct products of simple rings (obviously they are also Artinian). In a way, the commutative version of all this may be summed up in "a non-commutative way", as follows : the center of a semisimple ring is a direct product of fields.

An $M \in R$-mod is faithful if the canonical homomorphism $R \to \mathrm{End}_R(M)$ is injective. A ring R is primitive if there exists a simple faithful $M \in R$-mod (R is in fact "left" primitive). It is well-known that the primitive rings are exactly the dense rings of linear transformations of vector spaces over skew fields. Wedderburn's theorem yields that for a left Artinian ring, being simple is equivalent to being primitive. Furthermore, a ring is primitive if and only if it contains a maximal left ideal which contains no non-zero ideal.

The Jacobson radical J(R) of a ring R is defined to be the set of elements satisfying one of the equivalent properties listed in the following theorem.

I.1.4. Theorem. The following conditions for $r \in R$ are equivalent :

a. For each simple $M \in R$-mod, $rM = o$.

b. Each maximal left ideal of R contains r.

c. For all $x \in R$, $1-rx$ is left invertible in R.

a',b',c'. The right analogues of a,b,c.

The Jacobson radical is an ideal of R. If $J(R) = o$ then we say that R is semiprimitive and this is equivalent to the following statement: for each $r \neq o$ in R there is a simple $M \in R$-mod not annihilated by r i.e. $rM \neq o$. A useful application of these notions is :

I.1.5. Nakayama's Lemma. If $M \in R$-mod is finitely generated then $J(R)M = M$ implies $M = o$.

It is easily verified that the Jacobson radical is indeed a radical, viz. R/J(R) is semiprimitive for any ring R. Every left nilideal of R is contained in J(R). On the other hand J(R) need not be nilpotent, however we have :

I.1.6. Proposition. If R is left Artinian then J(R) is nilpotent. A left Artinian ring is semiprimitive if and only if the zero ideal is the unique nilpotent ideal.

I.1.7. Example. The ring $\begin{pmatrix} \mathbb{Q} & \mathbb{R} \\ 0 & \mathbb{R} \end{pmatrix}$ is right but not left Artinian. The ring k [[X]] of formal power series over the field k is a domain with non-zero Jacobson radical.

The Jacobson radical may be used to characterize left Artinian rings amongst left Noetherian rings :

I.1.8. Proposition. Let R be a left Noetherian ring, then R is left Artinian exactly then when R/J(R) is semisimple in R-mod and J(R) is nilpotent.

Since the center of a simple ring is a field, k say, it may be considered as a k-algebra. If a simple ring A has finite dimension over its center k then we say that A is a k-central simple algebra (or a c.s.a.). A criterion for checking whether a simple ring A is a c.s.a. will be provided in Section II.2, namely, a necessary and sufficient condition for this is that A satisfies some polynomial identities. We recall some almost-classical properties and theorems about c.s.a., without going into the theory of the Brauer group. For full detail on the subject we refer to [24], [49], [81], ...

I.1.9. Notation. If R is a ring with center C then R^o denotes the opposite ring of R while the enveloping ring of R is $R^o \otimes_C R = R^e$.

I.1.10. Proposition. If A is a k-central simple algebra then $A^e \cong M_n(k)$ where $n = [A : k]$.

I.1.11. Theorem. (Azumaya-Nakayama). If A is k-central simple then for each k-algebra B there is a lattice-isomorphism between the ideals of B and the ideals of $A \otimes_k B$.

I.1.12. Corollaries. If A is k-central simple and B is a simple k-algebra then so is $A \otimes_k B$.

The class of k-central simple algebras is closed under taking tensor products.

If ℓ is a field extension of k then $A \otimes_k \ell$ is ℓ-central simple.

It is true that every derivation δ of a c.s.a. A is an inner derivation i.e. $\delta(a) = xa - ax$ for some $x \in A$.

For automorphisms fixing the center a similar statement is true; this is contained in :

I.1.13. Theorem (Skolem-Noether). Let R be a simple ring with center k, let S be any finite dimensional simple k-algebra. If ρ_1, ρ_2 are morphisms from S to R then for some unit u of R we have $\rho_2(s) = u\rho_1(s)u^{-1}$ for all $s \in S$.

I.1.14. Corollary. Automorphisms of finite-dimensional semisimple k-algebras which leave the center elementwise fixed are inner automorphisms.

Simple subalgebras of a c.s.a. are isomorphic if and only if they are conjugated.

The simple subalgebras of a c.s.a. are accurately described by the "double centralizer theorem". If B is a subring of A then $Z_A(B) = \{a \in A,\ ab = ba \text{ for all } b \in B\}$ is called the centralizer of B in A.

I.1.15. Theorem (R. Brauer). Let A be k-central simple and suppose that B is a simple subalgebra with center ℓ then $Z_A(B)$ is simple with center ℓ and $Z_A(Z_A(B)) = B$, $Z_A(\ell) = B \otimes_k Z_A(B)$. The k-dimensions of A,B and $Z_A(B)$ are related by :

$[A : k] = [B : k] . [Z_A(B) : k]$ and if $[B : k] = n$ is finite then : $A \otimes_k B^\circ \cong M_n(Z_A(B))$.

I.1.16. Corollary. Let ℓ be a field extension of k contained in a k-central simple algebra A then the following statements are equivalent :

a. $Z_A(\ell) = \ell$.

b. $[A : k] = [\ell : k]^2 = [A : \ell]^2$

c. The field ℓ is maximal amongst the commutative subrings of A.

A splitting field of a k-central simple algebra A is an extension field of k, say ℓ, such that $A \otimes_k \ell \cong M_n(\ell)$. A maximal commutative subfield of A is a splitting field. Any splitting field contains a finite dimensional splitting field. For the theory of the Brauer group one may consult [81] or [168]. The theory of generic abelian crossed products, cf. Van Oystaeyen [172] and the generalization by Amitsur, Saltman [10], may be brought to bear on Sections II.3. and III2, but it would take us too far to include it here.

To end this introductory part let us mention :

I.1.17. Theorem.(Cartan-Hua-Brauer). Let A be a k-central simple algebra and let B be a subalgebra. If B is globally invariant under inner automorphisms of A then either $B \subset k$ or $B = A$.

The previous survey should make it clear that simple algebras, in particular c.s.a., have been extensively studied and that their structure is reasonably well-known. We have omitted the implications of the theory for the theory of representation of finite groups, cf. [49], [39], although this may be considered to be a main motivation for "embeddings" of non-commutative rings into simple rings and into central simple algebras.

The first problem is being treated in A. Goldie's theorems while the central simple algebra-case is the topic of Section II.2. i.e. Posner's theorem for P.I. rings. A major problem in the non-commutative theory is that the formation of total rings of fractions of certain integral domains need not be possible. Let us consider the following integral domains for which rings of fractions may still be constructed.

An integral domain R is said to be a left Ore domain if the intersection of any two non-zero left ideals is non-zero. Right Ore domains are defined in a similar way. There do exist integral domains which are neither a left or a right Ore domain; indeed, the free algebra generated by two symbols over a field has that property. Even better, (or worse?) there exist right Ore domains which are not left Ore domains e.g. the following ring of twisted polynomials. Let ℓ

be any field, σ an isomorphism of ℓ onto a proper subfield of ℓ, and consider the ring, $\ell[x,\sigma]$, of "polynomials" in the indeterminate x with multiplication defined by the rule : $rx^n = x^n\sigma^n(r)$ for $r \in \ell$. It is not too hard to verify that $\ell[x,\sigma]$ is a right Ore domain but not a left Ore domain.

<u>I.1.18. Theorem</u>. If R is an integral domain then the following statements are equivalent :

a. R is a left Ore domain.

b. There is a skew field containing R as a subring, S say, such that $S = \{b^{-1}a;\ a,b \in R,\ b \neq o\}$.

c. R has finite Goldie dimension in R-mod, i.e. R cannot contain an infinite direct sum of non-zero submodules.

We now try to extend this procedure to rings which are not necessarily integral domains.

A <u>regular</u> element in a ring R is an element r such that $rx \neq o$ and $xr \neq o$ for all non-zero $x \in R$. A <u>classical left quotient ring</u> for R is a ring Q containing R as a subring such that regular elements of R are units in Q and $Q = \{b^{-1}a;\ a,b \in R,$ b regular$\}$.

Let R be any ring and let S be a multiplicatively closed subset of R, then R is said to <u>satisfy the left Ore condition with respect to S</u> if, for every $s \in S$, $r \in R$ there exist $s' \in S$, $r' \in R$ such that $s'r = r's$. In a rather straightforward way one verifies :

<u>I.1.19. Theorem</u>. The ring R has a left quotient ring at S i.e. there exists a ring R_S and a ring homomorphism $j : R \to R_S$ such that $j(s)$ is invertible in R_S for each $s \in S$, every element of R_S is of the form $j(s)^{-1}\ j(r)$ with $s \in S$ and $j(r) = o$ if and only if $sr = o$ for some $s \in S$, exactly when R satisfies the left Ore condition with respect to S and S is left reversible ($rs = o$ with $s \in S$ implies $s'r = o$ for some $s' \in S$).

In particular, if S is the set of regular elements then the left Ore condition with respect to S is just called the left Ore condition. In this case S is clearly reversible and one easily derives from I.1.19. that the left Ore condition implies that R has a classical left quotient ring. Now Goldie's theorems provide a criterion for checking whether such a classical left quotient ring is semisimple or simple.

The left annihilator of an element $m \in M$, $M \in R$-mod, is the left ideal $\{r \in R, rm = o\}$. A left Goldie-ring R is a ring which has finite Goldie-dimension and such that left annihilators satisfy the ascending chain condition. An essential submodule N of $M \in R$-mod is a left R-module such that $N \subset M$ intersects all non-zero submodules of M non-trivially. An essential left ideal is then just an essential left submodule of R. The singular submodule of any $M \in R$-mod, denoted by $t_s(M)$, is defined to be the set $\{x \in M, Jx = o$ for some essential left ideal J of R$\}$. We define $t_s^{\ell}(R)$, $t_s^r(R)$ correspondingly; ℓ and r refer to left or right corresponding to whether R is considered as a left or a right R-module. Clearly, both $t_s^{\ell}(R)$ and $t_s^r(R)$ are ideals of R.

I.1.20. Lemma. If R satisfies the ascending chain condition on left annihilators, then $t_s^{\ell}(R)$ is a nilpotent ideal of R.

I.1.21. Corollary. A semiprime ring R is a left Goldie ring if and only if $t_s^{\ell}(R) = o$ and R has finite Goldie dimension.

I.1.22. Lemma. Let R be a ring with finite Goldie dimension such that $t_s^{\ell}(R) = o$. An $r \in R$ is regular if and only if its left annihilator is zero, and also if and only if Rr is an essential left ideal.

I.1.23. Proposition. In a semiprime left Goldie ring R the essential left ideals are exactly the left ideals of R which contain a regular element.

I.1.24. Theorem. (Goldie's second Theorem). The following conditions are equivalent :

a. R is a semiprime left Goldie ring.

b. R allows a classical left quotient ring which is semisimple.

I.1.25. Corollary. (Goldie's First Theorem). R allows a simple left quotient ring if and only if it is a prime left Goldie ring.

I.1.26. Proposition. If R is a semiprime left and right Goldie ring then R allows a semisimple ring Q for a classical left and right quotient ring. If an arbitrary ring R has a classical left quotient ring and also a classical right quotient ring, then these coincide.

I.1.27. Example. Let A be a k-central simple algebra and let C be a subring of k such that k is the field of fractions of C; let R be a C-algebra contained in A and containing a k-basis of A. Then R is a prime left and right Goldie ring (see also Section I.2.).

I.1.28. Comment. 1. Necessary and sufficient conditions for R to have a c.s.a. as left (and right) quotient ring are in II.2.

2. More general techniques of localization will be necessary in the sheaf theory of Chapter III, for these we refer to Sections I.3, I.4 and IV.

I.1.19. References. For the Jacobson radical and parts of the theory developed from and about it we refer to N. Jacobson's book [87]. Much of the basic theory of c.s.a. is of course in A. Albert's book [1], and also in B. Van der Waerden's [168]. Very nice compilations of the material we surveyed are I. Herstein [81] and P. Cohn [39]. A recent treatment of Goldie's theorems is given by K. Goodearl in [72]; of course one also may consult Goldie's papers [66],[68]. An easy example of a right but not left Goldie ring is a ring of twisted polynomials with respect to an injective (not bijective) endomorphism, more general results about these may be found in G. Cauchon's thesis, [33] .

I.2. Orders and Maximal Orders.

The arithmetical ideal theory developed for central simple algebras, cf. [49], is built upon the theory of commutative Dedekind rings and it is still very close to the latter. Once finite dimensionality is dropped from the hypotheses, Dedekind prime rings, Asano orders, hereditary orders and maximal orders, seem to lead rather independent lives from then on. We shall use some of the main properties of orders in studying the "geometry of P.I. rings" in particular in deriving a non-commutative version of Riemann-Roch's theorem. The elementary facts we need further on have been included in this section but we had to omit that part of the theory dealing with the more "number-theoretical" aspects, like primes, pseudo-valuations etc...., for which we refer to [171], [174].

Throughout the section, C will be a Noetherian integral domain with field of fractions k. Let V be any finite dimensional vectorspace over the field k.
A full C-lattice in V is a finitely generated C-module L in V such that $k.L = V$ where k.L stands for the k-space generated by L.
A C-order in a k-algebra A is a subring Λ which is a full C-lattice in A. A maximal C-order is a C-order which is not contained in any other C-order of A. Every element λ of a C-order Λ is integral over C and if C is integrally closed then the minimal and characteristic polynomial of λ over K have coefficients in C!
In what follows we assume that C is a Dedekind domain and A is a k-central simple algebra (although for many results quoted these assumptions may be weakened, cf. [144]). Then it is easy to see that C-orders are exactly subrings Λ of A which contain C, consist of C-integral elements and such that $k.\Lambda = A$. From this it is deduced that every C-order is contained in a maximal C-order of A (this uses the fact that A is a separable k-algebra). Recall that a ring is left hereditary if every left ideal is projective. Since C-orders are left and right Noetherian it will follow that a C-order Λ is left hereditary if and only if it is right hereditary. A C-order Λ is maximal if and only if for each prime ideal p of C the p-adic completion $\hat{\Lambda}_p$ is a maximal $\hat{C}_p$-order of $\hat{A}_p = \hat{k}_p \otimes_k A$. Local-global argumenta-

tion is particularly succesful in this theory and many properties may be proved by reduction to the local and complete local case.

I.2.1. Proposition. Let C be a Dedekind domain with field of fractions k and let A be a k-central simple algebra, then maximal C-orders of A are left and right hereditary.

From [144] we also recall the following theorem yielding a description of all maximal C-orders of A (stating in fact that maximal orders of A are Morita equivalent to maximal orders of the skewfield F for which $A \cong M_n(F)$).

I.2.2. Theorem. Let V be a left F-vectorspace of dimension n, let $A = \mathrm{Hom}_F(V,V)$ and let $Z(A) = Z(F) = k$. Let Δ be a maximal C-order of F and let L be any full left Δ-lattice in V. Then $\Lambda = \mathrm{Hom}_\Delta(L,L)$ is a maximal C-order of A. If Λ' is any maximal C-order of A then there exists a full left Δ-lattice L' in V such that $\Lambda' = \mathrm{Hom}_\Delta(L',L')$

I.2.3. Theorem. Let Λ be a maximal C-order of A. The non-zero prime ideals of Λ are maximal. For each prime ideal P of Λ, Λ/P is a finite dimensional simple algebra over $C/P \cap C$ (but the latter need not be equal to the center of Λ/P).

Next theorem states that maximal C-orders are Zariski central rings in the terminology of Section II.1.

I.2.4. Theorem. For each prime ideal P of Λ we have that $P = \mathrm{rad}\,\Lambda(P \cap C)$, where rad denotes the prime radical. Consequently there is a one-to-one correspondence between prime ideals P of R and prime ideals of C given by $P \leftrightarrow P \cap C$.

To a full C-lattice L in A there corresponds a full C-lattice $L^{-1} = \{a \in A,\ LaL \subset L\}$; note that L^{-1} is a right Λ-module if L is a left Λ-module.

I.2.5. Proposition. Let Λ be a maximal C-order of A, and let L be a full C-lattice and a left Λ-module, then :

a. $LL^{-1} = \Lambda$

b. $L^{-1}.L = O_r(L) = \{a \in \Lambda, Lx \subset L\}$

c. $(L^{-1})^{-1} = L$

d. $O_\ell(L^{-1}) = O_r(L)$; $O_r(L^{-1}) = O_\ell(L)$.

Recall that I is a normal Λ-ideal if I is a Λ-ideal such that $O_\ell(I)$ and $O_r(I)$ are maximal C-orders. A normal ideal I is two-sided if $O_\ell(I) = O_r(I)$.

I.2.6. Theorem. Let Λ be a maximal C-order in the k-central simple algebra A. The set of two-sided Λ-ideals of A is the free abelian group generated by the prime ideals of Λ. The group law is the usual multiplication of lattices, the identity element is Λ and the inverses are described by I.2.5.

Let us mention some "local" properties that will be useful, hence here we will assume that C is a discrete valuation ring of k. Our main objective is to state that maximal C-orders in this case are principal ideal rings i.e. one-sided ideals are one-sided principal.

In doing this we encounter the following :

I.2.7. Theorem. Let C be a discrete valuation ring of the field k and let A be a k-central simple algebra. Let Λ be a maximal C-order of A then :

a. If L_1 and L_2 are full C-lattices and left Λ-modules then $L_1 \cong L_2$ in Λ-mod if and only if they both have the same C-rank.

b. One-sided ideals of Λ are principal.

c. Every maximal C-order of A is a conjugate $u\Lambda u^{-1}$ of Λ for some unit u of A.

d. Let us consider completion with respect to the valuation v of C and denote this operation by $\hat{}$, then $\hat{A} = \hat{k} \otimes_k A \cong M_n(F)$ where F is a skewfield with center $\hat{k}$. Let Ω be the unique maximal $\hat{C}$-order in F (this uniqueness follows from the complete local theory cf. [144] since $\hat{C}$ is a discrete valuation ring of k which is complete with respect to the extension of v to $\hat{k}$), then :

$$\Lambda/J(\Lambda) \cong M_n(\Omega/J(\Omega)) ,$$

and $\Omega/J(\Omega)$ is a skewfield.

I.2.8. Theorem. Let Λ be a maximal C-order of A, then :

a. There is a one-to-one correspondence between maximal left ideals of Λ and maximal left ideals of $\Lambda/\mathrm{rad}\,\Lambda$ given by just taking the images under the canonical epimorphism $\Lambda \to \Lambda/\mathrm{rad}\,\Lambda$.

b. Maximal left ideals of Λ are conjugate under inner automorphisms by units of Λ.

c. If $\Lambda\lambda$ is a maximal left ideal of Λ then $\lambda\Lambda$ is a maximal left ideal of Λ.

If $[A : k]$ is not finite, then the theory becomes more complicated. Let us say that a ring Q is a quotient ring if every regular element of Q is invertible. A left order R in Q is a subring of Q such that every $q \in Q$ is of the form $b^{-1}a$ for some $a,b \in R$. If R is also right order in Q or if Q is left Artinian then every regular element of R is a regular element of Q. A left order S of Q is equivalent to the left order R if there exist regular elements a,b,c,d of Q such that $aRb \subset S$ and $cSd \subset R$. A left order R of Q which is not contained in an equivalent left order is called a maximal left order, the right and two-sided equivalents are defined similarly. Let R be a left order of Q. A left R-submodule L of Q is a (fractional) left R-ideal if L contains a regular element of Q and $Iq \subset R$ for some regular element $q \in Q$. If a left R-ideal L is contained in R then it is said to be an integral left R-ideal, the right and two-sided equivalents are defined in a left order in a simple (Artinian!) ring if and only if R is a prime left Goldie ring and in this case an integral left R-ideal is nothing but an essential left ideal of R; moreover, the integral R-ideals are just the non-zero ideals of R.

Exactly as in the first part of this section we can define the left and right order of a given left R-ideal L to be $O_\ell(L) = \{q \in Q,\ qL \subset L\}$ and $O_r(L) = \{q \in Q,\ Lq \subset L\}$ respectively. Each of these is a left order of Q equivalent to R. An R-ideal I is said to be invertible if there exists an R-ideal I' such that $II' = I'I = R$; this I' will then be denoted by I^{-1}.

A left order R in a quotient ring Q is called an Asano left order if the R-ideals form a group under multiplication, cf. J.C. Robson [147].

I.2.9. Theorem. For a left order R in a quotient ring Q the following statements are equivalent :

a. R is an Asano left order.

b. R is a maximal left order and integral R-ideals are projective left R-modules.

c. Integral R-ideals are invertible.

d. The R-ideals form an abelian group under multiplication.

Furthermore every R-ideal of a left Asano order is projective and finitely generated both as a left and right R-module. Every integral R-ideal is a unique product of maximal integral R-ideals and therefore, prime ideals are maximal.

A left hereditary, left Noetherian prime ring which is a maximal left order in its classical left ring of quotients (the latter exists by Goldie's Theorem) is a left Dedekind prime ring.

The left Dedekind prime rings are precisely the maximal left orders R in simple rings whose integral left R-ideals are projective. Clearly any left Dedekind prime ring is a left Asano order. A prime ring is said to be left bounded if each essential left ideal contains an ideal, an arbitrary ring is called a fully left bounded ring if each prime epimorphic image is left bounded.

A result of T. Lenagan, cf. [104], states that a left bounded left Noetherian prime ring is left Asano order in its classical ring of quotients if and only if it is a bounded Dedekind prime ring.

I.2.10. Proposition. If L is an essential left ideal of a Dedekind prime ring R then R/L is left Artinian. If I is a non-zero ideal of R then R/I is an Artinian principal ideal ring. If $L_1 \subset L_2$ are left ideals of R such that L_1 is an essential left R-submodule of L_2 then L_2/L_1 is a cyclic R-module; consequently, every left ideal of R can be generated by two elements, one chosen almost at random.

Closely related to Dedekind prime rings are hereditary Noetherian prime rings (HNP rings). As a matter of fact an HNP ring R is a Dedekind prime ring if and only if R has no non-trivial idempotent ideals cf. [148]. In general, it is still true for an HNP ring R that R/L is left Artinian for every essential left ideal L of R, cf. D. Webber [201].

It follows that non-zero prime ideals of an HNP ring are maximal. An ideal I of R for which some power I^n is idempotent is called an eventually idempotent ideal. Let us summarize some results of D. Eisenbud, J.C. Robson, [54] and collect them in a "theorem" :

I.2.11. Theorem . Let R be an HNP ring, then :

a. Maximal ideals are either idempotent or invertible.

b. Every proper invertible ideal is a product of maximal invertible ideals.

c. Let $I = M_1 \cap \dots \cap M_t$ where M_i is a maximal ideal of R for each $i \in \{1,\dots,t\}$, let J be an invertible ideal of R such that $J \not\subset M_i$, then $JI = J \cap I = IJ$.

d. A proper ideal not contained in a proper invertible ideal is eventually idempotent and conversely.

e. Every ideal is a product of an invertible ideal and an eventually idèmpotent ideal.

I.2.12. Comments. a. The extra assumption that R be a P.I. ring will reduce the general theory to the finite dimensional case. The infinite case will therefore not be very interesting for the geometry of P.I. rings, but it is included here mainly with an eye to some "along the way" applications.

b. The nicest maximal orders of a k-central simple algebra are those orders which are Azumaya algebras, we return to this in Section II.3.

I.2.13. References. The theory of orders in central simple algebras is well documented in the litterature, let us just mention Deuring [49] and Reiner [144]. Although not within the scope of this book we want to point towards Godement's papers on the zeta-function of a c.s.a. (theory partially exposed by Deuring in [49]), [63,1], [63,2]. The arithmetic of c.s.a.'s is brought to bear on the Riemann-Roch theorem for non-commutative curves and maybe it is not too far fetched to hope that the zeta-function of some c.s.a. will also relate to some curves. For Asano orders; HNP rings; and prime Dedekind rings one consults [104]; [105]; [147];

[48], [53], [54]. In surveying some of these results we benefitted a lot from E. Nauwelaerts' thesis, cf. [24].

I.3. Localization in Grothendieck Categories.

The generality of this section seems unnecessary at first sight. However, later we shall not only be interested in localization of rings and modules but we shall also need to know about localization of graded rings and graded modules, localization of two-sided modules and bimodules in M. Artin's sense, [17], localization of (pre) sheaves of modules over a fixed (pre) sheaf of rings over certain topological spaces. So with the inclusion of this "abstract" section we aim to kill at least five flies in one blow. For some elementary details about categories we refer to [109], [55].

Let $\mathcal{C}$ be a Grothendieck category. An injective hull of an object C of $\mathcal{C}$ is a maximal essential of C in $\mathcal{C}$ and this is unique up to isomorphism in $\mathcal{C}$. Any object C of $\mathcal{C}$ has an injective hull. A torsion theory for $\mathcal{C}$ is a pair of classes $(\mathcal{T},\mathcal{F})$ such that :

1°. $\mathrm{Hom}_{\mathcal{C}}(T,F) = o$ for all $T \in \mathcal{T}$, $F \in \mathcal{F}$.

2°. If $\mathrm{Hom}_{\mathcal{C}}(C,F) = o$ for all $F \in \mathcal{F}$ then $C \in \mathcal{T}$.

3°. If $\mathrm{Hom}_{\mathcal{C}}(T,C) = o$ for all $T \in \mathcal{T}$ then $C \in \mathcal{F}$.

$\mathcal{T}$ is called the torsion class and its objects are the torsion objects, whereas $\mathcal{F}$ is the torsion-free class consisting of torsion-free objects. The torsion theory $(\mathcal{T},\mathcal{F})$ is hereditary if and only if $\mathcal{T}$ is closed under taking subobjects. For a class $\mathcal{T}$ in $\mathcal{C}$ the following conditions are equivalent : i) $\mathcal{T}$ is the torsion class for some hereditary torsion theory : ii) $\mathcal{T}$ is closed under quotient objects, coproducts, extensions and subobjects.

A left exact subfunctor κ of the identity in $\mathcal{C}$ such that $\kappa(C/\kappa(C)) = o$ for every object C of $\mathcal{C}$ is called a kernel functor in $\mathcal{C}$. There is a one-to-one correspondence between hereditary torsion theories and kernel functors. If κ is a kernel functor then the corresponding torsion theory is given by $\mathcal{T}_\kappa = \{C \in \mathcal{C},\ \kappa(C) = C\}$, $\mathcal{F}_\kappa = \{C \in \mathcal{C},\ \kappa(C) = o\}$.

Let κ be a kernel functor in $\mathcal{C}$. An object E of $\mathcal{C}$ is said to be κ-injective if every exact diagram :

$$\begin{array}{ccccccccc} o & \longrightarrow & C' & \xrightarrow{i} & C & \longrightarrow & C'' & \longrightarrow & o \\ & & \downarrow f & \swarrow g & & & & & \\ & & E & & & & & & \end{array}$$

with $C'' \in T_\kappa$ may be completed by a morphism $g : C \to E$ such that $gi = f$. If this g is unique as such then E is said to be faithfully κ-injective. To be faithfully κ-injective it is necessary and sufficient that E be κ-injective and $\kappa(E) = o$. The class of all faithfully κ-injective objects in C form a full subcategory of C; it is denoted by $C(\kappa)$ and we call it the quotient category of C with respect to κ. For $C \in F_\kappa$ a κ-injective hull of C is defined to be an essential extension $C \to E$ such that E is κ-injective and $E/C \in T_\kappa$. Every $C \in F_\kappa$ has an essentially unique κ-injective hull which we shall denote by $E_\kappa(C)$. For arbitrary C we shall write $Q_\kappa(C)$ when we mean $E_\kappa(C/\kappa(C))$.

The inclusion functor $i_\kappa : C(\kappa) \to C$ has a left adjoint $\underline{\underline{a}}_\kappa$ and we put $Q_\kappa = i_\kappa \underline{\underline{a}}_\kappa$. For $C \in C$, $Q_\kappa(C)$ together with the canonical morphism $j_\kappa : C \to Q_\kappa(C)$ is called the C-object of quotients of C at κ.

Note that Q_κ is of course left exact but not necessarily right exact. The functor $\underline{\underline{a}}_\kappa$ is a reflector. A subcategory of a (complete) Grothendieck category which has a left exact reflector is called a Giraud subcategory.

I.3.1. Theorem (P. Gabriel, N. Popescu). Let C be a Grothendieck category with generator G and put $R = \mathrm{Hom}_C(G,G)^\circ$. Let $T : C \to R\text{-mod}$ be the functor defined by $T(C) = \mathrm{Hom}_C(G,C)$, $C \in C$; then :

1. T is full and faithful.
2. T induces an equivalence between C and R-mod(κ) where κ is the "largest" kernel functor in R-mod for which all modules T(C) are faithfully κ-injective.

(For this and related results, cf. [128] .)

Note the corollary: every Grothendieck category is complete.

Let us study the relations between localization in a Grothendieck category P and localization in a (strict) Giraud subcategory S of P with inclusion functor i and reflector $\underline{\underline{a}}$. (If our notation evokes presheaves and sheaves then that is exactly what one may bear in mind as the main example).

<u>I.3.2. Proposition.</u> An object E of S is injective in S if and only if iE is injective in P.

The injective hull in S resp. P will be denoted by $E^S(-)$ and $E^P(-)$ resp.

<u>I.3.3. Proposition.</u> Let $S \in S$ then $E^P(iS) = iE^S(S)$.

Let T stand for the class of objects C in P for which $\underline{\underline{a}}(C) = o$ and let F consist of the subobjects in P of objects of S, the latter objects are said to be <u>separated</u>. The pair (T,F) determines a torsion theory in P such that the quotient category in P is equivalent to S. As a corollary to I.3.3. one can deduce that $E^P(-)$ takes separated objects of P into separated objects.

<u>I.3.4. Proposition.</u> If $P \in P$ is separated then :

$$i\underline{\underline{a}}\ E^P(P) = E^P(i\underline{\underline{a}}P) = iE^S(\underline{\underline{a}}P) = E^P(P)\ .$$

If κ, κ' are kernel functors in P then κ' is said to be <u>Q_κ-compatible</u> if $\kappa' Q_\kappa = Q_\kappa \kappa'$ (note that we sometimes omit to write the canonical inclusion $i_\kappa : P(\kappa) \to P$ when this causes no ambiguities). It is easy enough to verify that for a Q_κ-compatible kernel functor κ', Q_κ is inner in $T_{\kappa'}$, as well as in $F_{\kappa'}$. We extend the terminology to reflectors of Giraud subcategories and we say that a kernel functor κ in P is <u>S-compatible</u> if and only if $i\underline{\underline{a}}\kappa = \kappa i\underline{\underline{a}}$. If $i\underline{\underline{a}}\kappa\, i\underline{\underline{a}} = \kappa\, i\underline{\underline{a}}$ i.e. κ maps objects of S to objects of S then κ is said to be <u>inner</u> in S. If κ is inner in S, then the functor κi will be denoted by κ^S; in general κ^S is not a kernel functor in S however :

<u>I.3.5. Proposition.</u> If κ is an S-compatible kernel functor in P then κ^S is a kernel functor in S.

An easy example of compatibility is obtained by considering kernel functors κ and κ' in P such that $\kappa' \geq \kappa$ $(T_{\kappa'} \supset T_\kappa)$; then κ' is $P(\kappa)$-compatible.

The relation between localization at κ in P and at κ^S in S may now be described :

I.3.6. Theorem. Let $\mathcal{S}$ be a strict Giraud subcategory of $\mathcal{P}$, κ an $\mathcal{S}$-compatible kernel functor in $\mathcal{P}$. Then an object S of $\mathcal{S}$ is (faithfully) κ^S-injective if and only if iS is (faithfully) κ-injective in $\mathcal{P}$.

I.3.7. Theorem. Let κ be an $\mathcal{S}$-compatible kernel functor in $\mathcal{P}$ and let S be a κ^S-torsion free object of $\mathcal{S}$, then $i\,E_{\kappa^S}(S) \cong E_\kappa(iS)$.

I.3.8. Corollary. If κ is an $\mathcal{S}$-compatible kernel functor in $\mathcal{P}$ then $i\,Q_{\kappa^S}(S) = Q_\kappa(iS)$ for any $S \in \mathcal{S}$.

I.3.9. Proposition. Let $\mathcal{S}$ be a strict Giraud subcategory of $\mathcal{P}$ and let α be the kernel functor in $\mathcal{P}$ such that $\mathcal{P}(\alpha) = \mathcal{S}$. Let κ be a kernel functor in $\mathcal{P}$ such that $\kappa \geq \alpha$ then $Q_\kappa(P) \in i\mathcal{S}$ for every $P \in \mathcal{P}$.

We include a construction of some very useful $\mathcal{S}$-compatible kernel functors following [188].
If P is a non-zero object of $\mathcal{P}$, let Kerf(P) be the class of kernel functors κ in $\mathcal{P}$ such that $P \in F_\kappa$. If P' is an essential extension of P in $\mathcal{P}$ then Kerf(P) = Kerf (P'), therefore, in describing Kerf(P) we may assume that P coincides with its injective hull $E^{\mathcal{P}}(P)$ in $\mathcal{P}$. We define $\kappa_P(Q) = \cap\{\mathrm{Ker}\, g,\ g \in \mathrm{Hom}_{\mathcal{P}}(Q, E^{\mathcal{P}}(P))\}$.

I.3.10. Proposition. κ_P is a kernel functor in $\mathcal{P}$ and $\kappa_P \in$ Kerf(P). If κ is a kernel functor in $\mathcal{P}$ then $\kappa \in$ Kerf(P) if and only if $\kappa \leq \kappa_P$.

I.3.11. Theorem. For every object S of $\mathcal{S}$, κ_{iS} is an $\mathcal{S}$-compatible kernel functor in $\mathcal{P}$. Conversely if $P \in \mathcal{P}$ is such that κ_P is $\mathcal{S}$-compatible then there is an $S \in \mathcal{S}$ such that $\kappa_P = \kappa_{iS}$; moreover $S = \underline{\underline{a}}\, P$.

Let us mention some facts about idempotent filters in a generator G for the Grothendieck category C.

I.3.12. Proposition. If $\mathcal{L}(G,\kappa)$ stands for the class of all subobjects of I in G such that G/I is κ-torsion then a necessary and sufficient condition for $E \in C$ to be κ-injective is that any morphism $I \to G$ with $I \in \mathcal{L}(G,\kappa)$ extends to a morphism $G \to E$.

The class $\mathcal{L}(G,\kappa)$ has the following properties :

1°. If $I \in \mathcal{L}(G,\kappa)$ and $I \subset J \subset G$ then $J \in \mathcal{L}(G,\kappa)$.

2°. If $I \in \mathcal{L}(G,\kappa)$ and $J \in \mathcal{L}(G,\kappa)$ then $I \cap J \in \mathcal{L}(G,\kappa)$.

3°. If $I \in \mathcal{L}(G,\kappa)$ and $\phi \in \mathrm{Hom}_C(G,C)$, then $\phi^{-1}(I) \in \mathcal{L}(G,\kappa)$.

4°. If $I \subset G$ and there exists $H \in \mathcal{L}(G,\kappa)$ such that for every $\phi \in \mathrm{Hom}_C(G,H)$ we have that $\phi^{-1}(I) \in \mathcal{L}(G,\kappa)$ then $I \in \mathcal{L}(G,\kappa)$.

Any class of subobjects of G satisfying the above properties is called an <u>idempotent filter with respect to G</u>. It is not true in general that idempotent filters $\mathcal{L}$ correspond bijectively to torsion theories in C. However any torsion theory κ in C may be completely determined by its filter $\mathcal{L}(G,\kappa)$. Moreover we have:

<u>I.3.13. Proposition</u>. There is a bijective correspondence between:

a. Kernel functors in C

b. Hereditary torsion theories in C

<u>I.3.14. References</u>. The general theory of localization in the framework of abelian categories emerged from P. Gabriel's thesis [61]. Readers not too familiar with category theory may prefer O. Goldman's presentation in R-mod, [70]. The Gabriel-Popescu theorem allows to connect both presentations, even the use of idempotent filters in some generator; for this one may consult A. Verschoren [192]. Compatibility of kernel functors is introduced by F. Van Oystaeyen in [176] and extended to the generality of this chapter in [188]. Actually this Section I.3 is a summary of Chapter II of [188].

I.4. Localization of Non-commutative Rings.

We apply the results of I.3 to $C = R$-mod, choosing for the generator G the ring R itself. If κ is a kernel functor in R-mod then $Q_\kappa(R)$ is a ring containing $R/\kappa(R)$ as a subring and the ring structure is uniquely determined by its R-module structure. If $M \in R\text{-mod}(\kappa)$ then M is in a natural way a left $Q_\kappa(R)$-module. In Section I.3 we have pointed out that the localization functor Q_κ is always left exact. Concerning right exactness :

I.4.1. Proposition. If κ is a kernel functor in R-mod then the following statements are equivalent :

a. Q_κ is exact and commutes with direct sums.
b. For any $M \in R$-mod : $Q_\kappa(M) \cong Q_\kappa(R) \otimes_R M$.
c. Every $Q_\kappa(R)$-module is faithfully κ-injective.
d. Let $\mathcal{L}(\kappa)$ be the idempotent filter in R associated with κ. Then for any $L \in \mathcal{L}(\kappa)$, $Q_\kappa(R) j_\kappa(L) = Q_\kappa(R)$, where $j_\kappa : R \to Q_\kappa(R)$ is the canonical ring morphism.
e. All $Q_\kappa(R)$-modules are κ-torsion free.

A kernel functor κ satisfying one of the above equivalent properties is called a t-functor (the corresponding torsion theory is usually said to be perfect then). If κ is a t-functor then $j_\kappa : R \to Q_\kappa(R)$ is left flat and an epimorphism in the category of rings. Conversely, every left flat epimorphism in the category of rings may be obtained as the canonical localization morphism j_κ for some t-functor κ. The relations between $Q_\kappa(R)$-mod and R-mod are amongst the main topics of interest in localization theory.

I.4.2. Lemma. Let $f : M \to M'$ be an R-linear map between $Q_\kappa(R)$-modules such that the $Q_\kappa(R)$-module generated by f(M) is κ-torsion free when considered in R-mod, then f is $Q_\kappa(R)$-linear.

I.4.3. Proposition. Let $M \in Q_\kappa(R)$-mod be κ-torsion free then M is injective in R-mod if and only if M is injective in $Q_\kappa(R)$-mod.

I.4.4. Proposition. Suppose that κ is a t-functor in R-mod. Then there is a one-to-one correspondence between kernel functors in $Q_\kappa(R)$-mod and the set of kernel functors σ in R-mod such that $\sigma \geq \kappa$. If $\sigma \geq \kappa$ let σ_κ be the corresponding kernel functor in $Q_\kappa(R)$-mod; then the idempotent filter $\mathcal{L}(\sigma_\kappa)$ of σ_κ in $Q_\kappa(R)$ is given by : $\mathcal{L}(\sigma_\kappa) = \{Q_\kappa(L),\ L \in \mathcal{L}(\kappa)\}$.

I.4.5. Proposition. For a kernel functor κ in R-mod we have :

a. If L is a left ideal of R then $j_\kappa^{-1}(Q_\kappa(R)j_\kappa(L))$ is contained in the set of elements $r \in R$ such that $Ir \subset L$ for some $I \in \mathcal{L}(\kappa)$ and equality holds trivially when κ is a t-functor.

b. If κ is a t-functor then $Q_\kappa(R)(L \cap j_\kappa(R)) = L$ for each left ideal L of $Q_\kappa(R)$. Moreover, for each left ideal H of R we have $Q_\kappa(H) = Q_\kappa(R)j\ (H)$.

I.4.6. Corollary. If κ is a t-functor then, if R is left Noetherian or left Artinian, so is $Q_\kappa(R)$.

Now we focus on ideals I of R and investigate whether the extensions $Q_\kappa(R)\ j_\kappa(I)$ are ideals of $Q_\kappa(R)$.

I.4.7. Proposition. Let κ be a t-functor, I an ideal of R. The following statements are equivalent :

a. For every $J \in \mathcal{L}(\kappa)$, I/IJ is κ-torsion.

b. $Q_\kappa(R)\ j_\kappa(I)$ is an ideal of $Q_\kappa(R)$.

A kernel functor κ is said to be geometric if it is a t-functor and if ideals of R extend to ideals of $Q_\kappa(R)$. A kernel functor κ is said to be symmetric if $\mathcal{L}(\kappa)$ has a filterbasis consisting of ideals of R. If κ and κ' are kernel functors then $\kappa \wedge \kappa'$ and $\kappa \vee \kappa'$ are defined by their respective filters $\mathcal{L}(\kappa \wedge \kappa') = \mathcal{L}(\kappa) \cap \mathcal{L}(\kappa')$; $\mathcal{L}(\kappa \vee \kappa')$ = the idempotent filter generated by $\mathcal{L}(\kappa)$ and $\mathcal{L}(\kappa')$. One may verify that the set of kernel functors in R-mod equiped with $\wedge$ and $\vee$, turns into a complete Brouwerian lattice. If κ and κ' are symmetric then so is $\kappa \wedge \kappa'$. On the other hand $\kappa \vee \kappa'$ need not be symmetric in general; if R is left Noetherian though then the sup. of symmetric kernel functors is symmetric!

I.4.8. Proposition. Let κ be a geometric kernel functor, then :

a. Maximal ideals of $Q_\kappa(R)$ correspond bijectively to ideals of R which are maximal amongst ideals of R which are not in $\mathcal{L}(\kappa)$.

b. If κ is moreover symmetric then proper prime ideals of $Q_\kappa(R)$ correspond bijectively to proper prime ideals of R which are not in $\mathcal{L}(\kappa)$.

If κ is an arbitrary kernel functor and R is a prime ring then $Q_\kappa(R)$ is a priori not necessarily prime. However if $\kappa(R) = o$ then this does hold; in particular, if R is a left Noetherian prime ring then $Q_\kappa(R)$ is prime for arbitrary κ. Furthermore, if κ is symmetric and R is a prime ring then $Q_\kappa(R)$ is a prime ring too.

I.4.9. Proposition. Let κ be a symmetric kernel functor and let M be a maximal ideal of R which is not in $\mathcal{L}(\kappa)$.
If $M^e = Q_\kappa(R)j_\kappa(M)$ is an ideal of $Q_\kappa(R)$ then $R/M \cong Q_\kappa(R)/M^e$ as rings and as left R-modules.

In the second part of this section we give some examples of kernel functors that will be useful in the sequel of the book. If S is a multiplicatively closed subset of R containing 1 then we may associate to it a kernel functor κ_S given by its filter $\mathcal{L}(\kappa_S) = \{L$ left ideal of R such that $[L:r] \cap S \neq \phi$ for all $r \in R\}$. If R is commutative then κ_S is a t-functor, in the non-commutative case this is not true in general. However if S is a left reversible left Ore set then κ_S is a t-functor and $\mathcal{L}(\kappa_S)$ has $\{Rs, s \in S\}$ for a filter-basis. Moreover, if S is a left reversible left Ore set then the left and right localizations at S are naturally isomorphic.

If P is a prime ideal of R then $C(P) = \{s \in R, rs \in P$ implies $r \in P\}$ is a multiplicatively closed subset of R. If R/P is a left Goldie ring then $s \in C(P)$ if and only if s mod P is regular in R/P, and then C(P) is easier to handle. In general we may associate a kernel functor κ_P in R-mod to P by putting $\mathcal{L}(\kappa_P) = \{L$ left ideal of R, $[L:r] \cap C(P) \neq \phi$ for all $r \in R\}$. Let Q_P be the localization functor associated to κ_P. J. Lambek, G. Michler studied this torsion theory rigourously in [102], in case R is a left Noetherian ring. A prime ideal P of R is said to be <u>left localizable</u> if C(P) is a left Ore set in R.

<u>I.4.10. Theorem.</u> Let P be a left localizable prime ideal of the left Noetherian ring R, then :

a. κ_P is a geometric kernel functor.
b. $P^e = Q_P(R)j_P(P)$ is the unique maximal ideal of the left Noetherian ring $Q_P(R)$; P^e is the Jacobson radical of $Q_P(R)$ and $Q_P(R)/P^e$ is isomorphic to the classical left ring of quotients of R/P, i.e. a simple Artinian ring.
c. The proper prime ideals of $Q_P(R)$ correspond bijectively to the proper prime ideals of R which are not in $\mathcal{L}(\kappa_P)$.

The following is result is due to A.G. Heinicke, [80] ;

<u>I.4.11. Proposition.</u> Let R be a left Noetherian ring and let κ_P be a t-functor. The largest ideal of $Q_P(R)$ contained in $Q_P(R)j_P(P)$ is the Jacobson radical of $Q_P(R)$ and this is the unique maximal ideal of $Q_P(R)$. If $Q_P(R)j_P(P)$ is an ideal then C(P) is a left Ore set.

In [102] J. Lambek, G. Michler proved that for a prime ideal P of a left Noetherian ring R, κ_P coincides with the kernel functor determined by the injective envelope $E_R(R/P)$ of R/P in R-mod, in the way described before I.3.9. This is still true for a prime ideal P of an arbitrary ring R provided R/P is a left Goldie ring, cf. B. Mueller [116]. In case R is commutative and prime Noetherian it is true that $\cap\{Q_P(R),\ P \in \text{Spec } R\} = R$ but the situation is not that nice in the non-commutative case. Since we are interested in the structure sheaf on Spec R it is reasonable to hope that the ring R will be the ring of global sections of the structure sheaf. Therefore the κ_P are not really that useful from the geometrical point of view and in order to remedy this D. Murdoch and F. Van Oystaeyen introduced in [112] another localization at a prime ideal. To a prime ideal P of

R we associate a symmetric kernel functor κ_{R-P} given by the filter $\mathcal{L}(\kappa_{R-P}) = \{L$ left ideal of R, L contains an ideal I of R such that $I \not\subset P\}$. In general $\mathcal{L}(\kappa_{R-P})$ is not idempotent (and the idempotent closure in Goldman's sense, [70], is not symmetric!) but if $\mathcal{L}(\kappa_{R-P})$ has a basis consisting of finitely generated left ideals then $\mathcal{L}(\kappa_{R-P})$ is idempotent (in particular if R is left Noetherian). The localization functor associated to κ_{R-P} is Q_{R-P}; when R is prime and $P = 0$ then we write Q_{sym} instead of Q_{R-0}.

<u>I.4.12. Proposition.</u> Let κ be a symmetric kernel functor in R-mod such that $\mathcal{L}(\kappa)$ has a filterbasis consisting of finitely generated left ideals. Then $\kappa = \Lambda\{\kappa_{R-Q}, Q \in C(\kappa)\}$, where $C(\kappa)$ is the set of ideals of R which are maximal amongst ideals of R not in $\mathcal{L}(\kappa)$.

<u>I.4.13. Corollary.</u> Let R be a prime ring and assume that ideals of R are finitely generated as left ideals. Then for any symmetric kernel functor κ in R-mod : $Q_\kappa(R) = \bigcap_{P \in C(\kappa)} Q_{R-P}(R)$. In particular : $R = \bigcap_{P \subset \Omega(R)} Q_{R-P}(R)$ where $\Omega(R)$ stands for the set of maximal ideals of R.

<u>I.4.14. Proposition.</u> Let P be a prime ideal of R such that R/P is a left Goldie ring then an ideal I of R is in $\mathcal{L}(\kappa_P)$ if and only if $I \not\subset P$. Consequently, we have that $\kappa_{R-P} < \kappa_P$.

If we strengthen our hypotheses somewhat we obtain a much stronger relation between κ_{R-P} and κ_P :

<u>I.4.15. Theorem.</u> Let P be a prime ideal of a prime Noetherian ring R. Then $Q_{R-P}(R) = Q_P(R) \cap Q_{sym}(R)$. More general, if κ is a kernel functor and κ^o is the symmetric kernel functor the idempotent filter of which is generated by the ideals in $\mathcal{L}(\kappa)$, then $Q_\kappa(R) \cap Q_{sym}(R) = Q_{\kappa^o}(R)$.

<u>I.4.16. Corollary.</u> If R is a prime Noetherian ring then $R = Q_{sym}(R) \cap (\bigcap_P Q_P(R))$, the intersection ranging over the maximal ideals of R.

In view of the sheaf theory expounded in Section V.1. the following kernel functors will be very important. Let R be a left Noetherian ring and let I be an ideal of R. The kernel functor κ_I is determined by the idempotent filter $\mathcal{L}(\kappa_I) = \{L$ left ideal of R, L contains some power of I$\}$. In the absence of the left Noetherian condition κ_I may be defined as being the smallest kernel functor such that R/I is a torsion module with respect to it.

<u>I.4.17. Proposition</u>. Let R be a left Noetherian ring and let P be a prime ideal of R, then :

$$\kappa_{R-P} = V\{\kappa_I, I \text{ an ideal of R such that } I \not\subset P\}.$$

From the definition it is clear that $\kappa_I = \kappa_J$ if and only if rad I = rad J and from I.4.12 we retain that $\kappa_I = \Lambda\{_{R-P}, P \not\supset I\}$. From this point of view it is interesting to know that the inf of t-functors need not be a t-functor but on the other hand, (cf. [120] for the proof) :

<u>I.4.18. Proposition</u>. If the κ_α, $\alpha \in A$ are t-functors then $\underset{\alpha}{V} \kappa_\alpha$ is a t-functor too.

Rephrasing I.3.8 we obtain that if κ' and κ are kernel functors in R-mod such that κ' is Q_κ-compatible (we denote by Q_κ the reflector R-mod$\to$R-mod(κ)) then $j_\kappa^\star\, Q_{\kappa'_\kappa} Q_\kappa = Q_{\kappa'} Q_\kappa$, where κ'_κ is the kernel functor induced by κ' in Q_κ(R)-mod and where $j_\kappa^\star$ is the restriction of scalars from Q_κ(R)-mod to R-mod via $j_\kappa : R \to Q_\kappa(R)$.

<u>I.4.19. Proposition</u>. If κ is a t-functor and κ' is Q_κ-compatible then $(\kappa' \vee \kappa)_\kappa = \kappa'_\kappa$ and moreover : $Q_{\kappa' \vee \kappa}(M) = Q_{\kappa'_\kappa} Q_\kappa(M) = Q_{\kappa'} Q_\kappa\ (M)$ for each $M \in$ R-mod. Furthermore $\mathcal{L}(\kappa'_\kappa) = \{Q_\kappa(L), L \in \mathcal{L}(\kappa')\}$.

The following criterion due to Van Oystaeyen [176] is useful in checking compatibility with a symmetric kernel functor :

<u>I.4.20. Theorem</u>. Let R be a left Noetherian ring, let κ be a symmetric kernel functor and let κ' be Q_κ-compatible. Then for any $I \in \mathcal{L}(\kappa')$, $J \in \mathcal{L}(\kappa)$ there exist

$I' \in \mathcal{L}(\ ')$, $J' \in \mathcal{L}(\kappa)$ such that $J'I' \subset IJ$. If both κ and κ' are symmetric then there is equivalence between :

a. κ' is Q_κ-compatible and κ is $Q_{\kappa'}$-compatible.

b. For $I \in \mathcal{L}(\kappa')$, $J \in \mathcal{L}(\kappa)$ there exist I', $I' \in \mathcal{L}(\kappa')$ and J', $J'' \in \mathcal{L}(\kappa)$, such that $J'I' \subset IJ$ and $I''J'' \subset JI$.

<u>I.4.21. Comments and References.</u> For details on torsion theory and localization in an abstract setting we refer to P. Gabriel [61], O. Goldman [70], J. Golan [64] or B. Stenström [167]. Geometric kernel functors appeared in Van Oystaeyen's paper [173] after symmetric kernel functors had been introduced by D. Murdoch, F. Van Oystaeyen in [120], [174] with an eye to the construction of "noncommutative schemes". For the localization at a prime ideal one needs to look up J. Lambek, G. Michler [101] and A.G. Heinicke [80], and B. Mueller [116], whereas for the systematic use and study of the symmetric localization at P we refer to F. Van Oystaeyen [173], [174], [183]. Compatibility of a kernel functor and a localization functor were first studied by F. Van Oystaeyen in [176] and put in the abstract setting of Grothendieck categories and Giraud subcategories by Van Oystaeyen, Verschoren in [188].

II. SOME NONCOMMUTATIVE ALGEBRA.

II.1. Birational Extensions of Rings.

Let A be a ring and let $X = \mathrm{Spec}(A)$ be the set of proper prime ideals of A. To an ideal I of A we associate $X(I) = \{P \in X,\ I \not\subset P\}$. Considering the sets X(I) as open sets in X we obtain the Zariski topology on X. It is clear that X(I) depends only on rad I. Put $V(I) = X - X(I)$.

Let A,B be rings and $f : A \to B$ a ring homomorphism. Then, following C. Procesi [136], we say that f is an extension of rings if $B = f(A)Z_B(A)$, where $Z_B(A) = \{b \in B,\ bf(a) = f(a)b$ for all $a \in A\}$. Unlike an arbitrary ringhomomorphism, a ring extension yields a map $\tilde{f} : \mathrm{Spec}(B) \to \mathrm{Spec}(A)$, which is given by $\tilde{f}(P) = f^{-1}(P)$ for $P \in \mathrm{Spec}(B)$.

An extension of rings $f : A \to B$ is called a birational extension if there exist ideals I of B and J of A such that $\tilde{f} : Y = \mathrm{Spec}(B) \to X = \mathrm{Spec}(A)$ restricts to a topological isomorphism $Y(I) \overset{\cong}{\to} X(J)$ such that if $f^{-1}(P) \in X(J)$ for some $P \in Y$ then $P \in Y(I)$. An extension of rings $f : A \to B$ is called a Zariski extension if there exist ideals I of B and J of A such that $\tilde{f}$ restricts to a topological isomorphism $Y(I) \overset{\cong}{\to} X(J)$ such that for every ideal $H \subset \mathrm{rad}(I)$, Y(H) corresponds to $X(H') \subset X(J)$ with $H' \subset f^{-1}(H)$.

II.1.1. Lemma. If $f : A \to B$ is a birational extension then $f : A \to f(A)$ and $f(A) \hookrightarrow B$ are birational extensions.

Proof : Let I,J , Y(I) , X(J) be as in the definition. It is easy to check that birationality of $f(A) \hookrightarrow B$ is obtained by considering Y(I) and Z(f(J)) where $Z = \mathrm{Spec}(f(A))$. On the other hand, birationality of $A \to f(A)$ may be verified by considering Z(f(J)) and X(J).

II.1.2. Remark. Since $\tilde{f}(Y(I)) \subset V(\mathrm{Ker}\ f)$ it is clear that Spec(A) cannot be connected unless $\mathrm{Ker}\ f \subset \mathrm{rad}(A)$. In the applications we will be dealing with semiprime or prime rings, therefore it is not restrictive to restrict attention to monomorphic f as the foregoing lemma already indicated. In the sequel f is assumed to be injective!

II.1.3. Lemma. The extension $f : A \hookrightarrow B$ is birational if and only if the homeomorphism $Y(I) \cong X(J)$ is such that for each ideal $H \subset rad(I)$, $Y(H)$ corresponds to $X(H') \subset X(J)$ with $H' \subset rad(H) \cap A$.

<u>Proof</u> : Assuming that f is birational, $\tilde{f}(Y(H)) = X(K)$ for some ideal K of A, $X(K) \subset X(J)$, $K \subset rad(J)$. If $P \in Y$, $P \supset H$ and $P \not\supset I$ then $P \cap A \not\supset J$ and $P \cap A \supset K$. If $P \in Y$, $P \supset H$ and $P \supset I$ then $P \cap A \supset J$ since $P \cap A \not\supset J$ would entail $P \not\supset I$ by birationality of f, hence $P \cap A \supset rad(J) \supset K$. Consequently $rad(H) \supset K$, i.e. we may take K for H'. ∎

<u>II.1.4. Lemma.</u> Let $f : A \to B$ be an injective extension of rings. The following properties are equivalent :

a. f is a Zariski extension described by ideals I,J of B,A resp. such that $Y(I) \cong X(J)$.

b. The homeomorphism $\tilde{f}|Y(I)$ is such that for every ideal $H \subset rad(I)$, $Y(H)$ corresponds to $X(H \cap A) \cap X(J)$.

c. $\tilde{f}(Y(I)) = X(J)$ and $rad(H) = rad\, B(H \cap A)$ for every ideal $H \subset rad(I)$.

<u>Proof</u> : a $\Rightarrow$ b. Let $H \subset rad\, I$. If $p \in X(H \cap A) \cap X(J)$ then there is a unique $P \subset Y(I)$ such that $P \cap A = p$. Clearly $P \not\supset H$. On the other hand if $Q \in Y(H)$ then a implies that $Q \cap A \in X(H')$ with $H' \subset H \cap A$ i.e. $Q \cap A \in X(H \cap A) \cap X(J)$.

b $\Rightarrow$ a. By definition.

b $\Rightarrow$ c. Let $H \subset rad(I)$. If $P \in Y$, $H \not\subset P$ then b. implies that $H \cap A \not\subset P \cap A$ thus $B(H \cap A) \not\subset P$. This means exactly that $rad(H) = rad\,(B\,(H \cap A))$.

c $\Rightarrow$ b. One easily checks that from c. it follows that if $P \in Y(I)$ and $K \cap A \subset P \cap A$ for some ideal K of B then $K \subset P$. Thus $\tilde{f}|Y(I)$ is injective. Let $H \subset rad\,(I)$ and consider $p \in X(H \cap A) \cap X(J)$. Since $\tilde{f}(Y(I)) = X(J)$ and $p \in X(J)$ we find $P \in Y(I)$ such that $P \cap A = p$. Obviously $H \not\subset P$. On the other hand, if $Q \in Y(H) \subset Y(I)$ then $H \cap A \subset Q \cap A$ would imply $rad(H) \subset Q$, contradiction, so $Q \cap A \in X(H \cap A) \cap X(J)$. ∎

Comparison of II.1.3 and II.1.4 shows that the notions of birational and Zariski extensions differ only in the following : if $X(H')$ corresponds to $Y(H)$ under $\tilde{f}$ then in the birational case $H' \subset rad\,(H) \cap A$ while in the Zariski case $H' \subset H \cap A$. The following proposition gives us sufficient conditions for both concepts to coincide.

II.1.5. Proposition. Let $f: A \to B$ be an extension of rings. If either A is commutative or B satisfies the ascending chain condition for ideals, then $\mathrm{rad}(H \cap A) = \mathrm{rad}(H) \cap A$ holds for every ideal H of B.

Proof. Since $A \hookrightarrow B$ is an extension, $\mathrm{rad}(H \cap A) \subset \mathrm{rad}(H) \cap A$. If A is commutative and $x \in A \cap \mathrm{rad}(H)$ then $x^n \in H \cap A$ for some $n \in \mathbb{N}$, thus $x \in \mathrm{rad}(H \cap A)$.
Now consider the case where B satisfies the a.c.c. for ideals. Then $(\mathrm{rad}(H))^n \subset H$ for some $n \in \mathbb{N}$, therefore : $(A \cap \mathrm{rad}(H))^n \subset A \cap (\mathrm{rad}(H))^n \subset \mathrm{rad}(A \cap H)$. Consequently, $A \cap \mathrm{rad}(H) \subset \mathrm{rad}(A \cap H)$. ∎

II.1.6. Remarks. Under the hypotheses of II.1.5 we also have that $\mathrm{rad}(B(H \cap A)) = \mathrm{rad}(B(A \cap \mathrm{rad}\, H))$. In case B satisfies the a.c.c. for ideals then each semiprime ideal of B is a finite intersection of prime ideals, and there is equivalence between :

a. $\mathrm{rad}(H) = \mathrm{rad}(B(A \cap H))$ for every ideal H of B, $H \not\subset \mathrm{rad}(I)$.

b. $Q \cap A \subset P \cap A$ implies $Q \subset P$ for every $Q \in Y$, $P \in Y(I)$.

For some theory on birational extensions in the non-Noetherian case cf. [182].
Since we will always be in the situation of II.1.5 we will here expound the theory of Zariski extensions. If $f: A \to B$ is a birational (Zariski) extension then we say that B is a birational (Zariski) A-algebra. It is clear that if B is a birational (Zariski) algebra then there exists a unique maximal open set in $Y = \mathrm{Spec}(B)$ which may be used as an open set of birationality. From hereon we shall always denote this maximal open set by $Y(I)$ and $X(J)$ is then the homeomorphic image of $Y(I)$ in X under $\tilde{f}$. If $Y(I) = Y$ then f is said to be a global Zariski extension. When relating to the center $Z(R)$ of a ring R it is usually advantageous to drop the topological considerations and we say that a ring R is Zariski central if for every ideal I of R : $\mathrm{rad}\, I = \mathrm{rad}(R(I \cap Z(R)))$.

II.1.7. Examples.

a. Every simple ring is a global Zariski extension of its center.

b. Every semisimple ring is a global Zariski extension of its center.

c. Let R be any ring, then the extension $R \hookrightarrow M_n(R)$ is a global Zariski extension.

d. Any Azumaya algebra (cf. II.3) is a global Zariski extension of its center.

e. (S. Amitsur, L. Small) If $A \hookrightarrow B$ is an extension such that there is a common ideal K of A and B then we may take $I = J = K$ i.e. B is a Zariski A-algebra if $K \not\subset \mathrm{rad}\, B$.

f. Semiprime P.I. algebras are Zariski extensions of their center.

II.1.8. Proposition. Let B be a Zariski A-algebra, then :

a. If B is semiprime then a non-trivial ideal H of B such that $H \subset \mathrm{rad}(I)$ intersects A non-trivially.

b. If B is prime then every non-trivial ideal of B intersects A non-trivially.

c. If B is semiprime and A is simple (or quasi-simple) then B is a quasi-simple ring i.e. B has no non-trivial ideals.

Proof : Straightforward from the foregoing.

II.1.9. Proposition. Let B be a Zariski A-algebra and let $\pi : B \to B_1$ be a surjective ring homomorphism. If $I \not\subset \mathrm{rad}(\mathrm{Ker}\, \pi)$ then B_1 is a Zariski $\pi(A)$-algebra.

Proof : cf. [182] or [183].

In order to avoid unnecessary Noetherian hypothesis we define a radical kernel functor to be a symmetric kernel functor on R-mod such that an ideal I of R is in $\mathcal{L}(\kappa)$ if and only if rad(I) is in $\mathcal{L}(\kappa)$.

If B is a Zariski A-algebra with open sets of birationality $Y(I) \subset Y = \mathrm{Spec}(B)$ and $X(I') \subset X = \mathrm{Spec}(A)$, let κ be a radical kernel functor on B-mod such that $I \in \mathcal{L}(\kappa)$. Then the set of ideals H' of A such that $X(H') \cong Y(H \cap I)$ for some $H \in \mathcal{L}(\kappa)$ is a basis for a filter defining a kernel functor κ' on A-mod. It is easy enough to verify that $I' \in \mathcal{L}(\kappa')$ and that κ' is a radical kernel functor.

II.1.10. Proposition. Let B be a Zariski A-algebra with open sets of birationality Y(I) and X(I'). Let κ be a radical kernel functor on B-mod such that $I\in\mathcal{L}(\kappa)$ and let κ' be the corresponding kernel functor on A-mod.
Then $\{BJ', J'\in\mathcal{L}(\kappa')\}$ is a basis for $\mathcal{L}(\kappa)$, consequently B-modules are κ-torsion if and only if they are κ'-torsion when viewed as A-modules by restriction of scalars.

Proof : Suppose H' is an ideal of A such that there exists an ideal $H\in\mathcal{L}(\kappa)$ such that $X(H')\cong Y(H\cap I)$. Then $\mathrm{rad}(BH')=\mathrm{rad}(H\cap I)$ and radicality of κ implies that $BH'\in\mathcal{L}(\kappa)$. Conversely, any $L\in\mathcal{L}(\kappa)$ contains an ideal H also in $\mathcal{L}(\kappa)$. Birationality yields that $Y(H\cap I)\cong X(H')$ for some ideal H' of A with $H'\subset H\cap I$. That $H'\in\mathcal{L}(\kappa')$ follows from the definition of κ'. ∎

II.1.11. Proposition. Let B be a Zariski A-algebra. With notation as before let $H\subset I$ be an ideal of B, then $(\kappa_H)'=\kappa_{H'}$, where H' is defined as usual by $X(H')\cong Y(H)$. (Note that neither κ_H nor $\kappa_{H'}$ need be idempotent, nevertheless both are radical pre-kernel functors in the sense of Van Oystaeyen, [183], and the proposition is indeed valid in the absence of any left Noetherian hypothesis on A or B, cf. loc. cit.) Furthermore if $P\in Y(I)$ then $(\kappa_{B-P})'=\kappa_{A-p}$ where $p=A\cap P$.

Proof : If $J\in\mathcal{L}((\kappa_H)')$ then J contains an ideal J" of A such that $X(J")\cong Y(L)$ for some ideal L of B such that $L\in\mathcal{L}(\kappa_H)$ i.e. $\mathrm{rad}(L)\supset H$. So, since $Y(L)\supset Y(H)$ the isomorphism $Y(L)\cong X(J")$ restricts to $Y(H)\cong X(H')\subset X(J")$ or $\mathrm{rad}(J")\supset H'$. The latter means that $J"\in\mathcal{L}(\kappa_{H'})$ and therefore $\mathcal{L}((\kappa_H)')\subset\mathcal{L}(\kappa_{H'})$. Since both filters contain I' it is sufficient to consider ideals contained in I' in order to compare the filters. If $J"\subset I'$ is an ideal of A which is in $\mathcal{L}(\kappa_{H'})$ then $\mathrm{rad}(J")\supset H'$. Moreover $J"\in\mathcal{L}(\kappa_{H'})$ means that $\mathrm{rad}(BJ")\in\mathcal{L}(\kappa_H)$ and $J"\in\mathcal{L}((\kappa_H)')$ follows. For the proof of the second statement, pick $J\in\mathcal{L}(\kappa'_{B-P})$. Then J contains an ideal J' of A such that $\mathrm{rad}(BJ')=\mathrm{rad}(J\cap I)$ and $J'\in\mathcal{L}(\kappa'_{B-P})$. Obviously $J'\not\subset P\cap A=p$.
Conversely, if J' is an ideal of A which is not in p then X(I'J') is an open set containing p and $X(I'J')\cong Y(IJ)$ for some ideal J of B. Therefore $P\in Y(IJ)$, $IJ\not\subset P$,

and hence $I'J' \in \mathcal{L}(\kappa'_{B-P})$ what entails that $J' \in \mathcal{L}(\kappa'_{B-P})$. ∎

II.1.12. Corollary. If B is a global Zariski A-algebra then, to every radical pre-kernel functor κ on B-mod there corresponds a radical pre-kernel functor on A-mod and κ is a kernel functor if κ' is.

To an arbitrary ring homomorphism $\varphi : A \to B$ there corresponds the restriction of scalars functor $\overset{\circ}{\varphi}$: B-mod→A-mod and also a lattice homomorphism φ_* : A-ker→B-ker (R-ker denoting the lattice of kernel functors on R-mod, cf. [64]). If $\kappa' \in$ A-ker then $\varphi_*(\kappa')$ is given by its class of torsion modules which consists of $M \in$ B-mod such that $\varphi^\circ(M)$ is κ'-torsion.

II.1.13. Lemma. Let B be a Zariski A-algebra and let κ be a radical kernel functor on B-mod such that $I \in \mathcal{L}(\kappa)$ and κ corresponds to a radical kernel functor κ' on A-mod. Then $f_*(\kappa') = \kappa$, where f is the extension $A \hookrightarrow B$.

Proof : If $J \in \mathcal{L}(\kappa)$ is an ideal then J contains an ideal J' of A such that rad(BJ')= rad($J \cap I$) and $J' \in \mathcal{L}(\kappa')$. Clearly, $\{BJ', J' \in \mathcal{L}(\kappa')\}$ is cofinal in $\mathcal{L}(f_*(\kappa'))$ and so the radicality of κ yields $\kappa = f_*(\kappa')$. ∎

We now aim to relate $Q_{\kappa'}(B)$ and $Q_\kappa(B)$ in some frequently occuring cases.

II.1.14. Theorem. Let $\varphi : A \hookrightarrow B$ be a monomorphic ring extension, let $\kappa' \in$ A-ker and write $\bar{\kappa}$ for $\varphi_*(\kappa)$, then :

a. If κ' is a t-functor such that $Q_{\kappa'}(A)$ is an extension of A, then $Q_{\bar{\kappa}}$ and $Q_{\kappa'}$ coincide on B-mod and $\bar{\kappa}$ is a t-functor.

b. If φ is an extension with central basis, i.e. B is a free A-module and has a basis which is a subset of $Z_B(A)$, then for any $\kappa' \in$ A-ker we have that $Q_{\bar{\kappa}}(B)$ and $Q_{\kappa'}(B)$ are isomorphic.

Proof : For the second statement we refer to Huang [85].

a. Since $Q_{\kappa'}(A)$ and B are A-bimodules in the sense of M. Artin [13] (see also section II.3.) there is a unique ringstructure on $Q_{\kappa'}(A) \otimes_A B$ such that the usual maps $Q_{\kappa'}(A) \to Q_{\kappa'}(A) \otimes_A B$ and $B \to Q_{\kappa'}(A) \otimes_A B$ are extensions. By property T for κ' we have that $Q_{\kappa'}(A) \otimes_A B = Q_{\kappa'}(B)$. Moreover since $Q_{\kappa'}(A) \otimes_A B$ and $B \otimes_A Q_{\kappa'}(A)$ are isomorphic A-bimodules and this isomorphism is compatible with the ring structure on $Q_{\kappa'}(A) \otimes_A B$ it follows that $Q_{\kappa'}(B) = Q_{\kappa'}(A) \otimes_A B \cong B \otimes_A Q_{\kappa'}(A)$ is a ring and the localization morphism $j_B' : B \to Q_{\kappa'}(B)$ is a ring homomorphism. All this, together with the fact that $Q_{\kappa'}(A)$ is a flat right A-module yields that $Q_{\kappa'}(B)$ is a flat right B-module. Let us now check whether $B \to Q_{\kappa'}(B)$ is epimorphic in the category of rings. So consider $B \to Q_{\kappa'}(B) \rightrightarrows R$ in Rings, such that $\varphi_1 j_B' = \varphi_2 j_B'$.

Then we obtain the following commutative diagram in Rings :

$$\begin{array}{ccccc} B & \xrightarrow{\;j_B'\;} & Q_{\kappa'}(B) & \overset{\varphi_1}{\underset{\varphi_2}{\rightrightarrows}} & R \\ \uparrow \varphi & & \uparrow Q_{\kappa'}(\varphi) & & \\ A & \xrightarrow[\;j_A'\;]{} & Q_{\kappa'}(A) & & \end{array}$$

Since j_A' is an epimorphism in Rings we get : $\varphi_1 Q_{\kappa'}(\varphi) = \varphi_2 Q_{\kappa'}(\varphi)$. Therefore, both φ_1 and φ_2 define in an unambiguous manner a left $Q_{\kappa'}(A)$-module structure in R and as κ' is a t-functor $\kappa'(R) = o$ follows. Suppose $x \in Q_{\kappa'}(B)$ is such that $\varphi_1(x) \neq \varphi_2(x)$: then, choosing $J \in \mathcal{L}(\kappa')$ such that $Jx \subset j_B'(B)$ we obtain $\varphi_1(Jx) = \varphi_2(Jx)$ and consequently $J.(\varphi_1(x) - \varphi_2(x)) = o$ by definition of the left A-module structure of R via $j_A' : A \to Q_{\kappa'}(A)$. However $\kappa'(R) = o$ then entails that $\varphi_1(x) = \varphi_2(x)$. It is well-known that a right flat ring epimorphism is obtained as the localization morphism at a t-functor cf. [64]. So there is a t-functor $\sigma \in B$-ker such that $Q_\sigma(B) \cong Q_{\kappa'}(B)$ and then $Q_\sigma(M) \cong Q_{\kappa'}(M)$ for all $M \in B$-mod follows from property T for σ and κ'.

If $J \in \mathcal{L}(\kappa')$ then B/BJ is κ'- torsion, hence $Q_\sigma(B/BJ) = Q_{\kappa'}(B/BJ) = o$, hence $BJ \in \mathcal{L}(\sigma)$. Conversely if $H \in \mathcal{L}(\sigma)$ then $Q_{\kappa'}(B/H) = o$ yields that $\kappa'(B/H) = B/H$ i.e. $J.1 \subset H$ for some $J \in \mathcal{L}(\kappa')$ and thus $BJ \subset H$ with $J \in \mathcal{L}(\kappa')$. Thus $\sigma = \overline{\varphi}(\kappa') = \overline{\kappa}$. ∎

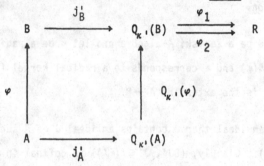

II.1.15. Corollary. Let B be a Zariski A-algebra and let κ be a radical kernel functor on B-mod such that $I\in\mathcal{L}(\kappa)$ and κ' is a t-functor on A-mod such that $A\to Q_{\kappa'}(A)$ is an extension (E.g. if κ' is a central localization of A) then $Q_{\kappa'}(B)\cong Q_{\kappa}(B)$ and $Q_{\kappa'}$ and Q_{κ} coincide on B-mod (in fact: up to natural equivalence of functors). Indeed, combine II.1.13 and II.1.14.

The usefulness of the above theorem resides in its application to κ_H and κ_{B-P}, since by Proposition II.1.11 $(\kappa_H)'=\kappa_{H'}$ and $(\kappa_{B-P})'=\kappa_{A-p}$ where $p=A\cap P$. In particular if A is a commutative ring, hence contained in the center of B, then both $\kappa_{H'}$ and κ_{A-p} are easily investigated e.g. κ_{A-p} has property T, so these properties established for these kernel functors are inherited by κ_H and κ_{B-P} with $H\subset I$, $P\in Y(I)$.

Let us now focus on commutative A, i.e. $A\subset Z(B)$.

II.1.16. Theorem. Let B be a Zariski A-algebra, $A\subset Z(B)$. Let κ be a radical kernel functor on B-mod such that $I\in\mathcal{L}(\kappa)$ and κ' is a t-functor, then :

a. Ideals of B extend to ideals of $Q_{\kappa}(B)$ under localization.

b. Write $\kappa^{\ell}=\kappa$ and let κ^{r} be the kernel functor on mod-B with filter generated by the ideals in $\mathcal{L}(\kappa)$ then $Q_{\kappa^{\ell}}(B)$ and $Q_{\kappa^{r}}(B)$ are isomorphic rings.

c. $Q_{\kappa'}(A)$ is contained in the center of $Q_{\kappa}(B)$. If $A=Z(B)$ and B is either prime or left Noetherian then $Q_{\kappa'}(A)\cong Z(Q_{\kappa}(B))$.

d. $Q_{\kappa}(B)$ is a global Zariski $Q_{\kappa'}(A)$-algebra.

e. If in the hypothesis of the theorem commutativity of A is replaced by the hypothesis that ideals of A localize to ideals of $Q_{\kappa'}(A)$ and ideals of B localize to ideals of $Q_{\kappa}(B)$ then $Q_{\kappa}(B)$ is still a Zariski $Q_{\kappa'}(A)$-algebra.

f. If B is prime and left bounded then B is right bounded and a left and right Goldie ring.

g. If B is fully left bounded and a global Zariski A-algebra, then B is fully fully right bounded and all prime factor rings are Zariski Goldie rings.

Proof : a. Let J be any ideal of B; $H\in\mathcal{L}(\kappa)$. There is an ideal H' of A such that $X(H')\cong Y(H\cap I)$ and clearly $BH'J\subset JH$ since H' commutes with B. Now Proposition I.4.7 finishes the proof.

b. From II.1.14. we retain :

$$Q_{\kappa_\ell}(B)\cong Q_{\kappa'}(A)\underset{A}{\otimes} B\cong B\underset{A}{\otimes} Q_{\kappa'}(A)\cong Q_{\kappa_r}(B) \ .$$

c. The first assertion is easily verified. For the second statement, consider $x\in Z(Q_\kappa(B))$. Then, for some $L'\in\mathcal{L}(\kappa')$: $L'x=j_B(L')x\subset j_B(B)$ and $L'x\; j_B(b)=j_B(b)L'x$ for any $b\in B$. Choose an $\ell\in L'$ and write $\ell x=j_B(c)$ with $c\in B$. Then $cb-bc\in\kappa(B)$ for all $b\in B$. If B is left Noetherian then $J'\kappa(B)=o$ for some $J'\in\mathcal{L}(\kappa')$. Consequently $J'cb=bJ'c$ for all $b\in B$ and thus $J'c\subset A$ because $A=Z(B)$. It follows that $j_B(J'L')x\subset j_B'(A)$ and $j_B'(A)=j_A'(A)$. The latter inclusion entails that $x\in Q_{\kappa'}(A)$ because κ' is a t-functor.

In the second case, if B is a prime ring then $\kappa(B)=o$. Hence $c\in A$ and $j_B(L')x\subset j(A)$ yields $x\in Q_{\kappa'}(A)$.

d. (a slight modification of the following proof will also establish the claim e, it is left to the reader to check this).

The assumptions yield that $Q_{\kappa'}(A)\hookrightarrow Q_\kappa(B)$ is an extension with $Q_{\kappa'}(A)\subset Z(Q_\kappa(B))$, and there is a one-to-one correspondence between the proper prime ideals of $Q_\kappa(B)$ (resp. $Q_{\kappa'}(A)$) and prime ideals of B (resp. A) not in $\mathcal{L}(\kappa)$ (resp. $\mathcal{L}(\kappa')$ (indeed see a. and I.4.8).

A proper prime ideal of $Q_{\kappa'}(A)$ has the form $Q_{\kappa'}(A)j(Q)$ for some prime ideal Q of A not in $\mathcal{L}(\kappa')$. Since $I'\in\mathcal{L}(\kappa')$ as $I\in\mathcal{L}(\kappa)$, $Q\in X(I')$ follows. Therefore there is a unique $Q\in Y(I)$ such that $Q\cap A=q$ and from $q\notin\mathcal{L}(\kappa')$, $Q\notin\mathcal{L}(\kappa)$ follows. Consequently $Q_\kappa(B)\; j_B(Q)$ is a proper prime ideal of $Q_\kappa(B)$. If $x\in Q_\kappa(B)\; j_B(Q)\cap Q_{\kappa'}(A)$ then, for some $J'\in\mathcal{L}(\kappa')$ we find that : $J'x\subset j_B(Q)\cap j_B(A)$ and since $j_B(Q)\cap j_B(A)=j_B(q)=j_A(q)$ it follows that $x\in Q_{\kappa'}(A)\; j_A(q)$, therefore $Q_\kappa(B)\; j_B(Q)\cap Q_{\kappa'}(A)=Q_{\kappa'}(A)\; j_A(q)$.

Let now H be an ideal of $Q_\kappa(B)$ and Q a proper prime ideal of $Q_\kappa(B)$. Assume $H\cap Q_{\kappa'}(A)\subset Q\cap Q_{\kappa'}(A)$.

Then $j_B^{-1}(H) \cap A \subset j^{-1}(Q) \cap A$ and $j_B^{-1}(Q)$ is a prime ideal of B not in $\mathcal{L}(\kappa)$. Moreover $I \in \mathcal{L}(\kappa)$ yields that $j_B^{-1}(Q) \in Y(I)$, consequently $j_B^{-1}(H) \subset j_B^{-1}(Q)$ and as κ is a t-functor, $H \subset Q$ follows.

Therefore $rad(H) = rad(Q_\kappa(B)(H \cap Q_{\kappa'}(A)))$ and it is clear that $Q_\kappa(B)$ is a Zariski $Q_{\kappa'}(A)$-algebra because of Lemma II.1.4. Note that the open sets of birationality are $Spec(Q_\kappa(B))$ and $Spec(Q_{\kappa'}(A))$.

f. If A is left bounded then $Q^{\ell}_{B-o}(B)$ is left bounded since κ_{B-o} is a t-functor, hence $Q^{\ell}_{B-o}(B)$ is simple Artinian, therefore $Q^{r}_{B-o}(B)$ is right bounded and B is right bounded, cf [124]. It is easily seen that $Q^{\ell}_{B-o}(B) = Q^{\ell}_{c\ell}(B)$ and $Q^{r}_{B-o}(B) = Q^{r}_{c\ell}(B)$, thus B is left and right Goldie. The remaining statements are obvious consequences of this. ■

Methods similar to the ones used above allow to derive the following result relating κ_{R-P} and κ_P by looking at the center of the localised rings.

<u>II.1.17. Proposition</u>. Let B be a prime Zariski A-algebra with $A \subset Z(B)$. If $Q_{B-P}(B)$ is identified with its canonical image in $Q_P(B)$, for some prime ideal P of B not containing I, then $Q_{B-P}(B)$ and $Q_P(B)$ have the same center.

<u>Proof</u> : Slight modification of Corollary 2.11.4 in [183]. □

<u>II.1.18. Comments and References.</u>

Birational extensions have been introduced by F. Van Oystaeyen in [182] and [183]. Classification of the birational algebras within certain classes of rings like asano orders, regular rings etc.... is the topic of [125] and E. Nauwelaerts' thesis, [124]. In particular, consequent use of the theory enables to extend and add to L. Small's results on localization of P.I. rings, cf. [165], as this has been carried out by E. Nauwelaerts in [121]. We also refer to the chapters on P.I. rings in this book. Birationality of graded rings has been studied in [182] and [121].

II.2. Rings Satisfying a Polynomial Identity. Generalities.

Let C be a commutative ring with unit. A non-empty class V of C-algebras is said to be a variety if it is closed under homomorphic images, products and taking subalgebras. Let $\underline{Alg}_C$ stand for the category of C-algebras. For any C-algebra R let $I(V,R) = I$ be the minimal ideal of R such that $R/I \in V$. The canonical inclusion functor $i : V \to \underline{Alg}_C$ has an adjoint $p : \underline{Alg}_C \to V$ which is given by $p(R) = R/I(V,R)$. It follows that any variety V considered as a category, contains free objects. If S is an arbitrary set then the free algebra over S in V may be constructed as follows. Let $F_S = C\{X_s, s \in S\}$ be the free associative algebra of S over C and put $p(F_S) = U$; we say that U is a free object of V. We have $U \in V$ and there is a map $r : S \to U : s \to p(X_s) = X_s \bmod I(V,F)$, such that for any R in V the set maps, $\sigma : S \to R$ correspond one-to-one to the C-algebra morphisms, $\pi : U \to R$ for which $\pi r = \sigma$. (One takes S to be an infinite set.) For $R \in \underline{Alg}_C$ we denote by $V(R)$ the variety generated by R. The set $I_S(R) = \{f(X_s) \in C\{X_s, s \in S\}, f(r_s) = o \text{ for all } r_s \in R\}$ is called the set (ideal!) of polynomial identities of R in S. A polynomial identity $f \in C\{X_s, s \in S\}$ of R is said to be non-trivial if $Rc(f) \neq o$, where $c(f)$ is the ideal of C generated by the coefficients of f. We say that f is a proper identity for R if $c(f)r = o$ with $r \in R$ yields $r = o$. A non-trivial identity of a prime C-algebra is proper! A P.I. algebra (we usually omit the C-) is a C-algebra satisfying a proper identity. A P.I. ring is a P.I. algebra over $\mathbb{Z}$.

An $f \in C\{X_s, s \in S\}$ is said to be linear in X_i, for some $i \in S$, if every monomial appearing in f is of degree 1 in X_i. An $f \in C\{X_s, s \in S\}$ is regular if it is linear in one of the variables appearing in it (up to renumbering we may assume then that f is linear in the "last" variable since only a finite number of X_s appear in f). If f is linear in each of its variables then it is called a multilinear polynomial. Any C-algebra which satisfies a non-trivial identity of degree d also satisfies a multilinear identity of degree at most d, cf. [136]. 3.12. Multilinear identities behave well under base change i.e. if f is a multilinear identity for $R \in \underline{Alg}_C$ and if $C \to D$ is a morphism of commutative rings then $R \otimes_C D$

satisfies the identity f. A very particular multilinear identity that may be satisfied is the standard identity of degree m :

$$S_m(X_1,\dots,X_m) = \sum_{\sigma\in S_m} (\operatorname{sgn}\sigma)\, X_{\sigma(1)} X_{\sigma(2)}\cdots X_{\sigma(m)}$$

where S_m is the permutation group on m elements. For example $S_2(X_1,X_2) = X_1X_2 - X_2X_1$, is the polynomial which characterizes the variety of all commutative algebras. More generally

II.2.1 Theorem (Amitsur-Levitzki). The polynomial $S_{2n}(X_1,\dots,X_{2n})$ is an identity for $M_n(C)$, and $M_n(C)$ does not satisfy any non-trivial polynomial identity of degree less than 2n.

II.2.2 Theorem (Kaplansky). Let R be a primitive C-algebra (i.e. possessing a faithful irreducible representation) satisfying a non-trivial polynomial identity of degree d. Then R is a central simple algebra of dimension n^2 over its center, where $2n < d$. Moreover $S_{2n}(X_1,\dots,X_{2n})$ is a polynomial identity for R.

II.2.3 Corollary. If $R \in \underline{\mathrm{Alg}}_C$ satisfies a polynomial identity f such that $c(f) = C$ then any primitive ideal of R is maximal.

II.2.4 Theorem (Amitsur). Let $R \in \underline{\mathrm{Alg}}_C$ satisfy a proper polynomial identity of degree d, then :

1. There exists a finite number of semiprime commutative C-algebras $C_1,\dots,C_s$ with $s < \frac{d}{2}$ and an embedding :

$$o \to R/\mathrm{rad}(o) \to \bigoplus_{i=1}^{s} M_i(C_i)$$

such that all stable identities, i.e. identities preserved under base change, satisfied by R are also satisfied by $\bigoplus_{i=1}^{s} M_i(C_i)$.

2. If R is a prime ring then there exists a field k and an embedding of rings:

$$o \to R \to M_s(k)$$

with $s<\frac{d}{2}$ and $M_s(k)=R.k$.

For the above theorems and detailed proofs we refer to [136]; let us mention however that, in proving Theorem II.2.4., one needs to know that in an arbitrary P.I. ring the upper nilradical coincides with the lower nilradical, whereas in a prime C-algebra satisfying a polynomial identity there are no non-zero nil left ideals.
Modification of II.2.4. also yields that a prime ring R satisfies a polynomial identity if and only if there exists a monomorphism $\lambda : R\to M_n(k)$ for some field k and some $n\in \mathbb{N}$, such that $M_n(k)=\lambda(R).k$ i.e. such that λ is a central extension of rings.

<u>II.2.5. Theorem (Amitsur)</u>. Let $R\in \underline{Alg}_C$ satisfy a proper polynomial identity, then R satisfies $S_n(X_1,\dots,X_n)^m$ for suitable $n,m\in \mathbb{N}$; i.p. R satisfies a multilinear identity with coefficients equal to 1 or -1.

In addition to the foregoing theorem one may prove that a semiprime ring R satisfying S_{2n}^m satisfies S_{2n} as well.

Note that in general, the vanishing of a product of polynomials f(X)g(X) on R does not imply that f or g is a polynomial identity on R, even if R is semiprime. If R is infinite and prime then fg is a polynomial identity of R if and only if f or g is such. The same statement holds if R is a semiprime algebra over an infinite field in view of the following remarks.

An algebra R is said to be <u>generic</u> for a variety V if R generates V. In this terminology, a semiprime algebra over an infinite field C=k is either generic in the variety of all C-algebras, or it generates the variety generated by $M_n(k)$ for some $n\in \mathbb{N}$. The <u>p.i. degree</u> of a P.I. algebra R is the smallest $n\in \mathbb{N}$ such that S_{2n}^k is an identity of R, for some $k\in \mathbb{N}$. It follows that the $n\in \mathbb{N}$ for which $V(R)=V(M_n(k))$ mentioned above is exactly the p.i. degree of R. It will be useful to pay special attention to the variety V_n generated by $M_n(K)$ where K is an arbitrary infinite field. The ideal of the free algebra which corresponds to the variety V_n is called the ideal of polynomial identities of n by n matrices.

Actually, if R is a semiprime P.I.-algebra then $R \in V_n$ where n is the p.i. degree of R. If V is any proper variety then there exists $n \in \mathbb{N}$ such that $V \supset V_n$ but $V \not\supset V_{n+1}$ and if R is any algebra in V then : p.i. deg $R \leq n$.

Next we consider certain polynomials which are just one step away from being identities : central polynomials.

A polynomial $f(X_1,\dots,X_m) \in C\{X_s, s \in S\}$ is said to be a central polynomial for $R \in \underline{Alg}_C$ if $f(r_1,\dots,r_m)$ is an element of the center $Z(R)$ for all $r_1,\dots,r_m$ in R but f is not an identity for R.

In [57], E. Formanek constructed for each $n \in \mathbb{N}$ a polynomial F_n over $\mathbb{Z}$ which is central for $M_n(C)$, where C is an arbitrary commutative ring with unit.

First consider

$$g(z_1,\dots,z_{n+1}) = \prod_{i=2}^{n}(z_1 - z_i)(z_{n+1} - z_i) \prod_{\substack{i=2 \\ i<j}}^{n} (z_i - z_j)^2 ,$$

in commuting variables $z_1,\dots,z_{n+1}$ over $\mathbb{Z}$. Let c_k be defined by $g(z_1,\dots,z_{n+1}) = \sum_k c_k z_1^{k_1} z_2^{k_2} \dots z_{n+1}^{k_{n+1}}$, $k = (k_1,\dots,k_{n+1})$. Now construct the following polynomial over $\mathbb{Z}$, in non-commuting indeterminates $x, y_1,\dots,y_n$:

$$G(x,y_1,\dots,y_n) = \sum_k c_k x^{k_1} y_1 x^{k_2} y_2 \dots y_n x^{k_{n+1}} .$$

The Formanek central polynomial is then given by :

$$F(x,y_1,\dots,y_n) = G(x,y_1,\dots,y_n) + G(x,y_2,\dots,y_n,y_1) + \dots + G(x,y_n,y_1,\dots,y_{n-1}).$$

Note that in any central polynomial the constant term may be deleted; then:

II.2.6. Proposition (C. Procesi) Let f be a central polynomial for $M_n(C)$, and suppose that f has constant term zero, then f is a polynomial identity for $M_{n-1}(C)$.

Proof : Embed $M_{n-1}(C)$ in $M_n(C)$ by $(a_{ij}) \mapsto \begin{pmatrix} (a_{ij}) & 0 \\ 0 & 0 \end{pmatrix}$. If we compute values of f for substitutions in $M_{n-1}(C)$ we obtain scalar matrices of $M_n(C)$. However the only scalar matrix of $M_n(C)$ in $M_{n-1}(C)$ is the zero matrix. ∎

II.2.7. Corollary. Let $C = k$ be an infinite field. If R and $M_n(k)$ satisfy the same identities then, for any semiprime ideal J of R either R/J satisfies the same identities of R or every central polynomial of R is an identity of R/J.

Proof : Since R/J is a semiprime P.I. algebra over an infinite field k, it satisfies the identities of $M_m(k)$ for some m.

If m=n then R and R/J satisfy the same identities. If $m \neq n$ then Proposition II.2.6. finishes the proof. ∎

The Formanek polynomial F_n is a central polynomial for any central simple algebra of dimension n^2 over its center, while F_n is an identity for central simple algebras of degree less than n.

II.2.8. Theorem (L. Rowen) A non-zero ideal of a semiprime P.I. algebra R intersects the center non-trivially. In particular, if the center of R is a field then R is simple.

II.2.9. Theorem (E. Posner) Let R be a prime P.I. algebra with center C and let K be the field of fractions of C. Then $Q = R \underset{C}{\otimes} K$ is a central simple algebra over K and Q is the classical left and right ring of quotients of R. The algebras R and Q satisfy the same identities and if p.i. deg $R = n$ then $\dim_K Q = n^2$.

Both of foregoing theorems are "classics" in P.I. theory, cf. [136] , [151] ,..., we pay some more attention to the following related results.

II.2.10. Proposition. Let $i : R \to S$ be a monomorphic extension of rings then:

1. $Z(R) \subset Z(S)$.
2. If S is prime then R is prime.
3. If S is semiprime then so is R.

Proof : Obvious. ∎

II.2.11. Theorem. Let $i : R \to S$ be a monomorphic extension of P.I. rings and let S be a prime ring, then :

1. R is a prime ring.
2. An $a \in R$ is regular in R if and only if a is regular in S.
3. Let $Q(R)$, $Q(S)$, denote the classical ring of quotients of R,S, resp.; there is a canonical commutative diagram of monomorphic extensions of rings :

$$\begin{array}{ccc} R & \longrightarrow & S \\ \downarrow & & \downarrow \\ Q(R) & \longrightarrow & Q(S) \end{array}$$

4. Put $U = Z(Q(R))$, $V = Z(Q(S))$. Then $U \subset V$. If $T = Z_{Q(R)}(Q(S))$, then $Q(S) = Q(R) \underset{U}{\otimes} T$.
5. If $Z_R(S) = Z(S)$ then $T = V$.

Proof : 1. Obvious.

2. If a is regular in R then Ra contains a non-zero ideal I of R. The non-zero part of I in $Z(R)$ extends then to an ideal of S contained in Sa and therefore a is regular in S.

3. Since each regular element of R becomes invertible under the morphism $R \to S \to Q(S)$, by 2., the universal property for the classical ring of quotients yields a commutative diagram as desired. Injectivity of those morphisms follows from the fact that R and S are prime rings. Note that the morphism $Q(R) \to Q(S)$ actually derives from :

$$\begin{array}{ccc} Q(R) & \longrightarrow & Q(S) \\ \cong \uparrow & & \uparrow \cong \\ Q(Z(R)) \underset{Z(R)}{\otimes} R & \longrightarrow & Q(Z(S)) \underset{Z(S)}{\otimes} S \end{array}$$

where the vertical arrows represent the canonical isomorphisms and the bottom morphism is the combination of $i : R \to S$ and the induced $Q(Z(R)) \to Q(Z(S))$ (cf. II.2.10.-1).

4. Clearly $U \subset V$, $j : Q(R) \to Q(S)$ is an extension of rings, and we have a morphism $\lambda : Q(R) \underset{U}{\otimes} T \to Q(S)$ which is even surjective. The kernel of λ is an ideal of $Q(R) \underset{U}{\otimes} T$, hence of the form $Q(R) \underset{U}{\otimes} J$ for some ideal J of T (by the Azumaya-Nakayama theorem of Chapter I). However since T is the centralizer of the central simple algebra Q(R) in Q(S), $J = o$ follows, consequently λ is an isomorphism.

5. If $Z_R(S) = Z(S)$ then $Z_R(S) \subset V$ and thus $Q(S) = RV$ follows. However, in this case $Q(S) = Q(R)V$ and $T = V$. ∎

II.2.12. Remark. Posner's theorem, II.2.9, and Rowen's theorem, II.2.8, imply that Q(R) is obtained from R by symmetric localization at the prime ideal o. Therefore most of the above theorem is a direct consequence of the localization results in Chapter I,

Returning to central polynomials we mention :

II.2.13. Proposition. The Formanek polynomial F_n is a central polynomial for each semiprime P.I. ring R of p.i. degree n.

Proof : If R is a prime P.I. ring of p.i. degree n then Q(R) is a simple algebra of dimension n^2 over its center, while R and Q(R) satisfy the same identities. Since F_n is a central polynomial and not an identity for Q(R), F_n is also a central polynomial and not an identity for R. Now let R be a semiprime P.I. ring of p.i. degree n. If P is a proper prime ideal of R then F_n is a central polynomial or an identity for R/P according to whether p.i. deg.R/P = n or not. Therefore $G = F_n(X_1,\dots,X_{n+1})X_{n+2} - X_{n+2}\, F_n(X_1,\dots,X_{n+1})$ is a polynomial identity for all prime factor rings of R and hence, as R is semiprime, G is an identity for R. Now G can only be an identity of R if F_n takes central values when evaluated in R. On the other hand, there always exists a prime ideal of R such that R/P has p.i. degree equal to n, so F_n cannot be an identity of R/P nor of R. ∎

II.2.14. Remarks. 1. The Formanek polynomial is a regular central polynomial for any semiprime P.I. algebra R with p.i. deg R = n.
Y. Razmyslov gave an explicite construction of multilinear central polynomials in [144]

2. A prime C-algebra with central polynomial is also a P.I. algebra.

3. As an example, recall that $X_1X_2 - X_2X_1$ is a central polynomial for all Grassman algebras.

II.2.15. Definitions.

1. The central kernel J of a C-algebra R is the additive subgroup of the center Z(R) of R generated by the evaluations of the regular central polynomials of R. If R has no regular central polynomial, then we put $J = o$. Observe that J is an ideal of Z(R) and that $J = Z(R)$ implies that R is a P.I. algebra.

2. Let R satisfy the identities over $\mathbb{Z}$ of n by n matrices. Let J_n be the additive subgroup of Z(R) generated by all evaluations of F_n in R; then J_n is an ideal of Z(R). We call J_n the n-kernel of R. Clearly $J_n \subset J$. Moreover, if P is a prime ideal of R such that R/P has p.i. degree n then F_n is not an identity of R/P, hence $J_n \not\subset P$ follows, and conversely.

II.2.16. Proposition. Let A be a generic commutative C-algebra, let R be a C-algebra satisfying the same identities (with coefficients in C) as $M_n(A)$. Then rad RJ_n = rad RJ.

Proof : Any identity for $M_n(A)$ is also an identity for $M_n(B)$ for any other commutative $B \in \underline{Alg}_C$, cf. [136] . Suppose $RJ_n \subset P$ for some prime ideal P of R. Then R/P has p.i. degree $t < n$ and R/P embeds into $M_t(B)$ for some commutative C-algebra B. A regular central polynomial for R is a central polynomial without constant term for $M_n(A)$ hence an identity for $M_t(A)$ i.e. for $M_t(B)$ i.e. for R/P. It follows that $J \subset P$, hence $RJ \subset P$. ■

<u>II.2.17. Corollary</u>. 1. If R is a prime P.I. algebra then rad RJ_n = rad RJ, where n = p.i. deg R.

2. If R is a semiprime P.I. algebra over an infinite field K then rad RJ_n = rad RJ.

<u>Proof</u> : 1. If R is finite then Z(R) is a field and R is simple. If R is infinite and $\varphi : C \to R$ defines the C-algebra structure of R, then R is a $\varphi(C) = C'$-algebra and C' is a domain contained in Z(R). By Posner's theorem R may be embedded in a simple algebra Q of dimension n^2 over its center $K \supset Z(R)$, while R and Q satisfy the same identities. Since K is an infinite field containing C', Q satisfies the same identities as $Q \otimes_K \overline{K} = M_n(\overline{K})$, where $\overline{K}$ is an algebraic closure of K. Clearly, $\overline{K}$ is a generic commutative C'-algebra and thus rad RJ_n = rad RJ by the proposition.

2. Following C. Procesi, [136], one may establish that R satisfies the same identities as $M_n(K)$. ■

If $R \in \underline{Alg}_C$ satisfies the identities of $M_n(A)$ for some generic commutative C-algebra, then the <u>Formanek center</u> F(R) of R is the subring of R obtained by evaluating in R <u>all</u> central polynomials of $M_n(A)$ which have constant term equal to zero. Clearly F(R) is contained in Z(R) and a proper prime ideal P of R contains F(R) if and only if R/P has p.i. degree less than n. This will be of some use further on.

<u>II.2.18. Theorem</u>. If the central kernel J of the C-algebra R is not contained in rad(o) then R is a Zariski Z(R)-algebra.

<u>Proof</u> : Put Y = Spec(R), X = Spec(Z(R)). In [152], L. Rowen pointed out that if $P \in X$ is such that $P \not\subset J$ then there is a $P' \in Y$ such that $P' \not\supset RJ$ and $P' \cap Z(R) = P$. Moreover if $P' \not\supset RJ$ then for any ideal H of R such that

$H \cap Z(R) \subset P' \cap Z(R)$ we have $H \subset P'$.
Hence $\mathrm{rad}(H) = \mathrm{rad}(R(H \cap Z(R)))$ for any ideal H of R such that $H \subset \mathrm{rad}(RJ)$. All this amounts to saying that R is a Zariski Z(R)-algebra in the sense of Chapter I.4., described on the non-empty open sets Y(RJ) and X(I).

<u>II.2.19. Remark.</u> There is a one-to-one correspondence between maximal ideals M in Y(RJ) and maximal ideals m in X(J), given by $M \to M \cap Z(R) = m$. In this case $Rm = M$. Also $J = Z(R)$ if and only if $RJ = R$.

<u>II.2.20. Corollary.</u> A semiprime P.I. algebra is a Zariski extension of its center.

A C-algebra R with central kernel $J \not\subset \mathrm{rad}(o)$ is thus a Zariski extension of its center. From Chapter I.4 we derive that there is a pair of "largest" Zariski open sets $Y(RI) \subset Y$, $X(I) \subset X$ for which the Zariski-extension property holds. That Y(RI) may be properly containing Y(RJ) is seen in the following easy example. Put $R = \mathbb{C}[X,-]$, the ring of twisted polynomials over $\mathbb{C}$ with respect to conjugation, i.e. $X\lambda = \bar{\lambda}X$ for every $\lambda \in \mathbb{C}$. Clearly R has rank 4 over its center $\mathbb{R}[X^2]$ hence it is a P.I. ring. Now (X) is a maximal ideal of $\mathbb{C}[X,-]$ such that $R/(X) \cong \mathbb{C}$ has p.i. degree 1, hence $(X) \notin Y(RJ)$.
On the other hand $\mathbb{C}[X,-]$ is a Zariski central ring and thus $Y(RI) = \mathrm{Spec}\, R$.
An even more simple example is $R = M_2(K) \oplus M_3(K)$, where K is an infinite field. Obviously R is a P.I. ring of P.I. degree 3 but $P = M_3(F) \subset R$ is a prime ideal of R such that R/P has p.i. degree 2. On the other hand R is a semisimple Artinian ring hence a global Zariski extension of its center.

Let $R \in \underline{\mathrm{Alg}}_C$ be such that its central kernel $J \not\subset \mathrm{rad}(o)$. Denote by $Y(RI) \subset \mathrm{Spec}\, R$, $X(I) \subset \mathrm{Spec}\, Z(R)$ the pair of largest open sets on which the Zariski-extension property holds (thus these contain at least Y(RJ), X(J)).
From I.4.1. it follows that :

$$Q_{R-p}(R) \cong Q_{Z(R)-p}(R) \cong Q_{Z(R)-p}(Z(R)) \underset{Z(R)}{\otimes} R \cong Z(R)_p \underset{Z(R)}{\otimes} R \cong R_p$$

where $p = P \cap Z(R)$ and all isomorphisms are isomorphisms of C-algebras.

Furthermore :

II.2.21. Proposition. Suppose $R \in \underline{Alg}_C$ such that $J \not\subset \mathrm{rad}(o)$ and $P \in \mathrm{Spec}(R) = Y$ such that $P \not\supset RJ$ then :

1. The central kernel of $Q_{R-P}(R)$ coincides with the center of $Q_{R-P}(R)$ and this is exactly $Q_{Z(R)-p}(Z(R))$.
2. $Q_{R-P}(R)$ is a P.I. algebra
3. Let $j_{R-P} : R \to Q_{R-P}(R)$ be the canonical localization morphism. The unique maximal ideal $Q_{R-P}(R)\, j_{R-P}(P) = P^e$ of $Q_{R-P}(R)$ is the Jacobson radical and $Q_{R-P}(R)/P^e$ is simple Artinian and isomorphic to the classical ring of quotients of R/P.

Proof : 1. cf. [112]

2. Follows from 1.

3. Kaplansky's theorem II.2.2 yields that primitive ideals of a P.I. algebra are maximal, therefore, by 2., P^e must be the Jacobson radical of $Q_{R-P}(R)$. Since $Q_{R-P}(R)/P^e$ is a simple P.I. algebra it must be Artinian and even finite dimensional over its center. Being a prime P.I. algebra, R/P has a classical ring of quotients which must therefore be isomorphic to $Q_{R-P}(R)/P^e$.

II.2.22. Remark. It is easy to derive from the results obtained so far that a P.I. algebra is a fully bounded ring, hence κ_{R-P} and κ_P coincide (cf. [177] [122]). Then II.2.21.3 is a consequence of the fact that κ_P is a central localization. More precisely :

II.2.23. Proposition. Let $R \in \underline{Alg}_C$ be such that $J \not\subset \mathrm{rad}(o)$ If $P \in Y(RI)$:

1. The elements of C(P) map onto invertible elements of $Q_{R-P}(R)$ under $j_{R-P} : R \to Q_{R-P}(R)$.
2. $\kappa_P = \kappa_{R-P}$.
3. The set C(P) is a left and right reversible Ore set of R.

<u>Proof</u> : 1. Pick $s\in C(P)$, then $s \bmod P=\overline{s}$ is left regular in R/P. Then $j_{R-P}(s)$ is left regular in $Q_{R-P}(R)/P^e$ i.e. a unit. Now P^e is the Jacobson radical of $Q_{R-P}(R)$, therefore $j_{R-P}(s)$ is a unit of $Q_{R-P}(R)$.

2. Since R/P is a prime P.I. algebra it is a Goldie ring (left or right) by Posner's theorem, hence $\kappa_{R-P}\leq\kappa_P$ follows from I.4.14. On the other hand, $L\in\mathcal{L}(\kappa_P)$ means $s\in L$ for some $s\in C(P)$ and by 1. this yields $Q_{R-P}(R)\, j_{R-P}(L)=Q_{R-P}(R)$ i.e. $L\in\mathcal{L}(\kappa_{R-P})$.

3. Easy from 1. Note that the fact that R/P is a Goldie ring entails $C(P)=\{s,\ sr\in P\Rightarrow r\in P\}$.

<u>II.2.24. Remark</u>. The use of symmetric localization made it possible to extend most of L. Small's results on localization of P.I. algebras cf. [165], at prime ideals in $Y(RJ)$ to prime ideals in the larger open set $Y(RI)$. There is still the problem of characterizing $Y(RI)$ intrinsically in P.I. terms or equivalently : determine the class of Zariski central P.I. algebras. Some results in this vein are in [122], [124].

Let us now turn to the study of identities of n by n matrices. Since each semiprime P.I. ring satisfies the identities of n by n matrices for a suitable $n\in\mathbb{N}$, the study of those identities presses forward.

Let R and S be algebras over a commutative ring D and suppose that S is a free D-module with basis $\{u_1,\dots,u_m\}$. Fix a set T and consider symbols $\{x_{t,i},\dots,t\in T, i=1,\dots,m\}$. We may consider the free associative D-algebra generated by the $x_{t,i}$, $D\{x_{t,i}\}$ and let $U=D\{\overline{x}_{t,i}\}=D\{x_{t,i}\}/I(V(R),D\{x_{t,i}\})$ be the free object in the variety generated by R in $\underline{Alg}_D$. In $U\otimes_D S$ we consider the following elements :

$$\xi_t=\sum_{i=1}^{m}\overline{x}_{t,i}\otimes u_i .$$

Next, consider a commutative ring C and a ring homomorphism $C\to D$. With these notations we formulate the following result, cf. C. Procesi [136].

II.2.25. Theorem. Let $\psi : C\{x_t, t\in T\} \to U \otimes_D S$ be defined by sending x_t to ξ_t. The kernel of ψ is the ideal of polynomial identities of $R \otimes_D S$ in $\underline{Alg}_C$. Consequently $C\{\xi_t\}$, the image of ψ, is isomorphic to the free algebra (in variables x_t, $t\in T$) in the variety of C-algebras generated by $R \otimes_D S$.

II.2.26. Corollary. If C is a commutative ring and $C[x_{t,i,j}]$ the polynomial ring in the variables $x_{t,i,j}$ $t\in T$, $i,j = 1,\dots,n$, then we may consider the $\xi_t \in M_n(C[x_{t,i,j}])$, where $\xi_t = (x_{t,i,j})$.
The morphism $\lambda : C\{x_t, t\in T\} \to M_n(C[x_{t,i,j}])$ which is defined by $\lambda(x_t) = \xi_t$ has for its kernel exactly the ideal of polynomial identities of $M_n(D)$ where D is any generic commutative algebra.

II.2.27. Theorem. Let C be an integral domain with field of fractions K. Take T such that it contains at least two elements, then :

1. $C\{\xi_t, t\in T\}$ has no zero divisors.
2. $C\{\xi_t, t\in T\}$ is an order in a division ring of dimension n^2 over its center.
3. $C\{\xi_t, t\in T\}$ is generic in the variety given by the identities of n by n matrices.

The ring $C\{\xi_t, t\in T\}$ is called the ring of generic matrices of dimension n (over T). The ring of fractions of $C\{\xi_t, t\in T\}$, denoted by $C<\xi_t, t\in T>$ is the generic division ring. The generic division ring is a subring of $M_n(K(x_{t,i,j}))$ and it is obtained from the ring of generic matrices by localization at the multiplicative set of non-zero central elements of R. Posner's theorem entails that the center of $C<\xi_t, t\in T>$ is the field of fractions of the center of $C\{\xi_t, t\in T\}$. Let us give a hint of how this center looks in case C is a field K of characteristic zero. In [130], C. Procesi has shown that the center of $K\{\xi_1,\dots,\xi_m\}$, in the case of generic 2 by 2 matrices, is a rational function field over K. In [59] E. Formanek proved that the same is true in the 3 by 3 case. In these cases the transcendence degree over K equals $mn^2-(n^2-1)$. Let us clarify this a little. Assume that K is an infinite field. A (polynomial) invariant of m-tuples of n by n matrices over K is a function, $\varphi : M_n(K)^m \to K$, such that $\varphi(A_1,\ldots,A_m)$ is a polynomial function in the entries of $A_1,\ldots,A_m$, and for any $P \in GL(n,K)$ we have :

$$\varphi(PA,P^{-1},\ldots,PA_mP^{-1}) = \varphi(A_1,\ldots,A_m)$$

A rational invariant is defined similarly but with φ being a rational function in the entries of $A_1,\ldots,A_m$. It may be shown that every rational invariant is a quotient of polynomial invariants. Polynomial invariants of m-tuples of n by n matrices over K form a commutative domain and the rational invariants yield its field of fractions. Let $R = K\{\xi_1,\ldots,\xi_m\}$ be a ring of generic n by n matrices and put $Q = K<\xi_1,\ldots,\xi_m>$. Each element of Z(R) is a scalar matrix pI where p is a polynomial in the entries of $\xi_1,\ldots,\xi_m$, and Z(Q) is the field of fractions of Z(R). Therefore Z(R) may be viewed as a subring of the ring of polynomial invariants and Z(Q) as a subfield of the field of rational invariants. It is clear that not all polynomial invariants belong to the center of R but, as E. Formanek points out in [59], it is conceivable that Z(Q) consists of all rational invariants. However the latter is known to be true, so far, only in the case where the field has characteristic zero, cf. C. Procesi [137]. Then the ring of invariants is generated by the polynomial functions :

$$\varphi(A_1,\ldots,A_n) = Tr\ (A_{i_1}\ldots A_{i_j})\ ,$$

where $A_{i_1}\ldots A_{i_j}$ is any monomial in $A_1,\ldots,A_m$, repetitions and omissions allowed. Moreover, Q is a central simple algebra of dimension n^2 over its center, thus the reduced trace of any element is in Z(Q) i.e. $Tr(\xi_{i_1}\ldots\xi_{i_j}).I \in Z(Q)$. The latter implies that rational invariant ly in Z(Q) and this yields C. Procesi's result, mentioned earlier.

Because all of this we may reformulate C. Procesi's and E. Formanek's results as follows. If K is a field of characteristic zero then the field of rational invariants of m-tuples of 3 by 3 matrices over K is a rational function field of transcendence degree 9(m-1)+1. (Recently E. Formanek studied the 4 by 4 case !).

Rings of generic matrices over a field K are examples of finitely generated P.I. algebras over a field; these are called affine K-algebras. Recall from [136] the following extension of Hilbert's Nullstellensatz :

II.2.28. Theorem. If R is an affine K-algebra then R is a Hilbert algebra.

Recall that R is a Jacobson ring if every prime ideal of R is the intersection of the primitive ideals containing it; and R is a Hilbert algebra if it is a Jacobson ring such that every primitive ideal of R has finite codimension. In the commutative case it is known that, if R is a Jacobson ring, then :

1°. R[X] is a Jacobson ring,

2°. For any maximal ideal M in R[X], $R \cap M$ is maximal in R and the dimension of R[X]/M over $R/R \cap M$ is finite,

3° If R is a Hilbert algebra, then so is R[X].

In the non-commutative case it is not hard to find counter-examples to each of these properties. On the other hand, these properties do generalize to the case of finitely generated extensions. If R is a subring of S then S is a finitely generated extension of R if there exist $x_1, \ldots, x_n \in Z_R(S)$ such that $\{x_1, \ldots, x_n\}$ generates S as an R-algebra.

II.2.29. Theorem. Suppose that S is a P.I. ring and a finitely generated extension of a Jacobson P.I. ring R then the following properties hold :

1. S is a Jacobson ring.

2. If M is a maximal ideal of S then $R \cap M$ is a maximal ideal of R and S/M has finite rank over $R/R \cap M$.

3. In case R is a Hilbert algebra over a field K then S is a Hilbert algebra over K.

For the proof of II.2.29, cf. [136], the following fundamental fact may be used :

<u>II.2.30. Theorem.</u> If $R \to S$ is a finitely generated extension of prime P.I. rings, then :

1. If R is semisimple then so is S.
2. If R is semisimple and S is simple then R is simple and S is finite dimensional over the center of R.

In the sequel of this section we recollect some results of a somewhat "geometric" nature, most of which are due to C. Procesi, cf. [136].

<u>II.2.31. Proposition.</u> Let $R \to S$ be a finitely generated extension of prime P.I. rings. Then :

1. There exists a regular $c \in R$ such that for every maximal ideal M of R not containing c, $MS \cap R = M$.
2. If R is a Jacobson ring, then for every prime ideal P of R with $c \notin P$ we have $PS \cap R = P$.

<u>II.2.32. Corollary.</u> Under the hypotheses of II.2.31 a regular $c \in R$ may be selected such that :

1. For a maximal ideal M of R such that $c \notin M$, there is a maximal ideal $\tilde{M}$ of S such that $\tilde{M} \cap R = M$.
2. If R is a Jacobson ring and $P \in \mathrm{Spec}(R)$ is such that $c \notin P$ then there is a $\tilde{P} \in \mathrm{Spec}(S)$ with $\tilde{P} \cap R = P$.
3. If R(hence S too) is a Jacobson ring, then the assumption that the minimal prime ideals of S are finite in number entails that there is a $c \in R$ which is not in any minimal prime ideal and such that $c \notin P$ with $P \in \mathrm{Spec}(R)$ implies that there exists a $\tilde{P} \in \mathrm{Spec}(S)$ with $\tilde{P} \cap R = P$. Rephrasing this in

geometric terms : Spec(S) decomposes into a finite number of irreducible components and for a given irreducible component V_i of Spec(R) there is an irreducible component $\tilde{V}_i$ of Spec(S) which maps generically onto V_i under the canonical continuous map Spec(S)→Spec(R). The element $c \in R$ defines a Zariski closed set V(c) in Spec(R) which does not contain any component of Spec(R); the result mentioned just states that the image of Spec(S)→Spec(R) contains the (dense) open set Spec(R)− V(c) = X(RcR). Again, this geometric interpretation may be rephrased in terms of birational extensions, introduced in Section II.2, but we leave this to the reader.

If R satisfies the ascending chain condition on semiprime ideals then every semiprime ideal of R is a finite intersection of prime ideals. Furthermore, the number of minimal prime ideals over an arbitrary ideal I of R is finite and, in particular, there is only a finite number of minimal prime ideals in this case. The usefulness of II.2.32.3 is thus established by :

II.2.33. Proposition. If R is a finitely generated P.I. algebra over a Noetherian ring then R satisfies the ascending chain condition on semiprime ideals. In particular, the statement holds for any affine algebra over a field.

The foregoing applies in proving the following generalization of Posner's theorem :

II.2.34. Theorem. Let R be a finitely generated semiprime P.I. algebra over a Noetherian ring, then R has a total ring of (left and) right quotients which is a semisimple algebra with descending chain condition and satisfying the same identities as R.

Assume now that R→S is a finitely generated extension of prime P.I. rings, and let Q(R) and Q(S) be the rings of quotients of R and S resp. We have observed before that Z(Q(R)) = Q(Z(R)) and Z(Q(S)) = Q(Z(S)) and also :

$$Q(R) \subset Q(S)\ ,\ Z(Q(R)) \subset Z(Q(S))$$

II.2.35. Proposition. $Z(Q(S))$ is finitely generated as a field over $Z(Q(R))$.

Proof : (cf. [136]). Assume that S is generated as an R-ring by $\{a_1,\dots,a_t\} \subset Z_R(S)$. Pick a basis $\{m_1,\dots,m_n\}$ of $T = Z_{Q(R)}(Q(S))$ over $Z(Q(S))$. It is clear that we may write :

$$a_i = \sum_{j=1}^{n} \alpha_{ij}\, m_j \ ;\ m_i m_j = \sum_{k=1}^{n} \gamma_{ij}^{k}\, m_k$$

with $\alpha_{ij} \in Z(Q(S))$ and $\gamma_{ij}^{k} \in Z(Q(S))$.

Consider $F = Z(Q(R))(\alpha_{ij}, \gamma_{ij}^{k})$. We claim : $F = Z(Q(S))$! Indeed $U = \sum_j Fm_j$ is an F-algebra and $T = U \otimes_F Z(Q(S))$. Consequently U has to be a simple algebra with center F and $Q(S) = Q(R) \otimes_{Z(Q(R))} T = (Q(R) \otimes_{Z(Q(R))} U) \otimes_F Z(Q(S))$.

It will be sufficient to prove that $Q(S) = Q(R) \otimes_{Z(Q(R))} U$, to establish our claim. Now, from $R \subset Q(R) \otimes_{Z(Q(R))} U$, $a_i \in Q(R) \otimes_{Z(Q(R))} U$ we gather that $S \subset Q(R) \otimes_{Z(Q(R))} U \subset Q(S)$. But since $Q(R) \otimes_{Z(Q(R))} U$ is simple, elements of S which become units of Q(S) must become units in $Q(R) \otimes_{Z(Q(R))} U$. Finally, since Q(S) is generated by S and inverses of regular central elements of S we obtain that $Q(R) \otimes_{Z(Q(R))} U = Q(S)$. ■

Making use of techniques developed by Lestman, [107], some of the foregoing results may be strengthened; this is included in Section II.4.

Let us conclude this section by stating some general facts about Krull dimension of P.I. algebras. Recall that for any prime ideal P of R, the rank of P, denoted by rkP, is the supremum of the lengths of chains $P = P_0 \supset P_1 \supset \dots \supset P_n$ of prime ideals of R. The Krull dimension, dim R, of R is then defined to be sup {rkP, P ∈ Spec R}.

II.2.36. Theorem. Let R be a prime P.I. algebra over a field K and let $P \neq o$ be a prime ideal of R. Put $\overline{R} = R/P$ and let Q, $\overline{Q}$ be the rings of quotients of R, $\overline{R}$ resp.. We have : $\text{tr.deg}_K\, Z(\overline{Q}) \leq \text{tr.deg}_K\, Z(Q)$. Moreover, if $\text{tr.deg}_K\, Z(Q) < \infty$, then we have that $\text{tr.deg}_K\, Z(\overline{Q}) < \text{tr.deg}_K\, Z(Q)$.

It is just technical know-how to derive from the above theorem that prime ideals of a P.I. algebra which is finitely generated over a Noetherian ring have finite rank (cf. [29]).

II.2.37. Corollary. Let $P_1, \ldots, P_n$ be the minimal prime ideals of the affine K-algebra R, then $\dim R = \max\{\dim R/P_i\}$. If R is prime as well then $\dim R \leq \text{tr.deg}_K\, Z(Q)$.

Proof : The first statement is obvious. By Theorem II.2.36 it is clear that any chain of prime ideals in R has length at most equal to $\text{tr.deg}_K\, Z(Q)$. That Z(Q) is finitely generated as a field over K follows from II.2.35. ∎

Representation theory of P.I. rings will provide means of establishing that $\dim R = \text{tr.deg}_K\, Z(Q)$ holds for every prime affine K-algebra. These techniques are essentially different from the ones in this section and we postpone accurate treatment of this topic to Chapter IX.

The following generalization of the Krull-Akizuki theorem may be useful. (cf. [133] for its proof).

II.2.38. Theorem. Let R be a finitely generated P.I. extension of a commutative Noetherian ring, then the following statements are equivalent :

1. $\dim R = o$.
2. R is left Artinian.
3. R is right Artinian.

II.2.39. Comments and References.

General P.I. theory is well-documented in C. Procesi's book [136], L. Rowen [152], N. Jacobson [68]. For the center of the ring of generic matrices one may look up E. Formanek's papers, [58],[59]. It is only just to mention some important papers by S.A. Amitsur, [6],[7],[12], C. Procesi [132],[133]. For further references see Section II.3, II.4.

II.3. Azumaya Algebras.

Throughout this section, C is a commutative ring with unit. If A is a C-algebra then A^o denotes the opposite algebra i.e. A^o has the same underlying additive group as A but with multiplication defined by $(a,b) \to ba$. The enveloping algebra A^e of A is the C-algebra $A \otimes A^o$. It is clear that A is a left A^e-module, by $(a \otimes b).a' = aa'b$ with $a,a' \in A, b \in A^o$. The multiplication map $m : A^e \to A$, given by $m(a \otimes b) = ab$ is left A^e-linear and Ker(m) is the left ideal of A^e generated by the elements of the form $a \otimes 1 - 1 \otimes a$.
The C-algebra A is called separable if A is a projective left A^e-module, or equivalently, if A^e contains an element e such that $m(e) = 1$ and $(\mathrm{Ker}(m))e = o$. Obviously any left A^e-module M may be regarded as a two-sided A-module and conversely, (two-sided A-modules are supposed the induce the same C-module structure). There is a covariant functor from the category of two-sided A-modules to C-mod constructed by associating to the two sided A-module M the C-module $\{m \in M,\ am = ma \ \text{ for all } a \in A\} = M^A$. The functors $(-)^A$ and $\mathrm{Hom}_{A^e}(A,-)$ are naturally equivalent and $\mathrm{Hom}_{A^e}(A,A) = Z(A)$. In these terms we may define separability as follows : a C-algebra A is separable if and only if $(-)^A$ is a right exact functor. Using this it is not hard to deduce :

II.3.1. Proposition. Let R_1, R_2 be commutative C-algebras. Consider a separable R_1-algebra (resp. R_2-algebra) A_1 (resp. A_2)
Then $A_1 \otimes_C A_1$ is a separable $R_1 \otimes_C R_2$-algebra provided that $A_1 \otimes_C A_2 \neq o$ and $R_1 \otimes_C R_2 \neq o$.

II.3.2. Corollary. Let A be a separable C-algebra and R any commutative C-algebra then $A \otimes_C R$ is a separable R-algebra.

If A is not faithful as a C-algebra, say $I.1 = o$ for an ideal I of C, $1 \in A$ then A is naturally an algebra over C/I and $A \otimes_C A^o = A \otimes_{C/I} A^o$. Consequently A is C-separable if and only if A is C/I-separable, therefore it is not really restrictive to assume that no ideal of C annihilates $1 \in A$.

II.3.3. Proposition. Let A be a separable C-algebra and let J be an ideal of A. Then A/J is a separable C-algebra, hence a separable $C/C \cap J$-algebra. Moreover $Z(A/J) = Z(A) \bmod J$.

Proof : A two-sided A/J-module M is naturally a two-sided A-module and $M^A = M^{A/J}$. Now right exactness of $(-)^A$ yields right exactness of $(-)^{A/J}$ and thus A/J is a separable C-algebra but, by the remark preceding this proposition, then also a separable $C/C \cap J$-algebra. Separability of A over C yields that $A^A \to (A/J)^A \to o$ is exact and since $A^A = Z(A)$, $(A/J)^A = Z(A/J)$ it follows that Z(A/J) is the image of Z(A) under the canonical epimorphism $A \to A/J$. ∎

II.3.4. Remark. Let C_1, C_2 be commutative rings and A_1 an algebra over C_1, A_2 an algebra over C_2. Then $A_1 \oplus A_2$ is a separable $C_1 \oplus C_2$-algebra if and only if both A_1 and A_2 are separable over C_1 and C_2 resp.

An Azumaya algebra A is an algebra separable over its center which is faithful as a $Z(\Lambda)$-module. Before turning to the study of Azumaya algebras let us mention a useful local-global property :

II.3.5. Proposition. Let $A \in \underline{Alg}_C$ be finitely generated in C-mod. Then the following statements are equivalent :

1. A is a separable C-algebra.
2. A_p is a separable C_p-algebra for all $p \in Spec(C)$
3. A_m is a separable C_m-algebra for all maximal ideals m of c.

Proof : cf. [48]. ∎

II.3.6. Proposition. Let A be an Azumaya algebra over $C = Z(A)$. Then C is a direct summand of A in C-mod.

Proof : Let $\varphi : A^e \to Hom_C(A,A)$ be given by : $\varphi(\alpha)(a) = \sum_i a_i a b_i$, where $\alpha = \sum_i a_i \otimes b_i$. Let $e \in A^e$ be the element introduced at the beginning of the section. Then $\varphi(e)$ is an idempotent and therefore the image of $\varphi(e)$ is a

C-direct summand of A, say E, such that $E \oplus \mathrm{Ker}\,\varphi(e) = A$. Utilizing the fact that $(a \otimes 1 - 1 \otimes a)e = o$ we see that, for any $a, b \in A$: $a(\varphi(e)(b)) = (a \otimes 1)e.b = (1 \otimes a)e.b = (\varphi(e)(b))a$.

Thus $\varphi(e)(b) \in Z(A) = C$ and this entails that the image of $\varphi(e)$ is exactly C. ∎

II.3.7. Theorem. Let A be a C-algebra; the following statements are equivalent:

1. A is an Azumaya algebra with $Z(A) = C$.
2. The functors $N \to A \otimes_C N$ and $M \to M^A$ define a category equivalence between C-mod and A^e-mod.
3. A is a faithful C-module of finite type and A/mA is a central simple C/m-algebra for each maximal ideal m of C.

Proof : cf. [95], Théorème 5.1. p.93. ∎

II.3.8. Proposition. Let A be an Azumaya algebra with $C = Z(A)$, then we have the following properties :

1. If I is an ideal of C then $AI \cap C = I$.
2. Every ideal of A is of the form AI for some ideal I of C.
3. For every two-sided A-module M we have that the mapping $A \otimes_C M^A \to M$ defined by $a \otimes m \to am$ for $a \in A$, $m \in M$, is an isomorphism in A-mod.
4. Every C-endomorphism of A is an automorphism.

Proof. 1. By Proposition II.3.6. : $A = C \oplus E$ for some C-submodule E of A. Obviously we obtain : $AI \cap C = (C \oplus E)I \cap C = (I \oplus EI) \cap C = I$.

2. Any ideal of A may be considered as an A^e-submodule M of A. By Theorem II.3.7.2 it follows then that M is generated by $M^A \subset Z(A) = C$.

3. Obvious.

4. If $f : A \to A$ is an endomorphism of C-algebras then $\mathrm{Ker}(f) = AI$ for some ideal I of C by II.3.8.2. However, as f is such that $f|C = 1_C$, f has to be injective. Considerations of rank (i.e. local rank) yields that the centralizer of f(A) in A is nothing but the center and $A = f(A)$ follows. ∎

II.3.9. Corollaries.

1. An Azumaya algebra satisfies a polynomial identity in view of Theorem II.3.7.3.
2. An Azumaya algebra is a global Zariski extension of its center in view of II.3.8.2. Therefore the results of Section II.1. apply. For any prime ideal P of A the rings $Q_{A-P}(A)$ and $Q_{C-p}(A)$ are isomorphic (where $p = P \cap C$, $P = A$) and κ_{A-P} is a t-functor. The local ring $Q_{A-P}(A)$ is an Azumaya algebra with center $Q_{C-p}(C)$ and therefore $Q_{A-P}(A)$ is a free $Q_{C-p}(C)$-module of rank r_p say. Since we have : $Q_{A-P}(A/P) = Q_{A-P}(A)/Q_{A-P}(P)$ we may use the first corollary to derive that p.i. deg $P = [A/P : C/P] = [Q_{A-P}(A) : Q_{C-p}(C)]$. The prime ideals P of A with the same $r = r_p$ form an open subset X_r of $X = \mathrm{Spec}(A)$ and we obtain a partition $X = \bigcup_{r=1}^{n} X_r$ where $n =$ p.i. deg A.
3. If an Azumaya algebra A has constant rank (i.e. $r_p = r$ for all prime ideals P of A) then every prime ideal P of A has p.i. degree r.
4. From the second Corollary we also retain that A satisfies the left and right Ore conditions with respect to C(P) for every prime ideal P of A.

We do not go into the theory of the Brauer group of a commutative ring here, nevertheless let us mention the following results :

II.3.10. Proposition. 1. If A and B are Azumaya algebras over C then $A \otimes_C B$ is an Azumaya algebra over C.

2. If A is an Azumaya algebra with center C and D is a commutative C-algebra then $A \otimes_C D$ is an Azumaya algebra with center D.

Proof : cf. [48]. ∎

II.3.11. Proposition. If A is an Azumaya algebra with center C then there exists a Noetherian subring C_o of C and an Azumaya algebra A_o with center C_o such that $A_o \otimes_{C_o} C = A$.

<u>Proof</u> : By "descent" to the Noetherian case for projective modules, see for example Proposition 2.8 in [95] p.8.

<u>II.3.12. Lemma</u>. Let A be a Noetherian Azumaya algebra with center C. Let $M,N \in A\text{-mod}$ be such that M is of finite type whilst N has finite representation. If $f : M \to N$ is A-linear such that $Q_{A-P}(f) : Q_{R-M}(M) \to Q_{R-P}(N)$ is an isomorphism for some prime ideal P of A then there is a t-functor κ_I such that $Q_I(f) : Q_I(M) \to Q_I(N)$ is an isomorphism for some ideal I of A.

<u>Proof</u> : Since $Q_{R-P}(N/f(M)) = o$ and $N/f(M)$ being of finite type, it follows that there is an ideal $J \in \mathcal{L}(\kappa_{A-P})$ for which $J.(N/f(M)) = o$. Choose $j \in J \cap C$, then κ_{A_j} is a t-functor such that $Q_j(f) : Q_j(M) \to Q_j(N)$ is surjective. Exactness of Q_j implies that $Q_j(M)$ and $Q_j(N)$ are of finite $Q_j(A)$-type and $Q_j(N)$ has finite representation. Therefore Ker $Q_j(f)$ is of finite type, and $Q_{A-P}(\text{Ker}(Q_i(f)) = 0$. As before we obtain κ_{A_i} with $i \in C$ such that $Q_{ij}(f) : Q_i(Q_j(M)) \to Q_i(Q_j(N))$ is an isomorphism. Taking $I = A_i \cap A_j$ yields the desired κ_I;. ■

<u>II.3.13. Theorem</u>. Let A be a Noetherian Azumaya algebra and let M be a two-sided A-module which is a projective left A-module of finite type. Let P be a prime ideal of A, then $Q_{A-P}(M)$ is a free $Q_{A-P}(A)$-module.

<u>Proof</u> : From II.3.9.2. it follows that $Q_{A-P}(M) = Q_{A-P}(A) \underset{A}{\otimes} M$ is a projective $Q_{A-P}(A)$-module of finite type, hence it has finite representation. It is easily seen that $P^e = Q_{A-P}(P)$ is the Jacobson radical of $Q_{A-P}(A)$ and that :

$$P^e = Q_{A-P}(A) \underset{Q_{C-p}(C)}{\otimes} p.Q_{C-p}(C) = p\, Q_{C-p}(C) \underset{Q_{C-p}(C)}{\otimes} Q_{A-P}(A) .$$

Therefore : $P^e \underset{Q_{A-P}(A)}{\otimes} Q_{A-P}(M) \cong p\, Q_{C-p}(C) \underset{Q_{C-p}(C)}{\otimes} Q_{A-P}(M)$.

Now, $Q_{A-P}(A)$ is $Q_{C-p}(C)$-free of finite rank and $Q_{A-P}(M)$ is projective of finite type, therefore $Q_{A-P}(M)$ is $Q_{C-p}(C)$-projective of finite type and so we obtain that $\psi : p\, Q_{C-p}(C) \underset{Q_{C-p}(C)}{\otimes} Q_{A-p}(M) \to Q_{A-p}(M)$, which is derived from the injection $p\, Q_{C-p}(C) \to Q_{C-p}(C)$, is again injective. Furthermore, since the localization Q_{A-p} is central, $Q_{A-p}(M)$ is a two-sided $Q_{A-p}(A)$-module.

Now, $Q_{A-p}(M)/P^e\, Q_{A-p}(M) = Q_{A-p}(A/P) \underset{Q_{A-p}(A)}{\otimes} Q_{A-p}(M)$ is a projective two-sided $Q_{A-p}(A/P)$-module of finite type, and as $Q_{A-p}(A/P)$ is simple, the latter entails that $Q_{A-p}(M)/P^e\, Q_{A-p}(M)$ a free $Q_{A-p}(A/P)$-module.

Summarizing this we obtain :

1°. $Q_{A-p}(M)$ has finite presentation as a left $Q_{A-p}(A)$-module.

2°. The $Q_{A-p}(A)$-linear $\psi : P^e \underset{Q_{A-p}(A)}{\otimes} Q_{A-p}(M) \to Q_{A-p}(M)$ is injective.

3°. The $Q_{A-p}(A)/P^e$-module $Q_{A-p}(M)/P^e\, Q_{A-p}(M)$ is free, and P^e is the Jacobson radical of $Q_{A-p}(A)$.

These conditions yield that $Q_{A-p}(M)$ is a free $Q_{A-p}(A)$-module (cf. [32]).

II.3.14. Corollary. If M is a two sided Λ module then M is locally free.

Proof : Combine Lemma II.3.12 and Theorem II.3.13. ∎

II.3.15. Remark. The restriction to the Noetherian case in II.3.12, II.3.13 and II.3.14. can be avoided if one utilizes Proposition II.3.11 and "descent" of the projective modules as in Proposition 2.8 of [95].

It is possible to give an intrinsic characterization of Azumaya algebras within the class of P.I. algebras. This is exactly the content of the celebrated Artin-Procesi theorem. Let us point out here how this can be achieved; following the set up used by M. Artin in [17] and assuming that we consider algebras which contain a field in the center. Along the way we introduce and study "bimodules in Artin's sense"; these will play a fundamental role in the construction of schemes over P.I. rings in Chapter V.

Let M be a two-sided R-module; the category of two-sided R-modules is a Grothendieck category and it will be denoted by $\underline{\underline{2}}(R)$. If $M \in \underline{\underline{2}}(R)$ then its R-center $Z_R(M)$ is defined to be the set $\{m \in M,\ rm = mr \text{ for all } r \in R\}$. We say that $M \in \underline{\underline{2}}(R)$ is an R-bimodule (from now on we drop the qualification "in Artin's sense" which one usually finds joined with the term bimodule defined here) if M is generated as an R-module (left or right!) by $Z_R(M)$. The category of R-bimodules is not a Grothendieck category but is a full subcategory of $\underline{\underline{2}}(R)$ which we will denote by $\underline{\underline{bi}}(R)$. A morphism in $\underline{\underline{bi}}(R)$ is a map carrying $Z_R(M)$ into $Z_R(N)$ and being left (hence right) R-linear; the set $\mathrm{Hom}_{\underline{\underline{bi}}(R)}(M,N)$ is a module over the center $Z(R)$ of R.

Clearly $\mathrm{Hom}_{\underline{\underline{2}}(R)}(R,M)$, for any $M \in \underline{\underline{2}}(R)$, corresponds bijectively to the set $Z_R(M)$, the correspondence being given by $f \mapsto f(1)$. A free bimodule is one isomorphic to $R^{(n)}$. Every $M \in \underline{\underline{bi}}(R)$ is a quotient of a free bimodule F i.e. there is an exact sequence $0 \to \kappa \to F \to M \to 0$ where $\kappa \in \underline{\underline{2}}(R)$, not necessarily in $\underline{\underline{bi}}(R)$! We have:

$\mathrm{Hom}_{\underline{\underline{bi}}(R)}(R \underset{Z(R)}{\otimes} N, M) \approx \mathrm{Hom}_{Z(R)}(N, Z_R(M))$ for all $N \in Z(R)$-mod, $M \in \underline{\underline{bi}}(R)$.

II.3.16. Remark. If R is an Azumaya algebra then $\underline{\underline{2}}(R) = \underline{\underline{bi}}(R)$, as indicated before. Moreover the categories of R-algebras (in the sense of C. Procesi i.e. ring extensions $R \to A$) is equivalent to $\underline{\mathrm{Alg}}_{Z(R)}$ via the functor $R \underset{Z(R)}{\otimes} -$.

It is easily seen that the tensor product $M \underset{R}{\otimes} N$ of bimodules is again a bimodule and that $M \underset{R}{\otimes} N \cong N \underset{R}{\otimes} M$ in $\underline{\underline{bi}}(R)$. Let us summarize the results of Section 4 of [17] in :

II.3.17. Proposition. Let $R \to A$ be an extension of rings and let $N \in \underline{\underline{bi}}(A)$. Then we have the following properties :

1°. $N \in \underline{\underline{bi}}(R)$ by restriction of scalars.

2. $R\,Z_A(N)$ is an R-bimodule and $Z_A(N) = Z_R(R.Z_A(N))$.

3. If $M \in \underline{\underline{bi}}(R)$ then $\mathrm{Hom}_{\underline{\underline{bi}}(A)}(A \underset{R}{\otimes} M, N) \cong \mathrm{Hom}_{\underline{\underline{bi}}(R)}(M, R\,Z_A(N))$.

4. If $R \to B$ is another ring extension then there is a unique ring structure on $A \underset{R}{\otimes} B$ such that $\alpha \otimes \beta = (\alpha \otimes 1)(1 \otimes \beta) = (1 \otimes \beta)(\alpha \otimes 1)$ for all $\beta \in Z_R(B)$, $\alpha \in Z_R(A)$, and such that the following diagram is a commutative diagram of ring extensions

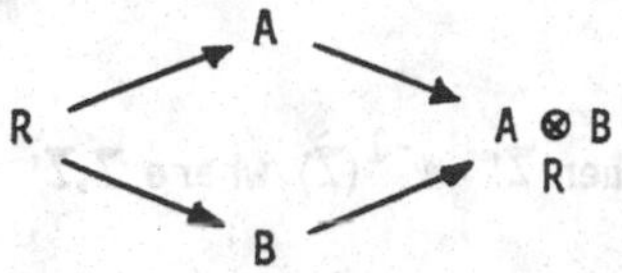

If either $b' \in Z_R(A)$ or $c \in Z_R(B)$ then $(b \otimes c)(b' \otimes c') = bb' \otimes cc'$.

Let $f : R \to A$ be an extension of rings. We say that f is <u>flat for bimodules</u> if an exact sequence $M' \to M \to M''$ in $\underline{bi}(R)$ yields an exact sequence:

$A \underset{R}{\otimes} M' \to A \underset{R}{\otimes} M \to A \underset{R}{\otimes} M''$

which is exact in $\underline{bi}(A)$.

<u>II.3.18. Proposition.</u> The following statements are equivalent :

1. A sequence $M' \to M \to M''$ in $\underline{bi}(R)$ is exact if and only if the sequence $A \underset{R}{\otimes} M' \to A \underset{R}{\otimes} M \to A \underset{R}{\otimes} M''$ is exact.
2. A is flat for R-bimodules and for every $M \in \underline{bi}(R)$ the map $M \to A \underset{R}{\otimes} M$ defined by $m \to 1 \otimes m$ is injective.

An extension having one of the equivalent properties of II.3.18 is said to be <u>faithfully flat</u> for bimodules. Faithful flatness of an extension entails many useful properties for the prime spectra of the rings involved. In particular if $f : R \to A$ is a faithfully flat extension for bimodules then the induced map $\tilde{f} : \mathrm{Spec}(A) \to \mathrm{Spec}(R)$ is surjective.

<u>II.3.19. Proposition.</u> Let $R \to R'$ be a flat extension for bimodules and let $f : R \to A$ be an arbitrary extension. Put $A' = A \underset{R}{\otimes} R'$ and let $f' : R' \to A'$ be the map induced by f. Put $K = \mathrm{Ker}(f)$, $K' = \mathrm{Ker}(f')$, then $K' = KA'$.

II.3.20. Corollaries.

1. Let $X = \mathrm{Spec}\, R \leftarrow X' = \mathrm{Spec}\, R'$ correspond to : $R \longrightarrow R'$

$$\begin{array}{ccc} X = \mathrm{Spec}\, R & \leftarrow & X' = \mathrm{Spec}\, R' \\ \uparrow{\scriptstyle \tilde f} & & \uparrow{\scriptstyle \tilde f'} \\ Y = \mathrm{Spec}\, A & \underset{\psi}{\leftarrow} & Y' = \mathrm{Spec}\, A' \end{array} \qquad \begin{array}{ccc} R & \longrightarrow & R' \\ {\scriptstyle f}\downarrow & & \downarrow{\scriptstyle f'} \\ A & \longrightarrow & A' \end{array}$$

and put $Z = \mathrm{Im}\, \tilde f$, $Z' = \mathrm{Im}\, \tilde f'$ then $\overline{Z'} = \psi^{-1}(\overline{Z})$ where $\overline{Z}, \overline{Z'}$ stand for the Zariski-closures of Z, Z' resp.

2. If $R \to R'$ is moreover faithfully flat for bimodules and $Y'' \underset{p_2}{\overset{p_1}{\rightrightarrows}} Y' \underset{\psi}{\longrightarrow} Y$ is the diagram corresponding to $R \to R' \rightrightarrows R' \underset{R}{\otimes} R'$. Then a closed set $C' \subset Y'$ is of the form $\psi^{-1}(C)$ for a closed set $C \subset Y$ if and only if $C' = p_1\, p_2^{-1}(C')$.

II.3.21. Lemma. There exist polynomials f and g in two variables over the prime field k. such that for $n \times n$ - matrices U and V the following equation holds identically : $f(U,V)\mathrm{Tr}\, U = g(U,V)$, where $\det(f(U,V))$ is not identically zero.

Now let k be an infinite field (take an infinite extension field of k_o) and let T be the ring of $n \times n$ matrices generated over k by the generic matrices ξ_i and by the traces $\mathrm{Tr} g$, $g \in k\{\xi_i\}$ (Trg is identified with the matrix Trg.I.). The ring T is a subring of $M_n(k(\xi^k_{ij}))$ and since $\mathrm{Tr} g \in Z(T)$ for all $g \in k\{\xi_i\}$, the inclusion $k\{\xi_i\} \hookrightarrow T$ is a central extension. Clearly $\mathrm{Tr}\, t \in T$ for all $t \in T$. Clearly the identities of $n \times n$-matrices hold in T and $T \hookrightarrow M_n(k(\xi^k_{ij}))$ is a central extension, therefore T is a prime ring.

II.3.22. Proposition. The inclusion $k\{\xi_i\} \hookrightarrow T$ induces a map $\psi : \mathrm{Spec}(T) \to \mathrm{Spec}(k\{\xi_i\})$ which restricts to a homeomorphism of the open subsets consisting of prime ideals of p.i. degree n in both spaces, denoted by $U' = \mathrm{Spec}_n(T)$, resp. $\mathrm{Spec}_n(k\{\xi_i\}) = U$.

Proof. Clearly $\psi^{-1}(U) = U'$. To show that U' maps onto U, consider $P \in U$. Then $k\{\xi_i\}/P$ embeds into a matrix algebra $M_n(K)$ for some field K containing k. The

map $k\{\xi_i\} \to M_n(K)$, defined by specializing the generic matrices ξ_i to ξ_i mod P , extends to a map on the traces i.e. to a map $T \to M_n(K)$ and the kernel of this map is in U' and in the pre-image of P. A prime ideal $P' \in U'$ corresponds to $\psi(P') = P = P' \cap k\{\xi_i\}$. We claim that $P' = \{t \in T,\ st \in P$ for some $s \in k\{\xi_i\}$ such that $s \bmod P$ is regular$\}$, so P' is uniquely determined by P and thus $\psi|U'$ is injective, i.e. bijective. Since $P' \in U'$, T/P' embeds into some $M_n(K)$ and the image of $k\{\xi_i\}$ generates $M_n(K)$ as a K-vector space. Since k is generic, it is clear that we can find $u,v \in k\{\xi_i\}$ such that $\det(f(u,v)) \neq o$, $\det f(\overline{u},\overline{v}) \neq o$ where f is a polynomial selected as in Lemma II.3.21., $\overline{u},\overline{v}$ being the images of u,v in $M_n(K)$. For any $t \in k\{\xi_i\}$ and sufficiently general $c \in k$ we get : $\det(f(u+ct,v)) \neq o$, $\det(f(\overline{u+ct},\overline{v})) \neq o$. Write $t \in T$ as $\sum_i s_j t_j$ where each t_j is a product of traces of elements of $k\{\xi_i\}$. By linearity of the trace we may rewrite Tr t as c^{-1} (Tru-Tr(u+ct)).
Applying Lemma II.3.21. to the traces appearing in the t_j and clearing denominators by multiplying t with enough elements f(u,v), f(u+ct,v) we obtain $t = s^{-1}z$ with $s,z \in k\{\xi_i\}$. Now the elements f(u,v), f(u+cz,v) have invertible images in $M_n(K)$ and thus they cannot divide o in $k\{\xi_i\}/P$, i.e. s is regular in $k\{\xi_i\}/P$. It is now easily seen that ψ is indeed a homeomorphism. ■

<u>II.3.23. Lemma.</u> $\mathrm{Spec}_n(T)$ is a union of spectra of the form $\mathrm{Spec}(T_s)$ where T_s is obtained from T by inverting $s \in Z(T)$ and T_s is an Azumaya algebra of rank n^2 over its center, which is also a left and right flat central extension of $k\{\xi_i\}$.

<u>Proof.</u> Let $P' \in U'$, i.e. T/P' is an order in a central simple algebra Λ of rank n^2 over its center which is a field. Choose elements $z_{ij} \in k\{\xi_i\}$ such that their images in Λ generate Λ over its center and such that $d = \det(\mathrm{tr}(z_{ij}\, z_{k\ell}))$ $\neq o$. The localised ring T_d may be viewed as a subring of $M_n(k(\xi^k_{ij}))$. Since d is invertible, the z_{ij} are linearly independent in $M_n(k(\xi^k_{ij}))$, thus also over $Z(T_d)$. We show that T_d is generated by the z_{ij} over $Z(T_d)$.

Given a matrix m then the n^2 equations : $c_{11}\,Tr(z_{11}\,z_{ij})+\ldots+c_{nn}\,Tr(z_{nn}\,z_{ij})=Tr(m\,z_{ij})$ have a unique solution for $c_{ij}=c_{ij}(m)$ which is an integral expression in the elements $Tr(mz_{ij})$, $Tr(z_{ij}\,z_{k\ell})$ and d^{-1}. Applied to $m\in T$ this yields that the $c_{ij}(m)$ are integral expressions in $Tr(m\,z_{ij})$, $Tr(z_{ij}\,z_{k\ell})$ and d^{-1} where these are now elements of $Z(T_d)$. Consequently T_d is a free $Z(T_d)$-module with basis $\{z_{ij}\}$. It is not hard to derive that T_d is an Azumaya algebra (since T_d specializes to a central simple algebra to of rank n^2 one may use the theory as set up in [25], or use results on unramified pseudo-places as in Van Oystaeyen [174]). Moreover it is obvious that $U'\cap \mathrm{Spec}\,T_d$ has a covering of the desired type and so we obviously obtain such a covering for U'. For the statement on flatness we refer to M. Artin [17], Lemma 10.10. ■

<u>II.3.24. Lemma</u>. Let $f: A\to B$ be a central extension which is faithfully flat for bimodules. Then B is an Azumaya algebra of rank n^2 over its center if and only if A is.

<u>Proof</u>. The "if" part is obvious. For the converse we refer to Lemma 11.1 of [17]; but it is also possible to modify the techniques of faithfully flat descent in [95] to this case. ■

<u>II.3.25. Theorem</u>. (M. Artin, C. Procesi). A k_o-algebra A is an Azumaya algebra of rank n^2 over its center Z(A) if and only if :

1. A satisfies the identities of nxn-matrices in characteristic char k_o.
2. Every $P\in \mathrm{Spec}(A)$ has p.i. deg. P = n.

<u>Proof</u>. By Lemma II.3.24. it will be sufficient to establish that $A\underset{k_o}{\otimes}k$ is an Azumaya algebra of constant rank n^2 over its center, where k is an infinite extension field of k_o. Hence we may as well suppose that A is a k-algebra, i.e. A is a quotient of the ring $k\{\xi_i\}$ of generic n x n-matrices, say $\pi: k\{\xi_i\}\to A\to o$. The induced map $\tilde{\pi}: \mathrm{Spec}(A)\longrightarrow \mathrm{Spec}(k\{\xi_i\})$ identifies Spec(A)

with a closed subset of $\mathrm{Spec}(k\{\xi_i\})$ which is contained in $U = \mathrm{Spec}_n(k\{\xi_i\})$. Since $\mathrm{Spec}(A)$ is quasi-compact it will be covered by the images of finitely many $\mathrm{Spec}(T_{s_j})$ $j = 1,\dots,N$, which yields open sets in U. Put $T_j = A \underset{k\{\xi_i\}}{\otimes} T_{s_j}$ and $\overline{T} = T_1 \times \dots \times T_N$. Then, from Lemma II.3.23 it follows that T_j is left flat over A and thus $\overline{T}$ is left flat over A. By Lemma II.3.24 it will therefore suffice if we establish that $\overline{T}$ is faithfully flat over A for left modules. By the covering property $\mathrm{Spec}\,\overline{T} \to \mathrm{Spec}\,A$ is surjective. If M is a simple left A-module then we have to show that $\overline{T} \underset{A}{\otimes} M \neq o$. However, A satisfies the identities of $n \times n$-matrices, hence by Kaplansky's theorem, the primitive quotients of A are central simple algebras. Now $A/\mathrm{ann}_A M$ is isomorphic to a sum of copies of M. On the other hand, surjectivity of $\mathrm{Spec}(\overline{T}) \to \mathrm{Spec}(A)$ yields that

$$\overline{T}/(\mathrm{ann}_A M).\overline{T} = \overline{T} \underset{A}{\otimes} A/\mathrm{ann}_A M \neq o.$$

Therefore $\overline{T} \underset{A}{\otimes} M \neq o$ and it follows that $A \to \overline{T}$ is faithfully flat. ∎

<u>II.3.26. Remark.</u> If $p \in \mathrm{Spec}(A)$ is a maximal ideal and $P = \pi^{-1}(p) \in \mathrm{Spec}(k\{\xi_i\})$ then it is not hard to see that $A/p = Q_{k\{\xi_i\}-P}(k\{\xi_i\}/P)$.
Since p.i.deg $p = n$, also p.i. deg $P = n$ and the $Q_{k\{\xi_i\}-P}$ is a central localization (we use the results on birationality of Section II.1. here). However, looking at the construction in Lemma II.3.24 and the nature of π in II.3.21 it is clear that d maps to an invertible element of A/P, consequently : $Q_{k\{\xi_i\}-P}$ extends to a localization of T such that certain s are inverted i.e. if P' is the prime ideal of T sitting over P then $Q_{T-P'}(T)$ is a localization of certain T_s (for some s which are not in P') and as such $Q_{T-P'}(T)$ is an Azumaya algebra. The obvious extension of π to T and to $Q_{T-P'}(T)$ is a specialization $Q_{T-P'}(T) \to Q_{A-p}(A) \to o$. Therefore the $Q_{A-p}(A)$ are Azumaya algebras of rank n^2 for each $p \in \mathrm{Spec}(A)$. However the fact that $\mathrm{Spec}_n(A) = \mathrm{Spec}(A)$ entails that A is a global Zariski extension of its center, hence the Q_{A-P} are in fact central localization $Q_{Z(A)-p^c}$, where $p^c = p \cap Z(A)$.
Now Proposition II.3.5 entails that A is an Azumaya algebra of constant rank

n^2.

Note that we have obtained another proof of the Artin-Procesi theorem in which some of the flatness-argumentation has been implicitely used in terms of exactness of certain (central) localizations. Moreover from this proof it is immediate that in formulating the Artin-Procesi theorem only maximal ideals have to be considered (that is of course a direct consequence of the formulation of the theorem but maybe less obvious in the proof given).

Applications of the above theorem will be encountered in the study of non-commutative varieties, in particular if these are to be linked to certain "central varieties".

II.3.27. Comments and References.

G. Azumaya introduced maximally central algebras in [25]; in N. Bourbaki [31] these algebras are called Azumaya algebras. The natural environement for these algebras is the theory of orders (over Dedekind rings) in the arithmetical ideal theory of central simple algebras, cf. [49], [144]. M-A. Knus and M. Ojanguren treated Azumaya algebras in the context of "descent" and Galoisian Cohomology in [95]. The use of Azumaya algebras over a fixed commutative ring i.e. the study of the Brauer group of a commutative ring is well documented by A. Grothendieck in [74], , and also by F. De Meyer, E. Ingraham in [48]. Azumaya algebras appear as certain "unramified pseudo-places" in some work of F. Van Oystaeyen [174]. If one thinks of the specialization $\pi : k\{\xi_i\} \to A/p$ encountered in this section, as being extended to the central localization at $\pi^{-1}(p)$ of $k\{\xi_i\}$, then the fact that p.i. deg $p = n$ will ensure that π is induced by an unramified pseudo-place of $M_n(k(\xi^k_{ij}))$ but we have not gone into this here. The Artin-Procesi theorem may be found in Procesi's book [136], where the theorem is stated for a ring A which need not be an algebra over some (prime) field. Since the assumption : A is an algebra over a field is perfectly compatible with our aims of studying the geometry of P.I. rings we have chosen for M. Artin's presentation, following

[17]. Let us point out that P.M. Cohn included a very neat proof of Artin-Procesi's theorem for prime rings in [39]; this proof uses Rasmyslovs multilinear central polynomial and is credited to W. Schelter.

II.4. Further Results on P.I. Rings.

In this section we have grouped together some results on rings with polynomial identities which have a particular geometrical meaning. Let us start by refining some properties mentioned in II.2; most of these refinements are due to Lestmann, [107].

II.4.1. Lemma. Let $R \to S$ be an extension of P.I. rings and let $P \subset P_1$ be prime ideals of R. Suppose that Q_1 is a prime ideal of S such that $Q_1 \cap R = P_1$. Put $W = \{Q_\alpha \in \mathrm{Spec}(S), Q_\alpha \cap R = P\}$ and $I = \cap\{Q_\alpha, Q_\alpha \in W\}$. If $W \neq \phi$ then a necessary and sufficient condition for the existence of a $Q \in \mathrm{Spec}\, S$ such that $Q \subset Q_1$, $Q \cap R = P$ is that I is contained in Q_1.

Proof. If such $Q \in \mathrm{Spec}(S)$ exists then obviously $I \subset Q_1$. Conversely suppose $I \subset Q_1$ i.e. $P \subset I \subset Q_1$. Let $Q \in \mathrm{Spec}(S)$ be contained in Q_1 and minimal over P. If $Q \cap R \neq P$ then in $\overline{S} = S/I$ we get that $\overline{Q} \cap \overline{R} \neq o$. Since $\overline{R} \to \overline{S}$ is again an extension of P.I. rings, while $I \cap R = P$, we see that $\overline{R}$ is a prime P.I. ring. Hence, by L. Rowen's results, we find $\overline{q} \neq o$ in $Z(\overline{R}) \cap \overline{Q}$. Now $Z(\overline{R}) \subset Z(\overline{S})$ and $\overline{Q} \cap Z(\overline{S})$ must consist of zerodivisors (by the minimality assumption on Q), therefore we can find $\overline{t} \neq o$ in $\overline{S}$ such that $\overline{qt} = o$. But $\overline{t} \neq o$ entails that there exists a Q_α such that $\overline{t} \notin \overline{Q}_\alpha$ and the latter forces $\overline{q} \in \overline{Q}_\alpha$ (otherwise $\overline{S}/\overline{Q}_\alpha$ is a prime P.I. ring with a non-zero central zero-divisor, which is of course unheard of). Now $Q_\alpha \cap R = P \subset I$ entails that $\overline{q} \in \overline{Q}_\alpha \cap \overline{R} = o$, a contradiction. Thus $\overline{Q} \cap \overline{R} = o$ or $Q \cap R = I \cap R = P$ follows. ∎

II.4.2. Theorem. Let $R \to S = R\{a_1, \ldots, a_k\}$ be a finitely generated extension of prime P.I. rings. There exists $c \neq o$ in $Z(R)$ such that, if $P \in \mathrm{Spec}(R)$ does not contain c, then $PS \cap R = P$,

Proof. Let $Q(R) \subset Q(S)$ be the rings of fractions of R and S and put $T = Q(R)\{a_1, \ldots, a_k\}$. Then $R \subset S \subset T \subset Q(S)$. Pick a maximal ideal N of T and write $U = T/N$. Since T is finitely generated as an algebra over $C = Z(Q(R))$ it follows that U is finite dimensional over C. Note that we may regard $R \subset Q(R) \hookrightarrow U$ as an isomorphic embedding. Put $\bar{S} = S \bmod N$. Observe that $P\bar{S} \cap R = P$ entails $PS \cap R = P$ for any $P \in \mathrm{Spec}(R)$. Therefore if $c \in Z(R)$ does satisfy the statements of the theorem when formulated in $\bar{S}$ then it will satisfy the statements in S. Henceforth we assume $R \subset S \subset U$ and this does not restrict generality. If $V = Z_{Q(R)}(U)$ then $Q(R) \otimes_C V = U$ follows from the centralizer theorem in Section I.1. The fact that V is finite dimensional over C allows to embed V in $M_m(C)$ for some suitable $m \in \mathbb{N}$. Consequently, both Q(R) and U embed into $Q(R) \otimes_C M_m(C) = M_m(Q(R))$. Since Q(R) is obtained from R by central localization we have that $M_m(Q(R)) = Q(M_m(R))$. It follows that we may choose $c \neq o$ in Z(R) such that $ca_i \in M_m(R)$ for each a_i, $i=1,\dots,k$. Thus for every $s \in S$ there exists an integer $r \in \mathbb{N}$ such that $c^r s \in M_m(R)$.

Suppose now that $P \in \mathrm{Spec}(R)$ is such that $c \notin P$, and write $b = \sum_j p_j\, a_{i_1} \dots a_{i_m} \in PS \cap R$, where $p_j \in P$ and $a_{i_j} \in \{a_1, \dots, a_k\}$, $m \in \mathbb{N}$.

Up to multiplication by powers of c we obtain :

$$c^q\, b = \sum_j p_j'\, (ca_{i_1}) \dots (ca_{i_m})$$

where $p_j' \in P$ and $ca_{i_j} \in M_m(R)$ for each j.

Clearly, the correspondence between ideals of R and ideals of $M_m(R)$ yields that $c^g\, b \in PM_m(R) \cap R = P$, hence $b \in P$ and $PS \cap R = P$ follows. ∎

In the foregoing proof we have used the following general but perhaps less known result :

<u>II.4.3. Lemma</u>. Let R be any ring and P a minimal prime ideal of R then $P \cap Z(R)$ consists of zero-divisors of R.

<u>Proof</u>. Let S be the multiplicatively closed subset of R generated by R-P and all regular elements of R which are central. An ideal of R maximal with the property of being disjoint from S is a prime ideal Q and since $R - Q \supset S \supset R - P$ it follows that $Q \subset P$ hence $Q = P$. Thus P is disjoint from the set of regular

elements of R which are central i.e. $P \cap Z(R)$ must consist of zero-divisors. ∎

For further use we state :

<u>II.4.4. Proposition</u>. Let $R \hookrightarrow S = R\{a_1,\dots,a_k\}$ be a finitely generated extension of prime P.I. rings and suppose that $\{P_\alpha, \alpha \in A\}$ is a family of non-zero prime ideals of R such that $\bigcap_{\alpha \in A} P_\alpha = o$. Then there exists a family of prime ideals: $\{Q_\beta, \beta \in \mathcal{B}\}$, in Spec(S) such that the following conditions hold :

1. $\bigcap_{\beta \in \mathcal{B}} Q_\beta = o$.
2. $Q_\beta \cap R \in \{P_\alpha, \alpha \in A\}$ for each $\beta \in \mathcal{B}$.

<u>Proof</u>. Put $T = Q(R)\{a_1,\dots,a_k\}$, this is a prime ring because $T \to Q(S)$ is an extension. Since T is a P.I. ring and finitely generated over $Z(Q(R))$ it follows that T is a Hilbert algebra, hence semisimple. Put $\mathcal{S} = \{Q_\beta \in \text{Spec}(S)$ for which there is an $\alpha \in A$ such that $Q_\beta \cap R = P_\alpha\}$.
By II.4.2. it is clear that $\mathcal{S}$ is not empty; indeed we may select a non-zero $c \in Z(R)$ such that each $P \in \text{Spec}(R)$ which does not contain c has the property $PS \cap R = P$. From $\cap P_\alpha = o$ it follows that some P_α do not contain c and for such a P_α we can choose Q to be an ideal of S maximal with the property $Q \cap R = P_\alpha$ and it will be easily checked that $Q \in \mathcal{S}$.
We only have to show that $I = \bigcap_\beta Q_\beta = o$. If I were non-zero then there exists a nonzero $b \in Z(S) \cap I$ i.e. a non-zero $b \in Z(T) \cap I$. However, the semisimplicity of T entails that there is a maximal ideal N of T such that $b \notin N$. The image $\bar{b}$ of b in $U = T/N$ is invertible in U. (Identifying Q(R) with its image in U, as in II.4.2).
We get that :

$$U = \{\sum_i q_i \bar{a}_{i_1} \dots \bar{a}_{i_j}, q_i \in Q(R), \bar{a}_{i_j} \text{ the image of } a_{i_j}\}$$

$$\bar{S} = \{\sum_i r_i \bar{a}_{i_1} \dots \bar{a}_{i_j}, r_i \in R, \bar{a}_{i_j} \text{ image of some } a_{i_j}\} .$$

By Theorem II.4.2. there is a non-zero $C_1 \in Z(R)$ such that $P \in \text{Spec}(R)$ which does not contain c_1 has a unique prime ideal of $\overline{S}$ lying over it. Obviously :

$$o \neq c_1 d = c_1 d.1 = c_1 d.(\overline{b}.u) = c_1 b\, du \in \overline{I},$$

where d is selected in Z(R) such that $du \in \overline{S}$. But $c_1 d \neq o$ implies that $c_1 d \notin P_\alpha$ for some P_α, a fortiori $c_1 \notin P_\alpha$, hence $P_\alpha \overline{S} \cap R = P_\alpha$. The same reasoning as in II.4.2. yields that there exists a prime ideal $\overline{Q}_\alpha$ of $\overline{S}$ such that $\overline{Q}_\alpha \cap R = P_\alpha$ and $c_1 d \notin Q_\alpha$. Finally, $c_1 d \in I + (N \cap S) \subset Q_\alpha$ yields a contradiction, thus $I = 0 = \bigcap_\beta Q_\beta$ holds. ∎

If R is a commutative ring and $\bigcap_\alpha P_\alpha = o$ then any minimal prime ideal of R is contained in the union $\bigcup_\alpha P_\alpha$. This is no longer true in the non-commutative case as we will point out later, but the following result(s) will be used in proving an analogue of a theorem of Grothendieck in the non-commutative geometry further on in this Section.

<u>II.4.5. Proposition</u>. Let R be a semiprime P.I. ring and let P be a minimal prime ideal of R. Given any finite subset $\{c_1,\ldots,c_k\} \subset Z(R) \cap P$ then amongst any set of $P_\alpha \in \mathrm{Spec}(R)$ such that $\bigcap_\alpha P_\alpha = o$ there can be found an α such that $P_\alpha \supset \{c_1,\ldots,c_k\}$.

<u>Proof</u>. By induction on k. If $k = 1$ and $c \in Z(R) \cap P$ then c is a zero-divisor i.e. $cb = o$ for some $b \neq o$ in R. From $\bigcap_\alpha P_\alpha = o$ it follows that $b \notin P_{\alpha_0}$ for some α_0 hence $c \in P_{\alpha_0}$ follows. Now suppose that the result holds for subsets consisting of k-1 elements and put :

$$I = \cap\{P_\alpha \,,\, \{c_1,\ldots,c_{k-1}\} \subset P_\alpha\}$$
$$J = \cap\{P_\alpha \,,\, \{c_1,\ldots,c_{k-1}\} \not\subset P_\alpha\}.$$

Clearly $I \cap J = o$ hence either $I \subset P$ or $J \subset P$. Suppose $J \subset P$ and consider the semiprime P.I. ring R/J. We have $o = \cap\{P_\alpha/J \,,\, \{c_1,\ldots,c_{k-1}\} \not\subset P_\alpha\}$ while P/J is again a minimal prime ideal.
The induction hypothesis yields the existence of some α_1 such that

$P_{\alpha_1}/J \supset \{c_1 \bmod J, \dots, c_k \bmod J\}$, but this contradicts $\{c_1,\dots,c_{k-1}\} \not\subset P_{\alpha_1}$. Thus $J \not\subset P$ and $I \subset P$.

Apply the result with k=1 to the semiprime P.I. ring R/I, then we may select an α such that P_α/I contains $c_k \bmod I$ while $\{c_1,\dots,c_{k-1}\} \subset P_\alpha$ i.e. $\{c_1,\dots,c_k\} \subset P_\alpha$. ■

At this point we may benefit from the introduction of the notion of the Formanek center of a P.I. ring. Let V_n be the variety of rings satisfying the identities of $n \times n$-matrices. If $R \in V_n$ then the Formanek center FR of R is the subring of R obtained by evaluating all central polynomials without constant term.

<u>II.4.6. Proposition</u>. With notations as above :

1. F is functor in V_n.
2. $FR \subset Z(R)$.
3. If $\psi : R \to S$ is surjective then $F\psi : FR \to FS$ is surjective.

<u>Proof</u>. 1. and 2. are obvious from the definition.

3. Let $f(x_1,\dots,x_s)$ be a central polynomial with $f(o,\dots,o) = o$ and let $a_1,\dots,a_s \in S$. Pick $b_1,\dots,b_s \in R$ such that $\psi(b_i) = a_i$, $i = 1\dots s$. Then $f(b_1,\dots,b_s) \in FR$ and $\psi f(b_1,\dots,b_s) = f(a_1,\dots,a_s)$, since the latter elements generate FS; 3. is proved. ■

Next let us look at absolutely irreducible representations. For a map $\alpha : R \to M_n(K)$, where K is a field, it is equivalent to say that α is a central extension or that for any field extension $L \supset K$ the representation $R \xrightarrow{\alpha} M_n(K) \to M_n(H)$ is irreducible. We call such representations : <u>absolutely irreducible representations of dimension n</u> (abbreviated : AI_n-representation). Given AI-representations $\alpha : R \to M_n(K)$, $\beta : R \to M_m(L)$ then these are said to be <u>equivalent representations</u> if $m = n$ and there exists a field H, $K \subset H$, $L \subset H$, together with an H-automorphism ψ of $M_n(H)$, such that the following diagram of extensions is commutative :

$$\begin{array}{ccccc} & & M_n(K) & \longrightarrow & M_n(H) \\ & \overset{\alpha}{\nearrow} & & & \\ R & & & & \downarrow \psi \\ & \underset{\beta}{\searrow} & & & \\ & & M_n(L) & \longrightarrow & M_n(H) \end{array} .$$

<u>II.4.7. Theorem. (Procesi [136]).</u> **The following statements are valid for any algebra R :**

1. **The kernel Ker α of an AI-representation $\alpha : R \to M_n(K)$ is a prime ideal and $\alpha(R)$ is a prime P.I. algebra.**
2. **AI-representations α, β of R are equivalent if and only if Ker α = Ker β.**
3. **There is a one-to-one correspondence between equivalence classes of AI-representations of R and prime ideals P of R for which R/P is a PI algebra.**

<u>Proof</u>. 1. The map $R/\mathrm{Ker}\ \alpha \to M_n(K)$ induced by α is an injective central extension hence 1. follows from the fact that $M_n(K)$ is a prime P.I. ring.

2. If α and β are equivalent then Ker α = Ker β is obvious. Conversely, let Ker α = Ker β = P and put S = R/P. Then α and β yield injective central extensions $\overline{\alpha} : S \to M_n(K)$, $\overline{\beta} : S \to M_m(L)$. Writing Q(S) for the ring of fractions of S, we obtain isomorphisms :

$$\tilde{\alpha} : Q(S) \underset{Z(Q(S))}{\otimes} K \to M_n(K) \ , \ \tilde{\beta} : Q(S) \underset{Z(Q(S))}{\otimes} L \to M_m(L)$$

and the following diagrams of central extensions are commutative :

$$\begin{array}{ccc} S & \longrightarrow & Q(S) \\ {\scriptstyle \overline{\alpha}} \downarrow & & \downarrow \\ M_n(K) & \overset{}{\underset{\tilde{\alpha}}{\text{———}}} & Q(S) \underset{Z(Q(S))}{\otimes} K \end{array} \qquad \begin{array}{ccc} S & \longrightarrow & Q(S) \\ {\scriptstyle \overline{\beta}} \downarrow & & \downarrow \\ M_m(L) & \underset{\tilde{\beta}}{\text{———}} & Q(S) \underset{Z(Q(S))}{\otimes} L \end{array}$$

Thus m=n follows immediately. Moreover if H is any common extension field of

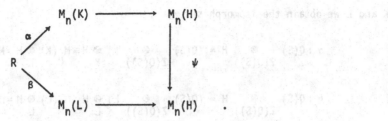

K and L we obtain the isomorphisms γ, δ :

$$\gamma : Q(S) \underset{Z(Q(S))}{\otimes} H = (Q(S) \underset{Z(Q(S))}{\otimes} K) \underset{K}{\otimes} H \cong M_n(K) \underset{K}{\otimes} H = M_n(H)$$

$$\delta : Q(S) \underset{Z(Q(S))}{\otimes} H = (Q(S) \underset{Z(Q(S))}{\otimes} L) \underset{L}{\otimes} H \quad M_n(L) \underset{L}{\otimes} H = M_n(H) .$$

Putting $\psi = \delta\gamma^{-1}$ we obtain a commutative diagram

$$\begin{array}{ccccc} & \nearrow & M_n(K) & \longrightarrow & M_n(H) \\ & & \uparrow \bar{\alpha} & & \\ R & \longrightarrow & S & & \downarrow \psi \\ & & \downarrow \bar{\beta} & & \\ & \searrow & M_n(L) & \longrightarrow & M_n(H) \end{array} .$$

3. It is enough to establish that if $P \in \mathrm{Spec}(R)$ is such that R/P is a P.I. ring then there is an AI representation $\alpha : R \to M_n(K)$ with $\mathrm{Ker}\, \alpha = P$. This however is a direct consequence of Posner's theorem cf. Section II.2. ∎

The effect of the above theorem is that $X = \mathrm{Spec}(R)$ decomposes into an infinite disjoint union $X = X_1 \cup X_2 \cup \ldots X_n \cup \ldots \cup X_\infty$. where $X_n = \mathrm{Spec}_n(R) = \{P \in \mathrm{Spec}(R),\ R/P$ is a P.I. ring, p.i. deg $R/P = n\}$. Put $\Sigma_m(R) = \bigcup_{1 \leq i \leq m} \mathrm{Spec}_i(R)$.

<u>II.4.8. Proposition.</u> 1. For any $m \in \mathbb{N}$, $\Sigma_m(R)$ is closed in $\mathrm{Spec}(R)$ for the Zariski topology.

2. If R satisfies a proper polynomial identity of degree d, then $\mathrm{Spec}(R) = \Sigma_{[d/2]}(R)$, where $[d/2]$ is the largest integer less or equal to d/2.

<u>Proof</u>. 1. If I is the minimal ideal of R such that $R/I \in V_n$ then $\Sigma_n(R) = \{P \in \mathrm{Spec}(R),\ I \subset P\}$ as is easily seen.

2. Follows from Amitsur's theorem, cf. Section II.2. Indeed if R satisfies a proper identity of degree d then $R/L(R)$ satisfies all identities of $n \times n$ matrices for $n = [d/2]$ and any R/P will satisfy the same identities. On the

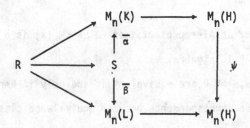

other hand, if $\alpha : R \to M_i(K)$ is an AI-representation with $\operatorname{Ker}\alpha = P$ then $M_i(K)$ has to satisfy all multilinear identities of $n \times n$ matrices, hence $i \leq n = [d/2]$.

■

II.4.9. Theorem. 1. A representation $\alpha : R \to M_n(K)$, is exactly then absolutely irreducible when $\alpha(FR) \neq o$.

2. R is an Azumaya algebra of rank n^2 over its center if and only if $R.FR = R$.

3. $\Sigma_{n-1}(R) = V(FR) = \{P \in \operatorname{Spec}(R), FR \subset P\}$.

Proof. 1. If α is an AI-representation then $\alpha(R)$ is a prime ring of p.i. degree n, so it is either a finite matrix ring or it satisfies exactly the identities of $n \times n$ matrices.

In either case we may evaluate a suitable central polynomial (e.g. the Formanek central polynomial) and obtain a non-zero value i.e. $\alpha(FR) \neq o$.

Conversely, suppose that $\alpha(R)K \neq M_n(K)$. It is clear that $\alpha(R)K$ is a finite dimensional K-algebra, with $\dim_K \alpha(R)K < n^2$. If J is the Jacobson radical of $\alpha(R)K$ then $\alpha(R)K/J$ satisfies all identities of $n-1 \times n-1$ matrices. However, all central polynomials for $n \times n$ matrices, without constant term are identities for $n-1 \times n-1$ matrices hence $\alpha(FR) \subset J$. On the other hand $\alpha(FR) \subset K$. Since J has to be nilpotent, $J \cap K = o$ i.e. $\alpha(FR) = o$.

2. If $RFR \neq R$ then there is a maximal ideal M of R containing FR. Then FR is sent to zero by $\alpha : R \to R/M$ i.e. M is not the kernel of an AI-representation by 1, hence $M \notin \operatorname{Spec}_n(R)$. The Artin-Procesi theorem, II.3.25, then states that R is not an Azumaya algebra.

Conversely if $RFR = R$ and M is a maximal ideal of R then the image of FR in R/M is non-zero, i.e. $\alpha : R \to R/M$ is an AI-representation and $M \in \operatorname{Spec}_n(R)$. Again, by the Artin-Procesi theorem, it follows that R is an Azumaya algebra.

3. This has been established along the way in the foregoing parts of this proof. ■

A prime ideal P of R is said to be identity faithful if $FR \not\subset P$. If R is semiprime then P is identity faithful if and only if $P \in \Sigma_{n-1}(R)$ i.e. p.i.deg $R/P = n$.

II.4.10. Proposition. Consider a collection $\{P_\alpha, \alpha \in A\}$ of prime ideals of a semiprime P.I. ring such that $\bigcap_\alpha P_\alpha = o$. If P is a minimal prime ideal of R which is identity faithful, then $P \subset \bigcup_\alpha P_\alpha$.

Proof. Since $P \not\supset FR$, pick $c \in FR - P$. Put $J_1 = \cap\{P_\alpha, c \in P_\alpha\}$, $J_2 = \cap\{P_\alpha, c \notin P_\alpha\}$. Then $J_1 \cap J_2 = o$, i.e. J_1 or J_2 is contained in P. However $J_1 \not\subset P$ is impossible, thus $J_2 \subset P$. Up to replacing R by R/J_2 we may assume that $J_2 = o$ and $c \notin P_\alpha$ for every $\alpha \in A$. Clearly this implies that c is regular in R, hence we may consider the localization R_c of R at $\{1, c, c^2, \ldots\}$ where $1 = cc^{-1} \in Z(R_c)FR$. It is easily seen that $R_c FR_c = R_c$, thus R_c is an Azumaya algebra. Write $P^e = PR_c$ and $P^e_\alpha = P_\alpha R_c$; then P^e and P^e_α are proper prime ideals of R_c such that $\cap P^e_\alpha = o$. Moreover, P^e is a minimal prime ideal of R_c.
By Proposition II.4.5. any chosen finite collection of elements of $Z(R_c)$ in P^e lies also in some P^e_α. Now the ideal P^e of R_c is generated by its central part; together with the foregoing statement, this yields : $P^e \subset \bigcup_\alpha P^e_\alpha$. So for any $p \in P$, there exists $i \in \mathbb{N}$, $\alpha \in A$, such that $pc^i \in P_\alpha$; consequently $P \subset P_\alpha$ for that $\alpha \in A$, and $P \subset \bigcup_\alpha P_\alpha$. ■

II.4.11. Proposition. Let R be a P.I. ring and let $I = \bigcap_\alpha P_\alpha$ for prime ideals $\{P_\alpha, \alpha \in A\}$ of R. If P is a prime ideal of R minimal over I and such that P/I is identity faithful in R/I then for any finite collection $\{p_1, \ldots, p_k\} \subset P$ there is an $\alpha \in A$ such that $\{p_1, \ldots, p_k\} \subset P_\alpha$.

Proof. An easy corollary of the foregoing. ■

II.4.12. Proposition. The set of identity faithful prime ideals of a semiprime P.I. ring R is contained in the open set of birationality of R.

Proof. An identity faithful prime ideal P is the union of all ideals I of R such that $I \cap Z(R) \subset P \cap Z(R)$. ■ (See also Section II.1).

II.4.13. Proposition. Let $f: R \to A$ be a central extension and P a prime ideal of A, then $f^{-1}(P)$ is a prime ideal of R and p.i. deg $R/f^{-1}(P)$ = p.i. deg A/P.

Proof. That $f^{-1}(P)$ is prime goes without saying. Clearly $R/f^{-1}(P)$ satisfies all identities satisfied by A/P since $R/f^{-1}(P) \to A/P$ is an injective extension. Moreover one easily verifies that A/P satisfies all multilinear identities satisfied by $R/f^{-1}(P)$. Since the p.i. degree is expressible in terms of the degrees of the standard identities satisfied by the rings in question, and since the standard identities are multilinear, the equality of p.i. degrees follows. ∎

II.4.14. Proposition. Let R be a P.I. ring. Let $\{P_\alpha, \alpha \in A\}$ be a collection of prime ideals of R and let P be a minimal prime ideal of R over $I = \cap_\alpha P_\alpha$, such that P/I is identity faithful in R/I. Let $S = R[X_1,\dots,X_k]$ be the ring of poly nomials over R in k central variables. To a prime ideal Q of S such that $Q \cap R = P$ there corresponds a collection $\{Q_\beta, \beta \in \mathcal{B}\}$ of prime ideals of S such that Q is minimal over $J = \cap_\beta Q_\beta$ and $Q_\beta \cap R \in \{P_\alpha, \alpha \in A\}$ for each $\beta \in \mathcal{B}$.

Proof. Put $I_1 = \cap \{P_\alpha, P_\alpha/I$ is identity faithful in $R/I\}$, as before, $I_1 \subset P$ follows. So $\tilde{I}_1 = I_1[X_1,\dots,X_k] \subset P[X_1,\dots,X_k] \subset Q$. Now $S/\tilde{I}_1 = R/I_1[X_1,\dots,X_k]$ and $Q/\tilde{I}_1$ is a prime ideal of $S/\tilde{I}_1$ lying over P/I_1 in R/I_1. Up to replacing S by $S/\tilde{I}_1$ and R by R/I_1 we may assume that $\{P_\alpha, \alpha \in A\}$ consists of identity faithful primes of R such that $\cap_\alpha P_\alpha = o$. Consider the set $\{Q_\gamma \in \mathrm{Spec}(S), Q_\gamma \cap R = P_\alpha$ for some $\alpha \in A\}$. By proposition II.4.13, it follows that each Q_γ is an identity faithful prime ideal of S. Since identity faithful prime ideals correspond in a one-to-one way to their central parts in a semiprime P.I. ring we obtain:

$$\cap_\alpha (P_\alpha \cap C) = (\cap_\alpha P_\alpha) \cap C = o$$

where $C = Z(R)$. Put $D = C[X_1,\dots,X_k] = Z(S)$.
Suppose that P_1' is a prime ideal of C such that $P_1' \subset P \cap C$. In that case

$FR \not\subset P_1'$ and therefore $P_1' = P_1 \cap C$ for some prime ideal P_1 of R. Thus $P_1 \subset P$ but then $P_1 = P$ by the minimality of P i.e. $P_1' = P \cap C$. Moreover, $(Q \cap D) \cap C = Q \cap C = (Q \cap R) \cap C = P \cap C = P'$. We have now reduced everything to the following situation: P' is a prime ideal of C which is minimal, $Q' = Q \cap D$ is a prime ideal of $D = C[X_1,\dots,X_k]$ lying over P' and we are given a collection $\{P_\alpha', \alpha \in A\}$ of prime ideals of C such that $\bigcap_\alpha P_\alpha' = o$. From [107] it follows that there is a collection $\{Q_\beta', \beta \in B\}$ of prime ideals of D such that Q' is minimal over $\bigcap_\beta Q_\beta'$ and such that $Q_\beta' \cap C \in \{P_\alpha', \alpha \in A\}$ for each $\beta \in B$.

Let $\{Q_\beta, \beta \in B\}$ be the set of prime ideals of S corresponding to $\{Q_\beta', \beta \in B\}$ in the open set of birationality in Spec(S); then $Q_\beta' = Q_\beta \cap D$. Each Q_β lies over some P_α and $(\bigcap_\beta Q_\beta) \cap D = \bigcap Q_\beta' \subset Q'$. The latter fact yields that Q contains $J = \bigcap_\beta Q_\beta$ because of II.4.12. If Q were not minimal over J then $J \subset Q_1 \subset Q$ for some $Q_1 \in \mathrm{Spec}(S)$ yields $J \cap C \subset Q_1 \cap C \subset Q'$ and $Q' = Q_1 \cap C$ by minimality of Q'. Both Q_1 and Q lie over Q', hence $Q_1 = Q$. It follows that the set $\{Q_\beta, \beta \in B\}$ is indeed the desired collection of prime ideals. ∎

<u>II.4.15. Proposition.</u> With assumptions as in II.4.14, if Q is a prime ideal of S with $Q \cap R = P$ then for any set $\{q_1,\dots,q_m\} \subset Q$ and $g \in S-Q$ we can select an index $\alpha \in A$ such that there is a prime ideal $Q_\alpha \in \mathrm{Spec}(S)$ such that $Q_\alpha \cap R = P_\alpha$, $\{q_1,\dots,q_m\} \subset Q_\alpha$ and $g \notin Q_\alpha$.

<u>Proof</u>. Let $\{Q_\beta, \beta \in B\}$ be as in II.4.14. Since P/I is identity faithful in R/I it follows that $Q/I[X_1,\dots,X_k]$ must be identity faithful in $S/I[X_1,\dots,X_k] = R/I[X_1,\dots,X_k]$. But $I[X_1,\dots,X_k] = \bigcap_{\alpha \in A} P_\alpha[X_1,\dots,X_k] \subset J$ entails that Q/J is identity faithful in S/J. Replacing R by S, P by Q, $\{P_\alpha, \alpha \in A\}$ by $\{Q_\beta, \beta \in B\}$ and I by J in Proposition II.4.14. the claim is established. ∎

The foregoing results relate to the following :

<u>II.4.16. Theorem (A. Grothendieck)</u>. Let $R \subset T$ be commutative rings and assume that T is finitely presented as a ring over R. If $R \hookrightarrow T$ has the "going down" property then the induced morphism $\mathrm{Spec}(T) \to \mathrm{Spec}(R)$ is an open map.

Recall that a finitely generated central extension $R \hookrightarrow T = R[a_1,\dots,a_n]$ is said to be a finitely presented extension if the kernel of the morphism $R[X_1,\dots,X_n] \to T$, which is defined by sending X_i to a_i while R remains fixed, is finitely generated as an ideal of $R[X_1,\dots,X_n]$. The equivalent of II.4.16 for P.I. rings fails as a close investigation of the following example due to G. Bergman and L. Small [29], will learn. Let R be a semiprime P.I. ring possessing a set of minimal prime ideals $\{P_\alpha, \alpha \in A\}$ and a prime ideal P such that $\cap_\alpha P_\alpha = o$ and $P \not\subset \cup_\alpha P_\alpha$. Let X be a commuting variable and consider $S = R[X]/(pX)$ where p is chosen arbitrarily in $P - \cup_\alpha P_\alpha$. The inclusion $R \to S$ is a finitely presented central extension of P.I. rings. To obtain an explicit example one may consider $R = \{(x_1,x_2,\dots) \in M_2(K)^{\mathbb{N}}, \exists n > o,\ a,b \in K$ such that $x_n = x_{n+1} = \dots = \begin{pmatrix} a & o \\ o & b \end{pmatrix}\}$. In that case the prime ideals of R are both maximal and minimal and therefore the condition of having the "going down" property will trivially be satisfied. We leave it to the reader to verify that the induced map $\mathrm{Spec}(S) \to \mathrm{Spec}(R)$ is not open. The following version of II.4.16. may be stated :

<u>II.4.17. Theorem.</u> Let $R \hookrightarrow R[a_1,\dots,a_k] = T$ be a finitely presented central extension of P.I. rings such that $R \hookrightarrow T$ satisfies the "going down" property. To an open set U in $\mathrm{Spec}(T)$ we may associate the ideal $I = \cap\{P \in \mathrm{Spec}(R), P \in f(U)\}$, and we let n denote the p.i. degree of R/I.
Let $f : \mathrm{Spec}(T) \to \mathrm{Spec}(R)$ be the induced map and put $\mathrm{Spec}_n^+(R) = \bigcup_{j \geq n} \mathrm{Spec}_j(R)$, then we have that $f(U) \cap \mathrm{Spec}_n^+(R)$ is an open subset of $\mathrm{Spec}_n^+(R)$ in the induced topology.

<u>Proof.</u> Let $U = \mathrm{Spec}(T) - V(H)$ for some ideal H of T, i.e. a prime ideal Q of T is in U if and only if $H \not\subset Q$. If $f(U) = \mathrm{Spec}(R)$ then the theorem becomes trivial; henceforth assume $f(U) \subsetneq \mathrm{Spec}(R)$. It will be sufficient to establish that $f(U) \cap \mathrm{Spec}_n^+(R) \cap V(I) = \phi$. If this were not true then we can find $Q \in U$ with $Q \cap R = P \in \mathrm{Spec}_n^+(R)$ and $I \subset P$. From p.i. deg $R/P \leq$ p.i. deg R/I it follows that p.i. deg $R/P \leq n$ whereas $P \in \mathrm{Spec}_n^+(R)$ means that p.i. deg $R/P \geq n$; consequently

$P \in Spec_n(R)$. Let $P_1 \subset P$ be a prime ideal of R which is minimal over I. Clearly $P_1 \subset P$ yields $P_1 \in Spec_n^+(R) \cap V(I)$. Moreover, the "going-down" property implies that $P_1 = Q_1 \cap R$ for some prime ideal Q_1 of T such that $Q_1 \subset Q$. Since $H \not\subset Q_1$, $Q_1 \in U$ hence $P_1 \in f(U)$ and $P_1 \in f(U) \cap Spec_n^+(R) \cap V(I)$. In other words, we may assume that P is minimal over I. Pick $g \in H-Q$. Let $\pi : R[X_1,\dots,X_k] \to T$ be defined by $\pi(X_j) = a_j$, $j=1,\dots,k$ and $\pi|R = 1_R$. Since $R \to T$ is finitely presented, Ker π may be generated by a finite set $\{q_1,\dots,q_m\} \subset S = R[X_1,\dots,X_k]$. Select $g_1 \in S$ such that $\pi(g_1) = g$. We have that $\{q_1,\dots,q_m\} \subset \pi^{-1}(Q)$ and $g_1 \notin \pi^{-1}(Q)$, whereas $\pi^{-1}(Q) \cap R = P$. Clearly P/I is identity faithful in R/I. Apply Proposition II.4.15. to find $Q' \in Spec(S)$ such that $\{q_1,\dots,q_m\} \subset Q'$, $g_1 \notin Q'$ and $Q' \cap R \notin f(U)$. Thus Ker $\pi \subset Q'$ and $\pi(Q') \in Spec(T)$, $g \notin \pi(Q')$. Therefore $\pi(Q') \in U$. But this contradicts $\pi(Q') \cap R = Q' \cap R \notin f(U)$. ■

Since $R \hookrightarrow T$ is a central extension of P.I. rings, f induces a map $f^* : Spec_n(T) \to Spec_n(R)$ where n = p.i. deg T.

<u>II.4.18. Corollary</u>. If $R \to T$ is a finitely presented central extension of P.I. rings which has the "going down" property, then f^* is an open map.

<u>Proof</u>. For any m, the open sets in $Spec_m(T)$ are exactly the sets of the form $U \cap Spec_m(T)$ with U open in Spec T. Note that we have, for each $m \in \mathbb{N}$:

$$(*)_m \quad f_m^*(U \cap Spec_m(T)) = f(U) \cap Spec_m(R) = Spec_m(R) \cap (f(U) \cap Spec_m^+(R)).$$

By Theorem II.4.17, if we take $m = m_o$ = p.i. deg R/I as in the theorem, these sets are open in $Spec_{m_o}(R)$. If $P \in Spec_{m_o}^+(R)$, let Q be a prime ideal of T maximal with respect to the property $Q \cap R \subset P$. Then :

$m_o \leq$ p.i. deg R/P $\leq$ p.i. deg R/Q $\cap$ R = p.i. deg T/Q $\leq$ n. It follows that $Spec_n R \subset Spec_{m_o}^+(R)$ and therefore by restriction of $(*)_{m_o}$ to $Spec_n(R)$ the corollary follows. ■

A finitely generated algebra R over a Noetherian ring A is not necessarily Noetherian but it does satisfy the ascending chain condition on semiprime ideals. Under that condition, the minimal prime ideals P_i over a semiprime ideal I of R are finite in number and $I = \cap P_i$ is an irredundant decomposition of I.

If $P_1,\dots,P_m$ are the minimal prime ideals of a ring R which satisfies the ascending chain condition on semiprime ideals then the rings R/P_i, $i = 1,\dots,m$, are called the <u>components</u> of R. If we endow $V(P_i)$ with the induced topology then there is a canonical homeomorphism $V(P_i) \cong \mathrm{Spec}(R/P_i)$, $i = 1,\dots,m$ mapping a prime ideal P containing P_i to P/P_i. Clearly $\mathrm{Spec}(R) = \bigcup_{i=1}^{m} V(P_i)$.

Next we develop some machinery, needed for the study of "products" of certain "varieties". Assume in the sequel of this section that R is affine over a field K i.e. R is a prime P.I. and finitely generated over K. Let Q be the ring of quotients of R and let us write $C = Z(R)$, $Z = Z(Q)$. If F is a field extension of K then $R \otimes_K F$ is affine over F, but if R is prime then $R \otimes_K F$ need not be prime. Recall the following result in this vein stemming from [26].

<u>II.4.19. Proposition (G. Bergman)</u> Let R and S be prime algebras over an algebraically closed field K then $R \otimes_K S$ is a prime algebra too.

If K is algebraically closed then $R \otimes_K F$ is prime for every field extension F over K. The center $D = C \otimes_K F$ of $R \otimes_K F$ is a Noetherian ring which is an order in a commutative Artinian ring, A say. We shall pay some attention to the ring $T = (Q \otimes_K F) \otimes_D A$. It is rather straightforward to verify the following, (cf [134]) :

<u>II.4.20. Proposition.</u> With notations as above :

1. T is a free A-module of dimension $[Q : Z]$.
2. T is an Artinian ring.
3. $R \otimes_K F$ is an order in T and $T = (R \otimes_K F).A$
4. The Jacobson radical of T is the extension of the Jacobson radical of A to T.
5. $J(R \otimes_K F) = (R \otimes_K F) \cap J(T)$

6. $R \underset{K}{\otimes} F/J(R \underset{K}{\otimes} F)$ is an order in $T/J(T)$.

7. A is the center of T and $A/J(A)$ is the center of $T/J(T)$.

8. If S is a component of $R \underset{K}{\otimes} F$ then :

   $\text{tr.deg}_F\, Z(Q(S)) = \text{tr.deg}_K\, Z(Q(R))$.

We aim to apply this in the study of affine rings R_1 and R_2 over a field K, with rings of quotients Q_1 and Q_2, $Z_1 = Z(Q_1)$, $Z_2 = Z(Q_2)$. The algebra $R_1 \underset{K}{\otimes} R_2$ is finitely generated over K.

<u>II.4.21. Theorem (Regev).</u> If R and S are P.I. algebras over a field F then $R \underset{F}{\otimes} S$ is a P.I. algebra.

If R,S satisfy identities of minimal degree p,q say, then $R \underset{F}{\otimes} S$ satisfies an identity of degree n, where n is the smallest integer for which :

$$n! > (3.4^{p-3}.3.4^{q-3})^n \ .$$

Putting $\ell = ((p-1)(q-1))^2 + 1$, $k = \frac{\ell^4}{4}$ then S_ℓ^k is an identity for $R \underset{F}{\otimes} S$, where S_ℓ is the standard identity of degree ℓ.

<u>II.4.22. Note.</u> If $p = q = 4$ then $\ell = 82$, $k = 11.403.044$ and $\deg S_\ell^k = 935.049.608$. The conclusion is left to the reader.

Let A be the ring of fractions of the Noetherian ring $Z_1 \underset{F}{\otimes} Z_2$ and put $T = (Q_1 \underset{F}{\otimes} Q_2) \underset{Z_1 \otimes Z_2}{\otimes} A$. Clearly, A is Artinian. With notations as before :

<u>II.4.23. Theorem.</u> 1. T is a free A-module of rank $[Q_1 : Z_1].[Q_2 : Z_2]$.

2. T is Artinian.

3. $R_1 \underset{F}{\otimes} R_2$ is an order in T and $T = (R_1 \underset{F}{\otimes} R_2).A$.

4. The Jacobson radical J(T) is generated by the Jacobson radical J(A) of A i.e. $J(T) = J(A).T$.

5. $J(R_1 \underset{F}{\otimes} R_2) = (R_1 \underset{F}{\otimes} R_2) \cap J(T)$.

6. $R_1 \underset{F}{\otimes} R_2/J(R_1 \underset{P}{\otimes} R_2)$ is an order in $T/J(T)$.

7. $A = Z(T)$ and $A/J(A)$ is the center of $T/J(T)$.

8. If U is a component of $R_1 \underset{F}{\otimes} R_2$ then we have :

$$\text{tr.deg}_F\, Z(Q(U)) = \text{tr.deg}_F\, Z(Q(R_1)) + \text{tr.deg}_F\, Z(Q(R_2)).$$

Proof. (Sketchy, for full detail cf. [130]):

1. Obvious.

2. Since A is Artinian 2. follows from 1.

3. For $q \in T$ there are $c \in Q_1 \underset{F}{\otimes} Q_2$, $b \in Z_1 \underset{F}{\otimes} Z_2$. Such that $q = cb^{-1}$.
It follows that a regular element in $R_1 \underset{F}{\otimes} R_2$ is also regular in T. Moreover $b = ad^{-1}$ for some $a, d \in R_1 \underset{F}{\otimes} R_2$, hence $q = cda^{-1}$ and $cd = pr^{-1}$ for some $p, r \in R_1 \underset{F}{\otimes} R_2$. Thus $q = p(ar)^{-1}$ where p and ar are in $R_1 \underset{F}{\otimes} R_2$. Now $T = (Q_1 \underset{F}{\otimes} Q_2).A$ with $Q_1 = R_1 Z_1$, $Q_2 = R_2 Z_2$ yields that $T = (R_1 \underset{F}{\otimes} R_2).A$.

4. Put $T_1 = (Q_1 \underset{F}{\otimes} Z_2) \underset{Z_1 \otimes Z_2}{\otimes} A$.

It is easily checked that $T_1 \underset{Z_2}{\otimes} Q_2 = T$.

If D is a central simple algebra over a field G, and S is any G-algebra then it is well-known that $J(S \underset{G}{\otimes} D) = J(S) \underset{G}{\otimes} D$. By Theorem II.4.20 :

$$J(T) = J(T_1 \underset{Z_2}{\otimes} Q_2) = J(T_1) \underset{Z_2}{\otimes} Q_2 = J(A)(T_1 \underset{Z_2}{\otimes} Q_2) = J(A)T.$$

5./6. Can be checked as in II.4.20. It is clear that 6. follows from 5. and 3. whereas 5. may be derived from the fact that $J(R_1 \underset{F}{\otimes} R_2)$ is nilpotent i.e. $J(R_1 \underset{F}{\otimes} R_2) \subset J(T)$, and nilpotency of $J(T)$ entails that $J(T) \cap (R_1 \underset{F}{\otimes} R_2) \subset J(R_1 \underset{F}{\otimes} R_2)$.

7. Clearly $Z(T) = Z(T_1 \underset{Z_2}{\otimes} Q_2) = Z(T_1) \underset{Z_2}{\otimes} Z_2 = Z(T_1) = A$, and similarly for $Z(T)J(T))$.

8. The components of $R_1 \underset{F}{\otimes} R_2$ are orders in the components of $T/J(T)$; the centers of the latter are exactly the components of $Z_1 \underset{F}{\otimes} Z_2$ and these have the right dimensions by the commutative theory. ∎

It is worthwhile to mention Jategaonkar's non-commutative version of the "principal ideal theorem".

II.4.24. Theorem (Jategaonkar). Let R be a prime Noetherian (i.e. two-sided Noetherian) P.I. ring and let P be a prime ideal of R which is minimal over Rc for some non-zero central element $c \in Z(R)$. Then P does not properly contain any non-zero prime ideal of R.

II.4.25. Comments. The proof of the above theorem relies on the positive answer to Jacobson's conjecture for fully bounded Noetherian rings (given by Jategaonkar in [89]) i.e. if R is an FBN ring then $\bigcap_{n \in \mathbb{N}} J(R)^n = o$.

If one has convinced oneself, using Rowen's theorem, that any semiprime P.I. ring is fully bounded, the proof of II.4.24 runs along the lines of Kaplansky's proof of the principal ideal theorem, cf. [93]. This non-commutative version of it will allow the development of some dimension-theory for the algebraic varieties introduced later on.

In [29] G. Bergman and L. Small show that if R is quasi-local with maximal ideal P then p.i. deg P divides p.i. deg R. More generally, if R is any prime P.I. ring, then p.i. deg R can be written as a linear combination, with non-negative integer coefficients, of the p.i. deg P, where P ranges over the maximal ideals of R. For any specified P_0, maximal in R, the expression for p.i. deg R may be chosen in such way that it involves p.i. deg P_0 with <u>positive</u> coefficient. It is known that, if R is a finite dimensional torsion free algebra over a commutative valuation ring C, and U is a prime ideal of C, then there do not hold any inclusion relations between the prime ideals of R lying over U; furthermore, only finitely many prime ideals of R lie over U. In this situation, every prime ideal of R containing U contains already a prime ideal of R lying over U. This allows to show that we have the "going down property" for $C \hookrightarrow R$. On the other hand, even if $C = Z(R)$ "going up" does not necessarily hold, cf. [29] for a counterexample.

II.4.26. Proposition [29]. Let R be a finite dimensional torsion free algebra over a commutative valuation ring C. Let $f : C \to \overline{C}$ be a ring morphism of commutative rings and let $g : R \to \overline{R} = R \otimes_C \overline{C}$ be the induced morphism, then :

1. For any $U \in \mathrm{Spec}(\overline{C})$, g induces a surjective, p.i. degree preserving map from the set of prime ideals lying over U to the set of prime ideals lying over $f^{-1}(U)$.
2. If $\tilde{f} : \mathrm{Spec}(\overline{C}) \to \mathrm{Spec}(C)$ is surjective then so is $\tilde{g} : \mathrm{Spec}(\overline{R}) \to \mathrm{Spec}(R)$.

Before applying this to P.I. rings we need the following results due to Procesi, [136].

II.4.27. Lemma. Let R be an order in a central simple algebra with center Z. Let P be a prime ideal of R and c a non-zero element of Z. Either $PR[c] \cap R = P$ or $PR[\frac{1}{c}] \cap R = P$.

II.4.28. Corollary. Let R be a prime P.I. ring and let Q be its total ring of fractions with $Z(Q) = Z$. If P is a prime ideal of R then there exists a valuation ring O_v in Z and a prime ideal $\hat{P}$ in RO_v such that $\hat{P} \cap R = P$.

Proof. Consider a subring O_v of Z which is maximal with respect to the property $PO_v \cap R = P$. If $M \supset PO_v$ is maximal with respect to $M \cap G(P) = \phi$ in RO_v, then M is a prime ideal of RO_v and $M \cap R = P$. If $c \in Z - O_v$ then $M[c] \cap RO_v = M$ cannot hold because then $PO_v[c] \cap R \subset M[c] \cap R = M \cap R = P$, what contradicts the maximality hypothesis on O_v.
By the foregoing lemma : $PO_v[c^{-1}] \cap R = P$, and thus $O_v[c^{-1}] = O_v$ or $c^{-1} \in O_v$ follows. From the fact that for each $c \in Z$ either c or c^{-1} is in O_v it follows that O_v is a valuation ring of Z. ∎

This "technique" enabled Bergman and Small to free their results, mentioned before, from valuation rings. Actually, it may be shown that, for any P.I. ring R and any pair of prime ideals $P_o \subset P_1$ of R, p.i. deg P_o - p.i. deg P_1 can be written as a linear combination, with non-negative coefficients, of the p.i. deg P with P running through the set of maximal ideals of R which contain P_o. More generally let R be an arbitrary ring and let I be an ideal of R such that $I = \cap\{P$ prime ideal of R, p.i. deg $P = n_o\}$ for some fixed $n_o \in \mathbb{N}$; if P_o is a prime ideal of R containing I then n-p.i. deg P_o can be written as a linear combination, with non-negative integer coefficients, in terms of the p.i. deg P with P ranging through all maximal ideals of R.

<u>II.4.29. Proposition</u>. Let $n \in \mathbb{N} - \{o\}$, $n \in N \subset \mathbb{N} - \{o\}$, and such that n is not in the additive group generated by N-{n}. Suppose that R is a semiprime P.I. ring such that p.i. deg P = n for all minimal prime ideals P of R, whereas p.i. deg $P \in N$ for all (other) maximal ideals P of R. Then R is an Azumaya algebra of constant rank n^2 over its center.

<u>Proof</u>. Writing n as $\sum_{m \in M} i_m m$ with $i_m > o$ in $\mathbb{Z}$, it follows that n has to be the only term appearing with non-zero coefficient. If P_1 is a maximal ideal of R and $P_o \subset P_1$ is a minimal prime ideal of R then n = p.i. deg P_o can be expressed as $\sum_{P \in \Omega(R)} m_P$ (p.i. deg P) and such that p.i. deg P_1 appears with positive coefficient. Hence p.i. deg $P_1 = n$ for every maximal ideal P_1. But, Artin-Procesi's theorem II.3.25, then entails that R is an Azumaya algebra of constant rank n^2. ∎

<u>II.4.30. Corollaries</u>. 1. If R is a prime P.I. ring with p.i. deg R = n, and we have that for all maximal ideals P of R, p.i. deg P > n/2, then R is an Azumaya algebra of constant rank n^2.

2. In the situation of 1. let $f(X_1,\dots,X_r)$ be a polynomial identity for every R/P whenever p.i. deg P < n/2. Assume that for some $x_1,\dots,x_r \in R$, f takes an invertible value $f(x_1,\dots,x_r)$, then R is an Azumaya algebra of constant

rank n^2.

3. (L. Small) If R is a prime P.I. ring and P is a maximal ideal of R such that $\bigcap_n P^n = o$ then p.i. deg R/P = p.i. deg P divides p.i. deg R.

II.4.31. Comments, References.

Fundamental results were drawn from C. Procesi [136], G. Bergman [28], L. Small [165], A. Regev [143] and Lestman [107], Jategaonkar [89]. Some of the results towards the end of the section could benefit by giving them a treatment in the framework of orders in central simple algebras.

III. GRADED RINGS.

III.1. Generalities on Graded Rings.

A ring R is $\mathbb{Z}$-graded if there is a family of additive subgroups $\{R_n, n\in\mathbb{Z}\}$ such that $R = \bigoplus_{n\in\mathbb{Z}} R_n$ and, for $i,j\in\mathbb{Z}$: $R_i R_j \subset R_{i+j}$.
We say that R is positively graded if $R_i = o$ for $i<o$. A graded left R-module may be defined to be a left R-module M such that $M = \bigoplus_{n\in\mathbb{Z}} M_n$ for some additive subgroups M_n of M satisfying : $R_i M_j \subset M_{i+j}$ for all $i,j\in\mathbb{Z}$. Put $h(R) = \bigcup_{n\in\mathbb{Z}} R_n$, $h(M) = \bigcup_{n\in\mathbb{Z}} M_n$; we shall refer to elements in these sets as being homogeneous elements of R (resp. of M). If M and N are graded left R-modules then a map $f : M\to N$ is said to be a graded morphism of degree p if it is a left R-linear map satisfying : $f(M_i)\subset N_{i+p}$ for all $i\in\mathbb{Z}$. Let R-gr be the category of graded left R-modules with morphisms of degree o for the homomorphisms between objects. It is fairly trivial to define graded submodules of a graded module, epimorphic images, $\varinjlim$ and $\varprojlim$ of graded left R-modules, the group $\mathrm{HOM}_R(M,N)$ of morphisms of arbitrary degree, which is a graded abelian group in the obvious way.
Short : R-gr is a Grothendieck category. However R itself is not a generator for this category. If $M\in$ R-gr then define $M(d)$, $d\in\mathbb{Z}$, to be the graded left R-module defined by putting $M(d)_i = M_{d+i}$. This yields a functor T_d : R-gr $\to$ R-gr and $\bigoplus_{d\in\mathbb{Z}} R(d)$ is a generator for R-gr.
Note also that if M is a finitely generated graded left R-module then $\mathrm{HOM}_R(M,N) = \mathrm{Hom}_R(\underline{M},\underline{N})$, where $\underline{M},\underline{N}$ denote the left R-modules underlying M and N. Injective and projective objects may be defined and studied in R-gr, we refer to [121] for full detail. Recall that if $P\in$ R-gr is a projective object then $\underline{P}$ is a projective R-module and vice versa. On the other hand if $E\in$ R-gr is such that $\underline{E}$ is injective then E is injective in R-gr but the converse is not true. If P is a prime ideal of a graded ring R then the ideal of R generated by all homogeneous elements of R lying in P, denoted by P_g, is a graded ideal and still a prime ideal. This is an essential fact in the ideal theory of

graded rings; another useful fact is that $M \in R$-gr is graded left Noetherian i.e. a Noetherian object of R-gr if and only if $\underline{M} \in R$-mod is Noetherian but in the Artinian case only the obvious implication remains valid.
If every non-zero homogeneous element of a graded ring R is invertible then R is said to be a graded division ring.

III.1.1. Proposition. If R is a graded division ring then either $R = R_o$ and R is a division ring, or $R \cong R_o[X, X^{-1}, \varphi]$ where $\varphi : R_o \to R_o$ is an automorphism, X an indeterminate of degree $t > o$ such that : $Xa = \varphi(a)X$, $X^{-1}a = \varphi^{-1}(a)X^{-1}$ for all $a \in R_o$.

Proof. If $R \neq R_o$, choose $a \in h(R)$ with deg $a > o$ and minimal as such. Since $R = Ra$, $R_t = R_o a$, thus $R_i = o$ if $i \notin \mathbb{Z}t$. Moreover $a^n \neq o$ yields $R_{nt} = R_o a^n$ for all $n \in \mathbb{Z}$. Pick $\lambda \in R_o$. There is a unique element $\varphi(a) \in R_o$ such that $a\lambda = \varphi(\lambda)a$. It is easily seen that $\varphi : R_o \to R_o$ is a field isomorphism and $a^n\lambda = \varphi^n(\lambda)a^n$, $a^{-1}\lambda = \varphi^{-1}(\lambda)a^{-1}$. The graded ring homomorphism $\alpha : R_o[X, X^{-1}, \varphi] \to R$ defined by $\alpha(X) = a$, $\alpha(X^{-1}) = \alpha^{-1}$ is an isomorphism of degree zero if we put deg $X = t$. ∎

III.1.2. Corollary. Graded division rings are (fields or) two-sided principal ideal rings.

Let us summarize some results of [121], II.7., about graded simple modules. Recall that $M \in R$-gr is said to be Gr-simple if o and M are its only graded submodules. Note that in the case where R is a positively graded ring, any Gr-simple $M \in R$-gr. is such that $M = M_j$ for some $j \in \mathbb{Z}$, i.e. $R_+ M = o$ where $R_+ = \bigoplus_{i > o} R_i$.

III.1.3. Theorem. ([121]). Let R be a graded ring and let $M \in R$-gr be Gr-simple. Then $D = HOM_R(M,M)$ is a graded division ring. If $D \neq D_o$ then $\underline{M}$ is a 1-critical module. Any R-submodule of $\underline{M}$ is principal and $\underline{M}$ is either simple in R-mod or 1-critical. Furthermore, the intersection of all maximal R-submodules of $\underline{M}$ is zero.

If $M \in R$-gr then M is a graded module over $S = END_R(M) = HOM_R(M,M)$. The graded ring $B^g(M) = END_S(M)$ is called the graded bi-endomorphism ring of M.

We refer to [121] for the graded version of the well-known density theorem :
If M is a semi-simple object of R-gr and $x_1,\dots,x_n \in h(M)$, $\alpha \in h(B^g(M))$, then there exists $r \in h(R)$ such that $\alpha(x_i) = rx_i$, $i = 1,\dots,n$. This may be applied to derive that the polynomial module $M[X]$ over a simple R-module M is a 1-cocritical R-module; or also to derive that, if $M \in R$-mod has finite length n, then the $R[X]$-submodules of $M[X]$ may be generated by n elements.
Let us also point out the graded version of Hopkin's theorem :
If R is a left Gr-Artinian ring then R is left Gr-Noetherian i.e. left Noetherian too. Note that $Q[x,x^{-1}]$ is Gr-Artinian but neither left or right Artinian.

A graded ring R is said to be Gr.-semisimple if R-gr is a semisimple category i.e. if any $M \in R$-gr is semisimple. Clearly R is Gr-semisimple if and only if (*) : $R = L_1 \oplus \dots \oplus L_n$, where the L_i, $i=1,\dots,n$, are minimal graded left ideals of R. The graded ring R is said to be Gr-simple if it has the decomposition (*) but with $HOM_R(L_i,L_j) \neq o$ for any $i,j = 1,\dots,n$.
A Gr-simple ring R is said to be uniformly simple if $R = L_1 \oplus \dots \oplus L_n$ with $L_i \cong L_j$ for any $i,j = 1,\dots,n$.

Consider $M_n(R)$ over the graded ring R. Fix $\underline{d} = (d_1,\dots,d_n) \in \mathbb{Z}^n$ and for every $\lambda \in \mathbb{Z}$ put :

$$M_n(R)(\underline{d})_\lambda = \begin{pmatrix} R_\lambda & R_{\lambda+d_2-d_1} & \cdots & R_{\lambda+d_n-d_1} \\ R_{\lambda+d_1-d_2} & R_\lambda & & R_{\lambda+d_n-d_2} \\ \vdots & \vdots & & \vdots \\ R_{\lambda+d_1-d_n} & R_{\lambda+d_2-d_n} & \cdots & R_\lambda \end{pmatrix}$$

It is obvious that $M_n(R) = \bigoplus_{\lambda \in \mathbb{Z}} M_n(R)(\underline{d})_\lambda$, and thus we have defined a gradation on $M_n(R)$; let us denote this graded ring by $M_n(R)(\underline{d})$ because the gradation depends on the choice of $\underline{d} \in \mathbb{Z}^n$.

As in [121], Lemma 9.1.1. and Proposition 9.1.3, p. 68, it is not hard to detect that the number of isomorphism classes of graded structures defined in the above way is actually finite and not hard to calculate.

III.1.4. Theorem. (Graded Version of Wedderburn's Theorem)

The following statements are equivalent, for a graded ring R :

1. R is Gr-simple (resp. uniformly simple).
2. There exist a graded division ring D and some $n \in \mathbb{N}$, $d \in \mathbb{Z}^n$, such that $R \cong M_n(D)(\underline{d})$ (resp. $R \cong M_n(D)$).

Let R be a graded ring and consider a multiplicatively closed set in R which consists of homogeneous elements, S say. Then R satisfies the left Ore conditions with respect to S if and only if :

1°. If $rs = o$ with $r \in h(R)$, $s \in S$ then there is an $s' \in S$ such that $s'r = o$.

2°. For any $r \in h(R)$, $s \in S$ there are $r' \in h(R)$, $s' \in S$ such that $s'r = r's$.

This new characterization of the left Ore condition with respect to a multiplicative set of homogeneous elements is easily verified; it will be useful in checking Ore conditions at graded prime ideals in III.2.

A graded ring having finite Goldie dimension in R-gr. and satisfying the ascending chain condition on graded left annihilators is called a graded Goldie ring (abreviated Gr-Goldie ring). Surprisingly(?) a semiprime Gr-Goldie ring need not have a Gr-semisimple ring of fractions. For example, let k be a commutative field and R a graded ring containing k such that $R_+ = k[X]$, $R_- = k[Y]$, $XY = YX = o$, i.e. $R_n = kX^n$, $R_o = k$, $R_{-n} = kY^n$ for $n > o$. Clearly R is a semiprime Gr-Goldie ring which cannot have a Gr-semisimple ring of graded fractions. Therefore some extra assumption is needed in order to prove the following graded "analogue" of Goldie's theorem :

III.1.5. Theorem. Let R be a semiprime Gr-Goldie ring satisfying one of the following conditions :

1. R has a central regular homogeneous element s of degree $\deg s > o$.

2. R is positively graded and the minimal prime ideals (these are graded!) of R do not contain R_+.
3. R is positively graded and R has a regular homogeneous element of positive degree.
4. The homogeneous elements of R of positive degree are nilpotent.

Then R admits a Gr-semisimple Gr-Artinian ring of graded fractions.

Proof. cf. [121]. Proposition 9.2.3. p.71.

III.1.6. Corollary. Let R be a Noetherian graded ring which is either positively graded or commutative. Then either $ht(P) = ht(P_g)$ or $ht(P) = ht(P_g)+1$ for any prime ideal P of R.

Utilizing this together with an easy induction argument we get :

III.1.7. Corollary. If R is a Noetherian commutative graded ring and P is a graded prime ideal with $ht(P) = n$, then there exists a chain of graded prime ideals $P_o \subsetneq P_1 \subsetneq \dots \subsetneq P_n = P$.

Since R-gr is a Grothendieck category, general techniques of localization at kernel functors apply to it. However, here there is one important difference when compared to localization theory in R-mod, namely the fact that R is not a generator for R-gr. If κ is a kernel functor in R-gr then there is equivalence between :

1°. if $\kappa(M) = M$ then $\kappa(M(n)) = M(n)$ for all $n \in \mathbb{Z}$.

2°. if $\kappa(M) = o$ then $\kappa(M(n)) = o$ for all $n \in \mathbb{Z}$.

A kernel functor satisfying one of these requirements is said to be rigid. Note that, with κ-torsion free M and κ-torsion N in R-gr, we always have that $\mathrm{Hom}_{R\text{-gr}}(M,N) = o$ but $\mathrm{HOM}_R(M,N) = o$ for all such M and N is equivalent to κ being rigid. It is easily seen that κ is rigid if and only if $\kappa(M(n)) = \kappa(M)(n)$ for all $n \in \mathbb{Z}$.

In general κ could be characterized by an idempotent filter of subobjects of the generator, however this is not always satisfactory. Therefore the fact that rigid kernel functors do correspond in a bijective way to graded filters in R is very useful. For full detail on graded rings of quotients at kernel functors in R-gr we refer to [181], we shall here give a down to earth treatment of those results necessary in the sequel.

An idempotent kernel functor $\underline{\kappa}$ on R-mod is said to be a graded kernel functor if the filter $\mathcal{L}(\underline{\kappa})$ of $\underline{\kappa}$ possesses a cofinal set of graded left ideals. If $\underline{\kappa}$ is a graded kernel functor and $M \in$ R-gr then $\underline{\kappa}(\underline{M})$ and $M/\underline{\kappa}(\underline{M})$ are graded left R-modules, the canonical morphism $M \to M/\underline{\kappa}(\underline{M})$ is graded of degree o. Furthermore, $\underline{\kappa}$ induces a kernel functor κ on R-gr which is rigid (and idempotent), the class of κ-torsion objects in R-gr consists of $N \in$ R-gr such that $\underline{N}$ is a $\underline{\kappa}$-torsion module. For rigid κ and $\{\kappa_i; i \in J\}$, $\kappa = \bigwedge_i \kappa_i$ is equivalent to $\mathcal{L}_g(\kappa) = \bigcap_i \mathcal{L}_g(\kappa_i)$, where $\mathcal{L}_g(\kappa)$ denotes the graded filter in R; note however that this fails for non-rigid κ on R-gr. Associating to a kernel functor κ on R-gr the graded kernel functor $\underline{\kappa}$ on R-mod generated by the torsion class of κ we see that every rigid κ is induced by a graded kernel functor $\underline{\kappa}$ or R-mod and if $\mathcal{L}$ is the filter in R generated by the graded filter $\mathcal{L}_g(\underline{\kappa})$ then $\mathcal{L} \cap L_g(R) = \mathcal{L}(\kappa)$. For example, if S is a multiplicatively closed subset of R consisting of homogeneous elements and not containing zero then let $\underline{\kappa}_S$ be the usual kernel functor corresponding to S on R-mod; $\underline{\kappa}_S$ is graded and it induces a rigid kernel functor κ_S on R-gr the (graded) filter of which is given by :

$$\mathcal{L}_g(\kappa_S) = \{L \text{ a graded left ideal of } R,\ (L:r) \cap S \neq \phi \text{ for all } r \in h(R)\}.$$

If R is a left Noetherian graded ring, I a graded ideal of R and P a graded prime ideal of R then consider :

$$\mathcal{L}_g(I) = \{L \in L_g(R),\ L \supset I^n \text{ for some } n \in \mathbb{N}\}.$$

$$\mathcal{L}_g(R-P) = \{L \in L_g(R),\ L \supset RsR \text{ for some } s \in R-P\}\ ,$$

where $L_g(R)$ stands for the set of graded left ideals of R.
The kernel functors κ_I, κ_{R-P} on R-gr. associated to $\mathcal{L}_g(I)$, $h_g(R-P)$ resp. are rigid kernel functors and they are induced on R-gr by $\underline{\kappa}_I$, $\underline{\kappa}_{R-P}$ resp. on R-mod which are given by their filters :

$\mathcal{L}(I) = \{L \text{ left ideal of } R,\ L \supset I^n \text{ for some } n \in \mathbb{N}\}$

$\mathcal{L}(R-P) = \{L \text{ left ideal of } R,\ L \supset RsR \text{ for some } s \in R-P\}$.

We omit the relations between kernel functors and injective objects in R-gr, and we refer to [111] Section 12.3 for this.

General methods (taking into account the fact that R does not generate R-gr) allow to define the module of quotients of $M \in$ R-gr at a kernel functor κ on R-gr as follows : $Q^g_\kappa(M) = \varinjlim_{L \in \mathcal{L}_g(\kappa)} \mathrm{HOM}_R(L, M/\kappa(M))$.

If κ is rigid, i.e. induced on R-gr by some kernel functor $\underline{\kappa}$ on R-mod, then we can relate $Q^g_\kappa(M)$ to $Q_{\underline{\kappa}}(\underline{M})$ in some way. For indeed if $\underline{\kappa}$ is a graded kernel functor on R-mod, define $gQ_{\underline{\kappa}}(M)$, for a graded $M \in$ R-mod as follows :

$$(g\, Q_{\underline{\kappa}}(M))_n = \{x \in Q_{\underline{\kappa}}(M),\ \text{there is an } L \in \mathcal{L}(\underline{\kappa}) \cap L_g(R) \text{ such that } L_m x \subset (M/\underline{\kappa}(M))_{m+n} \text{ for all } m \in \mathbb{Z}\}$$

$$g\, Q_{\underline{\kappa}}(M) = \bigoplus_{m \in \mathbb{Z}} (g\, Q_{\underline{\kappa}}(M))_m.$$

One checks that : $g\, Q_{\underline{\kappa}}(R)$ is a graded ring containing $R/\underline{\kappa}(R)$ as a graded subring and the graded ring structure of $g\, Q_{\underline{\kappa}}(R)$ is the unique ring structure compatible with its graded R-module structure. For every $M \in$ R-gr, $g\, Q_{\underline{\kappa}}(\underline{M})$ is a graded $g\, Q_{\underline{\kappa}}(R)$-module. It is easy to derive from the definition that :

$$g\, Q_{\underline{\kappa}}(\underline{M})_m = \varinjlim_{L \in \mathcal{L}(\underline{\kappa}) \cap L_g(R)} \mathrm{HOM}_R(L, \underline{M}/\underline{\kappa}(\underline{M}))_m.$$

Consequently, for a rigid kernel functor κ which is induced by some $\underline{\kappa}$ on R-mod, we find :

$$Q_{\kappa}^{g}(M) \cong g\, Q_{\underline{\kappa}}(\underline{M}) \quad \text{for every } M \in R\text{-gr}.$$

<u>III.1.8. Remark.</u> If κ is a rigid kernel functor such that $\underline{\kappa}$ has finite type (e.g. if R is left Noetherian) then

$$Q_{\kappa}^{g}(M) \cong Q_{\underline{\kappa}}(\underline{M}) \quad \text{for all } M \in R\text{-gr}. \;!$$

Recall that $\underline{\kappa}$ has <u>finite type</u> if for every $L \in \mathcal{L}(\underline{\kappa})$ there is an $L' \subset L$, $L' \in \mathcal{L}(\underline{\kappa})$ and L' is finitely generated. A rigid kernel functor κ on R-gr is said to be a <u>rigid t-functor</u> if Q_{κ}^{g} is exact and commutes with direct sums in R-gr.

<u>III.1.9. Theorem [184]</u>. Let κ be a rigid kernel functor or R-gr and let $\underline{\kappa}$ be the graded kernel functor on R-mod associated to κ, then the following statements are equivalent :

1. $Q_{\kappa}^{g}(R) j_{\kappa}(L) = Q_{\kappa}^{g}(R)$ for all $L \in \mathcal{L}_{g}(\kappa)$, where j_{κ} is the canonical graded morphism $R \to R/\kappa(R)$.
2. If $\underline{M} \in Q_{\kappa}^{g}(R)$-mod then $\underline{\kappa}(\underline{M}) = o$.
3. If $M \in Q_{\kappa}^{g}(R)$-gr. then $\kappa(M) = o$.
4. $Q_{\kappa}^{g}(-) = Q_{\kappa}^{g}(R) \underset{R}{\otimes} -$ (natural equivalence in R-gr)
5. $\underline{\kappa}$ is a t-functor in R-mod.
6. κ is a rigid t-functor in R-gr.

In the sequel of this section we assume that R is a left Noetherian positively graded ring. A graded kernel functor or the corresponding rigid kernel functor on R-gr will be said to be <u>projective</u> if $R_{+} \in \mathcal{L}_{g}(\kappa)$. Recall that Proj R is the set of graded prime ideals P of R which do not contain R and such that $P \not\supset R_{+}$; the Zariski topology of Spec R induces a topology in Proj R which is also referred to as being the Zariski-topology.

III.1.10. Theorem. [184] . Let $P \in \mathrm{Proj}\ R$, R a positively graded left Noetherian ring. Let $\underline{\kappa}_{h(P)}$ be the kernel functor on R-mod associated to the multiplicatively closed set $h(G(P))$, let $\underline{\kappa}_{h(R-P)}$ be the kernel functor associated to the m-system $h(R-P)$, then for the associated rigid kernel functors on R-gr we have : $\kappa_P = \kappa_{h(P)}$ and $\kappa_{R-P} = \kappa_{h(R-P)}$, where κ_P is constructed from the injective hull of R/P in R-gr i.e. the graded version of the Lambek-Micheler localization at P, cf. [181].

A useful corollary is :

III.1.11. Corollary. If R is a left Noetherian positively graded ring and $P \in \mathrm{Proj}\ R$ such that the left Ore conditions with respect to $G(P)$ hold, then R satisfies the left Ore conditions with respect to $h(G(P))$.

Note that $P \in \mathrm{Proj}\ R$ if and only if κ_{R-P} is projective.
We will return to the study of Proj when studying sheaves over it in Section V.3.

III.2. Graded Rings with Polynomial Identity.

III.2.1. Lemma. Let R be a graded ring which satisfies the ascending chain condition for graded left annihilators, then the left singular radical, $t_S(R)$, is nilpotent.

Proof. Put $J = t_S(R)$ and note that J is a graded ideal. Our hypotheses imply that the ascending chain of left annihilants $\ell(J) \subset \ell(J^2) \subset \ldots \subset \ell(J^n) \subset \ldots$ is stationary i.e. $\ell(J^n) = \ell(J^{n+1})$ for some $n \in \mathbb{N}$. If $J^{n+1} \neq o$ then there is $a \in h(R)$ such that $aJ^n \neq o$ and $\ell(a)$ maximal with respect to this property. If $b \in J \cap h(R)$ then $\ell(b) \cap Ra \neq o$ since $\ell(b)$ is a left essential ideal of R. Therefore there exists $c \in h(R)$ such that $ca \neq o$ and $cab = o$.
Consequently $\ell(a) \subsetneq \ell(ab)$ and by maximality of $\ell(a)$, $abJ^n = o$. Since J is generated by $J \cap h(R)$ it follows that $aJ^{n+1} = o$, or $a \in \ell(J^{n+1}) = \ell(J^n)$, contradiction. Thus $J^{n+1} = o$ follows.

III.2.2. Proposition. For a semiprime graded ring R the following statements are equivalent :

1. R is a Gr-Goldie ring
2. R is a Goldie ring.

Proof. 2. ⇒ 1. Easy.

1. ⇒ 2. By the lemma it follows that $t_S(R) = o$.

Since R has finite Goldie dimension in R-gr, R also has finite Goldie dimension; cf. [111].

Together with $t_S(R) = o$ this means that the injective envelope of R in R-mod is semisimple and therefore R satisfies the ascending chain condition for left annihilators.

III.2.3. Corollary. Let R be a prime graded Goldie ring and let $E^g(R)$ be an injective hull of R in R-gr. Then $E^g(R)$ is a graded simple Artinian ring such that the canonical ring morphism $R \to E^g(R)$ is a left flat ring epimorphism.

Proof. By Proposition III.2.2., the injective hull E(R) of R in R-mod is a simple Artinian ring and therefore $E^g(R)$ is a Goldie ring. From the graded injectivity of $E^g(R)$ one deduces that it is a graded regular ring. Techniques similar to the techniques used in the ungraded case may be used to prove that $E^g(R)$ is a graded simple Artinian ring. Since $E^g(R)$ is the ring of quotients with respect to the graded torsion theory associated to the filter generated by the graded left essential ideals of R, the fact that $E^g(R)$ is graded simple Artinian entails that the graded kernel functor is a t-functor and the last statement of the corollary follows from this.

III.2.4. Remark. $E^g(R)$ need not be a graded ring of fractions because graded essential left ideals of R need not contain homogeneous regular elements.

III.2.5. Corollary. Let $h(R)_{reg}$ be the multiplicative set of regular homogeneous elements of R. Put $E(R) = Q$, $E^g(R) = Q^g$ and let Q_h be the graded ring

of fractions of R obtained by inverting elements of $h(R)_{reg}$.
We have canonical ring homomorphisms : $R \to Q_h \to Q^g \to Q$.
Each localization functor involved is a t-functor and we have $Q_h = Q^g$ if and only if graded essential left ideals of R do contain homogeneous regular elements.

Proof. R satisfies the left Ore condition with respect to the set of regular elements of R. An easy extension of III.1.11 (cf. [184] for the general version) then entails that R satisfies the left Ore condition with respect to $h(R)_{reg}$. All assertions follow now easily.

After this preliminary section we turn to the study of graded P.I. algebras

III.2.6. Proposition. If R is a prime graded P.I. ring then $Q_h = Q^g$ i.e. Q^g is a graded simple Artinian ring of fractions of R.

Proof. Put $C = Z(R)$. If there exists $c \in h(C)$ with deg $c > o$ then Theorem III.1.5 applies. If there exists $c \in h(C)$ with deg $c < o$ then the graded ring of fractions of R at $\{1, c, c^2, \ldots\}$, Q_c^g say, embeds as follows :

$$R \to Q_c^g \to Q_h \to Q^g \to Q .$$

Obviously Q_c^g is again a prime graded Goldie ring and c^{-1} with deg $c^{-1} > o$ is in the center of Q_c^g. Therefore Q_c^g has a graded simple Artinian ring of fractions S^g which has to be contained in Q_h. However, then S^g must be the graded simple Artinian ring of fractions of R too.
The problem has now been reduced to the case $C = C_o$. By Theorem III.1.4. : $Q^g \cong M_n(D)(\underline{d})_n$ for some graded division ring D and some $\underline{d} \in \mathbb{Z}^n$.
Posner's theorem for P.I. rings entails that Q is finite dimensional over the field of fractions of C. Thus for $x \in h(D)$ we may find an equation $c_o x^m + \ldots + c_{m-1} x + c_m = o$ with $c_i \in C$ for some $m \in \mathbb{N}$.
If $C = C_o$ then deg $x = o$ follows but then $D = D_o$ entails that D is a skewfield and thus Q^g is simple Artinian. Therefore Q^g coincides with Q and it is the

ring of fractions with respect to $C^* = C_0^*$. Note that we also obtained that a graded prime P.I. ring has trivially graded centre if and only if $Q^g = Q$.

III.2.7. Theorem. (Graded version of Posner's theorem).
A graded prime ring is a P.I. ring if and only if R is a graded order of $M_n(K)(\underline{d})$ for some graded field K.

Proof. Obviously, graded orders of $M_n(K)(\underline{d})$ are P.I. rings. Conversely, if R is a graded prime P.I. ring then it has a graded simple Artinian ring of fractions Q^g because of Proposition III.2.6. We may write : $Q^g = M_{n'}(D)(\underline{d}')$ for some $n' \in \mathbb{N}$, $\underline{d}' \in \mathbb{Z}^{n'}$ and for some graded division ring D.
First suppose $D = D_0$, hence $C = C_0$ and $Q^g = Q$. Take a splitting field L of D, put $L = L_0$ and consider $Q^g \otimes L$ with the usual tensor product-gradation. Clearly $Q^g \otimes L \cong M_{n'}(M_m(L))(\underline{d}')$ is of the desired form. Assume now that $D \neq D_0$. By Proposition III.1.1., D is of the form $D_0[X,X^{-1},\varphi]$ where D_0 is a skewfield which is of finite dimension over its center and φ is an automorphism of D_0 such that φ^e is inner for some $e \in \mathbb{N}$ (here we use some results of G. Cauchon, [33]). The center of D is of the form $K_0[T,T^{-1}]$ for some field K_0 and variable T over K_0. Since D is a Zariski extension of $Z(D) = K_0[T,T^{-1}]$ it follows that $Q(D) = D \otimes_{Z(D)} K_0(T)$.

Consider a maximal commutative graded subring of D, say L. It is obvious that L is a graded field and that it is also maximal as a commutative subring of D. Write Q(L) for the ring of fractions $L \otimes_{Z(D)} K_0(T)$ which is a commutative subring of $Q(D) = D_0(X,\varphi)$. If $y \in D_0(X,\varphi)$ commutes with Q(L) then for some $s \in K_0[T,T^{-1}]$ we have that $sy \in L$, hence $y \in Q(L)$. Consequently Q(L) is a maximal commutative subfield of $D_0(X,\varphi)$ i.e. a splitting field.
Now $D \otimes_{Z(D)} L$ is a prime ring and by exactness of the central localization functor we obtain :

$$Q(D \otimes_{Z(D)} L) = Q(D) \otimes_{K_0(T)} Q(L) \simeq M_n(Q(L)) .$$

Argumentation similar to the ungraded case yields that $D \underset{Z(D)}{\otimes} L$ is a graded simple Artinian ring with center L, hence $D \underset{Z(D)}{\otimes} L \simeq M_k(\Delta)(\underline{d})$ for some $k \in \mathbb{N}$, $\underline{d} \in \mathbb{Z}^k$, and for some graded skewfield Δ with center L. Central localization at o yields : $Q(M_k(\Delta)(\underline{d})) = M_k(Q(\Delta)) \simeq M_n(Q(L))$.

From this one derives immediately that k=n, $Q(\Delta) \cong Q(L)$, hence Δ is commutative and thus $\Delta = L$.

Finally we have established that $M_n(L)$ and $D \underset{Z(D)}{\otimes} L$ are isomorphic as graded rings when the gradation on $M_n(L)$ is described by $\underline{d} \in \mathbb{Z}^n$. ∎

<u>III.2.8. Corollary</u>. The center of the field of fractions of the ring of generic $n \times n$-matrices contains a subfield over which it is purely transcendental of degree 1.

Recall that $\mathrm{Spec}_n(R)$ stands for the set of prime ideals P of R with p.i. deg P = n.

<u>III.2.9. Theorem</u>. Let R be a (left Noetherian) graded P.I. ring satisfying the identities of nxn matrices. A prime ideal P is in $\mathrm{Spec}_n(R)$ if and only if $P_g \in \mathrm{Spec}_n(R)$. Consequently the prime radical of the ideal generated by the central kernel of R is a graded ideal.

<u>Proof</u> : If $P \in \mathrm{Spec}_n(R)$ then obviously $P_g \in \mathrm{Spec}_n(R)$.

Conversely suppose that $P_g \in \mathrm{Spec}_n(R)$. If p.i. deg(R/P) were smaller than p.i. deg(R) = n then the multilinear standard polynomial S_n is an identity for R/P, hence P contains all evaluations of S_n in R. Since S_n is multilinear, it is clear that then also P_g contains all evaluations of S_n in R i.e. p.i. $\deg(R/P_g) < n$, contradiction. Consequently $P \in \mathrm{Spec}_n(R)$ follows.

<u>III.2.10. Remark</u>. Let R be a graded affine P.I. ring over k, satisfying the identities of nxn matrices; write G_n for the ring of generic nxn matrices, $G_n = k\{\xi_1, \ldots, \xi_m\}$, and write $R = k\{a_1, \ldots, a_m\}$ where the a_i, i=1, ..., m, are homogeneous generators for the ring R over k. Define a gradation on

$M_n(k[\xi_j^{ik}, j=1,\ldots,m;\ i,k = 1\ldots n])$ in such a way that $\deg\xi_j = \deg a_j$ and consider the gradation induced in G_n. The canonical ring homomorphism $\pi : G_n \to R$, defined by $\pi(\xi_j) = a_j$ and $\pi|k = 1_k$, is a graded ring homomorphism of degree o. The center $Z(G_n)$ is a graded ring. However it is well-known that $Z(G_n)$-k is contained in every prime ideal of G_n of p.i. degree less than n, i.e. $Z(G_n) = k \cup J_n$, where J_n is the central kernel of G_n. It follows easily from this that J_n is graded. However, for R we only know that the radical of the central kernel is graded, by III.2.9.

III.2.11. Corollary. If R is a graded P.I. ring satisfying the identity of nxn matrices then we have the following strengthening of the Artin-Procesi theorem, II.3. 25 ; R is an Azumaya algebra of rank n if and only if p.i. deg(P) = n for every graded prime ideal P of R.

III.2.12. Remark. The following theorem is of some importance in the study of the Brauer groups (graded and common) of a graded ring of the form $k[X,X^{-1}]$ (termed : graded field). Without going into detail let us just mention that deg(X) intervenes in the determination of $Br^g\, k[X,X^{-1}]$ (the classes of graded Azumaya algebras over $k[X,X^{-1}]$ with respect to graded equivalence i.e. isomorphism up to tensoring up by endomorphism rings of graded progenerator moduls). For example consider $\mathbb{R}[X,X^{-1}]$. If degX is odd then $Br^g\, \mathbb{R}[X,X^{-1}]$ $= \mathbb{Z}/2\mathbb{Z}$ is generated by the class of $\mathbb{H}[X,X^{-1}]$; if degX is even then $Br^g\, \mathbb{R}[X,X^{-1}] = Br\, \mathbb{R}[X,X^{-1}] = \mathbb{Z}/2\mathbb{Z} \times \mathbb{Z}/2\mathbb{Z}$ is generated by the classes of $\mathbb{H}[X,X^{-}]$ and $\mathbb{C}[X,-]$. We refer to F. Van Oystaeyen [185], [186].

III.2.13. Theorem. If R is a graded P.I. ring satisfying the identities of n x n matrices and such that Z(R) is a graded field then R is an Azumaya algebra.

Proof. If f is a multilinear central polynomial for R then it is clear that f cannot vanish on every homogeneous substitution i.e. there exist $u_1,\ldots,u_k \in h(R)$ such that $o \neq f(u_1,\ldots,u_k) \in h(C)$.

The hypothesis on C entails that $f(u_1,\dots,u_k)$ is a unit of C and therefore the central kernel of R coincides with C. By II.4.9. , R has to be an Azumaya algebra of constant rank.

III.2.14. Corollary. Iterated twisted polynomial rings of the type $D_o[X_1,\dots,X_n,X_1^{-1},\dots,X_n^{-1},\varphi_1,\dots,\varphi_n]$ are Azumaya algebras.

III.3. Graded Birational Extensions.

Consider graded rings A and B and a graded ring extension of degree o, $f: A \to B$. With notations as in Section II.1. , we put Spec B = Y, Spec A = X and if f is birational then we write U = Y(I) , V = X(I') open set of birationality in Y and X resp.

We say that B is a graded birational A-algebra if the ideals I of B and I' of A are graded ideals. A kernel functor κ is radical if it is symmetric and if an ideal $H \in \mathcal{L}(\kappa)$ if and only if $\mathrm{rad}(H) \in \mathcal{L}(\kappa)$. Obviously, if κ_H or κ_{B-P} is idempotent then it is radical for any ideal H of B, or prime ideal P of B.

A graded ring R with center C is said to be Gr-Zariski central if for every graded ideal M of R we have that $\mathrm{rad}(H) = \mathrm{rad}(R(H \cap C))$. This definition is actually weaker than R is graded and Zariski central; however in this section we shall give proofs only in the case of a graded Zariski central ring leaving the Gr-Zariski case to the reader.

III.3.1. Lemma. Let B be a graded birational A-algebra. To a radical kernel functor κ on B-mod such that $I \in \mathcal{L}(\kappa)$ there corresponds in a natural way a radical kernel functor κ' on A-mod (see Section II.1. for the definition of κ'). Now κ is a graded kernel functor (in the sense of III.1) if and only if κ' is graded.

Proof. If $L \in \mathcal{L}(\kappa)$ and κ is graded then $L \supset J$ with $J \in \mathcal{L}(\kappa)$ and J is a graded left ideal of B. Since both κ and κ' are radical kernel functors it is not hard to see that $J\cap A\in\mathcal{L}(\kappa')$ is a graded ideal of A and (because a basis for $\mathcal{L}(\kappa')$ consists of $J\cap A$ with $J\in\mathcal{L}(\kappa)$) therefore κ' is graded. Conversely, suppose κ' is a graded kernel functor and let L be an ideal of B which is in $\mathcal{L}(\kappa)$. Then $X(L\cap A)\cong Y(L)$ (we may restrict ourselves to the case $L\subset I$ because $I\in\mathcal{L}(\kappa)$ i.e. $L\in\mathcal{L}(\kappa)$ if and only if $L\cap I\in\mathcal{L}(\kappa)$!) yields that $L\cap A\supset J'$ where J' is a graded ideal of A and $J'\in\mathcal{L}(\kappa')$, $L\supset BJ'$. Now $\mathrm{rad}(BJ')\in\mathcal{L}(\kappa)$ because $J'\in\mathcal{L}(\kappa')$ hence, by radicality of κ, it follows that $BJ'\in\mathcal{L}(\kappa)$ i.e. L contains a graded ideal of $\mathcal{L}(\kappa)$ i.e. κ is graded.

<u>III.3.2. Remark.</u> If B is left Noetherian, graded, then every symmetric kernel functor on B-mod is radical. In the sequel we are concerned with Noetherian rings only so the radicality hypothesis will be implicit. We leave it to the reader to expound the general theory of graded birational extensions along the lines of II.1.8. and II.1.11., etc... For some geometrical applications we content ourselves with the following theorem.

<u>III.3.3. Theorem.</u> Let B be a graded left Noetherian Zariski central ring with center $Z(B)=A$. Let S be a central Ore set. Let $\underline{\kappa}$ be the graded kernel functor on B-mod associated with $h(S)=\bigcup_n (S\cap B_n)$. Then the following properties hold:

1. $\underline{\kappa}$ is a t-functor and the kernel functor κ on B-gr associated to $\underline{\kappa}$ is a t-functor.
2. For any graded ideal H of B, $Q^g_\kappa(B)H$ is a graded ideal of $Q^g_\kappa(B)$.
3. There is a one to one correspondence between graded prime ideals P of B such that $P\notin\mathcal{L}(\kappa)$ and proper graded prime ideals of $Q^g_\kappa(B)$. Consequently: $Q^g_\kappa(B)\,\mathrm{rad}\,I=\mathrm{rad}(Q^g_\kappa(B)I)$ holds for any graded ideal I of B.

Proof. Combination of graded techniques with the usual birationality results, cf.[182].

In view of the general theory mentioned in III.3.2. it is clear that the foregoing theorem applies to κ_{B-P} where P is a graded prime ideal of B (because Q_{B-P} is just central localization at the Ore set A-p where $p=P\cap A$, and Q^g_{B-P}

corresponds to graded localization at h(A-p), cf. II.1.11.). The foregoing theorem also applies to $\kappa = \kappa_a$ where κ_a is given by $\mathcal{L}(\kappa_a)$ = {L left ideal of B such that $a^k \in L$ for some positive k}, where a is homogeneous in A. Let us now come back for a moment to the well-known example : skew polynomial rings.

Let A be a simple (not necessarily Artinian)ring .

Let σ be an automorphism of A and let δ be a σ-derivation i.e. $\delta(ab) = \delta(a)b + \sigma(a)\delta(b)$ for all $a, b \in A$. Consider the ring $A[X,\sigma,\delta]$ obtained by adding a variable to A submitted to the multiplication rule $Xa = \sigma(a)X + \delta(a)$ for all $a \in A$, and such that A is a subring of $A[X,\sigma,\delta]$.

Elements of $R = A[X,\sigma,\delta]$ may be written as polynomials with coefficients in A on the left i.e. $f = a_o + a_1X + \ldots + a_nX^n$, $a_i \in A$. If the leading coefficient of f is invertible then deg fg = deg f + deg g holds for every $q \in R$ and also f is then a regular element of R. Given $g \in R$ we have that $g = qf + r$ with $r = o$ or deg r < deg f. It is not hard to calculate the ideals of R, cf. G. Cauchon [33] i.e. any ideal of R is of the form Rgf^m where $g \in Z(R)$ and f is a monic polynomial of minimal degree with the property that Rf is an ideal of R. If k is the fixed field of σ and δ in Z(A) then either Z(R) = k or Z(R) = k [T] with $T = uf^s + v$, where $s \in \mathbb{N}$, u is a unit of A and $v \in k$. If $Z(R) \neq k$ then σ^m is an inner automorphism for some $m \in \mathbb{N}$ and $A[X,\sigma,\delta]$ is not simple.

<u>III.3.4. Proposition</u>. If the center of $R = A[X,\sigma,\delta]$ is not k then R is a filtered Zariski central ring.

<u>Proof</u>. cf. [121] . If $\delta = o$ then R is a graded Zariski central ring.

<u>III.3.5. Proposition</u>. If $R = A[X,\sigma,\delta]$ has non-trivial center then it has the following properties :

1. Every non-trivial ideal of R is a product of maximal ideals.
2. Every ideal of R is a product of prime ideals and the product of ideals of R is commutative.
3. The fractional ideals of R in $Q = Q_{R-o}(R)$ form an abelian group under multiplication.

4. Every ideal of R is projective as a left R-module.
5. R is not properly contained in any subring S of Q such that $cS \subset R$ for some $0 \neq c \in Z(R)$, non-trivial prime ideals of R are maximal and R satisfies the ascending chain condition for ideals.
6. Non-zero maximal ideals of R are invertible in $Q_{R-0}(R)$ and R satisfies the ascending chain condition for ideals.

Proof. cf. [121] or [125]. The properties have been listed this way because, for a prime Zariski central ring they are equivalent to one another.

III.3.6. Corollary. If A is simple Artinian then R is a Dedekind prime ring. On the other hand $A[X,\sigma]$ is a Dedekind prime ring if and only if A is simple Artinian.

III.3.7. Remark. $R = A[X,\sigma,\delta]$ is a P.I. ring if and only if :

1°. A is a central simple algebra.

2°. σ^m is inner for some $m \in \mathbb{N}$.

3°. δ is algebraic in A.

Then $R[X^{-1}]$ is an Azumaya algebra (see the graded strengthening of the Artin-Procesi theorem).

If R is Zariski central, T a variable commuting with R then R[T] need not be Zariski central as the following easy example shows. Put $R = \mathbb{C}[X,-]$ i.e. $A = \mathbb{C}$, $\delta = 0$, σ = conjugation, then $(X,T-i)$ and $(X,T+i)$ are maximal ideals of R[T] sitting over the same central prime ideal (X^2,T^2+1). It is clear that these phenomena will occur whenever R contains non-central subfields which split certain central irreducible polynomials.

On the other hand we have the following :

III.3.18. Proposition. If R is a Zariski central ring then R[T] is Gr-Zariski central, where R[T] is positively graded by the degree in T.

Proof. Graded prime ideals of R[T] which contain T are obviously up to radical generated by their central parts. Grade prime ideals of R[T] not containing T are in Proj(R[T]) and by dehomogenization there correspond in a one-to-one way to prime ideals of R. Using the fact that T is central in R[T] the proposition follows easily from dehomogenization and homogenization of prime ideals.

III.3.19. Note. The counterexample given shows that central extensions of Zariski central rings need not be Zariski central. However it is conjectured that, if R_1, R_2 are Zariski central rings with the same center C, then $R_1 \otimes_C R_2$ is Zariski central.
A natural way to look for possible counterexamples is to try it out for maximal orders in central simple algebras but then the statement is true! In other cases it is hard to verify and the problem remains open.

III.3.20. References.

For III.1. cf. C. Năstăsescu, F. Van Oystaeyen [121]. For III.2. we refer to F. Van Oystaeyen [181]. It seems that the graded structure of the ring of generic matrices, in particular of its center relates to the problem of representing the center as a rational function field in some number of variables. This is suggested by Corollary III.2.8. and one may hope to find some inductive procedure to repeat that argument; this has up to now not been carried out. Most results from III.3. stem from G. Cauchon [33], E. Nauwelaerts, F. Van Oystaeyen [125] and E. Nauwelaerts [124]. The geometric features, i.e. the study of Proj as a "scheme" will be returned to in further chapters.

IV. THE FINISHING TOUCH ON LOCALIZATION.

IV.1. Relative Localization and Bimodules.

Although some torsion theorists recently developed nice theories about localization in abelian or even in barely additive categories, cf. a.o. J.Lambek [100], the applications we have in mind require other tools. So we are led to consider functors defined on some full subcategory of a Grothendieck category and taking values in that Grothendieck category. The "relative" theory thus obtained generalizes both localization in a Grothendieck category and localization in an abelian category.

As a special case we obtain the "pointwise" localization in presheaf categories encountered in [175], [188]. However the aim of this abstract setting is to apply it to the theory of bimodules in M.Artin's sense, cf. [17]. Since our techniques reveal the connections between localization in the category of two-sided modules (which is a Grothendieck category) and localization in the category of bimodules (a full subcategory), we also obtain refinements and generalizations of some results due to J.P. Delale, [46].

In the sequel of this section we consider full subcategories $\underline{C}$ of a Grothendieck category $\underline{D}$ satisfying :

RL1 : The canonical inclusion $i : \underline{C} \to \underline{D}$ has a right adjoint $c : \underline{D} \to \underline{C}$ i.e. for $C \in \underline{C}$, $D \in \underline{D}$ there is a functorial bijection $\mathrm{Hom}_{\underline{C}}(C, cD) \cong \mathrm{Hom}_{\underline{D}}(iC, D)$.

RL2 : $\underline{C}$ is closed under quotients in $\underline{D}$, i.e. if $iC \to D$ is surjective in $\underline{D}$ then $icD = D$.

It is not a real restriction if we assume that ic is a subfunctor of the identity in $\underline{D}$.

Note that RL2 entails that the zero-object $o \in \underline{D}$ is in $\underline{C}$.

Convention. We omit to write the functor i when it is redundant.

<u>IV.1.1. Definition.</u> A $\underline{\mathcal{D}}$-valued kernel functor on $\underline{C}$ is a left exact subfunctor σ of the inclusion functor, such that $\sigma(C/\sigma C) = o$ for any $C \in \underline{C}$. The triple $(\underline{C},\underline{\mathcal{D}},\sigma)$ is called a <u>situation</u>. An object C of $\underline{C}$ is σ-torsion if $\sigma C = C$, it is σ-torsion free if $\sigma C = o$.

<u>IV.1.2. Lemma.</u> Consider a situation $(\underline{C},\underline{\mathcal{D}},\sigma)$, then :

1. If C' is a subobject of C in $\underline{C}$ then C' is σ-torsion if C is and C' is σ-torsion free if C is.
2. A quotient of a σ-torsion object is again σ-torsion.
3. If C' is a subobject of C in $\underline{C}$ then $\sigma C \cap C' = \sigma C'$ (note : intersections are carried out in the Grothendieck category $\underline{\mathcal{D}}$).

<u>Proof.</u> 1. will follow from 3.; 2. is easy enough. 3. Obviously $\sigma C' \subset \sigma C \cap C'$. From the exact sequence $o \to C' \to C \to C/C' \to o$ in $\underline{C}$ we derive the following exact sequence in $\underline{\mathcal{D}}$:

$$o \to \sigma C' \to \sigma C \to \sigma(C/C').$$

We have a commutative diagram in $\underline{\mathcal{D}}$, with exact rows,

$$\begin{array}{ccccccccc}
o & \to & C' & \to & C & \xrightarrow{\pi} & C/C' & \to & o \\
 & & & & \uparrow \nu & & \uparrow & & \\
o & \to & \sigma C' & \to & \sigma C & \xrightarrow{\varphi} & \sigma(C/C') & & \\
 & & & & & \searrow & \uparrow & & \\
 & & & & & & \sigma C/\sigma C' & \to & o
\end{array}$$

This yields $\pi\nu(C' \cap \sigma C) = o$, hence $\varphi(C' \cap \sigma L) = o$ and $C' \cap \sigma C \subset \sigma C'$ follows. ■

<u>IV.1.3. Corollary.</u> Let $C_1, C_2 \in \underline{C}$ be such that C_1 is σ-torsion and C_2 is σ-torsion free then :

$$\mathrm{Hom}_{\underline{\mathcal{D}}}(C_1,C_2) = \mathrm{Hom}_{\underline{C}}(C_1,C_2) = o.$$

For $C \in \underline{C}$ we may consider an injective hull $E^d(C)$ of C in $\underline{\mathcal{D}}$, which is essentially unique. Put $E^c(C) = cE^d(C)$. Then $E^c(C)$ is much like an "injective hull" in $\underline{C}$ because of the following :

<u>IV.1.4. Proposition.</u> Let C' be a subobject of C in $\underline{C}$ and let there be $\varphi' \in \mathrm{Hom}_{\underline{C}}(C', cE)$ where E is an injective object of $\underline{\mathcal{D}}$, then there exists a $\varphi \in \mathrm{Hom}_{\underline{C}}(C, cE)$ such that the following diagram commutes :

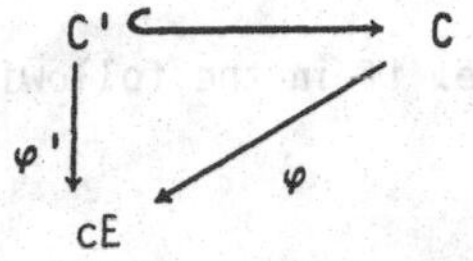

Moreover, if C' is a subobject of $E^c(C)$ in $\underline{C}$ then $C' \cap C = o$ (in $\underline{\mathcal{D}}$) if and only if $C' = o$.

<u>Proof.</u> Injectivity of E in $\underline{\mathcal{D}}$ yields a commutative diagram in $\underline{\mathcal{D}}$:

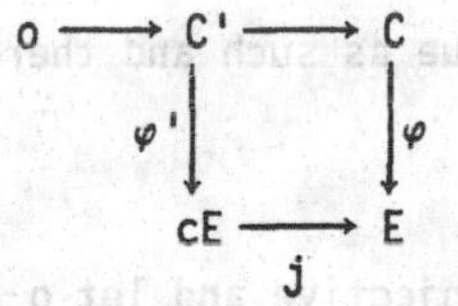

Now $\mathrm{Hom}_{\underline{\mathcal{D}}}(iC, E) = \mathrm{Hom}_{\underline{C}}(C, cE)$, thus $\varphi : C \to cE$ represents a $\underline{\mathcal{D}}$-homomorphism hence a C-homomorphism. The second statement is immediate from the fact that $E^d(C)$ is an essential extension of C in $\underline{\mathcal{D}}$. ∎

<u>IV.1.5. Remark.</u> A monomorphism $C \to E$ (E as above) extends to a morphism $\varphi : E^c(C) \to E$ with $\mathrm{Ker}\,\varphi \in \underline{\mathcal{D}}$. Since $\mathrm{Ker}\,\varphi \cap C = o$ if and only if $\mathrm{Ker}\,\varphi = o$, φ is monomorphic.

<u>IV.1.6. Comment.</u> It is clear that the properties mentioned in IV.1.4. and IV.1.5. characterize $E^c(C)$ up to isomorphism in $\underline{C}$. As mentioned before all intersections happen in $\underline{\mathcal{D}}$ however it is not unnatural to use $c\,(iC \cap iC')$ for

"intersection in $\underline{C}$". By means of this "intersection" one obtains a different notion of "essential extension" in $\underline{C}$ i.e. different "injective hulls". The exposition we have chosen avoids these tricky constructions in $\underline{C}$ completely.

Let $C \in \underline{C}$ be σ-torsion free, then we define a <u>σ-injective hull of C in $\underline{C}$</u> to be an object E of $\underline{C}$ such that the following properties hold :

IH_1 : E is σ-torsion free

IH_2 : E/C is σ-torsion

IH_3 : E is σ-injective in $\underline{C}$ i.e. if in the following diagram C" is a σ-torsion object of C :

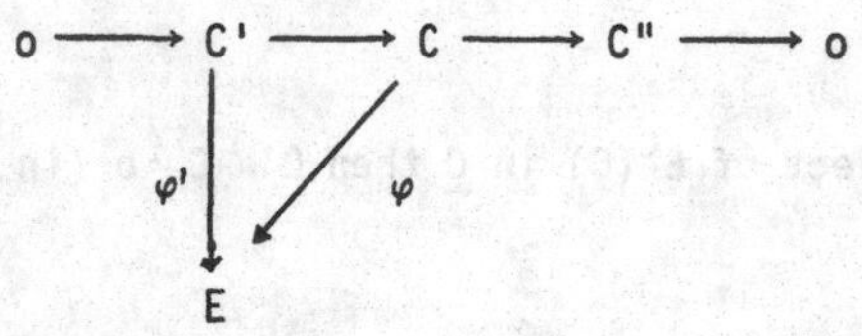

then any given $\underline{C}$-morphism $\varphi' : C' \to E$ extend to a $\underline{C}$-morphism $\varphi : C \to E$.

It follows from IV.1.3. that φ is unique as such and therefore we say that E is <u>faithfully σ-injective</u>.

<u>IV.1.7. Proposition.</u> Let $E \in \underline{C}$ be σ-injective and let $o \to E' \to E \to E" \to o$ be exact in $\underline{D}$. If E" is σ-torsion free then cE' is σ-injective in $\underline{C}$.

<u>Proof</u>. Let there be given a $\underline{C}$-morphism $\varphi : C' \to {}_cE'$. Consider the following exact sequence in $\underline{C}$:

$$o \to C' \to C \to C" \to o ,$$

where C" is σ-torsion.

Consider the following commutative diagram in $\underline{D}$ with exact rows,

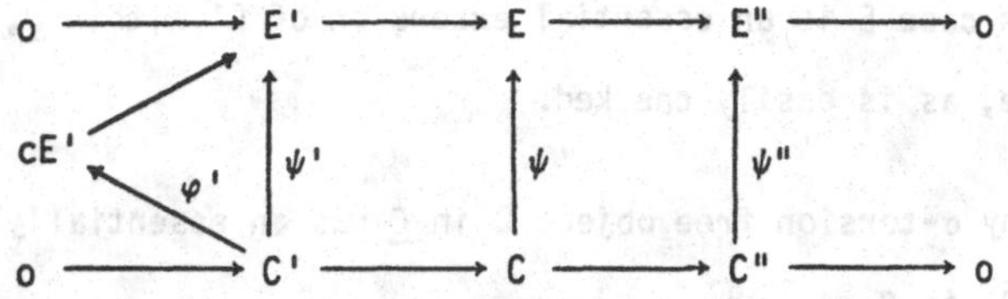

where ψ' is obtained from φ'; ψ is the extension of ψ' to C which exists by the σ-injectivity of E; ψ'' is the quotient map. By Corollary IV.13., ψ'' has to be the zero map i.e. ψ factorizes through E'. Now, because c is right adjoint to the inclusion i.e. $\mathrm{Hom}_{\underline{D}}(C,E') = \mathrm{Hom}_{\underline{C}}(C,cE')$ we find that $\psi : C \to cE'$ extends φ' to $\underline{C}$.

Let C' be a subobject of C in $\underline{C}$, then C is said to be a <u>$\underline{D}$-essential extension</u> of C' if for any nonzero subobject C" of C in $\underline{C}$, we have $C' \cap C \neq o$, where as always o and the intersection have to be interpreted in the larger Grothendieck category $\underline{D}$.

<u>IV.1.8. Lemma</u>. Let $o \to C' \to C \to C'' \to o$ be an exact sequence in $\underline{C}$ such that C is σ-torsion free and C" is σ-torsion, then C is a $\underline{D}$-essential extension of C' in $\underline{C}$.

<u>Proof</u>. Let C_1 be a subobject of C in $\underline{C}$ and consider the following commutative diagram in $\underline{D}$:

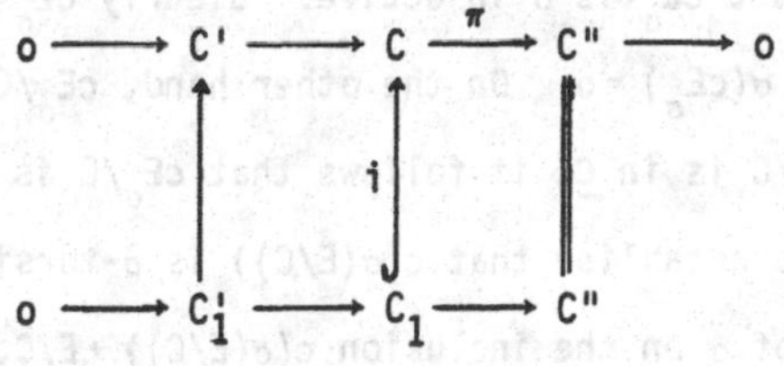

where $C_1' = \mathrm{Ker}(i\pi)$. If $C_1 \cap C' = o$, then $C_1' = o$, i.e. $C_1 \to C''$ is a monomorphism. By IV.1.2.1. we know that C_1 is σ-torsion; however, as a subobject of C in $\underline{C}$, it is also σ-torsion free, so $C_1 = o$. ∎

<u>IV.1.9. Remark</u>. Note that there may well exist $\underline{D}$-essential extensions E of a faithfully σ-injective E' in $\underline{C}$ such that E/E' is σ-torsion, which are not

isomorphic to E'. In case E is an essential extension of E' within $\underline{\mathcal{D}}$, then the above is not possible, as is easily checked.

IV.1.10. Theorem. Any σ-torsion free object C in $\underline{C}$ has an essentially unique σ-injective hull $E_\sigma(C)$ in $\underline{C}$.

Proof. Put $E = E^c(C) = cE^d(C)$. Using IV.1.2.3. and the fact that E is an essential extension in $\underline{\mathcal{D}}$ of C, it follows that E is σ-torsion free. It is obvious that E is also σ-injective. Consider the exact sequence in $\underline{C}$:

$$o \to C \to E \to E/C \to o$$

and put $E_\sigma = \pi^{-1}(\sigma(E/C)) \in \mathcal{D}$. Since E/E_σ is an object of $\underline{C}$ isomorphic to a subobject of $(E/C)/\sigma(E/C)$ in $\underline{C}$, it follows that E/E_σ is σ-torsion free. Consider the following commutative, exact diagram :

$$\begin{array}{ccccccccc} o & \longrightarrow & E_\sigma & \longrightarrow & E & \longrightarrow & E/E_\sigma & \longrightarrow & o \\ & & & & \downarrow{\scriptstyle\pi} & & \downarrow & & \\ o & \longrightarrow & \sigma(E/C) & \longrightarrow & E/C & \longrightarrow & (E/C)/\sigma(E/C) & \longrightarrow & o \end{array}$$

By IV.1.7. it follows that cE_σ is σ-injective. Clearly cE_σ is an essential extension in $\underline{\mathcal{D}}$ of C, so $\sigma(cE_\sigma) = o$. On the other hand, cE_σ/C is a subobject of $E_\sigma/C \cong \sigma(E/C)$ and as cE_σ/C is in $\underline{C}$, it follows that cE_σ/C is a subobject of $c(\sigma(E/C))$ in $\underline{C}$. Now, to establish that $c(\sigma(E/C))$ is σ-torsion, it suffices to use the left exactness of σ on the inclusion $c(\sigma(E/C)) \to E/C$, i.e.

$$\sigma(c\sigma(E/C)) = c\sigma(E/C) \cap \sigma(E/C) = c\sigma(E/C).$$

Therefore $c(\sigma(E/C))$ and also cE_σ/C are σ-torsion.
This proves that $E_\sigma(C) = cE_\sigma$ is a σ-injective hull of C.
Assume that E_1 and E_2 are both σ-injective hulls of C in $\underline{C}$, then we obtain the following diagram in $\underline{C}$

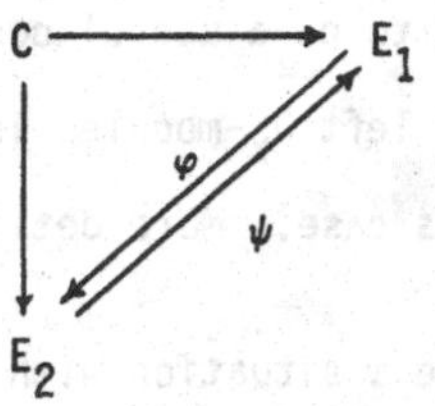

Since the identity of C extends in a unique way to a morphism $E_i \to E_i$, $i=1,2$, it follows that $\psi \circ \varphi = 1_{E_1}$ and $\varphi \circ \psi = 1_{E_2}$ and E_1 and E_2 are isomorphic in $\underline{\mathcal{D}}$ (and in $\underline{C}$!) ∎

IV.1.11. Definition. For an arbitrary object C in $\underline{C}$ the object of quotients of C at σ is defined to be the C-morphism $j_\sigma : C \to Q_\sigma(C)$, where $Q_\sigma(C) = E_\sigma(C/\sigma C)$, and where j_σ is the canonical morphism from $C \to C/\sigma C \to Q_\sigma(C)$.

IV.1.12. Proposition. Assume that $\underline{C}$ is closed under taking extensions, i.e. if $o \to C' \to C \to C'' \to o$ is an exact sequence in $\underline{\mathcal{D}}$, with C' and C'' in $\underline{C}$, then C lies in $\underline{C}$. Then $Q_\sigma(C)$ may be characterized by the following exact sequence in $\underline{C}$:

$$o \longrightarrow \overline{C} \longrightarrow Q_\sigma(C) \longrightarrow c\sigma(E^c(\overline{C})/\overline{C}) \longrightarrow o$$

where $\overline{C} = C/\sigma C$.

Proof. With notations as in IV.1.10, substituting $\overline{C}$ for C, we have that $Q_\sigma(C) = c\pi^{-1}(\sigma(E/\overline{C}))$, where $E = E^c(\overline{C})$ and $\pi : E \to E/\overline{C}$ the canonical projection. Since $cE_\sigma \in \underline{C}$, obviously π maps $cE_\sigma = Q_\sigma(C)$ into $c\sigma(E/\overline{C})$, while $\pi(Q_\sigma(C)) \in \underline{C}$. Conversely, $\pi^{-1}(c\sigma(E/\overline{C}))$ is in $\underline{C}$, because $\underline{C}$ is closed under extensions therefore $\pi^{-1}(c\sigma(E/\overline{C})) \subset cE_\sigma$, hence $cE_\sigma = \pi^{-1}(c\sigma(E/\overline{C}))$, yielding exactness of $\underline{C}$-sequence

$$o \longrightarrow \overline{C} \longrightarrow Q_\sigma(C) \longrightarrow c\sigma(E/\overline{C}) \longrightarrow o \quad .$$

IV.1.13. Remark and examples.

If $\underline{C}$ is the category of Artin R-bimodules and $\underline{\mathcal{D}}$ the category of twosided R-modules, then in general $\underline{C}$ is not closed under extensions as we shall see below. Yet, although this prevents us from using the characterization given in IV.1.12, bimodules of quotients will turn out to be a useful concept. On the other hand, the category of flabby presheaves of left $\underline{O}_X$-modules is closed under extensions and IV.1.12. may thus be used in this case. More details follow.

<u>IV.1.14. Proposition</u>. Let $(\underline{C},\underline{\mathcal{D}},\sigma)$ be a situation with $\sigma = \sigma' i$, where σ' is a (common) kernel functor in $\underline{\mathcal{D}}$. For any object C in $\underline{C}$ the object of quotients of C at σ is given by $Q_{\sigma'}(C) \cong cQ_{\sigma}(C)$ where $Q_{\sigma'}(C)$ is the object of quotients of C at σ' in the Grothendieck category $\underline{\mathcal{D}}$.

<u>Proof</u>. It is easy to verify that $cQ_{\sigma}(C)$ has all properties of a σ-injective hull of C in $\underline{C}$. Uniqueness arguments yield that $Q_{\sigma}(C) \cong cQ_{\sigma'}(C)$ holds. ∎

Let $(\underline{C},\underline{\mathcal{D}},\sigma)$ be a situation such that $\sigma = \sigma' i$ for some kernel functor σ' in $\underline{\mathcal{D}}$. We say that σ' is c-compatible if for all D in $\underline{\mathcal{D}}$ we have $\sigma'(D/cD) = D/cD$, i.e. D/cD is σ'-torsion. In this case we refer to $(\underline{C},\underline{\mathcal{D}},\sigma,\sigma')$ as a <u>compatibility situation</u>.

<u>IV.1.15. Proposition</u>. Let $(\underline{C},\underline{\mathcal{D}},\sigma,\sigma')$ be a compatibility situation, then $cQ_{\sigma'} = Q_{\sigma}c$ in $\underline{\mathcal{D}}$ (natural equivalence of functors).

<u>Proof</u>. We have the following exact commutative diagram :

$$\begin{array}{ccccccccc} 0 & \longrightarrow & \overline{D} & \longrightarrow & Q_{\sigma'}(D) & \longrightarrow & D_1 & \longrightarrow & 0 \\ & & \uparrow & & \uparrow & & \uparrow & & \\ 0 & \longrightarrow & \overline{D}\cap cQ_{\sigma'}(D) & \longrightarrow & cQ_{\sigma'}(D) & \longrightarrow & C_1 & \longrightarrow & 0 \end{array}$$

where $\overline{D} = D/\sigma' D$, D_1 is σ'-torsion and C_1 is σ-torsion because $C_1 \in \underline{C}$ and C_1 is a subobject of a σ'-torsion object in $\underline{\mathcal{D}}$. It is then clear that the bottom row gives another exact row :

$$0 \longrightarrow (\overline{D}\cap cQ_{\sigma'}(D))/c\overline{D} \rightarrow cQ_{\sigma'}(D)/c\overline{D} \longrightarrow C_2 \longrightarrow 0,$$

where $C_2 \in \underline{C}$ is isomorphic to C_1, i.e. σ-torsion. By compatibility $\overline{D}/c\overline{D}$ is σ'-torsion and since the class of σ'-torsion objects in $\underline{\mathcal{D}}$ is closed under extensions it follows that $cQ_{\sigma'}(D)/c\overline{D}$ is σ-torsion. That $cQ_{\sigma'}(D)$ is σ-torsion

free follows from the imposed relation on σ and σ'. Furthermore, it is straightforward to check that $cQ_{\sigma'}(D)$ is σ-injective in $\underline{C}$. Since $cQ_{\sigma'}(D)$ is a σ-injective hull of $c\overline{D}$, uniqueness arguments yield that $cQ_{\sigma'}(D) = Q_\sigma(c\overline{D})$.
From $\sigma(cD) = \sigma'(cD) = cD \cap \sigma'D$, we obtain the following exact commutative diagram in $\underline{\mathcal{D}}$:

$$\begin{array}{ccccccccc}
o & \longrightarrow & (cD) & \longrightarrow & cD & \longrightarrow & \overline{cD} & \longrightarrow & o \\
 & & \downarrow & & \downarrow & & \downarrow & & \\
o & \longrightarrow & \sigma'(D) & \longrightarrow & D & \longrightarrow & D & \longrightarrow & o \\
 & & & & \downarrow & & \downarrow & & \\
 & & & & D/cD & \relbar\joinrel\relbar & \overline{D}/\overline{cD} & & \\
 & & & & \downarrow & & \downarrow & & \\
 & & & & o & & o & &
\end{array}$$

Therefore $\overline{D}/\overline{cD}$ is σ'-torsion and since $\overline{cD} \subset c\overline{D} \subset \overline{D}$, it follows that $c\overline{D}/\overline{cD}$ is σ-torsion. Finally, this implies $Q_\sigma(\overline{cD}) = Q_\sigma(c\overline{D}) = Q_\sigma(cD)$, thus $cQ_{\sigma'}(D) = Q_\sigma(cD)$. ∎

Examples of compatibility situations will be encountered further on. Although c-compatibility seems to hold seldom in general, it is clear that the existence of c-compatible σ' in $\underline{\mathcal{D}}$ depends heavily on $\underline{\mathcal{D}}$, e.g. on the existence of certain generators for $\underline{\mathcal{D}}$.

For example, let $\underline{\mathcal{D}}$ be the category of twosided R-modules and $\underline{C}$ the category of R-bimodules, then for arbitrary R it is hard to construct a σ' in $\underline{\mathcal{D}}$ such that for every twosided R-module the quotient M/bi(M) is σ-torsion, cf. below. However, if R is an Azumaya algebra then every σ' in R-mod induces a compatibility situation, while $\underline{C} = \underline{\mathcal{D}}$ in this case! Checking compatibility may be simplified by the introduction of filters in some generators, we will not expound this in full depth, but just indicate some advantages of the use of filters.

The use of (idempotent) filters of subobjects of a generator for $\underline{\mathcal{D}}$ has been amply displayed in [192]. If G' is a generator for $\underline{\mathcal{D}}$, then the <u>filter associated to a</u>

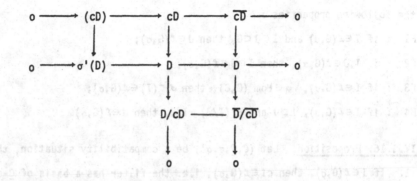

<u>kernel functor</u> σ' in $\underline{D}$ is given by the subobjects I of G', such that G'/I is σ'-torsion. Now, in case of a relative localization, let $(\underline{C},\underline{D},\sigma)$ be a situation such that $\underline{C}$ has a generator G, then the filter $\mathcal{L}(G,\sigma)$ is given by the $\underline{D}$-subobjects I of G such that G/I is σ-torsion.

If $(\underline{C},\underline{D},\sigma)$ is such that $\sigma=\sigma' i$ for some kernel functor σ' in $\underline{D}$, then $\mathcal{L}(G,\sigma)$ has the following properties :

F1 : if $I\in\mathcal{L}(G,\sigma)$ and $I\subset J\subset G$, then $J\in\mathcal{L}(G,\sigma)$;

F2 : if $I,J\in\mathcal{L}(G,\sigma)$, then $I\cap J\in\mathcal{L}(G,\sigma)$;

F3 : if $I\in\mathcal{L}(G,\sigma)$, $\varphi\in\mathrm{Hom}_{\underline{C}}(G,C)$, then $\varphi^{-1}(I)\in\mathcal{L}(G,\sigma)$;

F4 : if $I\in\mathcal{L}(G,\sigma)$, $I\supset J$ and $\sigma'(I/J)=I/J$, then $J\in\mathcal{L}(G,\sigma)$.

<u>IV.1.16. Proposition</u>. Let $(\underline{C},\underline{D},\sigma,\sigma')$ be a compatibility situation, then we have:

1. if $I\in\mathcal{L}(G,\sigma)$, then $cI\in\mathcal{L}(G,\sigma)$, i.e. the filter has a basis of $\underline{C}$-objects;
2. an object E in $\underline{C}$ is σ-injective if and only if for each $I\in\mathcal{L}(G,\sigma)$ and arbitrary $\varphi\in\mathrm{Hom}_{\underline{D}}(I,E)$ there is a $\bar{\varphi}\in\mathrm{Hom}_{\underline{D}}(G,E)$ extending φ.

<u>Proof</u>. Slight variation of Proposition 3.1. in [192]. ∎

These techniques will now be applied to bimodules. In the rest of this section R will always be an associative ring with unit. Let us recall some basic facts. An abelian group M, which is both a unitary left and right R-module is said to be a <u>twosided R-module</u>, when the left and right R-action are compatible, i.e. for every $m\in M$, r_1, $r_2\in R$, we have $r_1(mr_2)=(r_1m)r_2$. The category of two sided R-modules will be denoted by $\underline{\underline{2}}(R)$ morphisms being morphisms of abelian groups which are both left and right R-linear. The R-<u>center</u>, or simply <u>center</u> of a two-sided R-module M is the set

$$Z_R(M)=\{m\in M;\ rm=mr \text{ for all } r\in R\}\ .$$

Clearly $Z(R)=Z_R(R)$ is the center of R. Obviously $Z_R(M)$ is a twosided Z(R)-module. A twosided R-module M is said to be an R-<u>bimodule</u>, in the sense of M. Artin [17] , if M is generated as a left (or right) R-module by $Z_R(M)$. The category of R-bimodules, denoted by $\underline{\underline{bi}}(R)$, is the full subcategory of $\underline{\underline{2}}(R)$ consisting of all R-bimodules. Note that a map $M \to N$, with $M,N \in \underline{\underline{bi}}(R)$ mapping $Z_R(M)$ into $Z_R(N)$ is a morphism of R bimodules, exactly when it is left (or right) R-linear. Whereas $\underline{\underline{2}}(R)$ is a Grothendieck category, $\underline{\underline{bi}}(R)$ need not even be abelian! (Indeed, there is no reason why, in general, the kernel of a bimodule morphism should exist, i.e. in particular be a bimodule, when viewed as a morphism of two-sided modules.).

However, we may speak of exact sequences in $\underline{\underline{bi}}(R)$ in the following sense : a sequence of R-bimodules is <u>exact</u> when it is exact, if viewed as a sequence of twosided R-modules.

Define a functor $bi : \underline{\underline{2}}(R) \to \underline{\underline{bi}}(R)$, assigning to $M \in \underline{\underline{2}}(R)$ the largest R-bimodule contained in it, i.e. the R-bimodule generated by the R-center of M. One easily checks that bi is a right adjoint to the inclusion $\underline{\underline{bi}}(R) \to \underline{\underline{2}}(R)$. Since R is associative, we may associate to each left ideal I of R a largest twosided ideal contained in it, denoted by $2(I)$, the bound of I. Instead of $bi(2(I))$, we will simply write $bi(I)$.

<u>IV.1.17. Lemma</u>. Let I be a left ideal of R and let $x \in Z(R)$, then $2(I:x) = (2(I):x)$.

<u>Proof</u>. $2(I : x) = \{r \in (I:x);\ rR \subset (I:x)\}$

$= \{r \in R;\ rRx \subset I\}$

$= \{r \in R;\ rxR \subset I\}$

$= \{r \in R;\ rx \in 2(I)\}$

$= (2(I) : x)$. ∎

In order to study relative localization in $\underline{\underline{bi}}(R)$, we first study the Grothendieck category $\underline{\underline{2}}(R)$. Recall that a kernel functor σ in R-mod is symmetric if its idempotent filter $\mathcal{L}(\sigma)$ has a filterbasis consisting of ideals. An ideal I of R is said to be a <u>bi-ideal</u> of R if $bi(I) = I$. A kernel functor σ in R-mod is <u>hypersymmetric</u> if $\mathcal{L}(\sigma)$ has a filter basis consisting of bi-ideals. Unless otherwise indicated, all kernel functors will be assumed to be idempotent.

<u>IV.1.18. Proposition.</u> Let σ be a kernel functor in R-mod. Consider the following properties :

a. if I is a left ideal of R such that there is $J \in \mathcal{L}(\sigma)$ with the property that for all $x \in bi(J)$ we have $(I : x) \in \mathcal{L}(\sigma)$, then $I \in \mathcal{L}(\sigma)$,

b. σ is hypersymmetric;

c. if $I \in \mathcal{L}(\sigma)$, then $I/bi(I)$ is σ-torsion;

a'. if I is a left ideal of R such that for some $J \in \mathcal{L}(\sigma)$ with the property that for all $x \in \mathcal{Z}(J)$ we have $(I : x) \in \mathcal{L}(\sigma)$, then $I \in \mathcal{L}(\sigma)$,

b'. σ is symmetric;

c'. if $I \in \mathcal{L}(\sigma)$, then $I/\mathcal{Z}(I)$ is σ-torsion.

Then the following implications hold :

$$\begin{array}{ccccc} (a) & \Longleftrightarrow & (b) & \Longleftrightarrow & (c) \\ \downarrow & & \downarrow & & \downarrow \\ (a') & \Longleftrightarrow & (b') & \Longleftrightarrow & (c') \end{array} .$$

<u>Proof</u>. Straightforward. ∎

<u>IV.1.19. Remarks</u>. Every (idempotent) kernel functor σ in R-mod is "<u>inner</u>" in $\mathcal{Z}(R)$, i.e. if $M \in \mathcal{Z}(R)$, then $\sigma M \in \mathcal{Z}(R)$. Indeed, $\sigma M \in$ R-mod by definition, while, if $x \in \sigma M$, then we may find $L \in \mathcal{L}(\sigma)$ such that $Lx = o$. But then $LxR = o$, implying that $xR \subset \sigma M$, i.e. σM is a right R-module as well. Note also that in special cases the vertical arrows are equivalences too, e.g. when R is an Azumaya algebra, because in this case $\underline{\underline{bi}}(R) = \underline{\underline{\mathcal{Z}}}(R)$.

<u>IV.1.20. Theorem</u>. Let σ be a kernel functor in $\underline{\underline{\mathcal{Z}}}(R)$, let M be a σ-torsion free twosided R-module, then there exists an essentially unique twosided R-module $E^2_\sigma(M)$ in $\underline{\underline{\mathcal{Z}}}(R)$, which is called the σ-injective hull of M and which is faithfully σ-injective and such that $E^2_\sigma(M)/M$ is σ-torsion. It is characterized by the following exact sequence in $\underline{\underline{\mathcal{Z}}}(R)$

$$o \longrightarrow M \longrightarrow E^2_\sigma(M) \longrightarrow \sigma(E^2(M)/M) \longrightarrow o,$$

where $E^2(M)$ stands for an injective hull of M in $\underline{\underline{2}}(R)$.

Proof. This result is just a translation of the folklore of localization in general Grothendieck categories in terms of the particular Grothendieck category $\underline{\underline{2}}(R)$. ∎

The twosided R-module of quotients of M at σ is defined to be $Q^2_\sigma(M) = E^2_\sigma(M/\sigma M)$. Note that in general $Q^2_\sigma(R)$ is not a ring, due to the fact that except for some particular cases R is not a generator for the Grothendieck category $\underline{\underline{2}}(R)$.

To a kernel functor σ in $\underline{\underline{2}}(R)$ we may associate a filter $\mathcal{L}^2(\sigma)$ consisting of all ideals I of R such that R/I is σ-torsion. It is clear that $\mathcal{L}^2(\sigma)$ does not necessarily characterize σ, again by the fact that R is not necessarily a generator for $\underline{\underline{2}}(R)$. However, we have the following :

IV.1.21. Lemma. Assume that σ_1 and σ_2 are kernel functors in $\underline{\underline{2}}(R)$ such that $\mathcal{L}^2(\sigma_1) = \mathcal{L}^2(\sigma_2)$ and let M be an R-bimodule, then M is σ_1-torsion if and only if it is σ_2-torsion.

Proof. Pick a central element $x \in M$. If M is σ_1-torsion, then Rx is σ_1-torsion and thus $\mathrm{Ann}_R(x) \in \mathcal{L}^2(\sigma_1) = \mathcal{L}^2(\sigma_2)$, so Rx is σ_2-torsion. Since M is an R-bimodule, it is generated by its central elements and the lemma follows. ∎

Recall some well-known facts, cf. [189]. Let R^o denote the opposite ring of R then $R^e = R \underset{Z(R)}{\otimes} R^o$ is a generator for $\underline{\underline{2}}(R)$. If we define $\mathcal{L}^2(R^e,\sigma)$ as the set of twosided R-submodules J of R^e such that R^e/J is σ-torsion, then it is easily verified that $\mathcal{L} = \mathcal{L}(R^e,\sigma)$ has the well-known properties

a. if $I,J \in \underline{\underline{2}}(R)$, $I \subset J \subset R^e$ and $I \in \mathcal{L}$, then $J \in \mathcal{L}$;
b. if $I,J \in \mathcal{L}$ then $I \cap J \in \mathcal{L}$;
c. if $I \in \mathcal{L}$, $\varphi \in \mathrm{Hom}_{\underline{\underline{2}}(R)}(R^e,R^e)$, then $\varphi^{-1}(I) \in \mathcal{L}$.
d. if $I \in \underline{\underline{2}}(R)$, $I \subset R^e$ is such that there is a $J \in \mathcal{L}$ with the property that for every $\varphi \in \mathrm{Hom}_{\underline{\underline{2}}(R)}(R^e,J)$ we have $\varphi^{-1}(I) \in \mathcal{L}$, then $I \in \mathcal{L}$.

There is a one-to-one correspondence between kernel functors in $\underline{\underline{2}}(R)$ and filters in R^e with the above properties, this correspondence may be given as follows : $M \in \underline{\underline{2}}(R)$ is σ-torsion if and only if for all $\varphi \in \mathrm{Hom}_{\underline{\underline{2}}(R)}(R^e, M)$ we have $\mathrm{Ker}\,\varphi \in \mathcal{L}$. This is just an application to $\underline{\underline{2}}(R)$ of the results of I.3, or see [192].

In the next result we will implicitly use the techniques developed in I.3. Let $j : 2(R) \to R\text{-mod}$ be the canonical inclusion.

<u>IV.1.22. Theorem.</u> Let σ be a kernel functor in R-mod, let $\overline{\sigma}$ be the kernel functor in $\underline{\underline{2}}(R)$ induced by σ, i.e. $\overline{\sigma} = \sigma j$. Then for any twosided R-module M we have

1. $Q_\sigma(jM)$ is in a natural way endowed with a twosided R-module structure;
2. $Q_\sigma^2(M)$ and $Q_\sigma(jM)$ are naturally isomorphic with this structure.

<u>Proof.</u> For simplicity's sake, identify $\underline{\underline{2}}(R)$ with a subcategory of R-mod, this is an exact embedding; we thus omit writing j. Since $\overline{\sigma}M = \sigma M$ and $M/\sigma M$ are again twosided, we may reduce the proof to the case where M is torsion free. Let $E_\sigma(M)$ be the σ-injective hull of M in R-mod. Consider the following diagram

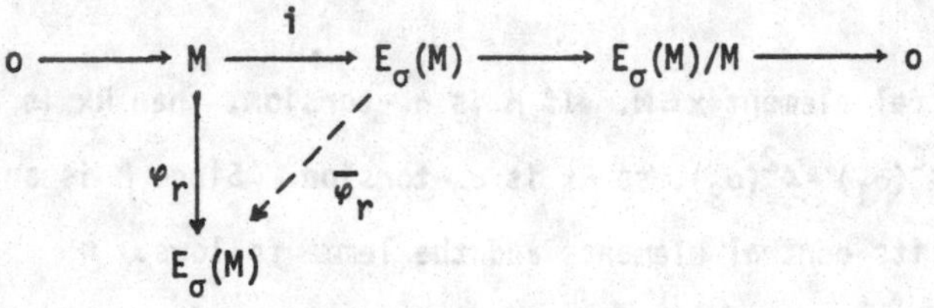

Here $i : M \to E_\sigma(M)$ denotes the canonical inclusion, and φ_r is the morphism which for some arbitrary, fixed $r \in R$ maps $m \in M$ to mr. Obviously φ_r is left R-linear, hence by the faithful injectivity of $E_\sigma(M)$ it follows that φ_r extends to a left R-linear map $\overline{\varphi}_r : E_\sigma(M) \to E_\sigma(M)$, which is the unique left R-linear map extending φ_r. The right action of R on $E_\sigma(M)$ is defined by $m.r = \overline{\varphi}_r(m)$ for $m \in E_\sigma(M)$ and $r \in R$. Then : $(sm)r = \overline{\varphi}_r(sm) = s(\overline{\varphi}_r(m)) = s(mr)$. Since $\overline{\varphi}_{r+r'}$ and $\overline{\varphi}_r + \overline{\varphi}_{r'}$ have equal restrictions to M, uniqueness arguments yield that $m(r+r') = mr + mr'$. The same uniqueness arguments and the commutativity of the following diagram of left R-linear morphisms :

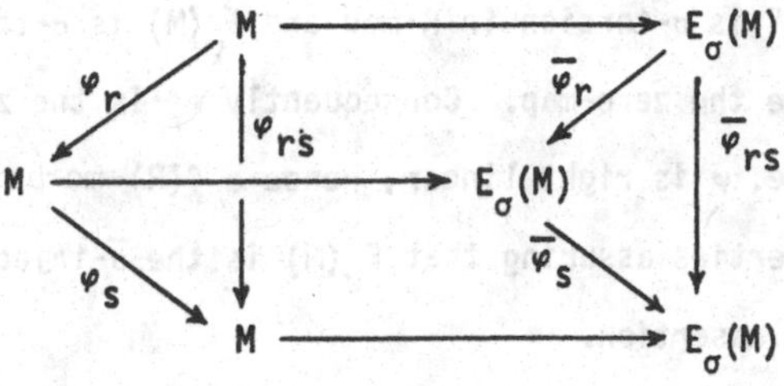

yield $\overline{\varphi}_{rs} = \overline{\varphi}_s\,\overline{\varphi}_r$, i.e. $m(rs) = (mr)s$ for all $m \in E_\sigma(M)$, $r,s \in R$. So $E_\sigma(M)$ is endowed with the structure of a twosided R-module which extends the twosided structure of M. Conversely, if $E_\sigma(M)$ has a twosided structure which extends that of M, then right multiplication by $r \in R$ in $E_\sigma(M)$ is left R-linear and induces φ_r on M. Again uniqueness arguments (on the left) yield that the twosided structure defined on $E_\sigma(M)$ in the above way is actually the unique $\underline{\mathcal{L}}(R)$-structure on $E_\sigma(M)$ extending the $\underline{\mathcal{L}}(R)$-structure of M.

Obviously $E_\sigma(M)/M$ is $\overline{\sigma}$-torsion and moreover $E_\sigma(M)$ is $\overline{\sigma}$-torsion free.

Consider an exact diagram in $\underline{\mathcal{L}}(R)$

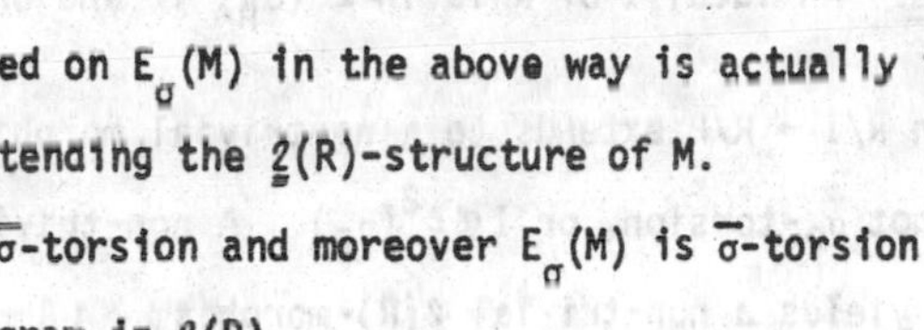

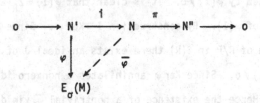

where N" is $\overline{\sigma}$-torsion. Now σ-injectivity of $E_\sigma(M)$ in R-mod entails that φ extends to a left R-linear map $\overline{\varphi}$ making the diagram commute.

Consider the map $\psi_r : N \to E_\sigma(M) : m \to \overline{\varphi}(mr) - \overline{\varphi}(m)r$. Since φ is $\underline{\mathcal{L}}(R)$-linear ψ_r vanishes on N'. Furthermore ψ_r is left R-linear as is easily seen so ψ_r factorizes through N":

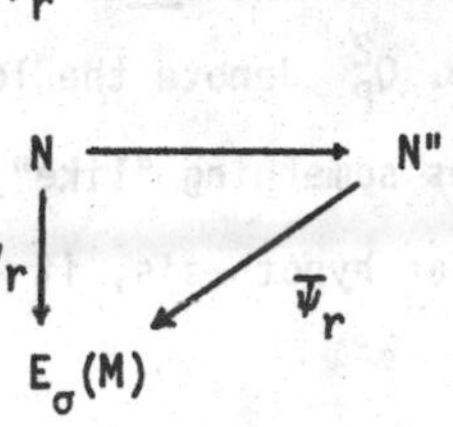

Since ψ_r is left R-linear N" is σ-torsion in R-mod and $E_\sigma(M)$ is σ-torsion free, it follows that $\overline{\psi}_r$ has to be the zero-map. Consequently ψ_r is the zero map and this holds for any $r \in R$, i.e. φ is right linear, hence a $\underline{\underline{2}}(R)$-morphism. We have established all of the properties assuring that $E_\sigma(M)$ is the $\overline{\sigma}$-injective hull of M in $\underline{\underline{2}}(R)$, which proves our assertion. ∎

<u>IV.1.23. Example</u>. If P is a prime ideal of R, then we may associate to it an endofunctor σ_P in $\underline{\underline{2}}(R)$, defined as follows : for any $M \in \underline{\underline{2}}(R)$ we put

$$\sigma_P(M) = \cap\{\mathrm{Ker}(f); f \in \mathrm{Hom}_{\underline{\underline{2}}(R)}(M, E^2(R/P))\} .$$

It is easily checked that σ_P is in fact a kernel functor in $\underline{\underline{2}}(R)$.

<u>IV.1.24. Proposition</u>. An ideal I of R is in $\mathcal{L}^2(\sigma_P)$ if and only if $I \not\subset P$.

<u>Proof</u>. If $I \subset P$ then $R/I \to R/P$ extends to a nontrivial morphism $R/I \to E^2(R/P)$ in $\underline{\underline{2}}(R)$, hence R/I is not σ_P-torsion, or $I \notin \mathcal{L}^2(\sigma_P)$. A non-trivial twosided R-linear map $\overline{\varphi} : R/I \to E^2(R/P)$ yields a non-trivial $\underline{\underline{2}}(R)$-morphism $\varphi : R \to E^2(R/P)$ which is completely determined by $\varphi(1) \neq o$. It is clear that $\varphi(1) \in Z_R(E^2(R/P))$ and Ker φ is the annihilator of $\varphi(1)$. By construction $I \subset \mathrm{Ker}\,\varphi$. Since $E^2(R/P)$ is an essential extension of R/P in $\underline{\underline{2}}(R)$ there exists an ideal J of R such that $J\varphi(1) \subset R/P$ and $J\varphi(1) \neq o$. Since Ker φ annihilates a nonzero ideal $J\varphi(1)$ of R/P clearly Ker $\varphi \subset P$. Hence the existence of a nontrivial $\overline{\varphi}$ yields $I \subset P$, which proves the converse of the first statement.

On the other hand, if R is left noetherian, then the symmetric kernel functor σ_{R-P} induces a kernel functor $\overline{\sigma}_{R-P}$ in $\underline{\underline{2}}(R)$ such that $\mathcal{L}^2(\overline{\sigma}_{R-P}) = \mathcal{L}^2(\sigma_p)$. Since σ_P is not characterized by $\mathcal{L}^2(\sigma_p)$, there are <u>no</u> a priori relations between $Q^2_{R-P}(R)$ and $Q^2_P(R)$, where Q^2_{R-P} resp. Q^2_P denote the localization at $\overline{\sigma}_{R-P}$ resp. σ_P. The construction of σ_P provides something "like" symmetric localization, even in the absence of the noetherian hypothesis; it will work extremely well in the bimodule case.

IV.1.25. Convention. The twosided R-module of quotients $Q^2_{\underline{\sigma}}(M)$ of M at $\overline{\sigma}$ will be denoted by $Q^2_\sigma(M)$ for any kernel functor σ in R-mod.

General relative localization theory may be restated as follows in the bimodule-case. A kernel functor in $\underline{\underline{bi}}(R)$ is a left exact subfunctor σ of the inclusion $i : \underline{\underline{bi}}(R) \to \underline{\underline{2}}(R)$ such that $\sigma(M/\sigma M) = o$ for any $M \in \underline{\underline{bi}}(R)$. If M is an R-bimodule then we define a σ-injective hull of M in $\underline{\underline{bi}}(R)$ to be an R-bimodule E with the following properties :

1. E is σ-torsion free, i.e. $\sigma E = o$;
2. E/M is σ-torsion, i.e. $\sigma(E/M) = E/M$;
3. E is σ-injective in $\underline{\underline{bi}}(R)$, i.e. if

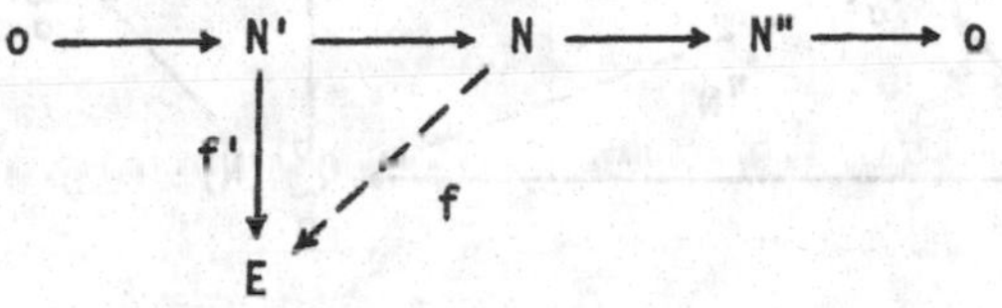

is an exact diagram in $\underline{\underline{bi}}(R)$, with $\sigma N" = N"$, then there is a map $f : N \to E$ in $\underline{\underline{bi}}(R)$ extending f'. This map is clearly unique, we say that E is faithfully σ-injective. The general theory now yields :

IV.1.26. Theorem. A σ-torsion free R-bimodule M has an essentially unique σ-injective hull, which we will denote by $E^{bi}_\sigma(M)$. ∎

The R-bimodule of quotients at σ of an R-bimodule M is by definition

$$Q^{bi}_\sigma(M) = E^{bi}_\sigma(M/\sigma M).$$

Let σ be a kernel functor in $\underline{\underline{bi}}(R)$, then $\tilde{\sigma} = \sigma i$ is a kernel functor in $\underline{\underline{bi}}(R)$. Application of IV.1.14. yields :

IV.1.27. Theorem. For any kernel functor σ in $\underline{\underline{2}}(R)$, the R-bimodule of quotients of an R-bimodule M at $\tilde{\sigma}$ is given by

$$Q^{bi}_{\tilde{\sigma}}(M) = \text{bi}\ Q^2_\sigma(M). \quad ∎$$

In other words, we have a natural isomorphism of functors in $\underline{\underline{bi}}(R)$

$$\eta : Q^{bi}_{\tilde{\sigma}} \to biQ^2_\sigma ,$$

i.e. for each $\varphi \in \mathrm{Hom}_{\underline{\underline{bi}}(R)}(M,N)$ the following diagram is commutative :

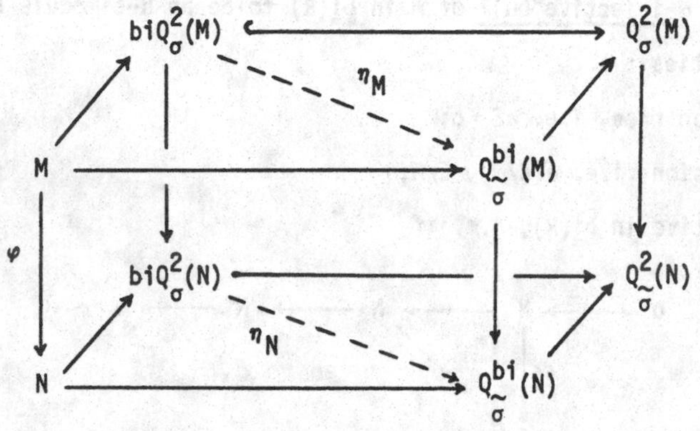

Since σ is not determined by its idempotent filter $\mathcal{L}^2(\sigma)$, we obviously cannot expect that σ-injectivity is detectable by studying filters. However, we have the following

IV.1.28. Proposition. Let σ be a hypersymmetric kernel functor in $\underline{\underline{2}}(R)$, then $E \in \underline{\underline{bi}}(R)$ is $\tilde{\sigma}$-injective in $\underline{\underline{bi}}(R)$ if and only if for each bi-ideal $I \in \mathcal{L}(\sigma)$ and any bilinear $\varphi : I \to E$ there is a bilinear $\overline{\varphi} : R \to E$ extending φ.

Proof. It suffices to mimic the proof of a similar statement on onesided modules, cf.[70] ∎

IV.1.29. Proposition. Let Q be an R-bimodule, then Q is faithfully σ-injective in $\underline{\underline{bi}}(R)$ if and only if Q is σ-torsion free and for each $I \in L^2(\sigma)$ and each $\underline{\underline{2}}(R)$-morphism $\varphi : I \to Q$, there exists $q \in Z_R(Q)$ such that $\varphi(i) = iq$ for each $i \in I$.

Proof. The "only if" part being obvious, let us prove the "if" part, mimicing an analogous result proved by Goldman in [70]. Consider the following diagram in $\underline{\underline{bi}}(R)$

$$\begin{array}{ccccccccc} o & \longrightarrow & M' & \xrightarrow{i} & M & \xrightarrow{\pi} & M'' & \longrightarrow & o \\ & & \downarrow{\scriptstyle g} & & & & & & \\ & & Q & & & & & & \end{array}$$

where M" is σ-torsion. Using Zorn's lemma, we may find a maximal extension $g' : N \to Q$ of g, where N is a sub-R-bimodule of M containing M'. We want to show that $M = N$. Assume the converse, i.e. there exists $m \in Z_R(M)$ not contained in N, and consider the following morphism of R-bimodules $h : R \to M : r \to rm$. Since M/N is σ-torsion and $R/h^{-1}(N)$ embeds into it, clearly $h^{-1}(N) \in \mathcal{L}^2(\sigma)$. Define $h_1 : I \to Q$ by assigning $g'(im)$ to $i \in I$, then by assumption there exists $q \in Z_R(Q)$ such that $h_1(x) = iq$ for all $i \in I$. Finally we define $g_1 : N + R_m \to Q$ by sending $n + rm$ to $g'(n) + rx$. This is easily verified to be well-defined and it properly extends $g' : N \to Q$; thus this contradicts the maximality assumption. This contradiction proves that $M = N$, as desired.

Assume now that g' and g" : $N \to Q$ both extend g, then $g' - g''$ factorizes through M", say $g' - g'' = \gamma \circ \pi$, where $\gamma : M'' \to Q$, as g' and g" coincide on M. But M" is σ-torsion and Q is σ-torsion free, hence $\gamma = o$ and $g' = g''$. ∎

IV.1.30. Convention. Instead of $Q^{bi}_{\underset{\sim}{\sigma}}$ we will always write Q^{bi}_{σ}.

IV.1.31. Theorem. Let σ be a kernel functor in $\underline{\underline{2}}(R)$, then $Q^{bi}_{\sigma}(R)$ is in a canonical way endowed with a unique ring structure such that $j_\sigma : R \to Q^{bi}_{\sigma}(R)$ is a central extension. If E is faithfully σ-injective in $\underline{\underline{bi}}(R)$, then it has a $Q^{bi}_{\sigma}(R)$-bimodule structure extending the R-bimodule structure. Each R-bimodule morphism $N \to M$, where N is a $Q^{bi}_{\sigma}(R)$-bimodule and M is faithfully σ-injective is a $Q^{bi}_{\sigma}(R)$-bimodule morphism.

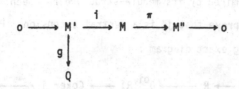

Proof. The proof runs along the lines of the well-known proofs of the similar statements in R-mod, cf. [70] . Since R is a generator for $\underline{\underline{bi}}(R)$, the use of idempotent filters presents no difficulties. That $Q^{bi}_{\sigma}(R)$ is a ring, the ringstructure being determined by its module-structure has been pointed out by Delale [45] [46]. Let us prove that j_σ is a central extension. Pick $c \in Z_R(Q^{bi}_\sigma(R))$ and consider the following exact diagram :

$$\begin{array}{ccccccccc} 0 & \longrightarrow & \sigma R & \longrightarrow & R & \longrightarrow & Q^{bi}_\sigma(R) & \longrightarrow & \text{Coker } j_\sigma \longrightarrow 0 \\ & & & \varphi_c \swarrow & & \swarrow \overline{\varphi}_c & & & \\ & & Q^{bi}_\sigma(R) & & & & & & \end{array}$$

where Coker j_σ is σ-torsion and $\varphi_c : R \to Q^{bi}_\sigma(R)$ is given by $r \mapsto rc$. By σ-injectivity of $Q^{bi}_\sigma(R)$ in $\underline{\underline{bi}}(R)$, the map φ_c extends in a unique way to a $\underline{\underline{bi}}(R)$-morphism $\overline{\varphi}_c : Q^{bi}_\sigma(R) \to Q^{bi}_\sigma(R)$. By the uniqueness assertion, $\overline{\varphi}_c$ has to coincide with right (and with left) multiplication by $c \in Q^{bi}_\sigma(R)$. By an earlier assertion $\overline{\varphi}_c$ has to be $Q^{bi}_\sigma(R)$-bilinear, i.e. $c \in Z_R(Q^{bi}_\sigma(R)) = Z(Q^{bi}_\sigma(R))$. ■

In [46] the following relative localization has been used. Let P be a prime ideal of R, and define for each $M \in \underline{\underline{bi}}(R)$

$$\sigma_P(M) = \cap \{\text{Ker}(f); f \in \text{Hom}_{\underline{\underline{bi}}(R)}(M, E^{bi}(R/P))\}$$

Here $E^{bi}(R/P) = \text{bi } E^2(R/P)$, the "injective hull" of R/P in $\underline{\underline{bi}}(R)$. One easily verifies that $\check{\sigma}_P = \tilde{\sigma}_P$ is the kernel functor in $\underline{\underline{z}}(R)$ defined in IV.1.23.
If P is a prime ideal of R, once would like to relate Q^{bi}_P to Q^{bi}_{R-P}, where Q^{bi}_P is the bimodule localization at $\tilde{\sigma}_P$ and Q^{bi}_{R-P} the bimodule localization at $\tilde{\sigma}_{R-P}$. The construction of Q^{bi}_{R-P} presupposes that σ_{R-P} is, in particular, idempotent. This is the case for all prime ideals of R if R is left noetherian, and it holds for almost all prime ideals of R (i.e. for an open set in the Zariski topology in Spec(R)) in case R is a semiprime P.I. ring, cf. [121] . We shall see in Section IV.3. that Q^{bi}_P and Q^{bi}_{R-P} are nicely related whenever R satisfies a polynomial identity.

<u>IV.1.32. Theorem.</u> Let P be a prime ideal of R. The following statements are equivalent :

1. σ_{R-P} is hypersymmetric;
2. σ_{R-P} is central;
3. if $p = P \cap Z(R) \in \text{Spec}(Z(R))$, then σ_{R-P} is induced by $\sigma_{Z(R)-p}$, the usual localization at p in Z(R)-mod;
4. $Q_{R-P}(R)$ and $Q_{Z(R)-p}(R)$ are ringisomorphic, and for any $M \in R$-mod the Z(R)-modules $Q_{R-P}(M)$ and $Q_{Z(R)-p}(M)$ are isomorphic.

<u>Proof.</u> The equivalence of the first three statements is immediate, if one considers σ' in Z(R)-mod given by $\mathcal{L}(\sigma') = \{I \subset Z(R); RI \in \mathcal{L}(R-P)\}$ and then shows that $\sigma' = \sigma_{Z(R)-p}$. The implication (3)$\Rightarrow$(4) follows from the fact $\sigma_{Z(R)-p}$ is a t-functor, i.e. $Q_{Z(R)-p}(R) \underset{R}{\otimes} M = Q_{Z(R)-p}(M)$ and from the well-known result that if σ is a kernel functor in R-mod induced by a kernel functor σ_1 in Z(R)-mod, such that σ_1 is a t-functor, then σ itself has property (T), whereas the rings $Q_\sigma(R)$ and $Q_{\sigma_1}(R)$ are isomorphic. For (4)$\Rightarrow$(1) apply (4) to the case where $M = R/RI$ for some $I \in \mathcal{L}(Z(R)-p)$. It is then easy to show that σ_{R-P} is hypersymmetric. ∎

<u>IV.1.33. Corollary.</u> If $P \in \text{Spec}(R)$ is such that σ_{R-P} is hypersymmetric, or equivalently, if $\mathcal{L}^2(R-P) = \mathcal{L}^2(\sigma_P)$ has a basis consisting of centrally generated ideals of R, then Q^{bi}_{R-P} and Q_{R-P} are isomorphic functors in $\underline{\underline{bi}}(R)$. ∎

<u>IV.1.34. Proposition.</u> Let $\tau \geq \sigma$ be kernel functors in $\underline{2}(R)$, then the kernel of the map $Q^2_\sigma(M) \to Q^2_\tau(M)$ is $Q^2_\sigma(\tau M/\sigma M)$.

<u>Proof.</u> Consider the following exact sequence in $\underline{2}(R)$:

$$o \longrightarrow \tau M/\sigma M \longrightarrow M/\sigma M \longrightarrow M/\tau M \longrightarrow o$$

yielding by left exactness an exact sequence

$$o \longrightarrow Q^2_\sigma(\tau M/\sigma M) \longrightarrow Q^2_\sigma(M/\sigma M) \longrightarrow Q^2_\sigma(M/\tau M).$$

As $\tau(M/\tau M) = o$, we get $\sigma(M/\tau M) = o$, hence $Q^2_\sigma(M/\tau M)$ is torsion free for τ and the canonical map $Q^2_\sigma(M/\sigma M) \to Q^2(M/\sigma M)$ is injective. Our assertion follows immediatly from the commutativity of the following diagram :

$$\begin{array}{ccccc} Q^2_\sigma(M/\sigma M) & \xrightarrow{\pi} & Q^2_\sigma(M/\tau M) & \longrightarrow & Q^2_\tau(M/\tau M) \\ \| & & & & \| \\ Q^2_\sigma(M) & & \longrightarrow & & Q^2_\tau(M) \end{array} .$$

<u>IV.1.35. Corollary</u>. If $\tau \geq \sigma$ are two kernel functors in $\underline{2}(R)$ and if for some two-sided R-module M we have $\tau M = \sigma M$, then the canonical map $Q^2_\sigma(M) \to Q^2_\tau(M)$ is injective. ∎

If τ and σ are symmetric kernel functors in R-mod, then IV.1.34. may be strengthened as follows :

<u>IV.I.36. Proposition</u>. Let $\tau \geq \sigma$ be arbitrary symmetric kernel functors in R-mod, then $Q^2_\sigma(\tau M) = \tau Q^2_\sigma(M)$ and $Q^2_\tau(Q^2_\sigma(M)) = Q^2_\tau(M)$ for any twosided R-module M.

<u>Proof</u>. From the exact sequence

$$o \longrightarrow \tau M \longrightarrow M \longrightarrow M/\tau M \longrightarrow o$$

one derives as above an exact sequence

$$o \longrightarrow Q^2_\sigma(\tau M) \longrightarrow Q^2_\sigma(M) \longrightarrow Q^2_\sigma(M/\tau M).$$

Let us show that $\tau Q^2_\sigma(M/\tau M) = o$. Take $x \in \tau Q^2(M/\tau M)$, then for some $I \in \mathcal{L}^2(\tau)$ we find $Ix = o$. On the other hand, we can pick some ideal $J \in \mathcal{L}^2(\sigma) \subset \mathcal{L}^2(\tau)$ such that $Jx \subset M/\tau M$ and as $IJ \subset I$. This yields $IJx = o$, i.e. $Jx \subset \tau(M/\tau M) = o$, hence x=o. From $\tau Q^2_\sigma(M/\tau M) = o$ one immediately deduces that $\tau Q^2(M) \subset Q^2(\tau M)$, the converse inclusion being obvious. Finally, from $M/\tau M \to Q^2_\sigma(M)/\tau Q^2_\sigma(M) \to Q^2_\tau(M)$ one deduces that $Q^2_\tau(Q^2_\sigma(M)) = Q^2_\tau(M)$. ∎

<u>IV.1.37. Lemma.</u> Let $\tau \geq \sigma$ be two kernel functors in $\underline{\mathcal{Q}}(R)$, then the unique R-linear morphism $Q_\sigma^{bi}(R) \to Q_\tau^{bi}(R)$ extending the localizing morphism $j_\tau : R \to Q_\tau^{bi}(R)$ to $Q_\sigma^{bi}(R)$ is a ringextension for the ring structure induced in $Q_\sigma^{bi}(R)$ and $Q_\tau^{bi}(R)$ by their R-module structure.

<u>Proof.</u> The diagram

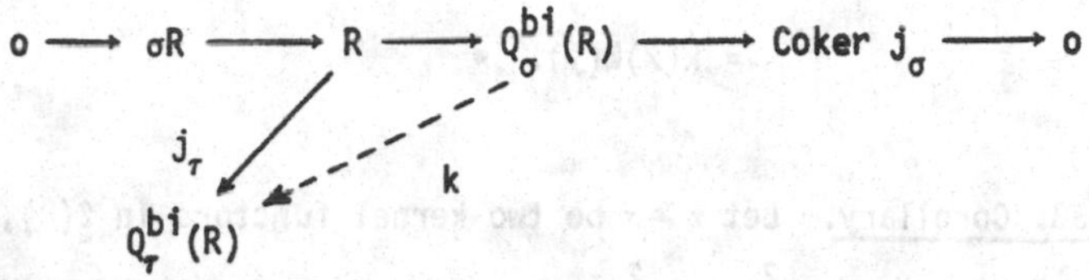

may be completed in a unique way by an R-linear map $k : Q_\sigma^{bi}(R) \to Q_\tau^{bi}(R)$, because $\tau \geq \sigma$. The R-module structure on $Q_\tau^{bi}(R)$ extends uniquely to a $Q_\sigma^{bi}(R)$-module structure which is obviously a $Q_\sigma^{bi}(R)$-bimodule structure, coinciding with the structure induced by ringmultiplication in $Q_\tau^{bi}(R)$, as is easily verified using uniqueness arguments. Choose ξ and η in $Z_R(Q_\sigma^{bi}(R)) =: C_\sigma$, then we may find $I \in \mathcal{L}^2(\sigma)$ such that $I\xi \subset R/\sigma R$, hence $k(I\xi\eta) = Ik(\xi\eta)$ and also $I\xi k(\eta) = k(I\xi\eta)$, by the definition of the $Q_\sigma^{bi}(R)$-module structure. So we find $I(\xi k(\eta) - k(\xi\eta)) = o$, i.e; $\xi k(\eta) - k(\xi\eta) \in \sigma(Q_\tau^{bi}(R)) \subset \tau(Q_\tau^{bi}(R)) = o$.

In other words, the restriction of k to C_σ is a ringmorphism. But from this one easily deduces that k is globally a ring morphism. Indeed take $x, y \in Q_\sigma^{bi}(R)$, then write these elements as

$$x = \Sigma r_i c_i \ , \quad y = \Sigma s_j d_j$$

where r_i, $s_j \in R$ and $c_i, d_j \in C$, then

$$k(x.y) = k(\sum_{i,j} r_i c_i s_j d_j) = k(\sum_{i,j} r_i s_j c_i d_j)$$

$$= \sum_{i,j} r_i s_j k(c_i d_j) \text{ (R-linearity)}$$

$$= \sum_{i,j} r_i s_j k(c_i) k(d_j) \quad \text{(since } k|C_\sigma \text{ is a ringmorphism)} .$$

$$= \sum_{i,j} r_i k(c_i) s_j k(d_j) \qquad (k(C_\sigma) \subset Z_R(Q_\tau^{bi}(R)))$$

$$= \sum_{i,j} k(r_i c_i) k(s_j d_j)$$

$$= k(\sum_i r_i c_i) k(\sum_j c_j d_j)$$

$$= k(x)k(y). \quad \blacksquare$$

IV.1.38. Corollary. Let $\tau \geqslant \sigma$ be two kernel functors in $\underline{\underline{2}}(R)$, then there is a unique R-linear map $Q_\sigma^2(M) \to Q_\tau^2(M)$ for each twosided R-module M, and for each R-bimodule M the following diagram is commutative :

$$\begin{array}{ccc} \mathrm{bi}\ Q_\sigma^2(M) & \longrightarrow & \mathrm{bi}\ Q_\tau^2(M) \\ \downarrow {\scriptstyle \eta_{\sigma,M}} & & \downarrow {\scriptstyle \eta_{\tau,M}} \\ Q_\sigma^{bi}(M) & \longrightarrow & Q_\tau^{bi}(M) \end{array}$$

the map $Q_\sigma^{bi}(M) \to Q_\tau^{bi}(M)$ is semilinear with respect to the ringmorphism $Q_\sigma^{bi}(R) \to$ $\to Q_\tau^{bi}(R)$. $\blacksquare$

We shall return to these results after expounding some technical detail concerning base change and ring extensions.

IV.2. Base Change and T-functors.

For some basic facts on bimodules and ring extensions we refer to Section II.3., or [17]. In particular recall the properties of "flatness for bimodules" e.g. II.3.18, II.3.19. etc...

If $\varphi : R \to S$ is a ring homomorphism and $M \in S\text{-mod}$, let ${}_RM$ be the image of M in R-mod obtained by restriction of scalars. Consider a kernel functor κ in R-mod, with torsion class T_κ. Then $T_{\varphi_*\kappa} = \{M \in S\text{-mod},\ {}_RM \in T_\kappa\}$ is the torsion class of a kernel functor in S-mod, which we denote $\varphi_*\kappa$.
In general modules of quotients at κ resp. $\varphi_*\kappa$ are poorly related. However if κ is symmetric and φ is an extension of rings then we obtain a much better description of $\varphi_*\kappa$.

IV.2.1. Lemma. Let $\varphi : R \to S$ be an extension of rings and let κ be a symmetric kernel functor on R-mod, then :

$$\mathcal{L}(\varphi_*\kappa) = \{L \text{ left ideal of } S,\ \varphi^{-1}(L) \in \mathcal{L}(\kappa)\}.$$

Proof. If $L \in \mathcal{L}(\varphi_*\kappa)$ then $S/L \in T_{\varphi_*\kappa}$ i.e. ${}_R(S/L) \in T_\kappa$ and vice versa. This means that for $\bar{1} \in S/L$, we may find a $J \in \mathcal{L}(\kappa)$ such that $J.\bar{1} = o$ i.e. $\varphi(J) \subset L$ or $J \subset \varphi^{-1}(L)$ and $\varphi^{-1}(L) \in \mathcal{L}(\kappa)$ follows. Conversely, suppose that L is a left ideal of S such that $\varphi^{-1}(L) \in \mathcal{L}(\kappa)$. If $z \in Z_R(S)$ then $\varphi^{-1}(L).z = \varphi(\varphi^{-1}(L))z = z\varphi(\varphi^{-1}(L)) \subset L$. If $z \in R$ and $J \in \mathcal{L}(\kappa)$ is a two-sided ideal such that $J \subset \varphi^{-1}(L)$, then $J.z = \varphi(J)z = \varphi(J)\varphi(z) \subset \varphi(J) \subset L$. Now, because φ is an extension, S is generated by $Z_R(S)$ and R, thus for any $s \in S$ there is a $J \in \mathcal{L}(\kappa)$ such that $J.s \subset L$ i.e. ${}_R(S/L) \in T_\kappa$ or $L \in \mathcal{L}(\varphi_*\kappa)$. ∎

IV.2.2. Corollary. Let $\varphi : R \to S$ be an extension of rings and let κ be a symmetric kernel functor on R-mod. For any $M \in S\text{-mod}$ we have : ${}_R(\varphi_*\kappa)(M) = \kappa({}_RM)$.

Proof. Straightforward, from the above lemma, if one notes that $J \in \mathcal{L}(\kappa)$ implies that $SJ \in \mathcal{L}(\varphi_*\kappa)$.

One further advantage of the consideration of extensions instead of arbitrary ring homomorphisms is that an extension $\varphi : R \to S$ yields an exact functor $\underline{\underline{bi}}(S) \to$ $\to \underline{\underline{bi}}(R)$ by restriction of scalars. Again, for $M \in \underline{\underline{bi}}(S)$ we write ${}_RM$ for its image in $\underline{\underline{bi}}(R)$. If σ is a kernel functor on $\underline{\underline{bi}}(R)$ then a kernel functor $\varphi_\star\sigma$ on $\underline{\underline{bi}}(S)$ may be defined by putting $(\varphi_\star\sigma)M$ equal to the S-module underlying $\sigma({}_RM)$ (that $\sigma({}_RM)$ has an S-module structure is exactly what IV.2.1. and IV.2.2. are about). Again from IV.2.2. it follows that the notation $\varphi_\star$ is unambiguous. Indeed, if the symmetric kernel functor κ on R-mod induces σ on bi(R) as described in Section IV.1. then the pull-back of σ under φ is induced on $\underline{\underline{bi}}(S)$ by the pull-back of κ. With notations as in Section IV.1. we have :

<u>IV.2.3. Proposition.</u> Let $\varphi : R \to S$ be an extension and $M \in \underline{\underline{bi}}(S)$, then :

1. If ${}_RM$ is faithfully κ-injective then M is faithfully $\varphi_\star\kappa$-injective.
2. If φ is a central extension then the converse of 1 holds.

<u>Proof.</u> 1. Look at the following diagram in $\underline{\underline{bi}}(S)$, where $(\varphi_\star\kappa)N'' = N''$:

$$\begin{array}{ccccccccc} o & \longrightarrow & N' & \longrightarrow & N & \longrightarrow & N'' & \longrightarrow & o \\ & & {\scriptstyle \psi'}\big\downarrow & \swarrow {\scriptstyle \psi} & & & & & \\ & & M & & & & & & \end{array}$$

Exactness of the restriction of scalars function yields a diagram in $\underline{\underline{bi}}(R)$:

$$\begin{array}{ccccccccc} o & \longrightarrow & N' & \longrightarrow & {}_RN & \longrightarrow & {}_RN'' & \longrightarrow & o \\ & & {\scriptstyle {}_R\psi'}\big\downarrow & \swarrow {\scriptstyle {}_R\psi} & & & & & \\ & & {}_RM & & & & & & \end{array}$$

where $\kappa({}_RN'') = {}_RN''$. Hence there exists a unique R-linear ${}_R\psi \in \mathrm{Hom}_{\underline{\underline{bi}}(R)}({}_RN, {}_RM)$ extending ${}_R\psi'$. Let us check that ${}_R\psi$ is actually a $\underline{\underline{bi}}(S)$-morphism. If $s \in \varphi(R)$, then for all $n \in N$ we get ${}_R\psi(sn) = s\,{}_R\psi(n)$. If $z \in Z_R(S)$ then, $\psi_S : N \to M :$ $n \to {}_R\psi(sn)$, and $\psi'_s : N \to M : n \to s\,{}_R\psi(n)$ are $\underline{\underline{bi}}(R)$-morphisms which coincide on N', hence on N. This yields that $s\,{}_R\psi(n) = {}_R\psi(sn)$ in this case too. Combining these

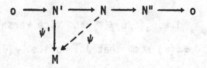

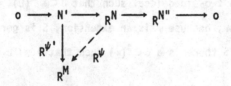

results yields that for all $s \in S$ and $n \in N$ we have $_R\psi(sn) = s_R\psi(n)$, and similar on
the right, proving that $_R\psi$ is actually a $\underline{bi}(S)$-morphism.

(2) Assume that φ is a central extension and that M is a faithfully $\varphi_*\kappa$-injec-
tive S-bimodule. Clearly $_RM$ is κ-torsion free. Consider the following diagram
in $\underline{bi}(R)$:

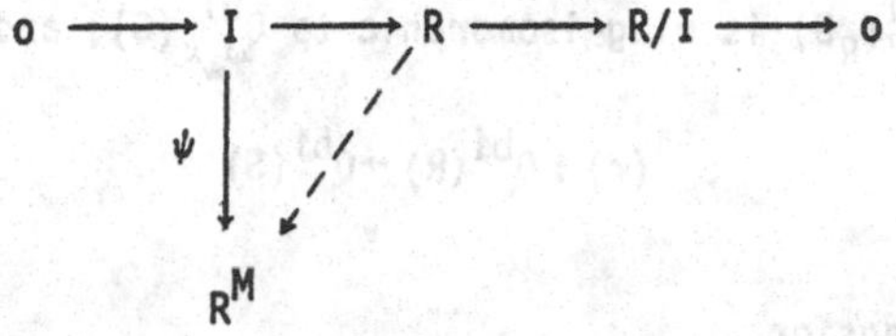

where $I \in \mathcal{L}^2(\kappa)$. Let us extend ψ to a $\underline{bi}(S)$-morphism $IS \to M$, then, as M is faithfully $\varphi_*\kappa$-injective, there exists a central element $m \in Z_R(M)$ such that $\overline{\psi}(j) = jm$
for all $j \in IS$, hence $\psi(i) = im$ for all $i \in I$. Since $Z_S(M) \subset Z_R(M)$ we also have
$m \in Z_R(M)$ and it suffices to apply IV.1.29. to derive the conclusion.
Let us now construct the map $\overline{\psi} : IS \to M$. Each $j \in IS$ can be written as $j = \Sigma i_\alpha s_\alpha$,
where $i_\alpha \in I$ and $s_\alpha \in Z(S)$, and we put

$$\overline{\psi}(j) = \Sigma\psi(i_\alpha)s_\alpha = \Sigma s_\alpha\psi(i_\alpha) \ .$$

For each $i \in I$ we have

$$i(\Sigma\psi(i_\alpha)s_\alpha) = \Sigma\psi(i)i_\alpha s_\alpha = \psi(i)(\Sigma i_\alpha s_\alpha)$$

and similar on the right. Now if $j = o$, then this implies that $\Sigma s_\alpha\psi(i_\alpha) = o$,
hence $\overline{\psi}$ is well-defined, and obviously left and right S-linear. ∎

IV.2.4. Remark.

If in the following commutative diagram of extensions φ, φ' and ψ_1 are central,
then so is ψ_2 :

$$
\begin{array}{ccc}
R & \xrightarrow{\ \psi_1\ } & S \\
\varphi \downarrow & & \downarrow \varphi' \\
R' & \xrightarrow[\ \psi_2\ ]{} & S'
\end{array}
$$
∎

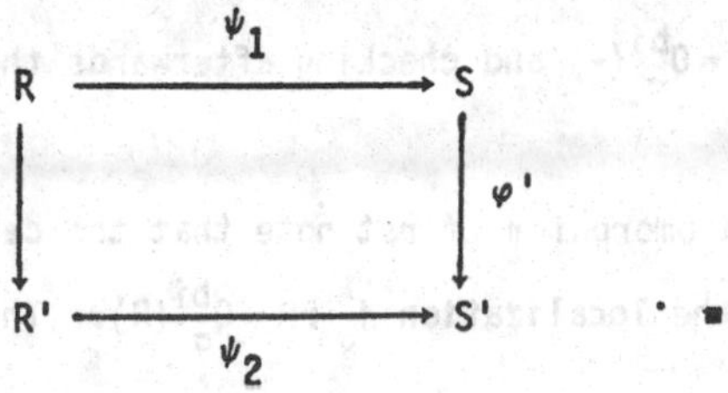

IV.2.5. Theorem. Let κ be a kernel functor in $\underline{\underline{bi}}(R)$ and assume that $\varphi: R \to S$ is a central extension, then for each S-bimodule M, there is a canonical isomorphism

$$_R[Q^{bi}_{\varphi_*\kappa}(M)] = Q_\kappa({}_RM).$$

In particular $Q^{bi}_\kappa({}_RS)$ is ring-isomorphic to $Q^{bi}_{\varphi_*\kappa}(S)$, and the induced morphism

$$Q^{bi}_\kappa(\varphi) : Q^{bi}_\kappa(R) \to Q^{bi}_\kappa(S)$$

is a central extension.

Proof. This follows immediately from IV.1. and the foregoing results, using the fact that localizing morphisms are central extensions. ∎

Let $\tau > \sigma$ be symmetric kernel functors in R-mod, then the kernel functor τ induces a kernel functor τ^e in $\underline{\underline{bi}}(Q^{bi}_\sigma(R))$ by putting

$$\tau^e\, M = \tau({}_RM)$$

for each $Q^{bi}_\sigma(R)$-bimodule M. In particular, the fact that $j_\sigma : R \to Q^{bi}_\sigma(R)$ is a central extension then implies that $Q^{bi}_{\tau^e}(Q^{bi}_\sigma(R)) = Q^{bi}_\tau({}_RQ^{bi}_\sigma(R))$.

With these notations we have :

IV.2.6. Theorem. We obtain ring isomorphisms

$$Q^{bi}_{\tau^e}(Q^{bi}_\sigma(R)) = Q^{bi}_\tau(Q^{bi}_\sigma(R)) = Q^{bi}_\tau(R) .$$

Proof. The first isomorphism follows from the foregoing, using the functorial isomorphism bi $Q^2_\sigma(-) = Q^{bi}_\sigma(-)$ and checking afterwards that the isomorphism is a ring isomorphism.
To check the second isomorphism first note that the central extension $Q^{bi}_\tau(R) \to Q^{bi}_\tau(Q^{bi}_\sigma(R))$ extends the localization $j_\sigma : R \to Q^{bi}_\sigma(R)$. The map $Q^{bi}_\tau(R) \to Q^{bi}_\tau(Q^{bi}_\sigma(R))$

may be viewed as induced by the isomorphism $Q_\tau^2(R) \to Q_\tau^2(Q_\sigma^2(R))$, hence it is injective. It finally suffices to check surjectivity.

Consider the following commutative diagram :

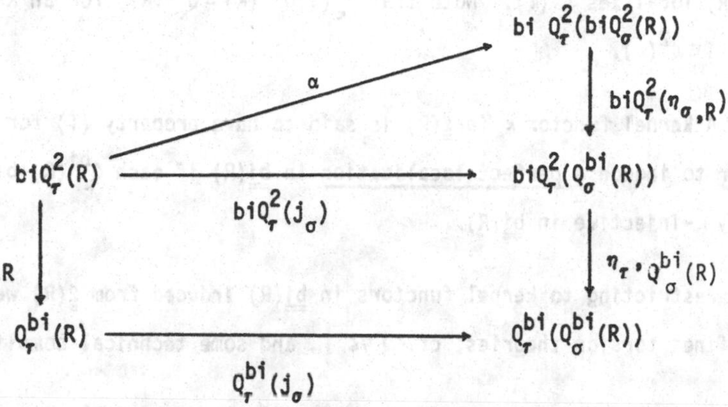

We have to show that α is surjective; clearly α is induced by the isomorphism

$$\beta : Q_\tau^2(R) \to Q_\tau^2\ Q_\sigma^2(R)$$

defined before. Take $z \in Z_R(Q_\tau^2(biQ_\sigma^2(R)))$, then there exists $z' \in Q_\tau^2(R)$ such that $\beta(z') = z$. Note that β being an isomorphism, $Z_R(Q_\tau^2(R))$ is mapped isomorphically onto $Z_R(Q_\tau^2(biQ_\sigma^2(R))$, hence $z' \in Z_R(Q_\tau^2(R)) \subset biQ_\tau^2(R)$.

Finally, pick an arbitrary element $x \in biQ_\tau^2(biQ_\sigma^2(R))$, i.e. $x = \Sigma r_\alpha z_\alpha$, when $r_\alpha \in R$ and $z_\alpha \in Z_R(Q_\tau^2(biQ_\sigma^2(R)))$, then we may find $z'_\alpha \in Z_R(Q_\tau^2(R))$, mapped onto z_α by β, i.e.

$$\beta(\Sigma r_\alpha z'_\alpha) = \Sigma r_\alpha \beta(z'_\alpha) = \Sigma r_\alpha z_\alpha = \alpha .$$

This proves that β induces a surjective map

$$\alpha : biQ_\tau^2(R) \to biQ^2(biQ_\sigma^2(R)),$$

hence, by the commutativity of the above diagram, $biQ_\tau^2(j_\sigma)$ is surjective, and we may then conclude that the extension $Q_\tau^{bi}(j_\sigma)$ is surjective too. ∎

Assume that $\mathcal{L}$ is a family of twosided ideals of a ring R. An extension $\varphi : R \to S$ is said to <u>localize</u> $\mathcal{L}$ if for all $I \in \mathcal{L}$ we have $S\varphi(I) = S$. In the rest of this paragraph we will be concerned with kernel functors κ in $\underline{\underline{bi}}(R)$ such that $j_\kappa : R \to Q_\kappa^{bi}(R)$ localizes $\mathcal{L}^2(\kappa)$. Note that $j_\kappa(I)Q_\kappa^{bi}(R) = Q_\kappa^{bi}(R)$ for an R-ideal I implies that $I \in \mathcal{L}^2(\kappa)$.

<u>Definition</u>. A kernel functor κ in $\underline{\underline{2}}(R)$ is said to have property (T) for <u>R-bimodules</u> or to induce a <u>perfect localization</u> in $\underline{\underline{bi}}(R)$ if each $Q_\kappa^{bi}(R)$-bimodule is faithfully κ-injective in $\underline{\underline{bi}}(R)$.

Note that by restricting to kernel functors in $\underline{\underline{bi}}(R)$ induced from $\underline{\underline{2}}(R)$ we avoid the use of "fine" torsion theories, cf. [46], and some technical complications.

<u>IV.2.7. Theorem</u>. Let κ be a kernel functor in $\underline{\underline{2}}(R)$, then the following assertions are equivalent :

1. κ has property (T) for R-bimodules.
2. each $Q_\kappa^{bi}(R)$-bimodule is κ-torsion free.
3. if $j_\kappa : R \to Q_\kappa^{bi}(R)$ denotes the localizing morphism, then $j_{\kappa,*}\,\kappa = \hat{o}$.
4. for each R-bimodule M the bimodule $Q_\kappa^{bi}(M)$ and the morphism $j_{\kappa,M} : M \to Q_\kappa^{bi}(M)$ are canonically isomorphic to $Q_\kappa^{bi}(R) \otimes_R M$ and $j_\kappa \otimes M : M \to Q_\kappa^{bi}(R) \otimes_R M$.

<u>Proof</u>. This is a straightforward adaptation of a similar result of Goldman's, cf.[40] using the machinery developed in the first part of this paragraph. ∎

<u>IV.2.8. Corollary</u>. If κ has property (T) for R-bimodules, then $j_\kappa : R \longrightarrow Q_\kappa^{bi}(R)$ localizes $\mathcal{L}^2(\kappa)$. The converse holds if κ is actually a symmetric kernel functor defined in R-mod.

<u>Proof</u>. Since each $Q_\kappa^{bi}(R)$-bimodule is κ-torsion free, by the foregoing result, the only κ-torsion $Q_\kappa^{bi}(R)$-bimodule is o, hence the next lemma will yield the first statement. Moreover, if κ is a kernel functor in R-mod and M a $Q_\kappa^{bi}(R)$-bimodule, then $x \in M$ lies in κM if and only if for some ideal $I \in \mathcal{L}^2(\kappa)$ we have $j_\kappa(T)x = o$.

But then $Q_\kappa^{bi}(R)j_\kappa(I)x = Q_\kappa^{bi}(R)x = o$, since j_κ is localizing, i.e. $x = o$. ∎

<u>IV.2.9. Lemma.</u> Let κ be a kernel functor in $\underline{\underline{2}}(R)$; the following statements are equivalent :

1. $j_\kappa : R \to Q_\kappa^{bi}(R)$ localizes $\mathcal{L}^2(\kappa)$;
2. the only κ-torsion $Q_\kappa^{bi}(R)$-bimodule is o;
3. $\mathcal{L}^2(j_{\kappa,*}\kappa) = \{Q_\kappa^{bi}(R)\}$.

<u>Proof</u>

(1) ⇒ (2) If $M \in \underline{bi}(Q_\kappa^{bi}(R))$ is κ-torsion and $m \in Z_R(M)$, then for some $I \in \mathcal{L}(\kappa)$, we have $Im = o$ and $Q_\kappa^{bi}(R)Im = o$. Hence $m = o$.

(2) ⇒ (3) Obvious.

(3) ⇒ (1) As $\mathcal{L}^2(j_{\kappa,*}\kappa) = \{K \text{ ideal of } Q_\kappa^{bi}(R);\ j_\kappa^{-1}(K) \in \mathcal{L}^2(\kappa)\}$, clearly $I \in \mathcal{L}^2(\kappa)$ implies $Q_\kappa^{bi}(R)j_\kappa(I) = Q_\kappa^{bi}(R)$. ∎

Note that the converse of IV.2.8. does not necessarily hold in general, which accounts for several mistakes in [45], where in particular III.4. Théorème 7 does not hold. We will come back to this later.

<u>Example</u>. Clearly $j_\kappa : R \to Q_\kappa^{bi}(R)$ localizes $\mathcal{L}^2(\kappa)$ iff $\mathcal{L}^2(j_{\kappa,*}\kappa) = \{Q_\kappa^{bi}(R)\}$ and κ has property (T) for bimodules iff $j_{\kappa,*}\kappa = \hat{o}$ and these are not equivalent. To give an example of a torsion theory $\nu \neq o$ such that $\mathcal{L}^2(\nu) = \{R\}$, define on $\underline{\underline{2}}(R)$

$$\nu = \sup\{\sigma \in \underline{\underline{2}}(R)\text{-Ker};\ \sigma(R/M) = o \text{ for all } M \in \Omega(R)\},$$

where $\Omega(R)$ denotes the set of all maximal ideals in R. We obviously have that $\mathcal{L}^2(\nu) = \{R\}$, for, indeed, if $I \in \mathcal{L}^2(\nu)$, then, if I is non-trivial, we get $I \subset M$ for some maximal ideal M of R, a contradiction. On the other hand $\nu \neq \hat{o}$, for otherwise $\nu(R/I) = o$ for all ideals I, which is not true in general. Example : take $R = \begin{pmatrix} k & o \\ k & k \end{pmatrix}$, where k is an arbitrary field, then $\nu(R) = \begin{pmatrix} o & o \\ k & o \end{pmatrix}$, so for $I = o$ we find $\nu(R/I) \neq o$.

IV.2.10. Corollary. Assume that $j_\kappa : R \to Q_\kappa^{bi}(R)$ localizes $\mathcal{L}^2(\kappa)$, then j_κ is an epimorphism of rings. Furthermore, for each left ideal J of $Q_\kappa^{bi}(R)$ we have $Q_\kappa^{bi}(R)j_\kappa(j_\kappa^{-1}(J)) = J$ and analogously for right ideals;

Proof. Assume that $\varphi, \psi : Q_\kappa^{bi}(R) \to S$ are ringmorphisms such that $\varphi j_\kappa = \psi j_\kappa$. It is clear that φj_κ localizes $\mathcal{L}(\kappa)$; it follows immediately from this that $bi\kappa(S) = o$. Now $\varphi - \psi$ induces a quotient morphism

$$\theta : Q/j_\kappa(R) \to S,$$

where $Q = Q_\kappa^{bi}(R)$, where $Q/j_\kappa(R)$ is a κ-torsion R-bimodule and $bi\kappa(S) = o$, hence $\theta = o$ and $\varphi = \psi$. In order to prove the second statement note that by construction, $j_\kappa(j_\kappa^{-1}(J)) = j_\kappa(R) \cap J = J^c$, hence the localizing morphism $j_\kappa : R \to Q_\kappa^{bi}(R)$ yields an injective left Q-morphism $J/J^c \to Q_\kappa^{bi}(R)/j_\kappa(R)$. Hence for all $j \in J$, we may find $I \in \mathcal{L}^2(\kappa)$ such that $Ij \subset J^c$. Indeed, $\overline{j} \in J/J^c$ may be identified with $\overline{q} \in Q_\kappa^{bi}(R)/j_\kappa(R)$, where $q = \Sigma j_\kappa(r_i)z_i$, $r_i \in R$, $z_i \in Z_R(Q_\kappa^{bi}(R))$. For each i we produce $I_i \in \mathcal{L}^2(\kappa)$ such that $I_i z_i \subset j_\kappa(R)$, and then $I = \cap I_i$ has the desired property. But $Ij \subset J^c$ implies $Q_\kappa^{bi}(R)j = Q_\kappa^{bi}(R)Ij \subset Q_\kappa^{bi}(R)J^c$; hence $j \in Q^{bi}(R)J^c$, i.e. $J \subset Q_\kappa^{bi}(R)\ j_\kappa(j_\kappa^{-1}(J))$. The converse inclusion being obvious, this concludes the proof. ∎

IV.2.11. Corollary. If κ has property (T) for R-bimodules, then

1. $j_\kappa : R \to Q_\sigma^{bi}(R)$ is epimorphic and flat for R-bimodules;
2. the lattice of (twosided) ideals of $Q_\kappa^{bi}(R)$ is isomorphic to the lattice of R-ideals I such that R/I is κ-torsion free. ∎

IV.2.12. Theorem. Let κ be a kernel functor in $\underline{\underline{2}}(R)$, which has property (T) for R-bimodules and let $\varphi : R \to S$ be an arbitrary ringextension. Denote $\varphi_*\kappa$ by κ' and assume that S is κ'-torsion free, then

1. $Q_{\kappa'}^{bi}(S)$ and $Q_\kappa^{bi}(S)$ are isomorphic rings;
2. $Q_{\kappa'}^{bi}(M)$ and $Q_\kappa^{bi}(M)$ are isomorphic R-bimodules for each $M \in \underline{\underline{bi}}(S)$ which is κ'-torsion free.

<u>Proof</u>. Clearly $Q_\kappa^{bi}(S) = Q_\kappa^{bi}(R) \otimes_R S \cong S \otimes_R Q_\kappa^{bi}(R)$ since κ has property (T) for bimodules, hence $Q_\kappa^{bi}(S)$ has a natural ring structure extending its $Q_\kappa^{bi}(R)$-module structure and such that $j : S \to Q_\kappa^{bi}(S) : s \mapsto 1 \otimes s$ is a ringmorphism. Moreover j is flat for R-bimodules since $Q_\kappa^{bi}(R)$ is. Consider the following diagram of rings

$$S \xrightarrow{\;j\;} Q_\kappa^{bi}(S) \overset{f}{\underset{g}{\rightrightarrows}} P$$

where f,g are ringmorphisms such that $fj = gj$. If $x \in Z(Q_\kappa^{bi}(S))$, then $I\,x \subset S$ for some $I \in \mathcal{L}^2(\kappa)$, hence $f(Ix) = g(Ix)$. Now $R \to Q_\kappa^{bi}(R)$ is an epimorphism, hence $fj|_R = gj|_R$, implying that $f|_{Q_\sigma^{bi}(R)} = g|_{Q_\sigma^{bi}(R)}$, since the following diagram is commutative

$$\begin{array}{ccccc} R & \longrightarrow & S & & \\ \downarrow & & \downarrow & & \\ Q_\kappa^{bi}(R) & \longrightarrow & Q_\kappa^{bi}(S) & \rightrightarrows & P \end{array} .$$

Hence on P we have induced unambiguously a $Q_\kappa^{bi}(R)$-bimodule structure via either $f|_{Q_\kappa^{bi}(R)}$ or $g|_{Q_\kappa^{bi}(R)}$. As κ has property (T), the ring P is κ-torsion free as an R-bimodule, so, $I(f(x)-g(x)) = o$ in the R-bimodule structure, thus we get $f(x) = g(x)$. Therefore $f = g$ follows, since $Z(Q_\kappa^{bi}(R))$ generates $Q_\kappa^{bi}(R)$.

Now, j being epimorphic and flat for bimodules, we get the existence of a kernel functor κ_1 in <u>bi</u>(S), having property (T) for bimodules and such that $Q_{\kappa_1}^{bi}(S) = Q_\kappa^{bi}(S)$ and then

$$Q_{\kappa_1}^{bi}(M) \cong Q_{\kappa_1}^{bi}(S) \otimes_S M \cong Q_\kappa^{bi}(S) \otimes_S M \cong (Q_\kappa^{bi}(R) \otimes_R S) \otimes_S M \cong Q_\kappa^{bi}(R) \otimes_R M \cong Q_\kappa^{bi}(M)$$

for each S-bimodule M (all isomorphisms in <u>bi</u>(R)).

Let us now check that $L^2(\kappa') = L^2(\kappa_1)$. Assume that S/I is κ_1-torsion, then $Q_{\kappa_1}^{bi}(S/I) \cong Q_\kappa^{bi}(S/I) = o$, hence S/I is κ-torsion in <u>bi</u>(R) hence κ'-torsion as an

S-bimodule; conversely, if S/I is κ'-torsion in $\underline{\underline{bi}}(S)$, then it is σ-torsion as an R-bimodule, i.e. it is κ_1-torsion, which proves our assertion.

Finally, let us show for an arbitrary $M \in \underline{\underline{bi}}(S)$ which is κ'torsion free, that $Q^{bi}_{\kappa}(M)$ is isomorphic to $Q^{bi}_{\kappa_1}(M)$. First, $Q^{bi}_{\kappa 1}(M)/M$ is κ'-torsion, since it is κ_1-torsion and $\mathcal{L}^2(\kappa') = \mathcal{L}^2(\kappa_1)$; next, consider an exact sequence of S-bimodules

$$\begin{array}{ccccccccc} o & \longrightarrow & N' & \longrightarrow & N & \longrightarrow & N'' & \longrightarrow & o \\ & & {\scriptstyle\varphi}\downarrow & \swarrow {\scriptstyle\overline{\varphi}} & & & & & \\ & & Q^{bi}_{\kappa 1}(M) & & & & & & \end{array}$$

where N" is κ'-torsion. Because N" is also κ_1-torsion, there exists a unique $\overline{\varphi} \in \mathrm{Hom}_{\underline{\underline{bi}}(S)}(N, Q^{bi}_{\kappa_1}(M))$, since $Q^{bi}_{\kappa_1}(M)$ is faithfully κ_1-injective; hence $Q_{\kappa_1}(M)$ is faithfully κ'-injective. Both assertions, together with the definitions yield the result. That the isomorphism thus described yields a ring isomorphism if $M = S$ is fairly trivial because the ring structure of $Q^{bi}_{\kappa_1}(S)$ resp. $Q^{bi}_{\kappa'}(S)$ is uniquely determined by its bimodule structure. ■

<u>IV.2.13. Corollary</u>. With assumptions as before, the kernel functor $\varphi_\star\kappa$ induces a perfect localization in $\underline{\underline{bi}}(S)$.

<u>Proof</u>. Consider the following commutative diagram of extensions

$$\begin{array}{ccc} R & \xrightarrow{\ \varphi\ } & S \\ {\scriptstyle j_\kappa}\downarrow & & \downarrow{\scriptstyle j_{\varphi_\star\kappa}} \\ Q^{bi}_{\kappa}(R) & \xrightarrow[\ \varphi_\kappa\ ]{} & Q^{bi}_{\varphi_\star\kappa}(S) \end{array}$$

Denote $j_{\varphi_\star\kappa}$ by j'_κ and take $M \in \underline{\underline{bi}}(Q^{bi}_{\varphi_\star\kappa}(S))$, then

$$((j'_\kappa)_\star\varphi_\star\kappa)M = ((j'_\kappa\varphi)_\star\kappa)M = ((\varphi_\kappa j_\kappa)_\star\kappa)M = ((\varphi_\kappa)_\star(j_\kappa)_\star\kappa)M$$

$$= ((j_\kappa)_\star\kappa)(_{Q_\kappa^{bi}(R)}M) = o$$

since κ has property (T) for bimodules. But $((j_\kappa')_\star\varphi_\star\kappa)M = o$ for every $M \in \underline{\underline{bi}}(Q^{bi}_{\varphi_\star\kappa}(S))$ implies that $\varphi_\star\kappa$ has property (T) for S-bimodules. ∎

An extension $\varphi : R \to S$ is said to be <u>absolutely torsion</u> free if and only if for all nonzero ideals I of R and non zero element $s \in S$ we have $\varphi(I)s \neq o$. An absolutely torsion free R-module is defined in a similar way. First as in the module case, the extension $\varphi : R \to S$ is said to be <u>torsion-free</u> if for all non trivial bilateral kernel functors σ in $\underline{\underline{bi}}(R)$ the ring S is torsion-free for the induced kernel functor $\varphi_\star\sigma$. (Here kernel functors may be non idempotent!).

<u>IV.2.13. Example</u>. If φ is injective and S is prime, then φ is absolutely torsion-free. If a ring is torsion-free over its center C, then the canonical inclusion $C \subset R$ is an absolutely torsion free central extension.

<u>IV.2.14. Lemma</u>.

1. An absolutely torsion-free extension is injective and torsion-free;
2. If R has a.c.c. on semiprime ideals, then an extension $\varphi : R \to S$ is torsion-free if and only if for all prime-ideals P of R the ring S is without $\varphi_\star\kappa_{R-P}$-torsion.
3. If κ is a symmetric kernel functor in R-mod and $j_\kappa : R \to Q_\kappa^{bi}(R)$ denotes the localizing morphism, then $\kappa R = o$, if j_κ is absolutely torsion-free. If $\kappa R = o$, then j_κ is torsion-free if and only if for all $\kappa \leq \tau$ the ring R is τ-torsion free.
4. If j_κ is torsion-free, then for all $\sigma \geq \kappa$ the ring $R/\kappa R$ is σ-torsion free.

<u>Proof</u>. (1) Let $\varphi : R \to S$ be absolutely torsion-free and take $r \in R$ such that $\varphi(r) = o$. Then $\varphi(RrR).1 = o$, hence $RrR = o$, so $r = o$ and φ is injective. Let κ be a non trivial symmetric kernel functor and $s \in (\varphi_\star\kappa)S$, then $s \in \kappa(_RS)$, i.e. there exists an ideal I in $\mathcal{L}^2(\kappa)$ such that $I.s = \varphi(I)s = o$, hence $s = o$, since κ is assumed to be non trivial.

(2) That the condition is necessary is obvious. Conversely, let σ be an arbitrary bilateral prekernel functor in R-mod, then

$$\mathcal{L} = \{L \text{ left ideal of } R;\ L \supset I \text{ for some ideal } I \text{ of } R \text{ such that there is } P \in \mathrm{Spec}(R) - \mathcal{L}(\sigma) \text{ for which } I \not\subset P\}.$$

is the filter of a prekernel functor σ', which is radical in the sense that rad $I \in \mathcal{L}$ implies $I \in \mathcal{L}$ and which has the property that $\sigma \leq \sigma'$. Since R satisfies the ascending chain condition on semiprime ideals, clearly each σ'-closed ideal of R is contained in a maximal σ'-closed ideal, for only chains of semiprime ideals have to be considered by the radicality of σ'. Let $C(\sigma')$ denote the set of maximal σ'-closed ideals of R. Its elements are prime ideals, for if I and J are ideals of R with $IJ \subset P$ for some $P \in C(\sigma')$ and if $I \not\subset P$ and $J \not\subset P$, then we have that I+P and J+P are in $\mathcal{L}(\sigma')$. We thus find $(I+P)(J+P) \subset$, yielding $P \in \mathcal{L}(\sigma')$ by the radicality of σ', a contradiction. Finally, this allows us to prove, exactly as in the noetherian case, cf. [174], Proposition 24, or I.4.12. that

$$\sigma' = \wedge\{\sigma_{R-P};P \in C(\sigma')\}.$$

Since $\sigma_{R-P}({}_RS) = o$, for all $P \in \mathrm{Spec}(R)$, we certainly have $(\varphi_*\sigma')S = \sigma'({}_RS) \subset \cap\{\sigma_{R-P}({}_SR);P \in C(\sigma')\} = o$, which proves that $(\varphi_*\sigma)S = o$, showing that the condition is also sufficient.

(3) This follows directly from (12.29.1.).

(4) Assume that there exists $\tau \geq \kappa$ such that $R/\tau R$ is not τ-torsion free; then there exists $o \neq \bar{x} \in R/\kappa R$ and an ideal $J \neq o$ in $\mathcal{L}(\tau)$ such that $J\bar{x} = \bar{o}$, i.e. $x \in R$ and $J \in \mathcal{L}(\kappa)$, both different from o, such that $j_\kappa(J)\bar{x} = o$, a contradiction.

(5) Assume now that $\kappa R = o$, then j_κ torsion-free implies by the foregoing that for all $\tau \geq \kappa$ we have $\tau R = o$. Conversely, assume that for all $\tau \geq \kappa$ the ring R is κ-torsion free. If $\tau \leq \kappa$ is a bilateral kernel functor in R-mod, then

$$j_{\kappa,*}\tau(Q_\kappa^{bi}(R)) \subset j_{\kappa,*}(Q_\kappa^{bi}(R)) = o,$$

and $Q_\kappa^{bi}(R)$ is $j_{\kappa,\star}\,\tau$-torsion free. If $\tau \not\leq \kappa$ is an arbitrary, nontrivial bilateral kernel functor then $\tau' = \widehat{\kappa \overset{\circ}{\vee} \tau}$, the largest bilateral kernel functor smaller than $\kappa \vee \tau$ is bilateral and non trivial. Assume $x \in j_{\kappa,\star}\tau'(Q_\kappa^{bi}(R))$, then there is an ideal $J \in \mathcal{L}^2(\tau')$ such that $j_\kappa(J)x = o$ and an ideal $I \in \mathcal{L}(\kappa)$ such that $j_\kappa(I)\,x \subset R$. If we view j_κ as an inclusion, this yields $o = (JI)x = J(Ix)$, hence $Ix \subset \tau' R = o$. But then $x \in \kappa(Q_\kappa^{bi}(R)) = o$. This proves that $j_{\kappa,\star}\tau'(Q_\kappa^{bi}(R)) = o$, but also that $j_{\kappa,\star}\tau(Q_\kappa^{bi}(R)) = o$, since $\tau \leq \tau'$. Hence j_κ is torsion-free central extension. ∎

A kernel functor σ in $\underline{\underline{bi}}(R)$ is said to be symmetric if it is induced by a symmetric kernel functor in R-mod.

IV.2.15. Corollary. If the ring R is prime, then all symmetric localizations in $\underline{\underline{bi}}(R)$ are torsion-free. ∎

Actually we may prove more :

IV.2.16. Proposition. If the ring R is semiprime, then an extension $\varphi : R \to S$ is absolutely torsion-free if and only if it is torsion-free.

Proof. One implication has been proved in IV.2.14. . Let us verify that the converse also holds. Assume that φ is torsion-free, let $o \neq I$ be an ideal of R and let $s \in S$. Since R is semiprime and I is nonzero, there exists a prime ideal $P \in \mathrm{Spec}(R)$ such that $I \not\subset P$. If $\varphi(I)s = o$, then $s \in \kappa_{R-P}({}_R S) = (\varphi_\star \kappa_{R-P})S = o$, as φ is torsion-free. This proves that φ is absolutely torsion-free. ∎

IV.2.17. Corollary. If the ring R is prime, then all symmetric localizations in $\underline{\underline{bi}}(R)$ are absolutely torsion-free and conversely. ∎

IV.2.18. Proposition. Let $\varphi : R \to S$ be an absolutely torsion-free extension and κ a symmetric kernel functor in $\underline{\underline{bi}}(R)$ which has property (T) for bimodules, then $Q_\kappa^{bi}(\varphi) : Q_\kappa^{bi}(R) \to Q_\kappa^{bi}(S) = Q_{\varphi_\star\kappa}^{bi}(S)$ is an absolutely torsion-free extension.

Proof. We will make use of the following diagram of extensions :

$$\begin{array}{ccc} R & \xrightarrow{\varphi} & S \\ \downarrow{\scriptstyle j_\kappa} & & \downarrow{\scriptstyle j'_\kappa} \\ Q^{bi}_\kappa(R) & \longrightarrow & Q^{bi}_\kappa(S) \end{array}$$

$$Q^{bi}_\kappa(\varphi) = \varphi_\kappa$$

Take an ideal I of $Q^{bi}_\kappa(R)$, different from zero and $s \in Q^{bi}_\kappa(S)$ such that $\varphi_\kappa(I)s = o$. Since Q^{bi}_κ is a perfect localization, we know that $I_1 = j_\kappa^{-1}(I) \neq o$ follows from $I = j_\kappa(j_\kappa^{-1}(I))Q^{bi}_\kappa(R)$; furthermore $j_\kappa(I_1) \subset I$, hence $\varphi_\kappa(j_\kappa(I_1))s \subset \varphi_\kappa(I)s = o$, i.e. $j_\kappa(\varphi(I_1))s = o$. As φ is absolutely torsion-free, it is torsion-free, hence $(\varphi_*\kappa)S = o$. In other words, we may view j'_κ as an inclusion and $\varphi(I_1)s = o$. Now, as $s \in Q^{bi}_\kappa(S)$, there exists an ideal $J \in \mathcal{L}^2(\varphi_*\kappa)$ such that $Js \subset S$, and by definition $\varphi^{-1}(J) = J_1 \in \mathcal{L}(\kappa)$, hence $J_1 \neq o$ and $\varphi(J_1) \subset J$, so $\varphi(J_1)s \subset S$. As I_1 is an ideal of R, we get $I_1J_1 \subset I_1$, so $\varphi(I_1)\varphi(J_1)s = \varphi(I_1J_1)s \subset \varphi(I_1)s = o$, i.e. $\varphi(J_1)s = o$ since φ is absolutely torsion-free. Since $\varphi^{-1}(S\varphi(J_1)) \supset J_1$, with $J_1 \in \mathcal{L}^2(\kappa)$, we get that the ideal $S\varphi(J_1)$ lies in $\mathcal{L}(\varphi_*\kappa)$ and as $Q^{bi}_\kappa(S)$ is by definition $\varphi_*\kappa$-torsion free, we finally find that $s = o$. ∎

<u>IV.2.19. Proposition</u>. Let $\varphi : R \to S$ be an extension which is epimorphic in the category of rings and flat for R-bimodules and let κ be a kernel functor in <u>bi</u>(S), then there exists a kernel functor $\varphi^*\kappa$ in <u>bi</u>(R), such that $\varphi_*\varphi^*\kappa = \kappa$.

<u>Proof</u>. For each R-bimodule M the extension φ induces a morphism $\varphi_M : M = R \otimes_R M \to S \otimes_R M$. We define $\varphi^*\kappa$ by $(\varphi^*\kappa)M = \varphi_M^{-1}(\kappa(S \otimes_R M))$. If $\varphi^*\kappa$ is a kernel functor in <u>bi</u>(R), then its definition and the flatness of φ immediately yield that $\varphi_*\varphi^*\kappa = \kappa$. So, let us check that we have indeed a kernel functor in <u>bi</u>(R). Functoriality and left exactness following from the flatness of φ, let us verify that $(\varphi^*\kappa)(M/(\varphi^*\kappa)M) = o$. Denote $M/(\varphi^*\kappa)M$ by M_1 and $S \otimes_R M/\kappa(S \otimes_R M)$ by M_2, then we obtain an injective bimodule morphism $\psi : M_1 \to M_2$ making the following diagram

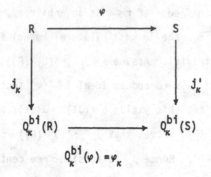

commutative :

$$\begin{array}{ccc} M_1 & \xrightarrow{\ \psi\ } & M_2 \\ \downarrow{\scriptstyle \varphi_{M_1}} & & \downarrow{\scriptstyle \varphi_{M_2}} \\ S \otimes_R M_1 & \xrightarrow[\ S \otimes \psi\ ]{} & S \otimes_R M_2 \end{array}$$

Since M_2 is an S-bimodule and φ is epimorphic, we get $S \otimes_R M_2 = S \otimes_R (S \otimes_S M_2) = (S \otimes_R S) \otimes_S M_2 = S \otimes_S M_2 = M_2$, hence φ_{M_2} is an isomorphism. Thus φ_{M_1} is injective and it suffices to check that $\kappa(S \otimes_R M_1) = o$ in order to have $\kappa M_1 = o$. But $S \otimes_R M_1$ injects in $S \otimes_R M_2 = M_2$ by the flatness of φ and $\kappa M_2 = \kappa(S \otimes_R M/\kappa(S \otimes_R M)) = o$, hence $\kappa(S \otimes_R M_1) = o$. ■

IV.2.20. Special case. Assume that κ has property (T) for R-bimodules, then the localizing morphism $j_\kappa : R \to Q_\kappa^{bi}(R)$ is an epimorphic extension which is flat for R bimodules. We then know that for each R-bimodule M we have $Q_\kappa^{bi}(R) \otimes_R M = Q_\kappa^{bi}(M)$ hence in this situation the map φ_M is the canonical morphism $M \to Q_\kappa^{bi}(M)$; for each kernel functor τ in $\underline{\underline{bi}}(Q_\kappa^{bi}(R))$ we get that $(\varphi^* \tau)M = \varphi_M^{-1}(\tau Q_\kappa^{bi}(M))$. Let $\mathcal{L}(\varphi,\kappa)$ be the idempotent filter in R-mod generated by the ideals I of R such that $Q_\kappa^{bi}(R) j_\kappa(I) \in \mathcal{L}(\kappa)$, then $\mathcal{L}(\varphi,\kappa)$ is the filter of a symmetric kernel functor which on R-bimodules coincides with $\varphi^* \kappa$. In other words :

IV.2.21. Corollary. Let κ be a symmetric kernel functor in $\underline{\underline{bi}}(R)$ having property (T) for R-bimodules, and τ a symmetric kernel functor in $\underline{\underline{bi}}(Q_\kappa^{bi}(R))$, then there exists a symmetric kernel functor τ' in $\underline{\underline{bi}}(R)$ such that $j_{\kappa,*}\tau' = \tau$. ■

IV.2.22. Corollary. Let $\varphi : R \to S$ be a torsion-free extension and κ a symmetric kernel functor in $\underline{\underline{bi}}(R)$ inducing a perfect localization, then $\varphi_\kappa = Q_\kappa^{bi}(\varphi) : Q_\kappa^{bi}(R) \to Q_\kappa^{bi}(S) = Q_{\varphi_*\kappa}^{bi}(S)$ is a torsion-free extension.

<u>Proof</u>. Let σ be a symmetric kernel functor on $Q_\kappa^{bi}(R)$-mod and consider the following commutative diagram

$$\begin{array}{ccc} R & \xrightarrow{\ \varphi\ } & S \\ {\scriptstyle j_\kappa}\big\downarrow & & \big\downarrow{\scriptstyle j'_\kappa} \\ Q_\kappa^{bi}(R) & \xrightarrow[\ \varphi_\kappa\ ]{} & Q_\kappa^{bi}(S) \end{array} .$$

By the foregoing corollary, there exists a symmetric kernel functor τ in <u>bi</u>(R) such that $j_{\kappa,\star}\tau = \sigma$. Take $s \in (\varphi_{\sigma,\star}\ \sigma)Q_\kappa^{bi}(S)$, then $s \in (\varphi_{\sigma,\star}\ j_{\sigma,\star}\ \tau)Q_\kappa^{bi}(S)$, i.e. $s \in (\varphi_\star\tau)({}_SQ_\kappa^{bi}(S))$. Since φ is torsion-free, j'_κ may be viewed as an inclusion, hence there exists an ideal $I \in \mathcal{L}(\tau)$ such that $\varphi(I)s = o$. On the other hand, there is an ideal J in $\mathcal{L}(\kappa)$ such that $\varphi(J)s \subset S$. Moreover, since I is an ideal, we obtain $IJ \subset I$. Hence

$$\varphi(I)\varphi(J)s = \varphi(IJ)s \subset \varphi(I)\, s = o$$

and from $\varphi(I)\varphi(J)s = o$, it follows that $\varphi(J)s = o$ since $I \neq o$ and τ is nontrivial. But then $s \in \sigma({}_SQ_\kappa^{bi}(S)) = \varphi_\star\sigma(Q_\kappa^{bi}(S)) = o$. As σ and s were arbitrary, the proposition follows. ∎

In most applications R will be a prime ring. The last corollary is then just a restatement of IV.2.18.

<u>IV.2.23. Lemma</u>. Let κ be a kernel functor in <u>bi</u>(R) which has property (T) for R-bimodules, and let τ be a kernel functor in $\underline{bi}(Q_\kappa^{bi}(R))$, then $j_\kappa^\star\tau > \kappa$.

<u>Proof</u>. First note that, if we denote by $\hat{o}$ the zero-kernel functor, then $j_\kappa^\star\hat{o} = \kappa$. Indeed, if $M \in \underline{bi}(R)$, then $(j_\kappa^\star\hat{o})M = j_{\kappa,M}^{-1}(\hat{o}Q^{bi}(M)) = \operatorname{Ker} j_{\kappa,M} = \kappa M$. Now, certainly $\tau > \hat{o}$, hence $j_\kappa^\star\tau > j_\kappa^\star\hat{o} = \kappa$. ∎

IV.2.24. Proposition. Let Q be a prime ideal of R and assume that κ_{R-Q} has property (T) for R-bimodules. Take a prime ideal T of $R_Q = Q^{bi}_{R-Q}(R)$ and call its inverse image $j_Q^{-1}(T) = P$, where $j_Q : R \to R_Q$ is the localizing morphism. Then $j_Q^{\star}\kappa_{R_Q-T} = \kappa_{R-P}$.

Proof. Let us first introduce some notation. If M is an R-bimodule, then M_Q will denote the R-bimodule of quotients of M at κ_{R-Q}, while $j_{Q,M} : M \to M_Q$ denotes the localizing morphism.

Since T is an ideal of R_Q, we have $T = R_Q j_Q(j_Q^{-1}(T)) = R_Q j_Q(P) = P_Q$. Now, for each R-bimodule M, by definition

$$(j_Q^{\star}\kappa_{R_Q-P_Q})M = j_{Q,M}^{-1}(\kappa_{R_Q-P_Q} M_Q)$$

as noted before. Hence $x \in (j_Q^{\star}\kappa_{R_Q-P_Q})M$ if and only if $j_{Q,M}(x) \in \kappa_{R_Q-P_Q} M_Q$ iff there exists an ideal $I \not\subset P_Q$ such that $Ij_{Q,M}(x) = o$.

Write J for the inverse image $j_Q^{-1}(I)$ of I, then $J \not\subset P$. Indeed, otherwise $J \subset P$ would imply $I = R_Q j_Q(j_Q^{-1}(I)) = R_Q j_Q(J) \subset R_Q j_Q(P) = P_Q$, a contradiction. As $j_Q(I) \subset J$, we get $o = j_Q(J) j_{Q,M}(x) = j_{Q,M}(Jx)$, i.e. $Jx \subset \sigma_{R-Q} M$. But as $P \subset Q$ we have $\kappa_{R-Q} < \kappa_{R-P}$; indeed, if $P \not\subset Q$, then $P \in \mathcal{L}(\kappa_{R-Q})$, i.e. $j_Q(P)R_Q = R_Q$ and the first member equals $T = P_Q$, which was supposed to be a proper ideal. Hence we derive a contradiction. Thus $Jx \subset \kappa_{R-P} M$, hence $J\overline{x} = \overline{o} \in M/\kappa_{R-P} M$, i.e. $\overline{x} \in \kappa_{R-P}(M/\kappa_{R-P} M) = \overline{o}$. In other words : $x \in \kappa_{R-P} M$.

Conversely, assume $X \in \kappa_{R-P} M$, then there exists an ideal $J \not\subset P$ such that $Jx = o$, and so $J_Q j_Q(x) = R_Q j_Q(J) j_Q(x) = o$. But $J_Q \not\subset P_Q$, for otherwise $J \subset j_Q^{-1}(J_Q) \subset j_Q^{-1}(P_Q) = P$, a contradiction. Hence $j_Q(x) \in \kappa_{R_Q-P_Q} M$, i.e. $x \in j_Q^{-1}(\kappa_{R_Q-P_Q} M) = (j_Q^{\star}\kappa_{R_Q-P_Q})M$. ∎

IV.2.25. Lemma. Let $\varphi : R \to S$ be an extension which is epimorphic in the category of rings and flat for R-bimodules and let κ be a kernel functor in $\underline{\underline{bi}}(S)$. Denote $\varphi^{\star}\kappa$ by κ, then for each R-bimodule M the bimodules $Q^{bi}_{\kappa}(M)$ and ${}_R Q^{bi}_{\kappa}(S \otimes_R M)$ are canonically isomorphic.

<u>Proof</u>. First note that φ is a <u>central</u> extension because it is an epimorphism cf. Delale [46], hence for each S-bimodule N the R-bimodules ${}_RQ^{bi}_{\kappa'}(N)$ and $Q^{bi}_{\varphi_*\kappa'}({}_RN)$ are canonically isomorphic. But as $\varphi_*\kappa' = \kappa$ by (IV.2.19.), it suffices to apply this to $N = S \otimes_R M$ to finish the proof. ∎

<u>IV.2.26. Corollary</u>. Let $\varphi : R \to S$ be an extension which is epimorphic in the category of rings and flat for R-bimodules, and let κ be a kernel functor in $\underline{\underline{bi}}(S)$, then κ induces a perfect localization in $\underline{\underline{bi}}(S)$ if and only if $\varphi^*\kappa$ induces are in $\underline{\underline{bi}}(R)$.

<u>Proof</u>. Denote $\varphi^*\kappa = \kappa'$. Assume first that κ has property (T) for S-bimodules, and take M $\underline{\underline{bi}}(R)$, then

$$Q^{bi}_{\kappa'}(M) \cong Q^{bi}_{\kappa}(S \otimes_R M) \cong Q^{bi}_{\kappa}(S) \otimes_S (S \otimes_R M) \cong Q^{bi}_{\kappa}(S) \otimes_R M$$

$$Q^{bi}_{\kappa}(S \otimes_R M) \otimes_R M \cong Q^{bi}_{\kappa'}(R) \otimes_R M ,$$

hence κ' induces a perfect localization in $\underline{\underline{bi}}(R)$.

Conversely, assume that κ' has property (T) for R-bimodules, and take $N \in \underline{\underline{bi}}(S)$, then

$$Q^{bi}_{\kappa}(N) \cong Q^{bi}_{\kappa'}(K) \cong Q^{bi}_{\kappa}(R) \otimes_R N \cong Q^{bi}_{\kappa}(R) \otimes_R (S \otimes_S N)$$

$$\cong (Q^{bi}(R) \otimes_R S) \otimes_S N \cong Q^{bi}(S) \otimes_S N \cong Q^{bi}(S) \otimes_S N,$$

which shows that κ induces a perfect localization in $\underline{\underline{bi}}(S)$. ∎

<u>IV.2.27. Corollary</u>. Let Q be a prime ideal of a ring R such that κ_{R-Q} has property (T) for R-bimodules, then

1. If for all prime ideals P of R the symmetric kernel functor κ_{R-P} has property (T) for R-bimodules, then the same property holds for $R_Q = Q^{bi}_{R-Q}(R)$.
2. If for all prime ideals T of R_Q the symmetric kernel functor κ_{R_Q-T} has property (T) for R_Q-bimodules, then the same property holds for all prime

ideals $P \subset Q$ of R.

<u>Proof</u>. Let T be a prime ideal of R_Q and let $j_Q^{-1}(T) = P$ be its inverse image in R. Then by IV.2.24 we know that $j_Q^* \kappa_{R_Q-T} = \kappa_{R-P}$, which induces a perfect localization under the first assumption. The foregoing corollary then yields that κ_{R_Q-T} has the same property, which proves the assertion.

Conversely, assume that 2 holds, and take $P \subset Q$ in R. Then $P_Q \subset Q_Q$ is a proper prime ideal in R_Q. Indeed, if for some ideals I and J in R_Q we have $IJ \subset P_Q$, then $j_Q^{-1}(I) j_Q^{-1}(J) = j_Q^{-1}(IJ) \subset j^{-1}(P_Q) = P$, hence $j_Q^{-1}(I) \subset P$ or $j_Q^{-1}(J) \subset P$. Assume the former holds then $I = R_Q j_Q(j_Q^{-1}(I)) \subset R_Q j_Q(P) = P_Q$, which shows that P_Q is a prime ideal of R_Q. By assumption $\kappa_{R_Q-P_Q}$ induces a perfect localization, hence by the foregoing corollary, so does $j_Q^* \kappa_{R_Q-P_Q}$. But by IV.2.24 this is exactly κ_{R-P}. ∎

<u>IV.2.28. Corollary</u>. If $\varphi : R \to S$ is a torsion-free extension with the property that for all prime ideals P (resp. maximal ideals M) of R the kernel functor κ_{R-P} (resp. κ_{R-M}) induces a perfect localization in $\underline{\underline{bi}}(R)$, then the induced morphisms $\varphi_P : R_P \to S_P$ (resp. $\varphi_M : R_M \to S_M$) are torsion-free extensions having the same property. ∎

The property stated in the last two results holds for Azumaya algebras or more generally, for Zariski central rings, many examples of which arise when one considers rings of twisted polynomials or power series, see section II.3.

Let us finish this section with some notes on the Amitsur-complex.
Assume that $\varphi : R \to S$ is a ring morphism, then the Amitsur-complex of φ is the cosimplicial object $S = (S^n, d^i, s^i)$ in the category of twosided R-modules

$$S \rightrightarrows S \otimes_R S \mathrel{\substack{\longrightarrow\\\longrightarrow\\\longrightarrow}} S \otimes_R S \otimes_R S \mathrel{\substack{\longrightarrow\\\longrightarrow\\\longrightarrow\\\longrightarrow}} \cdots$$

so that $S^n = S \otimes_R \ldots \otimes_R S$ (n-fold tensor product), and where

$$d^i(b_o\otimes\ldots\otimes b_n) = b_o\otimes\ldots\otimes 1\otimes\ldots\otimes b_n \in S^{n+1} \quad \text{(1 in the i th position)}$$

$$s^i(b_o\otimes\ldots\otimes b_n) = b_o\otimes\ldots\otimes b_i b_{i+1}\otimes\ldots\otimes b_n \in S^{n-1}\ .$$

The complex is augmented by the map $\varphi : R \to S$. If S is an R-algebra, then so are the objects S^n, and the face maps are R-extensions. The degeneracies are not ring morphisms in general. We have the following :

<u>IV.2.29. Proposition</u>. Let $\varphi : R \to S$ be an extension which is faithfully flat for R-bimodules. Then the Amitsur complex for φ is a resolution of R, i.e. the sequence

$$o \longrightarrow R \xrightarrow{\varphi} S \xrightarrow{\delta^1} S\otimes_R S \xrightarrow{\delta^2} \ldots$$

is exact, where $\delta^n = \Sigma_{i=o}^n (-)^i d^i$. In particular, φ maps R onto the subring of S which consists of elements b such that $b\otimes 1 = 1\otimes b$ in $S\otimes_R S$. More generally, if M is an R-bimodule (resp. if M is a left R-module and if φ is left faithfully flat), then the sequence

$$o \longrightarrow M \longrightarrow S\otimes_R M \longrightarrow S\otimes_R S\otimes_R M \longrightarrow \ldots$$

obtained from the above sequence by tensoring with M is exact.

<u>Proof</u>. It suffices to show that the sequence in question becomes exact after tensoring on the left with S, and the resulting complex is homotopically trivial, the homotopy being multiplication of the first two entries in a tensor, i.e. $h : S^{\otimes(n+2)} \to S^{\otimes(n+1)}$, defined by

$$b\otimes b_o\otimes\ldots\otimes b_n \longrightarrow bb_o\otimes\ldots\otimes b_n\ .$$

The case of a module may be treated in a similar way. ∎

This proposition is the basis for Grothendieck's theory of flat descent for modules, and his theory extends without difficulty to the case of left modules with respect to a left flat extension $\varphi : R \to S$.

IV.3. Bimodule Localization of P.I. Rings.

In this Section IV.3. kernel functors are idempotent. For the application we have in mind it is not restrictive to assume that the ring R is left Noetherian; but we stick to the first, weaker hypothesis.

IV.3.1. Lemma. If $P \in \operatorname{Spec} R$ then $\overline{\kappa}_{R-P} < \kappa_P$ in $\underline{\underline{2}}(R)$.

Proof. Since $\overline{\kappa}_{R-P}$ is induced in $\underline{\underline{2}}(R)$ by κ_{R-P} it is clear that, for any $M \in \underline{\underline{2}}(R)$, $m \in \overline{\kappa}_{R-P} M$ if there is an $s \in R-P$ such that $sRm = o$. If $m \notin \kappa_P M$ then there exists $f \in \operatorname{Hom}_{\underline{\underline{2}}(R)}(M, E^2(R/P))$ such that $f(m) \neq o$. Therefore $f(RmR) \neq o$ is a subobject of $E^2(R/P)$ in $\underline{\underline{2}}(R)$, hence $f(RmR) \cap R/P \neq o$. Select $r, r' \in R$ such that $o \neq f(rmr') \in R/P$. Then $o = f(sRrmr') = sRf(rmr')$. But $f(rmr') \neq o$ and $s \notin P$ contradicts $sRf(rmr') = o$.

IV.3.2. Theorem. If M is a κ_P-torsion free R-bimodule, then there is an isomorphism f_M: $Q_P^{bi}(M) \to Q_{R-P}^{bi}(M)$ in $\underline{\underline{2}}(R)$. Moreover if R is κ_P-torsion free then $Q_P^{bi}(R)$ and $Q_{R-P}^{bi}(R)$ are isomorphic rings. Upon identification of these, f_M becomes an isomorphism of $Q_{R-P}^{bi}(R)$-modules.

Proof. Since $\mathcal{L}^2(\kappa_P) = \mathcal{L}^2(\kappa_{R-P})$ it is clear that $M \in \underline{\underline{bi}}(R)$ is κ_P-torsion if and only if it is $\overline{\kappa}_{R-P}$-torsion. Being a bimodule, $Q_P^{bi}(M)/M$ is $\overline{\kappa}_{R-P}$-torsion whereas $\overline{\kappa}_{R-P} < \kappa_P$ entails that $Q_P^{bi}(M)$ is $\overline{\kappa}_{R-P}$-torsion free. The existence of the isomorphism f_M will follow immediately if $Q_P^{bi}(M)$ turns out to be faithfully R_{R-P}-injective. Therefore, consider the following exact diagram of R-bimodules :

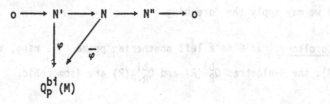

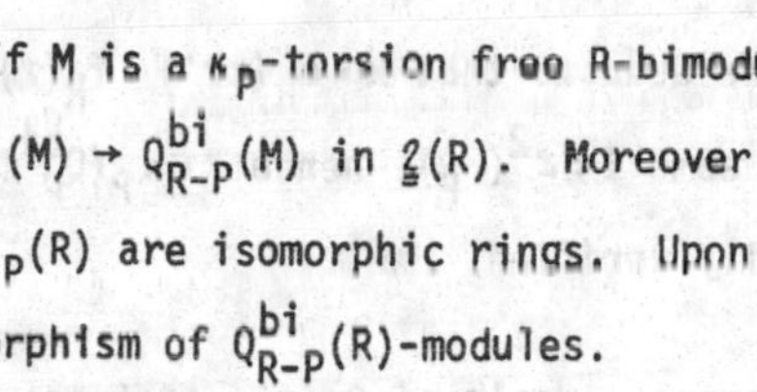

where N" is $\overline{\kappa}_{R-P}$-torsion, hence κ_P-torsion.

By the κ_P-injectivity of $Q_P^{bi}(M)$ in $\underline{\underline{bi}}(R)$, φ extends to a $\underline{\underline{2}}(R)$-morphism $\overline{\varphi}$ and thus

$Q_P^{bi}(M)$ is $\bar{\kappa}_{R-P}$-injective. Uniqueness arguments yield that $Q_P^{bi}(M)$ and $Q_{R-P}^{bi}(M)$ are canonically isomorphic. The case $M=R$ yields a $\underline{\underline{2}}(R)$-morphism $f_R : Q_{R-P}^{bi}(R) \to Q_P^{bi}(R)$. To prove that f_R is an isomorphism of rings as well, it suffices to check that $f_R(zz') = f_R(z)f_R(z')$ for <u>central</u> elements $z,z' \in Z_R(Q_{R-P}^{bi}(R))$. If z is such an element, then it is known that $\bar{z} = z \bmod R$ is a central element of the $\bar{\kappa}_{R-P}$-torsion bimodule $Q_{R-P}^{bi}(R)/R$, therefore $I\bar{z} = o$ for some $I \in \mathcal{L}^2(R-P)$. Thus $Iz \subset R$ for some $I \in \mathcal{L}^2(\kappa_{R-P})$. We have

$$If_R(zz') = f_R((Iz)z') = Iz\ f_R(z')$$

and

$$If_R(z)f_R(z') = f_R(Iz)f_R(z') = Izf_R(1)f_R(z') = Izf_R(z')\ .$$

Note that $f_R(1) = 1$, as f_R extends the identity of R to an isomorphism $Q_P^{bi}(R) \cong Q_{R-P}^{bi}(R)$ in $\underline{\underline{bi}}(R)$. From this one deduces that $x = f_R(zz') - f_R(z)f_R(z') \in Q_P^{bi}(R)$ is annihilated (on both sides) by some $I \in \mathcal{L}^2(\kappa_P)$, hence $x \in \kappa_P(Q_P^{bi}(R)) = o$. The remaining statement is now easily verified. ∎

<u>IV.3.3. Theorem</u>. Assume that proper ideals of R intersect the center of R non-trivially. If $P \in \mathrm{Spec}(R)$ is such that R is κ_{R-P} torsion free, then $Q_{R-P}^{bi}(R)$ and $Q_P^{bi}(R)$ are isomorphic rings.

<u>Proof</u>. Since $\kappa_P R$ is an ideal of R, it is nonzero if and only if it intersects the center of R nontrivially. If $z \in \kappa_P R$ is a central element, then $Iz = zI = o$ for some $I \in \mathcal{L}^2(\kappa_P) = \mathcal{L}^2(\sigma_{R-P})$, hence $z \in \kappa_{R-P} R = o$. This implies that R is κ_P-torsion free, and we may apply the foregoing. ∎

<u>IV.3.4. Corollary</u>. If R is a left noetherian prime P.I. ring, then for all $P \in \mathrm{Spec}(R)$, the R-algebras $Q_P^{bi}(R)$ and $Q_{R-P}^{bi}(R)$ are isomorphic.

<u>Proof</u>. This follows immediately from a result of Rowen's [151], which states that in a semiprime p.i. algebra each nonzero ideal intersects the center non trivially. That R is κ_{R-P}-torsion free for each $P \in \mathrm{Spec}(R)$ is obvious. ∎

Note that the same result holds for the enveloping algebra of a nilpotent Lie algebra.

<u>IV.3.5. Propositon.</u> If R is a left noetherian semiprime P.I. ring then $Q_P^{bi}(R)$ and $Q_{R-P}^{bi}(R)$ are isomorphic R-algebras for all $P \in \mathrm{Spec}(R)$.

<u>Proof</u>. If we show that $\kappa_P(R/\kappa_{R-P}R) = o$ and that $R/\kappa_{R-P} = R/\kappa_P R$ is semiprime, then (IV.3.3.) may be applied again, using Rowen's result.
Let $\pi : R \to R/\kappa_{R-P}$ be the canonical projection and let κ be a kernel functor in $\underline{\underline{bi}}(R)$. The torsion theory $\pi_\star\kappa$ in $\underline{\underline{bi}}(R/\kappa_{R-P}R)$ associated to κ in the usual way has the property that :

$$Q_{\pi_\star\kappa}^{bi}(R/\kappa_{R-P}R) \cong Q_\kappa^{bi}(R/\kappa_{R-P}R)$$

as R-algebras. Applying this to $\kappa = \kappa_P$ and $\kappa = \overline{\kappa}_{R-P}$ shows that (IV.3.3.) yields the result. We still have to prove that $\kappa_P(R/\kappa_{R-P}R) = o$ and that $\kappa_{R-P}R$ is a semiprime ideal.
Since R is left noetherian, there is an ideal $I \in \mathcal{L}^2(\kappa_{R-P})$ such that $I\kappa_{R-P}R = o$, hence $\kappa_I R = \kappa_{R-P}R$, where κ_I is the symmetric kernel functor which is given by $\mathcal{L}^2(I) = \{J$ ideal of R; rad $J \supset I\}$.
Since R is semiprime, we obtain that $\cap\mathrm{Spec}(R) = o$, hence $\mathrm{rad}(I) \cap \cap\{P \in \mathrm{Spec}(R); P \not\supset I\} = o$ and therefore $\cap\{P; P \not\supset I\} = L$ is κ_I-torsion, since $(\mathrm{rad}(I)) \cap L \subset \mathrm{rad}(I \cap L) = o$.
Conversely, if $x \in \kappa_I R$, then $I^n x = o$ for some $n \in \mathbb{N}$, therefore $x \in P$ if $P \not\supset I$; i.e. $x \in L$. Thus $\kappa_{R-P}R = \kappa_I R = L$ is a semiprime ideal. Now $\kappa_P(R/\kappa_{R-P}R)$ is an ideal of a semiprime P.I. ring hence it is nonzero if and only if it intersects the center nontrivially. However a central element $z \in \kappa_P(R/\kappa_{R-P}R)$ is annihilated by some $I \in \mathcal{L}^2(\kappa_P) = \mathcal{L}^2(R-P)$, hence $z \in \kappa_{R-P}(R/\kappa_{R-P}R) = o$ follows.
So $\kappa_P(R/\kappa_{R-P}R) = o$ and $\kappa_P R = \kappa_{R-P}R$. ∎

Let R be a ring, P a prime ideal of R and $\pi : R \to R/P$ the canonical projection. It is then fairly easy to show

<u>IV.3.6. Proposition.</u> The following statements are equivalent

1. there is an extension $R/P \hookrightarrow S$, where S is simple;
2. there is an extension $R/P \rightarrow S$ which localizes $\mathcal{L}^2(\kappa_P)$, but not the ideal P;
3. the composed morphism $R \rightarrow R/P \hookrightarrow Q_P^{bi}(R/P)$ localizes $\mathcal{L}^2(\kappa_P)$;
4. $Q_P^{bi}(R/P)$ is simple;
5. $\pi_\star \kappa_P$ has property (T) for R/P-bimodules.

<u>Proof.</u> The demonstration runs along the lines of a similar result in [64] and may be found in [46] . ■

<u>Remark.</u> In [45] Delale states that the assertions in the foregoing theorem are also equivalent to

6. κ_P has property (T) for R-bimodules,

which is definitely untrue. We will give a counterexample below.

<u>IV.3.7. Corollary.</u> Let P be a prime ideal in Rwith the property that κ_P induces a perfect localization in $\underline{\underline{bi}}(R)$, then $Q_P^{bi}(R)$ is a local ring with unique maximal ideal M such that $j_P^{-1}(M) = P$, where $j_P : R \rightarrow Q_P^{bi}(R)$ denotes the localizing morphism. Moreover, $Q_P^{bi}(R)/M$ is then canonically isomorphic to the simple ring $Q_P^{bi}(R/P)$.

<u>Proof.</u> Let $M = j_P(P)Q_P^{bi}(R)$, then M is a two sided ideal of $Q_P^{bi}(R)$, since j_P is an extension. Moreover, $j_P^{-1}(j_P(P)Q_P^{bi}(R)) = P$; indeed, by definition $\kappa_P(R/P) = o$, so we get a commutative diagram

$$\begin{array}{ccc} R/P & \xrightarrow{\; j_P \otimes R/P \;} & Q_P^{bi}(R) \underset{R}{\otimes} (R/P) \\ & \searrow \quad \swarrow {\scriptstyle \pi} & \\ & Q_P^{bi}(R/P) & \end{array}$$

hence $j_P \otimes R/P : R/P \rightarrow Q_P^{bi}(R) \underset{R}{\otimes} (R/P) = Q_P^{bi}(R)/M$ is injective, which proves the assertion. Next, if I is an ideal of $Q_P^{bi}(R)$, then $I = j_P(j_P^{-1}(I))Q_P^{bi}(R)$, hence $I \subset M$ implies $j_P^{-1}(j_P(P)Q_P^{bi}(R)) = P$, i.e. $j_P^{-1}(I) \in \mathcal{L}^2(\kappa_P)$. But then $Q_P^{bi}(R) =$

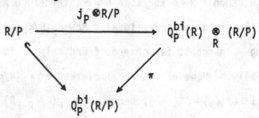

$j_P(j_P^{-1}(I))Q_P^{bi}(R) = I$. The other assertions now follow trivially. ∎

IV.3.8. Proposition. Let R be a prime ring with center C. The following statements are equivalent :

1. $Q_o^{bi}(R)$ is simple and $Z(Q_o^{bi}(R)) = Q(C)$;
2. each nonzero (two sided) ideal of R intersects C non-trivially;
3. $j : R \to Q(C) \otimes_C R$ localizes $\mathcal{L}(\kappa_o)$;
4. $\theta : Q(C) \otimes_C R \to Q_o^{bi}(R)$ is an isomorphism and $Q_o^{bi}(R)$ is simple.

Proof. (1) ⇒ (2) Let I be a nonzero ideal of R, then by (1) we can write $1 = \Sigma i_\alpha z_\alpha$, where $i_\alpha \in I$ and $z_\alpha \in Z(Q_o^{bi}(R)) = Q(C)$. Choose $o \neq c \in C$ such that for each α we have $z_\alpha c \in C$, then $o \neq c = \Sigma\, i_\alpha(z_\alpha)\in I \cap C$.

(2) ⇒ (3) Take $I \in \mathcal{L}^2(\kappa_o)$ and choose $c \in C \cap I$, $c \neq o$, then c is invertible in $Q(C) \otimes_C R$.

(3) ⇒ (4) Obviously $Q_o^{bi}(R)$ is simple and there exists a universal morphism $k : Q_o^{bi}(R) \to Q(C) \otimes_C R$, which is an inverse for θ.

(4) ⇒ (1) Obvious, since $Z(Q(C) \otimes_C R) = Q(C)$. ∎

IV.3.9. Corollary. The following statements are equivalent :

1. $Q_P^{bi}(R/P)$ is simple and $Z(Q_P^{bi}(R/P)) = Q(Z(R/P))$;
2. each nonzero ideal of R/P intersects Z(R/P) non-trivially.

Proof. In view of the foregoing and the well-known base change result with respect to central extensions for localizations, it clearly suffices to verify the obvious fact that $\pi_* \kappa_P = \kappa_o$ in bi(R/P), where $\pi : R \to R/P$ denotes the canonical projection.

If a prime ideal P of R satisfies one (and hence all) of the conditions in IV.3.6., then it is said to be a point of R. By the foregoing corollary, we derive in particular that for a P.I. algebra every prime ideal is a point. In particular, we derive from Posner's theorem

IV.3.10. Theorem. If R is a P.I. ring, then

1. each prime ideal of R is a point;
2. for every prime ideal P of R the ring $Q_P^{bi}(R/P)$ is a central simple algebra over its center $Q(Z(R/P))$;
3. $Q_P^{bi}(R/P)$ coincides with the classical ring of quotients of R/P.

Proof. (1) and (2) follow from (3). To prove (3), first note that if we denote by $\pi : R \to R/P$ the canonical projection, then $\pi_* \kappa_P = \kappa_o$ implies that

$$Q_P^{bi}(R/P) = Q_o^{bi}(\overline{R}),$$

where $\overline{R} = R/P$. So, it suffices to check whether for a prime P.I. ring R the ring $Q_o^{bi}(R)$ coincides with the classical ring of quotients Q(R) of R. This being the case we have that $Q_o^{bi}(\overline{R}) = Q_o(R)$ and from both Posner's theorem and the well-known fact that $Q_o(R) = Q(R)$. ∎

Assume that R is a prime noetherian p.i. ring and let P be a prime ideal of R. Suppose, as Delale states in [45], that IV.3.8.1 implies IV.3.8.6., then in particular the foregoing implies that κ_P has property (T) for R-bimodules. Hence $Q_P^{bi}(R)$ is a left noetherian local ring with unique maximal ideal $j_P(P)Q_P^{bi}(R) = M$. Clearly $Q_P^{bi}(R)$ is a prime P.I. ring too, (embed $Q_P^{bi}(R)$ into $Q_o^{bi}(R) = Q_{cl}(R)$, whereas the fact that $Q_P^{bi}(R)$ is prime follows from the fact that κ_P has property (T) for R-bimodules). Finally, let us note that the localizing morphism $j_P : R \to Q_P^{bi}(R)$ is injective which may be seen by using IV.3.3., or directly as follows : by definition Ker $j_P = \kappa_P R$ and if $o \neq r \in \kappa_P R \cap Z(R)$, which is nonzero if and only if $\kappa_P R$ is nonzero, then there exists $I \in \mathcal{L}^2(\kappa_P)$ such that $Ir = o$, but then $I = o$ or $r = o$, since R is assumed to be prime, a contradiction. Actually, we now just proved

IV.3.11. Proposition. Let R be a left noetherian prime P.I. ring and P a prime ideal of R such that κ_P has property (T) for R-bimodules. Then $Q_P^{bi}(R)$ is a left Noetherian local prime P.I. ring with maximal ideal $PQ_P^{bi}(R)$. ∎

To continue, since $Q_P^{bi}(R)$ is a P.I. ring, the unique maximal ideal of $Q_P^{bi}(R)$ is the Jacobson radical of $Q_P^{bi}(R)$, which under the noetherian prime P.I. hypothesis is known to have the following property, by a well known result of Jategaonkar's:

$$\bigcap_{n=o}^{\infty} J(Q_P^{bi}(R))^n = o.$$

Now, $P \subset PQ_P^{bi}(R)$, hence $\bigcap_{n=o}^{\infty} P^n = o$. So Delale's result would imply for each prime ideal P of a prime left noetherian P. I. ring R to have the property $\cap P^n = o$. Take a commutative discrete valuation ring D with maximal ideal M. The ring

$$R = \begin{pmatrix} D & D \\ M & D \end{pmatrix}$$

is a subring of a matrix ring hence is a P.I. algebra. Obviously R is a left noetherian prime ring. The ideal

$$P = \begin{pmatrix} M & D \\ M & D \end{pmatrix}$$

is a maximal hence prime ideal of R. Since it is idempotent, this shows that Delale's result leads to a contradiction.

Note that many results in this section remain valid for enveloping algebras of finite dimensional solvable Lie-algebras over a field of characteristic o. Details are left to the reader.

V. STRUCTURE SHEAVES AND SCHEMES.

This chapter is devoted to constructing structure sheaves on Spec (R) and Proj(R) by making use of the localization techniques available to us now. Two types of sheaves are of interest here: the "module"-type and the "bimodule"-type corresponding to whether the localization used is localization in R-mod or the relative localization in $\underline{\underline{bi}}(R)$. It should be clear from attentive reading of the technical section on "bimodule" localization of P.I. rings that the sheaf of "bimodule"-type is most likely to behave in a nice functorial way with respect to ring extensions. On the other hand the sheaf of "module"-type contains more information than the former, therefore we consider both sheaves equally interesting.

V.I. Presheaves, Sheaves and Localization.

First some generalities. Let X be a topological space and let $\underline{C}$ be an arbitrary category. Let Open(X) denote the category of open sets of X (with canonical inclusions for the morphisms). A <u>presheaf</u> over X with values in $\underline{C}$ is an object of the functor category $\underline{\text{Hom}}(\text{Open}(X)^{\text{opp}},\underline{C})$.
Thus a presheaf P over X with values in $\underline{C}$ assigns to each open subset U of X an object P(U) of $\underline{C}$ and to each inclusion of open subsets $V \subset U$ a morphism $P^U_V : P(U) \to P(V)$ satisfying : 1. For all $U \in \text{Open}(X)$, $P^U_U = 1_{P(U)}$, 2. For each triple $W \subset V \subset U$ in Open(X), $P^V_W \circ P^U_V = P^U_W$. A <u>morphism of presheaves</u> $P \to P'$ over X is thus given by morphisms $f(U) : P(U) \to P'(U)$ for all $U \in \text{Open}(X)$ such that for each couple $U \subset V$ the following diagram is commutative in $\underline{C}$.

$$\begin{array}{ccc} P(V) & \xrightarrow{f(V)} & P'(V) \\ {\scriptstyle P^V_U}\downarrow & & \downarrow{\scriptstyle P'^V_U} \\ P(U) & \xrightarrow[f(U)]{} & P'(U) \end{array}$$

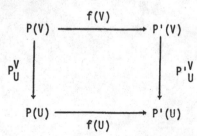

If $U \in \mathrm{Open}(X)$, let $\mathrm{Cov}_X(U)$ be the directed set of all families $\hat{U} = \{U_i, i \in I\}$ with $U_i \in \mathrm{Open}(X)$ for each $i \in I$ and $U = \bigcup_{i \in I} U_i$.
Depending on whether $\underline{C}$ is Ens, R-mod, Rings,... we speak of a presheaf, a presheaf of modules, a presheaf of rings,... .
If $\hat{U} \in \mathrm{Cov}_X(U)$, $U \in \mathrm{Open}(X)$ and P a presheaf over X then write $p_i : \prod_{i \in I} P(U_i) \to P(U_i)$ for the ith canonical projection. We have a morphism :

$$j : P(U) \to \prod_{i \in I} P(U_i)$$

such that $p_i j = P^U_{U_i}$ and we also have morphisms :

$$p, q : \prod_{i \in I} P(U_i) \rightrightarrows \coprod_{(j,k) \in I \times I} P(U_j \cap U_k)$$

where the (j,k)-component of p is $P^{U_j}_{U_j \cap U_k} p_j$ and the (j,k)-component of q is $P^{U_k}_{U_j \cap U_k} p_k$.

Thus we have obtained a diagram :

$$(*) : P(U) \xrightarrow{j} \prod_{i \in I} P(U_i) \underset{q}{\overset{p}{\rightrightarrows}} \coprod_{(j,k) \in I \times I} P(U_j \cap U_k) \quad .$$

The presheaf P over X is said to be a sheaf over X if and only if (*) is an equalizer diagram. Now we have defined the category of presheaves over X with values in $\underline{C}$, $P(X,\underline{C})$, and a full subcategory of it, $S(X,\underline{C})$ consisting of all sheaves in $P(X,\underline{C})$. Objects of $S(X,\underline{C})$ are called <u>sheaves over X with values in $\underline{C}$</u>.

<u>V.1.1. Theorem</u>. If $\underline{C}$ is a Grothendieck category then :

1. $P(X,\underline{C})$ is a Grothendieck category.
2. $S(X,\underline{C})$ is a Giraud subcategory of $P(X,\underline{C})$
3. If we agree to call a presheaf <u>separated</u> when it is a subobject in $P(X,\underline{C})$ of a sheaf in $S(X,\underline{C})$, then P is separated if and only if j is a monomorphism.

Let us recall some further basic facts in the terminology of Chapter I. The construction of the reflector $\underline{\underline{a}} : P(X,\underline{C}) \to S(X,\underline{C})$ is executed in two steps. First define a functor $L : P(X,\underline{C}) \to P(X,\underline{C})$ as follows. If $U \in \mathrm{Open}(X)$ then $\mathrm{Cov}_X(U)$ may be given the structure of a category i.e. if $\mathcal{U} = \{U_i, i \in I\}$ and $\mathcal{V} = \{V_j, j \in J\}$ are in $\mathrm{Cov}_X(U)$ then a morphism $\mathcal{U} \to \mathcal{V}$ is given by a map $\varepsilon : I \to J$ such that $U_i \subset V_{\varepsilon(i)}$ for all $i \in I$. Let $P \in P(X,\underline{C})$ and define $[P,U]$ for each $U \in \mathrm{Open}(X)$ by its action on $\mathcal{U} \in \mathrm{Cov}_X(U)$ as follows :

$$[P,U](\mathcal{U}) = \mathrm{Ker}\Big(\prod_{i\in I} P(U_i) \rightrightarrows \coprod_{(j,k)\in I\times I} P(U_j \cap U_k)\Big) .$$

Since $[P,U] : \mathrm{Cov}_X(U) \to \underline{C}$ is a contravariant functor we may define $LP \in P(X,\underline{C})$ as a covariant functor :

$$LP : \mathrm{Open}(X)^{\mathrm{opp}} \to \mathrm{Ens} : U \mapsto \varinjlim_{\mathcal{U} \in \mathrm{Cov}_X(U)} [P,U](\mathcal{U})$$

i.e. $LP(U) = \varinjlim_{\mathcal{U} \in \mathrm{Cov}_X(U)} \varprojlim_{V \in \mathcal{U}} P(V)$.

Hence if $\underline{C}$ admits $\varinjlim$ and $\varprojlim$ then the assignment $P \to LP$ yields a left exact endofunctor of $Q(X,\underline{C})$.

<u>V.1.2. Theorem</u>. Let $F(X,\underline{C})$ be the class of separated presheaves in $P(X,\underline{C})$, then the following properties hold :

1. If $P \in F(X,\underline{C})$ then the canonical $P \to LP$ is monomorphic and $LP \in S(X,\underline{C})$.
2. If $P \in P(X,\underline{C})$ then $LP \in F(X,\underline{C})$.
3. If $P \in S(X,\underline{C})$ then $LP \simeq P$ and vice versa.

Next, define $\underline{\underline{a}}$ by $i\underline{\underline{a}} = LL$ where $i : S(X,\underline{C}) \to P(X,\underline{C})$ is the canonical inclusion. It is easy enough to prove that $\underline{\underline{a}}$ is a left adjoint of i and then we may substitute $P(X,\underline{C})$ and $S(X,\underline{C})$ for P and S resp. in I.3. We say that $LLP = i\underline{\underline{a}}P$ is the sheaf associated to P. In order to clarify this abstract nonsense let us present these concepts in a more "visual" way. First, the definition of a sheaf may be rephrased in a "down to earth" way as follows : a presheaf P is a sheaf if and only if the following properties hold : Sh.1. Let U be an open subset of X which is covered by open sets U_i, $i \in I$, let f and g be in P(U) such that for all $i \in I : P^U_{U_i}(f) = P^U_{U_i}(g)$, then f = g i.e. elements of P(U) are recognizable on a covering of U, (this states exactly that P is separated). Sh.2. Let U be an open subset of X covered by $\{U_i, i \in I\}$ and let there be given elements $f_i \in P(U_i)$ for all $i \in I$, such that for each pair $(i,j) \in I \times I : P^{U_i}_{U_i \cap U_j}(f_i) = P^{U_j}_{U_i \cap U_j}(f_j)$ then there is an $f \in P(U)$ such that $f_i = P^U_{U_i}(f)$ for all $i \in I$, i.e. the "pieces" of f on a covering of U may be glued together if they coincide on intersections.

A sheaf-space over X is given by a triple (E,X,p) where E and X are topological spaces and $p : E \to X$ is a local homeomorphism. If Y is a subset of X then any continuous map $s : Y \to E$ such that $p \circ s = 1_Y$ is said to be a section (of the sheaf-space) over Y. The set of sections of (E,X,p) over Y is denoted $\Gamma(Y,E)$ and the assignment $U \to \Gamma(U,E)$ for $U \in \text{Open}(X)$ is easily seen to define a presheaf. This presheaf is actually a sheaf and it is equally easy to verify that $p^{-1}(x)$ may be identified with $\varinjlim_{U \ni x} \Gamma(U,E)$ for each $x \in X$.

More generally, if P is a presheaf over X then we define the stalk of P at $x \in X$ to be the object P_x of $\underline{C}$ given by $P_x = \varinjlim_{x \in U} P(U)$.

From the foregoing it follows that the stalk in $x \in X$ of the (pre)sheaf $\Gamma(-,E) : \text{Open}(X)^{opp} \to \text{Ens}$ is exactly the fibre $E(x) = \varinjlim_{x \in U} P(U)$

Let us show that all sheaves arise as a sheaf of sections of some sheaf-space in the above way. Let P be a presheaf over X. For $x \in X$ define $\tilde{P}(x) = \varinjlim_{x \in U} P(U)$ where U runs through all neighbourhoods of x in X.

Put $\tilde{P}$ equal to the disjoint union of the $\tilde{P}(x)$, $x \in X$, and let $p : \tilde{P} \to X$ be the mapping given by sending $\tilde{P}(x)$ to x. If $x \in U$ then there exists a canonical morphism $P(U) \to \tilde{P}(x) : s \mapsto \tilde{s}(x)$. In this way we associate to each $s \in P(U)$ a morphism $\tilde{s} : U \to \tilde{P} : x \mapsto \tilde{s}(x)$, which satisfies $p\tilde{s} = 1_U$. Clearly the restriction morphisms

behave as they should. Endow $\widetilde{P}$ with the coarsest topology which makes all maps $\widetilde{s}$ ($s \in P(U), U \in \mathrm{Open}(X)$) continuous then $(\widetilde{P},X,p)$ becomes a sheaf-space, called the sheaf-space associated to P. Furthermore, we obtain a presheaf morphism $P \to \Gamma(-,\widetilde{P})$ given on $U \in \mathrm{Open}(X)$ by $P(U) \to \widetilde{P}(U) = \Gamma(U,\widetilde{P}) : s \longmapsto \widetilde{s}$.
One easily verifies that this morphism is injective, resp. bijective if Sh.1., resp. Sh.1 and Sh.2, hold. We say that sheaf-spaces (E,p,X) and (E',p',X) are isomorphic if there is a homeomorphism $\varphi : E \to E'$ such that $p' \circ \varphi = p$, so:

V.1.3. Theorem. Any sheaf (of sets) over X is isomorphic to the sheaf of sections of a sheaf-space associated to it. The latter sheaf-space is uniquely determined (up to isomorphism) by the given sheaf.

A word on base change. Consider a continuous map $f : X \to Y$. To a presheaf $P \in \mathcal{P}(X,\underline{C})$ we associate a presheaf $f_* P \in \mathcal{P}(Y,\underline{C})$ by putting $f_* P(V) = P(f^{-1}(V))$ for any $V \in \mathrm{Open}(Y)$. This construction yields a functor $f_* : \mathcal{P}(X,\underline{C}) \to \mathcal{P}(Y,\underline{C})$ which restricts to a functor $f_* : \mathcal{S}(X,\underline{C}) \to \mathcal{S}(Y,\underline{C})$. Note that, whereas f_* is exact in $\mathcal{P}(X,\underline{C})$, it is in general only left exact in $\mathcal{S}(X,\underline{C})$.
Consider another continuous map $g : Y \to Z$, then $(gf)_* = g_* f_*$. The functor f_* : $\mathcal{P}(X,\underline{C}) \to \mathcal{P}(Y,\underline{C})$ has a left adjoint $f^o : \mathcal{P}(Y,\underline{C}) \to \mathcal{P}(X,\underline{C})$ which is given by

$$(f^o P)(U) = \varinjlim_{f(U) \subset V} P(V)$$

for $U \in \mathrm{Open}(X)$, $V \in \mathrm{Open}(Y)$. If we write i_Y for the canonical inclusion $\mathcal{S}(Y,\underline{C}) \to \mathcal{P}(Y,\underline{C})$ and $\underline{\underline{a}}_X : \mathcal{P}(X,\underline{C}) \to \mathcal{S}(X,\underline{C})$ for the reflector with respect to $\mathcal{S}(X,\underline{C})$ then $f^* = \underline{\underline{a}}_X f^o i_y$ is a left adjoint for f_* on sheaves. If $g : Y \to Z$ is continuous then $(gf)^* = f^* g^*$.
We say that f is relative if the topology on X is initial for f.

V.1.4. Lemma. If $f : X \to Y$ is a relative map then f_* is full. If $h : P \to Q$ is a presheaf morphism in $\mathcal{P}(X,\underline{C})$ such that $f_*(h)$ is an isomorphism in $\mathcal{P}(Y,\underline{C})$ then h is an isomorphism in $\mathcal{P}(X,\underline{C})$.

V.1.5. Corollary. If f is relative and $S \in \mathcal{S}(X,\underline{C})$ then there is a sheaf $S \in \mathcal{S}(Y,\underline{C})$ such that $f^* S' = S$. In particular if f is the inclusion of an open subset U of

X in X and if $S \in S(X,\underline{C})$ then f^*S is called the sheaf over U induced by S and we denote it by $S|U$.

Let us mention without much detail some results stemming from [188]; as a matter of fact we merely mention the point of view behind most results obtained in [188] and refer to loc. cit. for details, examples and applications.

Let X be some fixed topological space and let R be a (pre) sheaf of rings over X. Then, $(\pi(R))$, $\sigma(R)$ will be the category of left (pre-) R-Modules where a (pre) sheaf of abelian groups M is said to be a left (pre) R-Module if for each $U \in \text{Open}(X)$, $M(U)$ has an $R(U)$-module structure which is compatible with restriction morphisms. It is easily seen that $\pi(R)$ is a Grothendieck category containing $\sigma(R)$ as a strict Giraud subcategory. This allows us to apply the machinery of localization to the categories $\pi(R)$ and $\sigma(R)$ and in fact all constructions of ringed spaces in algebra known to the authors (Spec, Gam_k cf. [42], Prim_k, [171], R-Sp in [65], Prim as in [171]) are examples of localizations within the category of presheaves. Many interesting examples are obtained by "gluing together" kernel functors $\kappa(U)$ on $R(U)$-mod for $U \in \text{Open}(X)$, these are the so-called local kernel functors on $\pi(R)$, cf. [188]. An important remark is that $\underline{\underline{a}}$ itself is in fact a localization functor in $\pi(R)$ with quotient category $\sigma(R)$. So in order to verify whether a certain localization Q_κ in $\pi(R)$ takes sheaves to sheaves one is lead to compatibility conditions of the type considered in chapter I between $\underline{\underline{a}}$ and Q_κ. Of course, viewing the construction of a structure sheaf, Spec for example, as the construction of a localization in $\pi(R)$, these compatibility conditions yield a criterion for finding whether the usual presheaf on the prime spectrum of a ring is already a sheaf. Another feature, also very evident in algebraic geometry, is the technique of representing rings as rings of sections of sheaves. This can be carried out in a formal way by considering a relative localization in $\pi(R)$ with respect to the subcategory of flabby presheaves (presheaves such that the restriction morphisms are surjective), which yields the "pointwise localization" introduced in [175] and this in turn can be related to the sectional representation theory of J. Dauns, K. Hofmann a.o. In view of the bimodule constructions we introduced over P.I. rings, there is still another relative localization of interest, namely the relative localization in $\pi(R)$ with respect to the R-Bimodules. However we shall treat the sheaves of bimodules over the spectra of P.I. rings in a more constructive way and thus we leave the abstract theory for what it is worth.

After these generalities, we return now to the construction of structure sheaves, first over the prime ideal spectrum.

Throughout R is a left Noetherian ring with unit. Put X = Spec(R) and for each ideal I of R put $X(I) = \{P \in X, P \not\supset I\}$. The X(I) are the open sets of the Zariski topology on X. A $P \in X$ is closed if and only if P is a maximal ideal of R. A subset S of X is said to be <u>irreducible</u> if it is not the union of closed sets properly contained in S. A <u>generic point</u> for an irreducible set S is a point $P \in X$ such that $V(P) = S$, where $V(I) = X - X(I)$ for each ideal I. We know that the irreducible closed subsets of X are all of the form V(P) for some prime ideal P of R, and then P is the unique generic point for V(P). Obviously, X is covered by some open sets $X(I_\alpha)$, $\alpha \in A$, if and only if $1 \in \sum_\alpha I_\alpha$, hence X is compact but not necessarily Hausdorff. Moreover, since R is left Noetherian every open set X(I) is compact too.

On the other hand, to each ideal I of R there corresponds a kernel functor κ_I as in I.4.17. given by its filter $\mathcal{L}(I) = \{$L left ideal of R containing an ideal J such that $\mathrm{rad}(J) \supset I\}$.

Let $Q_I(R)$ be the ring of quotients of R at κ_I.

<u>V.1.6. Lemma</u>. Associating $Q_I(R)$ to any open subset X(I) of X = Spec(R) defines a presheaf of rings on X. The same construction applied to any left R-module M yields a presheaf of left modules over this presheaf of rings.

<u>Proof</u>. cf.[120], [174] or [168]. Note that both X(I) and κ_I depend on I up to taking the radical.

Note also that $I \not\subset \text{rad } o$ is equivalent to $X(I) \neq \phi$.

The restriction morphisms of the above presheaves may be explicitely defined as follows. Let $X(J) \subset X(I)$, then $\mathcal{L}(I) \subset \mathcal{L}(J)$ and $\kappa_I < \kappa_J$ follows. Consider $M \in R\text{-mod}$ and let $j_I = j_{I,M}$ be the canonical localization morphism $M \to Q_I(M)$.
The canonical projection $\pi : j_I(M) = M/\kappa_I(M) \to M/\kappa_J(M) = j_J(M)$ yields a commutative diagram in R-mod :

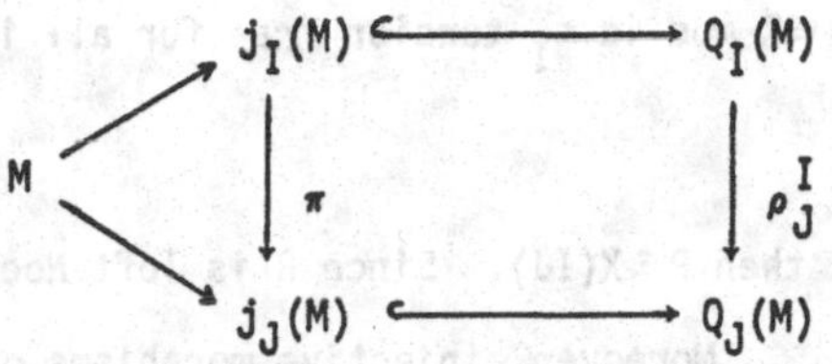

One easily checks that the maps ρ^I_J satisfy the conditions for being restriction homomorphisms.

For any $M \in R\text{-mod}$ let $\underline{Q}^o_M$ be the presheaf just constructed.

<u>V.1.7. Proposition.</u> For any $M \in R\text{-mod}$, the presheaf $\underline{Q}_M$ is separated.

<u>Proof</u>. cf. [174]. In general one considers $\underline{Q}_M = \underline{\underline{a}}\, \underline{Q}_M$ as the structure sheaf.

<u>V.1.8. Corollary.</u> If R is left Noetherian and $M \in R\text{-mod}$ such that $\kappa_I(M) = o$ for every ideal I of R then $\underline{Q}_M$ is actually a sheaf i.e. $\underline{Q}_M = \underline{\underline{a}}\, \underline{Q}_M$.

<u>V.1.9. Note.</u> If R is a left Noetherian prime ring then $\underline{Q}_R$ is a sheaf.

<u>V.1.10. Proposition</u>. For any $M \in R\text{-mod}$, $\underline{Q}_M(X) = M$.

<u>Proof</u>. To the open set X of X we have associated κ_R. But since any $M \in R\text{-mod}$ is κ_R-torsion free, the proposition follows.

<u>V.1.11. Corollary</u>. If M is κ_I-torsion free for all ideals I of R then $\underline{Q}_M(X) = M$.
In particular if R is a left Noetherian prime ring then $\underline{Q}_R(X) = R$.

V.1.12. Lemma. If $P \in \mathrm{Spec}(R)$ then $\kappa_{R-P} = V\{\kappa_I, P \in X(I)\}$.

Proof. If $P \in X(I)$ then $I \not\subset P$ i.e. $I \in \mathcal{L}(R-P)$ and thus $\kappa_I \leq \kappa_{R-P}$. Conversely, if $J \in \mathcal{L}(R-P)$ then $J \not\subset P$ and $P \in X(J)$. ■

V.1.13. Remark. Since R is left Noetherian we may apply Proposition I.4.12, i.e. $\kappa_I = \Lambda\{\kappa_{R-P}; P \in X(I)\}$.

V.1.14. Corollary. If $M \in R$-mod is κ_I-torsion free for all ideals I of R then $Q_{R-P}(M) = \varinjlim_{P \in X(I)} Q_I(M)$.

Proof. If $P \in X(I) \cap X(J)$ then $P \in X(IJ)$. Since R is left Noetherian it is easily verified that $\kappa_{IJ} = \kappa_I \vee \kappa_J$. Moreover, injective morphisms of $Q_I(M)$ resp. $Q_J(M)$ into some $Q_\kappa(M)$ extend to $Q_{IJ}(M)$ (the morphisms are supposed to coincide on M of course). In this case it is clear that the inductive limit of the system $\{Q_I(M), \rho^I_J;$ rad $0 \neq J \subset I, P \in X(J)\}$ may be obtained as the module of quotients with respect to the symmetric kernel functor $V\{\kappa_I, I \not\subset P\} = \kappa_{R-P}$. ■

The obvious geometric meaning of the module $Q_{R-P}(M)$ is that $\underline{O}_{M,P} = Q_{R-P}(M)$ is the stalk of $\underline{O}_M$. Note also that V.1.14. remains true for noetherian M, cf. V.3.9.

An open set X(I) of Spec(R) is said to be a T-set if κ_I is a T-functor, similarly, the stalk at $P \in \mathrm{Spec}(R)$ is a T-stalk if κ_{R-P} is a T-functor. A symmetric T-functor κ with the property that, for every ideal J of R, $Q_\kappa(R)J$ is an ideal of $Q_\kappa(R)$, then κ is said to be geometric. If κ_I is geometric then X(I) is called a geometric open set of X. Obviously the geometric open sets are generalizations of the basic open sets $X(Rf) = \{P \in X; f \notin P\}$ in the commutative case. If R is a Zariski central ring then Spec(R) has a basis of geometric open sets and each stalk is a geometric stalk, cf. Theorem II.1.16.1 and other results in II.1. The most familiar situation is where Spec(R) has a T-basis i.e. a basis for the Zariski topology consisting of T-sets (a geometric basis would be even better, but it seems that this happens only to rings which are very close to being Zariski central).

V.1.15. Proposition. Suppose that R is a left Noetherian prime ring such that Spec(R) has a T-basis. Let $M \in R$-mod be κ_I-torsion free for all ideals I of R, then we have : $\underline{O}_M = \underline{a}(\underline{O}_R \underset{R}{\otimes} M)$.

Proof. On a T-set X(I) we get $Q_I(M) = Q_I(R) \underset{R}{\otimes} M$. A section of $\underline{O}_R \underset{R}{\otimes} M$ over X(I) may be identified with a section of $\underline{O}_M$ over X(I). Since $\Gamma(X(I), \underline{O}_M) = Q_I(M)$ and since $\underline{O}_M$ and $\underline{O}_R \underset{R}{\otimes} M$ coincide on a basis for the topology on X it follows that $\underline{O}_M$ is isomorphic to the sheaf associated to $\underline{O}_R \underset{R}{\otimes} M$. ■

V.1.16. Proposition. Let κ be a geometric kernel functor, then there is one-to-one correspondence between proper prime ideals of $Q_\kappa(R)$ and κ-closed prime ideals of R i.e. prime ideals of R not in the filter $\mathcal{L}(\kappa)$.

Proof. Since κ is geometric $Q_\kappa(P)$ is an ideal of $Q_\kappa(R)$. Since κ is a t-functor $Q_\kappa(P) = Q_\kappa(R) j_\kappa(P) := P^e$.
If $I, J \subset P^e$ for some left ideals I,J of $Q_\kappa(R)$ then $j_\kappa^{-1}(I) j_\kappa^{-1}(J) \subset P$ i.e. $j_\kappa^{-1}(I)$ or $j_\kappa^{-1}(J) \subset P$.
However $I = Q_\kappa(R) j_\kappa^{-1}(I)$ or $J = Q_\kappa(R) j_\kappa^{-1}(J)$ by property T for κ, hence $I \subset P^e$ or $J \subset P^e$.■

V.1.17. Corollary. If R is a left Noetherian prime ring, then a geometric stalk is a "local" ring i.e. a left Noetherian prime ring with a unique maximal ideal. If R is also fully left bounded (e.g. P.I.) then the stalk is a local ring in the sense of [194] .

With the following lemma we return to a very general setting. Let $\tau > \sigma$ be symmetric kernel functors such that σ is a t-functor. It is clear that $\mathcal{L}(\tau^e) = \{Q_\sigma(R)I, I \in \mathcal{L}(\tau)\}$ is an idempotent filter for a kernel functor τ^e on $Q_\sigma(R)$-mod.

V.1.18. Lemma. With notations as above : $Q_{\tau^e}(Q_\sigma(R)) = Q_\tau(Q_\sigma(R)) = Q_\tau(R)$.

Proof. Via the canonical $R \to Q_\sigma(R)$ we consider $Q_\sigma(R)$-modules as R-modules.

Clearly, if $M \in Q_\sigma(R)$-mod then $x \in \tau^e M$ if and only if $Ix = o$ with $I \in \mathcal{L}(\tau)$ i.e. $x \in \tau M$. The left $Q_\sigma(R)$-module $Q_\tau e(Q_\sigma(R))/Q_\sigma(R)$ is τ^e-torsion hence τ-torsion and therefore we obtain an R-linear injective map $\varphi : Q_\tau e(Q_\sigma(R)) \to Q_\tau(Q_\sigma(R))$. Now, $Q_{\tau^e}(Q_\sigma(R))$ is τ^e-injective, $Q_\tau(Q_\sigma(R))$ is τ^e-torsion-free and $Q_\tau(Q_\sigma(R))/Q_{\tau^e}(Q_\sigma(R))$ is τ^e-torsion, therefore $Q_{\tau^e}(Q_\sigma(R))$ and $Q_\tau(Q_\sigma(R))$ turn out to be isomorphic as as $Q_\sigma(R)$-modules. Verifying τ-injectivity of $Q_{\tau^e}(Q_\sigma(R))$ one easily establishes that φ is in fact a ring morphism.∎

<u>V.1.19. Theorem</u>. If $X(I)$ is a geometric open set in Spec(R) then $X(I)$ is homeomorphic to $\mathrm{Spec}(Q_I(R))$ and the structure sheaf $\underline{O}_{Q_I(R)}$ on $X' = \mathrm{Spec}(Q_I(R))$ is homeomorphic to the sheaf $\underline{O}_R | X(I)$.

<u>Proof</u>. A prime ideal P of Spec(R) is in X(I) exactly then when $P \notin \mathcal{L}(I)$ therefore, by V.1.16, there is a one-to-one correspondence between the sets X(I) and $\mathrm{Spec}(Q_I(R))$. Any open subset of X(I) is of the form X(IJ) i.e. of the form X(K) with $K \subset I$. Obviously X(K) corresponds to $X'(K^e)$, where $K^e = Q_I(K)$, under the restriction of the correspondence between X(I) and $\mathrm{Spec}(Q_I(R))$. This establishes the homeomorphism $X(I) \cong \mathrm{Spec}(Q_I(R))$. By Lemma V.1.18. we have : $Q_{\kappa^e}(Q_I(R)) = Q_\kappa(Q_I(R)) = Q_\kappa(R)$ and this proves the sheaf isomorphism $\underline{O}_{Q_I(R)} \cong \underline{O}_R | X(I)$.∎

<u>V.1.20. Theorem</u>. Let $X(K) \subset X(I)$ be t-sets and suppose that X(I) is geometric then :

1. $X(K^e)$ is a t-set in $\mathrm{Spec}(Q_I(R))$ if and only if X(K) is a t-set.
2. Proper t-sets in $\mathrm{Spec}(Q_I(R))$ correspond bijectively to t-sets properly contained within X(I).

<u>Proof</u>. A nice gumbo of the foregoing results ∎

<u>V.1.21. Proposition</u>. If X(I) and X(J) are t-sets in Spec(R), where R is a left Noetherian prime ring, then $X(IJ) = X(I) \cap X(J)$ is a t-set.

<u>Proof</u>. We have to show that $\kappa = \kappa_I \vee \kappa_J = \sup\{\kappa_I, \kappa_J\}$ is a t-functor. The filter $\mathcal{L}(\kappa_I \vee \kappa_J)$ has a basis consisting of finite products of ideals in $\mathcal{L}(\kappa_I) \cup \mathcal{L}(\kappa_J)$.

Let $K = K_1 \ldots K_r$ be such a product. Since $Q_\kappa(R)$ contains both $Q_I(R)$ and $Q_J(R)$ as subrings it is easy enough to check that $Q_\kappa(R)K = Q_\kappa(R)$. Consequently, for every $K \in \mathcal{L}(\kappa)$ we have $Q_\kappa(R)K = Q_\kappa(R)$, showing that κ is indeed a t-functor. ∎

V.1.22. Proposition. Let R be a prime left Noetherian ring such that Spec R has a t-basis then each stalk of $\underline{O}_R$ is a t-stalk.

Proof. From $\kappa_{R-P} = \sup\{\kappa_I; P \in X(I)\}$ it follows that $\kappa_{R-P} = \sup\{\kappa_I, P \in X(I), X(I)$ a t-set$\}$. Straightforward modification of the proof of the foregoing proposition yields that κ_{R-P} is a t-functor i.e. the stalk at P of $\underline{O}_R$ is a t-stalk.

In connection with geometric open subsets of Spec(R) let us recall some results of [174] which may come in handy in recognizing whether an open set is geometric or not.

V.1.23. Definition. If κ is a t-functor then an ideal I of R is said to be a κ-ideal if for each $J \in \mathcal{L}(\kappa)$ there exists a $J' \in \mathcal{L}(\kappa)$ such that $J'I \subset IJ$.

V.1.24. Lemma. Let R be a left Noetherian prime ring and κ a symmetric t-functor, let I be an ideal of R. The following statements are equivalent :

1. I is a κ-ideal.
2. $Q_\kappa(R)I = I^e$ is an ideal of $Q_\kappa(R)$.

V.1.25. Lemma. Let κ_1 and κ_2 be symmetric t-functors and put κ equal to $\kappa_1 \vee \kappa_2$. If I is an ideal of R which is both a κ_1-ideal and a κ_2-ideal then I is a κ-ideal.

V.1.26. Corollary. If X(I) and X(J) are geometric then X(IJ) is geometric too.

Proof. Combine Proposition V.1.21 and Lemma V.1.25. ∎

V.1.27. Corollary. If Spec(R) possesses a geometric basis then each stalk of $\underline{O}_R$ is a geometric stalk.

Proof. Combine V.1.14 and the foregoing corollary. ∎

V.1.28. Proposition. Let $X(I)$ be a geometric open set in $\mathrm{Spec}(R)$ and let $X(K) \subset X(I)$ be a t-set, then $X'(K^e)$ is geometric in $X' = \mathrm{Spec}(Q_I(R))$ if and only if $X(K)$ is geometric. There is a bijective correspondence between proper geometric subsets of $\mathrm{Spec}(Q_I(R))$ and geometric subsets of $X(I)$.

Proof. By Theorem V.1.20, 2., it follows that t-sets contained in $X(I)$ correspond bijectively to proper t-sets in X'. Now suppose that $X(K)$ is geometric and let J be an ideal of $Q_I(R)$ such that $J \notin \mathcal{L}(\kappa_{K^e})$. Obviously $J^c = J \cap R$ is a κ_I-ideal. Since $J^c \notin \mathcal{L}(\kappa_K)$ it follows that $Q_K(R)J^c$ is an ideal of $Q_K(R)$ i.e. for every $L \in \mathcal{L}(\kappa_K)$ there is an $L' \in \mathcal{L}(\kappa_K)$ such that $L'J^c \subset J^c L$. Extension to $Q_I(R)$ yields $(L')^e J \subset JL^e$ and it is easily seen that this implies that J is a κ_{K^e}-ideal. Conversely, if J is an ideal of R then $Q_I(R)J$ is an ideal of $Q_I(R)$ since $X(I)$ is geometric. Thus $Q_{K^e}(Q_I(R))J$ is an ideal of $Q_{K^e}(Q_I(R)) = Q_K(R)$ and that finishes the proof because the bijective correspondence is now immediate from $X(K) = X(I) \cap X(K')$ where $K' = \{x \in R;\ \exists L \in \mathcal{L}(I), Lx \subset K\}$. ∎

V.2. Presheaves and Sheaves on Proj(R)

Throughout this section R is a left Noetherian graded ring and we will assume that R is positively graded (although many results still hold for arbitrary $\mathbb{Z}$-graded rings, these restrictions keep us within the framework of the geometrical applications we have in mind).

V.2.1. Theorem. Fix $n_0 \in \mathbb{N}$. For each $n \geq n_0$ let there be given an additive subgroup p_n of R_n. The following statement are equivalent :

1. There is a unique $P \in \mathrm{Proj}(R)$ such that $P \cap R_n = p_n$
2. a. For all $n, k \in \mathbb{N}$ and $m \geq n_0$: $R_n\, p_m\, R_k \subset p_{n+m+k}$
   b. If $r \in R_n$, $t \in R_m$, where $n, m \geq n_0$ then $rR_k t \subset p_{n+m+k}$ for all $k \in \mathbb{N}$, implies $r \in p_n$ or $t \in p_m$
   c. There is an $n \geq n_0$ such that $p_n \neq R_n$.

Proof. 1$\Rightarrow$2. Obviously a and b hold. Assume $p_n = R_n$ for all $n \geq n_o$. Then $aR_{\ell_1} a \dots aR_{\ell_r} a \subset R_{k(r+1)+\sum_I \ell_i}$ for all $(\ell_1,\dots,\ell_r) \in \mathbb{N}^r$ and all $a \in R_k$. Choose r such that we have $k(r+1) \geq n_o$, then $aR_{\ell_1} a \dots aR_{\ell_r} a \subset P$ for all $(\ell_1,\dots,\ell_r) \in \mathbb{N}^r$, hence $a \in P$ i.e. $a \in P_k$, contradicting $P \not\supset R_+$.

2$\Rightarrow$1. By c. we may fix $n \geq n_o$ and $a \in R_n - p_n$. For each $m \in \mathbb{N}$, $m \neq o$, put $P'_m = \{x \in R_m$, for almost all $t \in \mathbb{N}$, for all $(\ell_1,\dots,\ell_t) \in \mathbb{N}^t$, $aR_{\ell_1} a \dots aR_{\ell_t} x \subset p_{m+t_n+\sum_i \ell_i}$. Obviously, $m \geq n_o$ yields $P'_m = p_m$ because of b. For any m, P'_m is additively closed. Put $P = \oplus_m P'_m$. Then P is an ideal of R and $P_n \neq R_n$. If $s \in R_\lambda$, $w \in R_\mu$ satisfy $sR_k w \subset P'_{\lambda+\mu+k}$ for all $k \in \mathbb{N}$, then RsRw is generated as a left R-module by a finite number of elements of the form $sr_k w$ (R is left Noetherian!) and therefore we may choose $t \neq o$ large enough such that for all $(\ell_1,\dots,\ell_t) \in \mathbb{N}^t$ and for all of the chosen generators $sr_k w$ we have that $aR_{\ell_1} a \dots aR_{\ell_t} sr_k w \subset p_{\lambda+\mu+k+tn+\sum \ell_i}$. For any t' larger than t, fixed as above, we obtain :

$$((RaR)^{t'})_j \; sR_k w \subset p_{j+\lambda+\mu+k},$$

for all $j,k \in \mathbb{N}$.

Write A for the ideal RaR. Then, for all $j,k \in \mathbb{N}$:

$$(A^{t'})_j \; s(A^{t'})_k \; w \subset p_{j+\lambda+\mu+k}.$$

Now, $(A^{t'})_j \neq o$ implies that $j \geq nt'$, because $\deg(a) = n \neq o$; with $n \geq n_o$. Further, $A^{t'} \neq o$ because otherwise $a \in p_n$ would follow from (b.). Again from (b.) it follows now that either $(A^{t'})_j \, s$ is in $p_{j+\lambda}$, or $(A^{t'})_k \, w$ is in $P_{k+\mu}$, where both $j+\lambda$ and $k+\mu$ may be taken to be larger than $t'n > n_o$. Hence by construction of P, either s or w must be in P and therefore we have proved that P is a prime ideal of R.

Assume that Q is a graded prime ideal of R such that $P \neq Q$, while Q satisfies the conditions of (2.) too. Pick an element b of Q-P and suppose that b is homogeneous. If $d = \deg(b) > o$, then for all $m \geq n_o$ and all $(\ell_1,\dots,\ell_m) \in \mathbb{N}^m$ we have

$$b\,R_{\ell_1} b \ldots b\,R_{\ell_{m-1}} b \subset Q \cap R_\lambda \subset p_\lambda \subset P,$$

where $\lambda = \Sigma\, \ell_i + md$. So $(Rb)^m \subset P$, what $b \notin P$. If $d = o$ then, for all $\lambda \in \mathbb{N}$ we obtain a $R_\lambda\, b \subset R_{n+\lambda} \cap Q \subset P$. However $a \notin P$ then entails $b \in P$, contradiction. ∎

V.2.2. Remark. Note the deviations from the commutative case, where the noetherian hypothesis is not needed. We do not know, whether the above result holds in the noncommutative situation, if the noetherian hypothesis is dropped.

V.2.3. Theorem. Let I be a graded ideal of R and let $\underline{C}$ be the set of all ideals J of R which are maximal with respect to the property $J \notin \mathcal{L}(I)$, where $\mathcal{L}(I)$ is the idempotent filter generated by the powers of I. Put $\underline{C}^g = \{gJ; J \in \underline{C}\}$, then $\underline{C}^g$ consists of graded prime ideals and $\mathcal{L}(I) = \cap\{\mathcal{L}(\kappa^+_{R-P}); P \in \underline{C}^g\}$, where κ^+_{R-P} denotes the symmetric graded kernel functor associated to $(R-P)^+$ as in III.1.

Proof. Just as in the nongraded case one checks that $\underline{C}$ consists of prime ideals, hence that $\underline{C}^g$ consists of graded prime ideals. Obviously an ideal H belongs to $\mathcal{L}$ (I) if and only if $H \not\subset P$ for all $P \in \underline{C}$. However, since the filters we have to compare are both graded, the foregoing remark reduces to saying that a basis for $\mathcal{L}(I)$ is obtained by taking the graded ideals gH, which are not contained in any element $gP \in \underline{C}$, proving that $\mathcal{L}(I)$ is the intersection of the filters $\mathcal{L}(\kappa_{R-P})$, for $P \in \underline{C}^g$. ∎

Note that with the foregoing notations, we also obtain that if $I \subset R_+$, then the elements of $\underline{C}^g$ are in Proj(R). Indeed, if $I \subset R_+$, then for any $P \in \underline{C}^g$ we have $P \not\supset I$ and a fortiori $P \supset R_+$, whence $P \in \mathrm{Proj}(R)$ follows. ∎

V.2.4. Proposition. Let I be a graded ideal of R, then $R_+ \cap \mathrm{rad}(I_+) = R_+ \cap \mathrm{rad}(I)$.

Proof. This follows immediately from the following fact. If P is a graded prime ideal containing I_+, then either $P \supset I$ or $P \supset R_+$. Indeed, if $P \not\supset R_+$, then take $y \in I-P$. Clearly, $\deg(y) = o$, while for some $d > o$ there exists $x \in R_d - P_d$. So we get that $yRx \subset I_+ \subset P$ and this yields $y \in P$, contradiction. ∎

V.2.5. Corollary. Put $V_+(I) = \{P \in \mathrm{Proj}(R); P \supset I\}$, then $V_+(I) = V_+(I_+) = V_+(\mathrm{rad}(I)) = V_+((\mathrm{rad}(I))_+) = V_+(\mathrm{rad}(I_+))$. ∎

Denote by κ_I the symmetric graded kernel functor associated to $\mathcal{L}(I)$; let Q_I^g be the corresponding graded localization functor in R-gr and let $j_I^+ : R \to Q_I^g(R)$ be the corresponding canonical graded ring morphism (of degree o). If $I \subset R_+$, then we have a property for κ_I^+ which is something like being faithfully graded, cf. [181].

V.2.6. Proposition. Let I be a graded ideal of R, with $I \subset R_+$, then $(Q_I^g(R))_m$ consists of all $x \in Q_I(R)$ with the property that there exists $J \in \mathcal{L}^+(I)$ such that $Jx \subset j_I^+(R)$ and $J_n x \subset (j_I^+(R))_{n+m}$ for all $n \geq n_o$, where n_o is arbitrary but fixed.

Proof. We shall write S_m for the latter set. It is obvious that $(Q_I^g(R))_m$ is contained in S_m. Conversely, pick $x \in S_m$ and let $J \in \mathcal{L}^+(I)$ be such that $J_n x \subset (j_I^+(R))_{n+m}$ for all $n \geq n_o$. If $J_k x = o$ for all $k < n_o$, then there is nothing to prove. If $J_k x \neq o$, fix this k momentarily and choose $y \neq o$ in $J_k x$; moreover, choose $z \in (j_I^+)^{-1}(y)$.
If $j_I^+(R_N)J_k x = o$ for all $N \geq n_o - k$, then $R_N z$ maps to o in $j_I^+(R)$, hence $RR_N z \in \kappa_I(R)$ for all $N \geq n_o - k$. Therefore we obtain :

$$\Big(\sum_{N \geq n_o - k} RR_N\Big) z \in \kappa_I^+(R).$$

Now, $T = \sum_{N \geq n_o - k} RR_N$ is the graded ideal $\bigoplus_{n \geq n_o - k} R_n$ of R and if $T \notin \mathcal{L}^+(I)$, then $T \subset P$ for some $P \in C^g(I)$.
Thus, $C^g(I) \subset \mathrm{Proj}(R)$. However $P \supset T$ yields $P \supset R_+$, because of V.2.1 and this contradicts $P \in \mathrm{Proj}(R)$. Thus $T \in \mathcal{L}^+(I)$; So $Tz \in \kappa_I^+(R)$ implies $z \in \kappa_I^+(R)$ or $y = o$, contradiction. Hence we may assume that $j_I^+(R_N)J_k x \neq o$ for some $N \geq n_o - k$. Thus,

$$o \neq j_I^+(R_N)J_k x \subset J_{N+k} x \subset (j_I^+(R))_{N+k+m},$$

because $N+k \geq n_o$. Since $J_k x \subset j_I^+(R)$, the above inclusions imply that $J_k x \subset j_I^+(R)_{k+m}$, hence, since this has now been deduced for all possible k, we find that $x \in (Q^g_I(R))_m$. ∎

V.2.7. Remarks.

1. The similar statement for graded localization at κ^+_I holds for graded left R-modules too;
2. Since $I \subset R_+$, it is easily seen that $J \in \mathcal{L}^+(I)$, implies $J_+ \in \mathcal{L}^+(I)$,
3. One easily shows that $R_+ \cap \mathrm{rad}(I) = R_+ \cap \mathrm{rad}(\bigoplus_{n \geq n_o} I_n)$, for each graded ideal I of R and each positive integer $n_o \in \mathbb{N}$. This entails that the set S_m defined in the statement of the foregoing proposition is well defined. Indeed, if J_1 and J_2 are in $\mathcal{L}^+(I)$, such that $J_\nu x \subset j^+_I(R)$ and $(J_\nu)_n x \subset (j^+_I(R))_{n+m_\nu}$ for $\nu = 1,2$, and for all $n \geq n_o$, then either $m_1 = m_2$, as follows directly from $(J_1 \cap J_2)_n x \subset (j^+_I(R))_{n+m_1} \cap (j^+_I(R))_{n+m_2}$ or else $(J_1 \cap J_2)_n x = o$ for all $n \geq n_o$. Since $R_+ \cap \mathrm{rad}(\bigoplus_{n \geq n_o} (J_1 \cap J_2)_n) = R_+ \cap \mathrm{rad}(J_1 \cap J_2) \in \mathcal{L}^+(I)$, it follows that $\mathrm{rad} \bigoplus_{n \geq n_o} (J_1 \cap J_2)_n$ is in $\mathcal{L}^+(I)$ and since R is left noetherian, it also follows that $\bigoplus_{n \geq n_o} (J_1 \cap J_2)_n \in \mathcal{L}^+(I)$. But then $(J_1 \cap J_2)_n x = o$, for all $n \geq n_o$, meaning that $x \in \kappa^+_I(Q^g_I(R)) = o$.

Let us now endow Proj(R) with the topology induced by the Zariski topology of Spec(R) as follows : for any ideal I of R we put

$$V_+(I) = \{P \in \mathrm{Proj}(R); P \supset I\} = V(I) \cap \mathrm{Proj}(R)$$

and

$$X_+(I) = X(I) \cap \mathrm{Proj}(R) = \mathrm{Proj}(R) - V_+(I).$$

Note that in these definitions we may replace I by the smallest graded ideal of R containing it, so, from now on, if we write $V_+(I)$ or $X_+(I)$, then we assume silently that I is chosen to be graded. By the foregoing we may assume that $I \subset R_+$, and even, by the foregoing remarks, that $V_+(I) = V_+(\bigoplus_{n \geq n_o} I_n)$ for some positive integer n_o. It is also clear that $V_+(I)$ remains unaltered under taking radicals and positively graded parts of I in any order. One easily checks the following relations :

$$V_+(I+J) = V_+(I) \cap V_+(J)$$

$$V_+(IJ) = V_+(I \cap J) = V_+(I) \cup V_+(J),$$

which shows us that the collection of the sets $X_+(I)$, where I ranges through the graded ideals (or those contained in R_+) of R, exhausts the open subsets of the topology induced in Proj(R). We will speak of the Zariski topology on Proj(R).

To an open set $X_+(I)$ we associate the kernel functor κ_I^+, where κ_I^+ is given by its filter $\mathcal{L}(\kappa_I^+)$, which consists of all left ideals of R, containing a graded ideal J of R with the property that $\mathrm{rad}(J) \supset I_+$. The following theorem may be verified, step by step, by mimicing the proof of the corresponding properties for Spec(R), mentioned before, taking care to use the established graded theory where necessary.

<u>V.2.7. Theorem.</u> Assigning $Q_I^g(R)$ to $X_+(I)$ for each graded ideal I of R defines a presheaf of graded rings $\underline{Q}_R^g$ on Proj(R), endowed with the Zariski topology. Since R is a left noetherian graded ring every open subset is quasicompact and the presheaf $\underline{Q}_R^g$ is separated.

To prove the last statement, one uses the fact that the noetherian hypothesis entails that every open subset is quasicompact, exactly as in the nongraded case. If $\underline{Q}_R^g$ is separated, then it may be embedded in a sheaf of rings.
It is very easy to verify that the sheaf thus obtained is a sheaf of graded rings. Moreover, applying the same type of construction to a graded left R-module M yields a presheaf $\underline{Q}_M^g$ of left $\underline{Q}_R^g$-modules, which is canonically endowed with the structure of a presheaf of <u>graded</u> left $\underline{Q}_R^g$-modules, and the associated sheaf of (graded) left modules over $\underline{\underline{a}}\, \underline{Q}_R^g = L\underline{Q}_R^g$.

<u>V.2.8. Theorem.</u> If R is a left noetherian graded prime ring (resp. M a left Noetherian, torsion free graded left R-module), then $\underline{Q}_R^g$(resp. $\underline{Q}_M^g$) is a sheaf.

Proof. It suffices to mimic the demonstration of the nongraded, similar statement, see also V.1.8, V.1.9.■

V.2.9. Lemma. For any open subset $X_+(I)$ of Proj(R) we have that

$$\mathcal{L}(\kappa_I^+) = \cap\{\mathcal{L}(\kappa_{R-P}^+);\ P\in X_+(I)\}\ .$$

Proof. If $P'\in X_+(I)$, then $P'\notin\mathcal{L}^+(\kappa_I^+)$ hence $P'\subset P$ for some $P\in C^g(I_+)$. Now $P'\subset P$ yields that $\mathcal{L}(\kappa_{R-P}^+)\subset\mathcal{L}(\kappa_{R-P'}^+)$. We apply V.2.3 and obtain the result :

$$\mathcal{L}(\kappa_I^+) = \bigcap_{P\in C^g(I)} \mathcal{L}(\kappa_{R-P}^+) = \bigcap_{P'\in X_+(I)} \mathcal{L}(\kappa_{R-P'}^+).\quad ■$$

V.2.10. Theorem. For any $P\in\mathrm{Proj}(R)$ we have the following statements :

1. $\kappa_{R-P}^+ = \sup\{\kappa_I^+; P\in X_+(I)\}$
2. if R is a (left noetherian) prime graded ring, then the stalk of $\underline{Q}_R^+$ at P is exactly $Q_{R-P}^g(R)$.

Proof. (1) Since $P\in X_+(I)$ is equivalent to $P\not\supset I_+$, it follows at once that $\kappa_I^+\leq\kappa_{R-P}^+$ for every graded prime ideal I such that $P\in X_+(I)$. Conversely, if $J\in\mathcal{L}^+(\kappa_{R-P}^+)$, then J contains a nonzero graded ideal I such that $I_+\not\subset P$, hence $P\in X_+(I)$ and also $J\in\mathcal{L}(\kappa_I^+)$.

(2) The stalk of the sheaf $\underline{Q}_R^g$ at $P\in\mathrm{Proj}(R)$ is defined as

$$S = \varinjlim_{P\in X_+(I)} Q_I^g(R)\ .$$

For every graded ideal I such that $P\in X_+(I)$, we have a monomorphism f_I : $Q_I^g(R)\longrightarrow Q_{R-P}^g(R)$, which is graded of degree o. Therefore we obtain a monomorphism of degree o, say $f: S\to Q_{R-P}^g(R)$, which is the inductive limit of the morphisms f_I, as I runs through the graded ideals with the property that $X_+(I)\ni P$. Pick $x\in(Q_{R-P}^g(R))_m$, then, by definition, we may represent x by a graded morphism of degree m, say $\mu_x: I\to R$, for some $I\in\mathcal{L}^+(\kappa_{R-P}^+)$. Now μ_x also represents an element y_I of $Q_I^g(R)$. By construction of f_I it is obvious that $f_I(y_I) = x$. Hence, the image y of y_I in S has the property that $f(y) = x$ and this entails that f is an isomorphism of degree o; linearity with respect to the R-module structure is easily checked. ∎

We will need an easy general result on sheaves now. Let X be any topological space and let $\underline{R}$ be a presheaf of rings on X, the restriction maps being graded ring morphisms of degree o. Define a subpresheaf of rings $\underline{R}_o$ of $\underline{R}$ by $\underline{R}_o(U) = (\underline{R}(U))_o$, for each $U \in \mathrm{Open}(X)$.

V.2.11. Lemma. If $\underline{R}$ is a sheaf of graded rings, then $\underline{R}_o$ is a subsheaf of rings of $\underline{R}$.

Proof. Let $\underline{R}_V^U$ denote the restriction map of $\underline{R}$ with respect to the open subsets $V \subset U$ in X. Since $\underline{R}_V^U$ is graded of degree o, the restriction $(\underline{R}_V^U)_o$ of $\underline{R}_V^U$ to $\underline{R}(U)_o$ maps $\underline{R}(U)_o$ into $\underline{R}(V)_o$. Obviously $\underline{R}_o$ is separated, since it is a subpresheaf of a sheaf (= $\underline{R}$). Moreover, if $\mathcal{U} = \{U_i, i \in I\}$ is an open covering of $U \in \mathrm{Open}(X)$, and if $r_i \in \underline{R}(U_i)$ are elements of degree o satisfying the relations

$$\underline{R}^{U_i}_{U_i \cap U_j}(r_i) = \underline{R}^{U_j}_{U_i \cap U_j}(r_j)$$

for all $i,j \in I$, then there is $r \in \underline{R}(U)$ such that $\underline{R}^U_{U_i}(r) = r_i$, since $\underline{R}$ is a sheaf. Again, the fact that $\underline{R}_V^U$ has degree o implies $r \in \underline{R}(U)_o$; consequently $\underline{R}_o$ is a sheaf. ∎

The sheaf of rings $(\underline{\underline{a}}\, \underline{Q}_R^g)_o = (L\underline{Q}_R^g)_o$ defined over Proj(R) is called the structure sheaf of Proj(R), it will be denoted simply by $\underline{O}_R^+$. Since direct limits of graded morphisms of degree o respect taking homogeneous parts of degree o, we have

$$\underline{O}^+_{R,P} = (Q^g_{R-P}(R))_o ,$$

for every $P \in \mathrm{Proj}(R)$. If R is a graded prime ring then we obtain that

$$\Gamma(X_+(I), \underline{O}_R^+) = (Q_I^g(R))_o ,$$

for every graded ideal I of R. Anyhow, one easily shows that for c homogeneous and central in R (or more generally : for κ_I^+ perfect for graded R-bimodules), we obtain

$$\Gamma(X_+(Rc), \underline{O}_R^+) = (Q_c^g(R))_o ,$$

as is easily verified. Similar constructions and results hold for arbitrary graded left R-modules M, yielding a sheaf $\underline{O}_M^+$ of left $\underline{O}_R^+$-modules.
In the commutative case graded prime ideals may be related to common prime ideals in some of the rings appearing in the structure sheaf. This relation is fully expressed by saying that Proj(R) is a scheme i.e. that Proj(R) has a covering by open subsets $X_+(I)$, such that $\text{Proj}(R)|X_+(I) \cong \text{Spec}(Q_I^g(R))_o$. It suffices for example to consider $X_+(I)$, where I runs through the set of ideals generated by an homogeneous element of R. The general noncommutative case is hard to solve, and will not be treated here. Different techniques will be given below, while a most satisfactory theory holds for the class of Zariski central rings, as we will see further on.

<u>V.2.12. Lemma</u>. Let C be the center of R. Let a be an ideal of C, then for any $b \in \mathcal{L}(a)$, we have $Rb \in \mathcal{L}(Ra)$.

<u>Proof</u>. Obvious. ∎

Define the filter $\mathcal{L}'$ consisting of those left ideals of R containing Rb for some $b \in \mathcal{L}(a)$. By the lemma, $\mathcal{L}' \subset \mathcal{L}(a)$. If $\mathcal{L}(a)$ is an idempotent filter, then $\mathcal{L}'$ is idempotent too, as is straightforward to verify.

<u>V.2.13. Lemma</u>. If R is left noetherian, then $\mathcal{L}'$ is an idempotent filter which coincides with $\mathcal{L}(Ra)$.

<u>Proof</u>. Let $L \in \mathcal{L}(Ra)$, i.e. $Ra^n \subset L$ for some $n \in \mathbb{N}$; since $a^n \in \mathcal{L}(a)$, it follows that $L \in \mathcal{L}'$. Together with the foregoing remarks, it is thus shown that $\mathcal{L}' =$

$\mathcal{L}(Ra)$ and that $\mathcal{L}(Ra)$ is idempotent. ∎

Since we have assumed from the start that R is left noetherian, for any ideal a of C and any left R-module M there is equivalence between the following statements :

1. M is a κ_a-torsion C-module;
2. M is a κ_{Ra}-torsion R-module.

It is clear that $\kappa_a(C) = C \cap \kappa_{Ra}(R)$ and therefore the canonical inclusion $C \hookrightarrow R$ yields a canonical injective morphism $C/\kappa_a(C) \to R/\kappa_{Ra}(R)$ and therefore also a C-linear morphism $Q_a(C') \hookrightarrow Q_{Ra}(R)$ in the category of C-modules.

<u>V.2.14. Proposition.</u> Let R be a left noetherian ring with center C and let a be an ideal of C such that κ_a is a t-functor, then the following statements hold:

1. $Q_a(R)$ is a ring and the canonical C-linear map $R/\kappa_a(R) \hookrightarrow Q_a(R)$ is a ring monomorphism;
2. $Q_a(R)$ and $Q_{Ra}(R)$ are isomorphic rings;
3. for any left R-module M the R-modules $Q_a(M)$ and $Q_{Ra}(M)$ are isomorphic;
4. the center $Z(Q_a(R))$ of $Q_a(R)$ is $Q_a(C)$.

<u>Proof</u>. Using the foregoing remarks, this is an easy modification of the proof of similar properties in Section II.1. ∎

<u>V.2.15. Corollary</u>. Under the assumptions of V.2.14 , the kernel functor κ_{Ra} is a t-functor, hence $Q_a(R) = Q_a(C) \otimes_C R$, and the canonical morphism $R \to Q_{Ra}(R)$ is a central extension. The latter fact entails that ideals of R localize to ideals of $Q_{Ra}(R)$. Note also that in this case $Q_{Ra}(R)$ is left noetherian. ∎

Let $j_a : R \to Q_{Ra}(R)$ be the canonical morphism. If the correspondence $\mathrm{Spec}(Q_{Ra}(R)) \to \mathrm{Spec}(R) : P \to j_a^{-1}(P)$ is a one-to-one correspondence between $\mathrm{Spec}(Q_{Ra}(R))$ and the Zariski open set $X(a) = \{P \in \mathrm{Spec}(R);\ a \not\subset P\}$, then $Q_{Ra}(\mathrm{rad}(J)) = \mathrm{rad}(Q_{Ra}(J))$ for every ideal J of R. The latter is obviously the case when σ_a has property (T).

Let us now assume that R is a positively graded prime left noetherian ring satisfying the following conditions :

PS1. $R_+ \subset \mathrm{rad}(RC_+)$

PS2. C is generated by C_1 as a C_0-algebra.

Some comments. Condition PS1 amounts to saying that Proj(R) may be covered by open sets $X_+(Rc)$, where c is homogeneous in C_+, and that the assignment $P \to P \cap C$ yields a map Proj(R) → Proj(C). Note that condition PS1 does not hold for generic $n \times n$-matrix rings (these rings are of course not left noetherian). As for the second condition, instead of (PS2), actually it would be sufficient to assume that the following condition holds :

PS2'. if c is homogeneous of degree t in C_+, then there is a one-to-one corresponding (in fact a homeomorphism) between $\mathrm{Spec}(Q_C^g(R))$ and $\mathrm{Spec}(Q_C^g(R)^{(t)})$, where as usually, for any graded ring S we have

$$S^{(t)} = \bigoplus_{i \in \mathbb{Z}} S_{it}.$$

Anyhow, if we assume (PS2) from the start many proofs simplify a lot. Since one of the main aims of our construction is to present a projective theory explaining at least the constructions of Artin and Schelter in [17], and since in this context (PS2) is not restricting, we will do so from now on. Note however that (PS2) does not hold for most rings of twisted polynomials, e.g. $\mathbb{C}[X,-]$ has center $\mathbb{R}[X^2]$. Yet, these rings of twisted polynomials satisfy (PS2'), and moreover, since they are also Zariski central, the theory expounded in the sequel applies to them.

The rest of this paragraph will be devoted to showing that under the conditions cited above, the ringed space $(\mathrm{Proj}(R), \underline{O}_R^+)$ is locally of the form $(\mathrm{Spec}(S), \underline{O}_S)$, where S is prime noetherian, and where $\underline{O}_S$ is the structure sheaf on Spec(S) constructed in V.1. Now, our assumptions imply that Proj(R) may be covered by open sets $X_+(Rc)$, where c is homogeneous in C_+ of degree 1. As observed above, the graded kernel functor κ_{Rc}^+ has property (T) and $\mathcal{L}(\kappa_{Rc}^+)$ is generated by the ideals $R\alpha$, where α belongs to the filter of κ^{+}_{Cc}. We also know that Q^g_{Rc} is just central localization Q^g_c and that graded ideals I of R extend to graded ideals $Q^g_c(I) = Q^g_c(R)I$ of $Q^g_c(R)$. Finally, if C is the center of R, then $Z(Q^g_c(R)) = Q^g_c(C)$.

Let c be a homogeneous element of degree 1 in C. Given an ideal I of $(Q^g_c(R))_o$, we construct a graded ideal I^D of $Q^g_c(R)$ with the property $I^D \cap (Q^g_c(R))_o = I$, as follows. For $n \in \mathbb{Z}$, put

$$J_n = \{x \in (Q^g(R))_n;\ c^{-n}x \in I\}$$

and $J = \bigoplus_{n\in\mathbb{Z}} J_n$. It is not hard to verify that J is as required, thus we put $I^D = J$. If J is an ideal of $Q^g_c(R)$, then write J^D for $(J_o)^D$. Because of (PS2) we find $J = J^D$, while the weaker (PS2'), implies that rad J^D = rad J , and develop the theory from that!

<u>V.2.16. Lemma</u>. For any prime ideal P of $Q^g_c(R)$, the ideal P_o is prime in $Q^g_c(R)_o$.

<u>Proof</u>. Suppose that $x, y \in Q^g_c(R)_o$ are such that $xQ^g_c(R)_o y \subset P_o$, then for any $z \in Q^g_c(R)$, i.e. $z = z_1 + \ldots + z_s$ with z_i homogeneous in $Q^g_c(R)$, we have $xzy = xz_1y + \ldots + xz_sy$. Now, putting deg $z_i = t_i$, we obtain $c^{-t_i}xz_iy \in P_o \subset P$, and as c and c^{-1} do not belong to P, it follows that $xz_iy \in D$ for all $1 \leq i \leq s$, i.e. $xzy \in P$. Thus $x\,Q^g_c(R)y \subset D$, implying that x or y must belong to P_o, i.e. that P_o is prime. ■

<u>V.2.17. Corollary</u>. $Q^g_c(R)_o$ is a left noetherian prime ring.

<u>Proof</u>. That $Q^g_c(R)_o$ is prime follows immediately from the lemma because $Q^g_c(R)$ is prime. If $L_{(1)} \subset L_{(2)} \subset \ldots$ is an ascending chain of left ideals of $Q^g_c(R)_o$, then $L^D_{(1)} \subset L^D_{(2)} \subset \ldots$ is an ascending chain of left ideals of $Q^g_c(R)$, hence $L^D_{(n)} = L^D_{(n+1)} = \ldots$ for some $n \in \mathbb{N}$. Since $(L^D_{(i)})_o = L_{(i)}$, it follows that the original chain terminates, i.e. that $Q^g_c(R)_o$ is left noetherian.

<u>V.2.18. proposition</u>. There is a bijective correspondence between the sets $X_+(Rc)$ and $\mathrm{Spec}(Q^g_c(R))_o$.

Proof. Define $\psi : X_+(Rc) \to \mathrm{Spec}(Q_C^+(R)_o)$ by $\psi(P) = [Q_C^g(R)P]_o$. Conversely, we want to define $\varphi : \mathrm{Spec}(Q_C^g(R)_o) \to X_+(Rc)$ by $\varphi(Q) = R \cap Q^D$. It is easily checked that ψ and φ are well-defined maps (since Q^D is obviously a prime ideal of $Q_C^g(R)$ and the morphism $R \to Q_C^g(R)$ is an extension). Now $\varphi\psi(P) = ((P^e)_o)^D \cap R = P^e \cap R = P$, while on the other hand $\psi\varphi(Q) = ((R \cap Q^D)^e)_o$ and $(R \cap Q^D)^e$ is just Q^D, by property (T) for Q_C^g, implying $\psi\varphi(Q) = (Q^D)_o = Q$. ∎

V.2.19. Proposition. Endow $X_+(Rc)$ with the topology induced by the Zariski topology on Proj(R) and $\mathrm{Spec}(Q_C^g(R)_o)$ with its Zariski topology. Then ψ is a homeomorphism.

Proof. Let I be a graded ideal of R, then $\mathrm{rad}(Ic) = \mathrm{rad}\,(I \cap Rc)$ and $I^e = Q_C^g(R)I = Q_C^g(R)Ic = (Ic)^e$. Therefore the proposition will follow, if we can show that the open subset $X_+(Ic)$ of $X_+(Rc)$ corresponds bijectively with the open set Y $Y((I^e)_o)$ of $\mathrm{Spec}(Q_C^g(R)_o)$ under the map ψ. Since a prime ideal P of R is such that $P \supset Ic$ if and only if $P^e \supset Q_C^g(Ic)$, which is equivalent to $(P^e)_o \supset Q_C^g(Ic)_o = Q_C^g(I)_o = (I^e)_o$, the latter because $(P^e)_o \supset (I^e)_o$ yields $P^e \supset I^e$ and $P \supset I$, we find that ψ does map $X_+(Ic)$ to $Y((I^e)_o) \subset \mathrm{Spec}(Q_C^g(R)_o)$. ∎

V.2.20. Proposition. Let I be a graded ideal of R and put $I' = (Q_C^+(R)I)_o$, then $Q_{I'}(Q_C^g(R)_o) = Q_{Ic}^g(R)_o$.

Proof. An element $g \in Q_{Ic}^g(R)_o$ may be viewed as an R-linear graded morphism of degree o, say $\hat{g} : c^n I^n \to R$. Then $\hat{g}$ extends to a morphism $Q_C^g(\hat{g}) : Q_C^g(I^n c^n) = Q_C^g(I^n) \to Q_C^g(R)$, which is also of degree o. This $Q_C^g(\hat{g})$ in turn restricts to a $Q_C^g(R)_o$-linear map $(I')^n \to Q_C^g(R)_o$, which represents an element of $Q_{I'}(Q_C^g(R)_o)$. It is straightforward to check that we obtain a well-defined map

$$\theta : Q_{Ic}^g(R)_o \to Q_{I'}(Q_C^g(R)_o),$$

which is also a ring morphism. In order to verify that θ is injective, consider g and h in $Q_{Ic}^g(R)_o$ and suppose that $\theta(g) = \theta(h)$ i.e. there exists $m \in \mathbb{N}$ such that $Q_C^g(\hat{g})$ and $Q_C^g(\hat{h})$ restrict to the same $Q_C^g(R)_o$-linear map $(I')^m \to Q_C^g(R)_o$. Since I' generates I^e as a left $Q^g_C(R)$-module, it is a direct consequence of the $Q^g_C(R)$-linearity of $Q^g_C(\hat{g})$ and $Q^g_C(\hat{h})$, that $Q^g_C(\hat{g})$ and $Q^g_C(\hat{h})$ coincide on $(I^e)^m$ and thus we obtain that for some positive integer $m \in \mathbb{N}$ the maps $\hat{g}$ and $\hat{h}$ coincide on $c^m I^m$, i.e. $g = h$. Conversely, an element $y \in Q_{I'}(Q^g_C(R)_o)$ may be viewed as a $Q^g_C(R)_o$-linear morphism $\hat{y} : (I')^n \to Q^g_C(R)_o$, and thus as a graded $Q^g_C(R)$-linear map $(I^e)^n \to Q^g_C(R)$, which corresponds uniquely to an R-linear graded morphism $I^n \to Q^g_C(R)$, i.e. an element of $Q^g_I\, Q^g_C(R) = Q^g_{Ic}(R)$. Note that the fact that deg $c = 1$ is used heavily, since otherwise the construction should have been carried out in $Q^g_C(R)^{(d)}$. Now, since it is clear that all graded morphisms appearing here are of degree o, we finally obtain a morphism $\theta' : Q_{I'}(Q^g_C(R)_o) \to Q^g_{Ic}(R)_o$. Straightforward verification yields that θ' is inverse to θ. ∎

Note that the ring $Q^g_{Ic}(R)_o$ is exactly the ring of sections over $X^+(Ic)$ for the structure sheaf $\underline{O}^+_R$ of Proj(R), whereas $Q_{I'}(Q^g_C(R)_o)$ is the ring of sections of the structure sheaf $\underline{O}_{Q^g_C(R)_o}$ of Spec($Q^g_C(R)_o$) over Y(I'), where Y(I') denotes the open subset of $Y = \text{Spec}(Q^g_C(R)_o)$ corresponding to $X_+(Ic)$ under the isomorphism ψ. This shows that the ringed space (Proj(R), $\underline{O}^+_R$) is of the form announced in the remarks preceeding V.2.16.

In the next paragraph we will see how similar results may be obtained for Zariski central rings. The main difference with the set-up in this paragraph will be that conditions PSI and PS2 may be omitted in the Zariski central case. Moreover, we will even show that the ringed space (Proj(R),$\underline{O}^+_R$) will then have a <u>basis</u> $\mathcal{B}$ of open subsets (and not only an open covering), such that each induced ringed space $(U, \underline{O}^+_R|_U)$ for $U \in \mathcal{B}$ is of the form (Spec(S),$\underline{O}_S$), for some left noetherian prime ring S.

For details on Zariski algebras we refer to Section II.1, and [124], [180]; the reader should recall Theorem II.1.14, Corollary II.1.15. Theorem II.1.16. Now a positively graded ring is said to be a GZ-ring if it is Zariski central, it is said to be a ZG-ring if for every graded ideal I of R we have $I^n \subset R(I \cap C)$ for some $n \in \mathbb{N}$ Although we prefer to formulate the results we are about to deduce for GZ-rings, most of these remain valid for ZG-rings.

V.2.21. Theorem. Let R be a left Noetherian GZ-ring and let S be a central multiplicative subset of R not containing o. Let $\underline{\kappa}$ be the kernel functor on R-mod associated to $S^+ = \bigcup_n (S \cap R_n)$. Then : 1. $\underline{\kappa}$ is a graded kernel functor and it is a graded t-functor.

2. If j_κ denotes the canonical graded morphism $R \to Q^g_{\underline{\kappa}}(R)$ and I is a graded ideal of R then $Q^g_{\underline{\kappa}}(R)j_{\underline{\kappa}}(I)$ is a graded ideal of $Q^g_{\underline{\kappa}}(R)$.

3. Graded prime ideals P of R such that $P \notin \mathcal{L}(\underline{\kappa})$ correspond in a bijective way to proper graded prime ideals of $Q^g_{\underline{\kappa}}(R)$. Consequently, if I is a graded ideal of R then rad I is graded and $Q^g_{\underline{\kappa}}(R)j_{\underline{\kappa}}(\text{rad } I) = \text{rad}(Q^g_{\underline{\kappa}}(R)j_{\underline{\kappa}}(I))$.

V.2.22. Corollary. If P is a graded prime ideal of R then Q^g_{R-P} is just central localization Q^g_{C-p}, hence the foregoing applies to $\underline{\kappa} = \kappa_{R-P}$ if P is graded. The foregoing also applies to the case where $\underline{\kappa} = \kappa_c$, $c \in C$ and $\mathcal{L}(\kappa_c) = \{L$ left ideal of R, $\exists k \in \mathbb{N}$, $c^k \in L\}$.

V.2.23. Lemma. If R is a left Noetherian GZ-ring then Proj(R) has a basis for the Zariski topology consisting of open sets $X_+(Rc)$ where c runs through the homogeneous elements of C_+.

The graded kernel functor κ^+_{Rc} associated to $X_+(Rc)$ is a t-functor and it is obtained from κ^+_{Cc} in the usual way. Let Q^g_c be the localization functor corresponding to κ^+_{Rc} and let $j^+_c : R \to Q^g_c(R)$ be the canonical graded ring morphism. Then $Q^g_c(R)$ is Zariski central over its center $Q^g_{Cc}(C)$, where Q^g_{Cc} is the localization functor associated to κ^+_{Cc} on C-gr.

V.2.24. Lemma. If R is a left Noetherian GZ-ring then R_o is Zariski central over C_o; so if $p \in \text{Spec}(C_o)$ is such that $C_o \cap R_o p = p$ then rad $R_o p$ is the unique prime ideal of R_o lying over p.

Proof. That $C_o \subset Z(R_o)$ is obvious. Let I be an ideal of R_o and consider $I + R_+$.

Then $(I+R_+)^n \subset R(C\cap(I+R_+))$ because R is Zariski central. But $(I+R_+)\cap C = (I\cap C_0)\oplus C_+$, hence $I^n = (I^n)_0 = I_0^n \subset [R(C_+\oplus(C_0\cap I))]_0$, and as R is positively graded : $I^n \subset R_0(C_0\cap I)$ follows. The second statement is easily derived from the fact that R_0 is a Zariski extension of C_0. ∎

V.2.25. Proposition. Let R be a left noetherian GZ-ring with center C. Let Q^g be the localization functor corresponding to some graded central kernel functor in R-mod. Denote $Q^g(R)$ (resp. $Q^g(C)$) by S resp. D. Suppose that g is a graded prime ideal of C and define $q^{(e)}$ by $(q^{(e)})_m = q_{me}$ for a fixed $e\in\mathbb{N}$. Then we have

$$\mathrm{rad}(Sq) = \mathrm{rad}(Sq^{(e)}) .$$

Proof. The inclusion $\mathrm{rad}(Sq)\supset \mathrm{rad}(Sq^{(e)})$ is clear. Conversely, since q is central it follows that for any $x\in q_m$ we have that $x^e \subset q_{me} \in q^{(e)}$, whence $x \subset \mathrm{rad}(Cq^{(e)})$ and $q \subset \mathrm{rad}(Cq^{(e)})$ follows. Now, R is Zariski central, hence $Rq \subset \mathrm{rad}(Rq^{(e)})$. By V.2.21.,3, it follows that

$$Q^g(R)\ \mathrm{rad}(I) = \mathrm{rad}(Q^g(R)I)$$

for any ideal I of R. Applying this to $\mathrm{rad}(Rq^{(e)})$ yields that $Sq \subset \mathrm{rad}(Sq^{(e)})$ and it is then clear that $\mathrm{rad}(Sq) = \mathrm{rad}(Sq^{(e)})$ holds. ∎

V.2.26. Theorem. If R is a prime left noetherian GZ-ring then Proj(R) is a scheme, in the following sense. There exists a basis $\mathcal{B}$ for the Zariski topology on Proj(R), consisting of open sets $X_+(I)$ such that $X_+(I)$ endowed with the induced topology and sheaf is isomorphic to $\mathrm{Spec}(Q_I^g(R)_0)$, with its usual topology and structure sheaf. If $X_+(I)$ and $X_+(J)$ are in $\mathcal{B}$, then the ring $(Q_{IJ}^g(R))_0$ is generated as a ring by the restrictions of the rings $(Q_I^g(R))_0$ and $(Q_J^g(R))_0$.

Proof. For the set $\mathcal{B}$ we choose the set of all $X_+(Rc)$, where c runs through the homogeneous elements in C_+. Since R is Zariski central, this is a basis for the topology in Proj(R) as well as a covering for it. The proof splits into three parts : A. The settheoretic bijective map;

B. The topological homeomorphism;

C. The (pre)-sheaf isomorphism.

Bijective correspondence between the sets $X_+(Rc)$ and $\mathrm{Spec}(Q_c^g(R)_o)$. Let $q_o \in \mathrm{Spec}(Q_c^g(R)_o)$; $Q_c^g(R)_o$ is Zariski $Q_c^g(C)_o$ central, so, if $p_o = q_o \cap Q_c^g(C)_o$, then p_o is prime and q_o is the unique prime ideal of $Q_c^g(R)_o$, lying over p_o, i.e. $q_o = \mathrm{rad}(Q_c^g(R)_o p_o)$. Define q' by

$$q'_m = \{d \in C_m ; d^e c^{-m} \in q_o\} .$$

where $e = \deg(c)$. It is clear that q' is unaltered if one substitutes p_o for q_o in this definition! First, let us establish that $q'_m = \{d \in C_m ; \exists N, M \in \mathbb{N},\ eM = mN$ and $d^N c^{-M} \in q_o\}$. That q'_m is contained in this set is clear. Conversely, suppose $N \geq e$, then $M \geq m$ and $(d^e c^{-m}).(d^{N-e} c^{-M+m}) \in q_o$. Now, since q_o is prime and both factors are central, it follows that either $d^e c^{-m} \in q_o$, and then we are done or else $d^{N-e} c^{-M+m} \in q_o$. In this case we repeat the procedure, and in the end we have to consider the case $N < e$, $M < m$ with $eM = Nm$. In this situation, we have that $(d^{e-N} d^N)(c^{M-m} c^{-M}) = (d^{e-N} c^{M-m})$. $(d^N c^{-M})$ is an element of $Q_c^g(C)_o q_o \subset q_o$ and thus $d^e c^{-m} \in q_o$. It is easily derived from the latter characterization of $q' = \Sigma q'_m$ that it is a graded ideal of C and moreover that it is a graded prime ideal of C.

Consider $C \cap Rq'$ and pick $c_1 = \Sigma'_i x_i y_i \in C \cap Rq'$, with $x_i \in R$, $y_i \in q'$ and as $C \cap Rq'$ is graded, we may suppose that c_1, r_i and y_i are homogeneous. Since y_i commutes with r_j for all couples (i,j), we may choose N large enough such that

$$c_1^N = \Sigma' r_{i_1} \ldots r_{i_N} y_1^{\nu_1} \ldots y_i^{\nu_i - e} y_i^e ,\ \nu_i > e$$

i.e. each term in the sum has the form given above with respect to at least one index i appearing in the expression for c_1. We may assume that $N = e.\nu$, enlarging it if necessary. Thus $c_1^N c^{-\nu m}$, with $m = \deg(c_1)$, is in $(Q_c^g(R))_o q_o \subset q_o$, because each term in $c_1^N c^{-\nu m}$ may be written in the form :

$$(c^{-\nu m+\deg(y_i)} . r_{i_1}\dots r_{i_N} y_1^{\nu_1}\dots y_i^{\nu_i - e})(y_i^e c^{-\deg(y_i)})$$

which is obviously in q_o, since $y_i \in q'$.

It follows that $Rq' \cap C = q'$, hence $\mathrm{rad}(Rq') = Q$ is a prime ideal of R, since R is assumed to be Zariski central. Furthermore, if we write Q^{ex} for $Q_c^g(R)Q$, then, we find $Q^{ex} = \mathrm{rad}(Q_c^g(R)q')$ and by (V.2.25.) we obtain $Q^{ex} = \mathrm{rad}(Q_c^g(R)q'^{(e)})$. Clearly Q is a graded prime ideal of R which does not contain c, because $c \in Q$ yields $c \in q'$ and $1 \in q_o$, therefore $Q \in X_+(Rc)$. So we have obtained a well-defined assignment

$$\psi : \mathrm{Spec}(Q_c^g(R)_o) \to X_+(Rc)$$

$$q_o \mapsto Q \quad .$$

Conversely, given $Q \in X_+(Rc)$, consider $(Q^{ex})_o \cap (Q_c^+(C))_o$. If y belongs to this set, then $c^n g \in Q \cap C_{ne}$, for some $n \in \mathbb{N}$. Put $q' = Q \cap C$ and form $q'^{(e)}$. Then

$$p_o = \{d \in Q_c^g(C)_o ;\ \exists n \in \mathbb{N},\ c^n d \in q'_{ne}\}$$

is a prime ideal of $Q_c^g(C)_o$, so the relation $c^n y \in Q \cap C_{ne}$ translates to $y \in p_o$, i.e. $(Q^{ex})_o \cap Q_c^g(C))_o = p_o$, and by the Zariski $Q_c^g(C)_o$-centrality of $Q_c^g(R)_o$, it follows that $\tilde{q}_o = \mathrm{rad}((Q^{ex})_o)$ is a prime ideal of $Q_c^g(R)_o$. It is trivial to verify that Q corresponds to $\tilde{q}_o$ in the way described first, hence the constructed map ψ is clearly surjective. Injectivity of ψ will follow from $q_o = \tilde{q}_o = \mathrm{rad}((Q^{ex})_o)$, where $\psi(q_o) = Q$ and $\psi(\tilde{q}_o) = Q$. It has been established that $Q^{ex} = \mathrm{rad}(Q_c^+(R)q'^{(e)})$. Now, if $x \in (Q^{ex})_o$, then $Q_c^g(R)x \subset Q^{ex}$, and as $Q_c^g(R)$ is left noetherian, it follows that

$$Q_c^g(R)x\, Q_c^g(R)x \dots Q_c^g(R)x \subset Q_c^g(R)q'^{(e)}.$$

Since $q'^{(e)}$ is positively graded, taking parts of degree o in the foregoing inclusion yields

$$Q_c^g(R)_o \times Q_c^g(R)_o \times \dots Q_c^g(R)_o \times \subset Q_c^g(R)_o\, q'_o + \sum_i Q_c^g(R)_{-ie}(q')_{ie} \; .$$

However, $x\in Q_c^g(R)_{-ie}$ means that $c^i x\in Q_c^g(R)_o$, hence $Q_c^g(R)_{-ie}=Q_c^g(R)_o\ c^{-1}$, whence, for each index i we get

$$Q_c^g(R)_{-ie}(q')_{ie} = Q_c^g(R)_o\ c^{-1}(q')_{ie}.$$

Now $y\in(q')_{ie}$ means that $yc^{-1}\in p_o$, so, substituting $q_o'\subset p_o$ above, we derive

$$Q_c^g(R)_o \times Q_c^g(R)_o \times \ldots\ Q_c^g(R)_o \times \subset Q_c^g(R)_o\ p_o .$$

Thus $x\in \mathrm{rad}(Q_c^g(R)_o\ p_o)=q_o$. Finally, we obtain $p_o\subset(Q^{ex})_o\subset q_o$, and $q_o=\mathrm{rad}((Q^{ex})_o)$ implying that $q_o=\mathrm{rad}((Q^{ex})_o)$, because q_o is prime and $p_o=q_o\cap(Q_c^+(C))_o$.

<u>The homeomorphism</u> $X_+(Rc)\to \mathrm{Spec}(Q_c^g(R)_o)$.

Since $X_+(Rcd)=X_+(Rc)\cap X_+(Rd)$, for any $c,d\in C_+$, the open sets $X_+(Rcd)$ form a basis for the topology induced on $X_+(Rc)$, if we allow d to vary through C_+. Let

$$\psi : X_+(Rc)\to \mathrm{Spec}(Q_c^g(R)_o)$$
$$P\mapsto \mathrm{rad}((P^{ex})_o),$$

be the bijective map constructed before.

Let $d\in C_m$, then $d^e\ c^{-m}\in Q_c^g(R)_o$. If $P\in X_+(cd)$, then $P\in \mathrm{Proj}(R)$ has the property that $cd\notin P$. This implies that $d^e\ c^{-m}\notin \mathrm{rad}((P^{ex})_o)$. Indeed, $(d^e\ c^{-m})^N\in(P^{ex})_o$ would yield $d^{eN}\in P^{ex}$ or $c^M\ d^{eN}\in P$ for some $M\in\mathbb{N}$. Since both c and d are central, $c\notin P$ implies $d\in P$, hence $cd\in P$, a contradiction. Conversely, if $\psi(P)$ lies in the open set

$$\{q\in \mathrm{Spec}(Q_c^g(R)_o);\ d^e\ c^{-m}\notin q\} ,$$

then it is easily seen that $cd\notin P$. That ψ is a homeomorphism of topological spaces is now evident.

<u>The sheaf isomorphism.</u>

Let $c,d\in C_+$ and put $\deg(c)=e$, $\deg(d)=m$, $c'=c^m\ d^{-e}\in Q_d^g(C)_o$, $d'=d^e\ c^{-m}\in Q_c^g(C)_o$.

To the open set $X_+(Rcd)$ in $X_+(Rc)$, there corresponds the open set

$$Y_{d'} = \{q \in \mathrm{Spec}(Q_c^g(R)_o); d' \notin q\}$$

in $\mathrm{Spec}(Q_c^g(C)_o)$. In the (pre) sheaf structure of $X_+(Rc)$, the ring of sections over $X_+(Rcd)$ is $Q_{cd}^g(R)_o$, while the presheaf on $\mathrm{Spec}(Q_c^g(R)_o)$ has the ring $Q_{d'}(Q_c^g(R)_o)$, sitting over $Y_{d'}$, where $Q_{d'}$ denotes the localization functor in $Q_c^g(R)_o$-mod, corresponding to the central multiplicatively closed subset generated by d'.

If R is a prime ring, as we have assumed, then the presheaf induced on $X_+(Rc)$ is actually a sheaf. However $Q_c^g(R)_o$ need not necessarily be a prime ring, but, as it is certainly left noetherian, the presheaf on $\mathrm{Spec}(Q_c^g(R)_o)$ is separated. To prove the sheaf isomorphism it will thus be sufficient to find a basis of open sets for the topology, such that over basic open sets the ring of sections of the sheaf on $X_+(Rc)$ is isomorphic to the ring sitting over the image of the same open set in the presheaf on $\mathrm{Spec}(Q_c^g(R)_o)$ because then sheafification yields the isomorphism of the corresponding sheaves. So it will suffice to prove

$$Q_{d'}(Q_c^g(R)_o) = Q_{cd}^g(R)_o = Q_{c'}(Q_d^g(R)_o).$$

Since c and d commute, the second isomorphism will follow from the first by a symmetry argument.

Define a ring morphism $\varphi : Q_{d'}(Q_c^g(R)_o) \to Q_{cd}^g(R)_o$ by $\varphi((d')^{-N}(Xc^{-f})) = (Xc^{(e+m)}d^f).(cd)^{-(f+eN)}$, where $x \in Q_c^g(R)_{fe}$. The inverse homomorphism $\theta : Q_{cd}^g(R)_o \to Q_{d'}(Q_c^g(R)_o)$ will then be expressed by $\theta(y(cd)^{-f}) = (d')^{-M}(yd^{f(e-1)})(c^{-f(m+1)})$, where $y \in Q_{cd}^g(R)_{f(e+m)}$. Everything is easily checked; indeed this part of the proof is identical to the proof of the commutative analogue, since all localizations occuring are central localizations. Also the fact that $Q_{cd}^g(R)_o$ is generated by the images of $Q_c(R)_o$ and $Q_d^g(R)_o$ is now mere verification. ∎

If the ring R is not prime, then the presheaves of $X_+(I)$ and $\mathrm{Spec}(Q_I^g(R)_o)$ are isomorphic over the basic open sets $X_+(J)$ contained in $X_+(I)$. In this case the scheme structure is checked on the associated sheaf. Actually, the same proof may be repeated, taking care to write $j_c(R)$ instead of R, whenever needed. But since the torsion part behaves nicely, the whole proof carries over. So, finally one obtains, for a basis of the topology isomorphisms (over the basic open sets) between the rings occuring in the presheaves defined over $X_+(Rc)$ and $\mathrm{Spec}(Q_c^g(R)_o)$. The result follows then immediatelyfrom the well-known properties of the associated sheaf.

<u>V.2.27. Remark.</u> If in the situation of the theorem we have that $R_o \subset C$, i.e. $R_o = C_o$, then it follows that $Q_c^g(R)_o$ is an extension of R_o and therefore to the canonical ring morphism $R_o \to Q_c^g(R)_o$ there corresponds a presheaf morphism $\mathrm{Spec}(Q_c^g(R)_o) \to \mathrm{Spec}(R_o)$ In this case Proj(R) may be viewed as a $\mathrm{Spec}(R_o)$-scheme.

<u>V.2.28. Remark.</u> Let R be a left noetherian GZ-ring with center C and suppose that C is generated as a ring by C_1 over the subring C_o. Then, with the notations of the foregoing theorem, we have

1. if $P \in X_+(Rc)$, then $(P^{ex})_o$ is a prime ideal of $Q_c^g(R)_o$, i.e. the bijection ψ is given by $\psi(P) = (P^{ex})_o$;
2. if R is a prime ring, then $Q_c^g(R)_o$ is prime, i.e. the presheaf $\mathrm{Spec}(Q_c^g(R)_o)$ is a sheaf.

<u>Proof.</u> The second statement follows from the first, applied to the prime ideal o. Now, pick $c \in C_+$. It has been noted that $X_+(R\ )$ depends only on $\mathrm{rad}((Rc)_+)$, hence the fact that $C_m = C_1^m$ implies that we get a basis for the Zariski topology of Proj(R) by taking $\{X_+(Rc); c \in C_1\}$.

If I,J are ideals of $Q_c^g(R)_o$ such that $IJ \subset (P^{ex})_o$ for some $P \in X_+(Rc)$, then pick $i \in I$, $j \in J$. For any $q \in Q_c^g(R)_m$ we see that $c^{-m}q \in Q_c^g(R)_o$, hence $iqj \in c^m(P^{ex})_o \subset (P^{ex})_m$. Since P^{ex} is graded prime ideal of $Q_c^g(R)$, it follows that i or j is in P^{ex}, hence i or j is in $(P^{ex})_o$. ∎

The foregoing remark includes Azumaya rings over polynomial rings, but fails to include most rings of twisted formal power series. Indeed, the fact that R is generated by R_1 as an R_0-algebra, does not imply that C is generated by C_1 over C_0, as the following well-known example shows. Consider $R = \mathbb{C}[[t]]$ with $at = t\bar{a}$ $a \in \mathbb{C}$, then R is Zariski central with center $\mathbb{R}[[t^2]]$.

V.3. Bimodules and Structure Sheaves over Spec and Proj.

In this section we aim to remedy certain shortcomings of the ringed spaces constructed before, in particular, we use the localization techniques of V.1. in order to obtain new structure sheaves which behave functorially with respect to extensions of rings.

Throughout R is a left Noetherian ring and we write X for Spec(R), X_+ for Proj(R).

V.3.1. Proposition. If we assign the ring $Q_I^{bi}(R)$ to the open set X(I) of X where $I \not\subset \mathrm{rad}(o)$ then we obtain a presheaf of rings on Spec(R), denoted by $\underline{Q}^{bi}(R)$

Proof. The presheaf properties are easily verified.

V.3.2. Corollary. Let $M \in \underline{\underline{2}}(R)$. Assigning $Q_I^2(M)$ to X(I) for each ideal $I \not\subset \mathrm{rad}(o)$ of R, yields a presheaf $\underline{Q}^2(M)$ of two-sided R-modules over X. Choosing M = R, the presheaf $\underline{Q}^{bi}(R)$ is Module-isomorphic to a subpresheaf of $\underline{Q}^2(R)$.

V.3.3. Proposition. For any $M \in \underline{\underline{2}}(R)$, the presheaf $\underline{Q}^2(M)$ is separated.

Proof. Let the open set X(I) be covered by open sets $\{X(I_\alpha)\}_\alpha$. We have to show that the only $g \in Q_I^2(M)$ such that $\rho(I,I_\alpha)g = o$ for each I_α is $g = o$ (here $\rho(I,I_\alpha) : Q_I^2(M) \to Q_{I_\alpha}^2(M)$ are the restriction morphisms for the presheaf $\underline{Q}^2(M)$). Since R is left Noetherian, every open set of X is compact i.e. we may assume the covering $X(I) = \bigcup_\alpha X(I_\alpha)$ to be finite. There is an ideal $J \in \mathcal{L}(\kappa_I)$ such that

$Jg \subset M/\kappa_I(M)$. Hence if $g \in \bigcap_\alpha \operatorname{Ker} \rho(I, I_\alpha)$ then $Jg \subset \kappa_{I_\alpha}(M)/\kappa_I(M)$ for all α. Since $\kappa_{I_\alpha} \geq \kappa_I$ for each I_α and since the $X(I_\alpha)$ cover $X(I)$ it follows that $\bigcap_\alpha \kappa_{I_\alpha}(M)/\kappa_I(M) = o$. Thus $Jg = o$ but then $g \in \kappa_I(Q_I^2(M)) = o$. ∎

V.3.4. Corollary. The presheaf $\underline{Q}^{bi}(R)$ is separated too.

V.3.5. Theorem. If $M \in \underline{\underline{2}}(R)$ is torsion free with respect to every symmetric kernel functor then $\underline{Q}^2(M)$ is a sheaf.

Proof. It is well-known that $\underline{\underline{2}}(R)$ may be identified with R^e-mod where $R^e = R \underset{Z(R)}{\otimes} R^o$. Via this identification a kernel functor κ on $\underline{\underline{2}}(R)$ corresponds to a kernel functor κ^e on R^e-mod, moreover, the identification also identifies $Q_\kappa^2(M)$ and $Q_{\kappa^e}(M)$. From here on the proof reduces to the module case, treated in foregoing sections.

V.3.6. Remark. The analogues of V.3.1., V.3.4. remain valid if R is replaced by an arbitrary $M \in \underline{\underline{bi}}(R)$ i.e. we obtain a separated presheaf $\underline{Q}^{bi}(M)$. Let us write $\underline{O}_M^{bi}$ for the associated sheaf of $\underline{Q}^{bi}(M)$ and $\underline{O}_M^2$ for the sheaf associated to $\underline{Q}^2(M)$.

Recall that the inductive limit of R-bimodules is an R-bimodule. Moreover we have :

V.3.7. Lemma. Let $(E_\alpha, f_{\alpha\beta} \in \operatorname{Hom}_{\underline{\underline{2}}(R)}(E_\alpha, E_\beta))$ be an inductive system of two-sided R-modules. Assume there exists an index γ such that for all $\alpha \geq \gamma$ the maps $f_\alpha : E_\alpha \to E = \varinjlim E_\alpha$ are injective, then :

$$\varinjlim bi(E_\alpha) = bi(\varinjlim E_\alpha) = bi(E) .$$

Proof. Exactness of $\varinjlim$ entails that $\varinjlim bi(E_\alpha)$ injects into E and thus $\varinjlim bi(E_\alpha) \subset bi(E)$. Conversely, let us establish that $Z_R(\bigcup_\alpha f_\alpha(E_\alpha)) \subset \bigcup_\alpha f_\alpha(Z_R(E_\alpha))$. Since $\bigcup_\alpha f_\alpha(E_\alpha)$ is inductive, we may assume $\alpha \geq \gamma$ and take $r = f_\alpha(a_\alpha) \in Z_R(\bigcup_\alpha f_\alpha(E_\alpha))$. Hence for all $x \in R$ we obtain $xf_\alpha(a_\alpha) - f_\alpha(a_\alpha)x = f_\alpha(xa_\alpha - a_\alpha x) = o$. Now for $\alpha \geq \gamma$ we have: $\operatorname{Ker}(f_\alpha) = o$, thus $xa_\alpha - a_\alpha x = o$ for all $x \in R$ i.e. $a_\alpha \in Z_R(E_\alpha)$.

V.3.8. Theorem. Suppose that $M \in \underline{\underline{bi}}(R)$ is finitely generated as a left R-module (i.e. M is a left Noetherian R-bimodule). Then for each prime ideal P of R we obtain :

$$\varinjlim_{P \in X(I)} Q_I^{bi}(M) = Q_{R-P}^{bi}(M) .$$

Proof. Clearly $\kappa_{R-P}(M)$ is finitely generated as a left R-submodule of M and so we can find an ideal $I \in \mathcal{L}(R-P)$ such that $I\,\kappa_{R-P}(M) = o$. Consequently $\kappa_{R-P}(M) = \kappa_I(M)$ (note : $\kappa_{R-P} = V\{\kappa_I,\ P \in X(I)$!) and this entails that for all J such that $P \in X(J) \subset X(I)$, the canonical maps $Q_J(M) \to Q_{R-P}(M)$ and $Q_J^{bi}(M) \to Q_{R-P}^{bi}(M)$ are injective. First, one easily checks that $\varinjlim_{P \in X(J)} Q_J(M) = Q_{R-P}(M)$.

Now, applying Lemma V.3.7. yields :

$$\varinjlim_{P \in X(I)} Q_I^{bi}(M) = \varinjlim_{P \in X(I)} bi(Q_I(M)) = bi(\varinjlim_{P \in X(I)} Q_I(M)) = bi\ Q_{R-P}(M) = Q_{R-P}^{bi}(M).$$

V.3.9. Remark. If $M \in \underline{\underline{2}}(R)$ is finitely generated on the left then for every prime ideal P of R : $Q_{R-P}^2(M) = \varinjlim_{P \in X(I)} Q_I^2(M)$.

V.3.10. Corollary. Let $M \in \underline{\underline{2}}(R)$ (resp. $M \in \underline{\underline{bi}}(R)$) be finitely generated on the left, then for any prime ideal P of R the stalk of $\underline{O}_M^2$ at P (resp. of $\underline{O}_M^{bi}$ at P) is given by $Q_{R-P}^2(M)$ (resp. $Q_{R-P}^{bi}(M)$).

It is cristal clear that for each prime ideal P of R the left R-module R/P is torsion free for each symmetric kernel functor κ on R-mod. With this in mind we prove :

V.3.11. Theorem. Let $M \in \underline{\underline{bi}}(R)$ be left Noetherian and absolutely torsion free (i.e. torsion free for all symmetric kernel functors), then $\Gamma(X, \underline{O}_M^{bi}) = M$.

Proof. That $\Gamma(X, Q^{bi}(M)) = M$ is easy enough. Since the presheaf $Q^{bi}(M)$ is separated we have an injective morphism $\underline{Q}^{bi}(M) \hookrightarrow \underline{O}_M^{bi}$, such that for each prime ideal P of R the following diagram is commutative :

$$\begin{array}{ccc} \Gamma(M) = \Gamma(X, \underline{O}^{bi}_M) & \searrow & \\ \uparrow & & Q^{bi}_{R-P}(M) \\ \Gamma(X, Q^{bi}_{\_}(M)) & \nearrow & \end{array}$$

Because M is absolutely torsion free, all arrows in the diagram represent monomorphisms of bimodules. Now $Q^{bi}_{R-P}(M)/M$ is κ_{R-P}-torsion hence so is $\Gamma(M)/M$, but since this holds for every $P \in \mathrm{Spec}(R)$ it follows that $\Gamma(M)/M = o$ i.e. $\Gamma(M) = M$.

<u>V.3.12. Corollary</u>. If R is a left Noetherian prime ring then $\Gamma(X, \underline{O}^{bi}_R) = R$.

<u>V.3.13. Proposition</u>. Let R be a left Noetherian prime ring. For any open subset X(I) of X : $\Gamma(X(I), \underline{O}^{bi}_R) = \bigcap_{P \in X(I)} Q^{bi}_{R-P}(R)$.

<u>Proof</u>. Pick $s \in \Gamma(X(I), \underline{O}^{bi}_R)$. Then s may be viewed as a section $s : X(I) \to \underline{\tilde{O}}^{bi}_R$ in the topological sense, where $\underline{\tilde{O}}^{bi}_R$ is the sheaf space (espace étalé) associated to $\underline{O}^{bi}_R$ in the usual way. Thus s(X(I)) may be written as $\{r_P, P \in X(I)\}$ with $r_P \in Q^{bi}_{R-P}(R)$ for each $P \in X(I)$. For each $X(J) \subset X$ and each $t \in \Gamma(X(J), Q^{bi}_{\_}(R))$, the set $\{P \in \mathrm{Spec}\, R,\ t(P) = r_P\}$ is open, where t(P) is obtained from $Q^{bi}_J(R) \to Q^{bi}_{R-P}(R)$: $t \mapsto t(P)$. Since $Q^{bi}_{\_}(R)$ is separated, we obtain for each $P \in X(I)$ a commutative diagram :

$$\begin{array}{ccc} \Gamma(X(I), \underline{O}^{bi}_R) & \searrow & \\ \uparrow & & \\ \Gamma(X(I), Q^{bi}_{\_}(R)) = Q^{bi}_I(R) & \longrightarrow & Q^{bi}_{R-P}(R) \end{array}$$

Thus $\Gamma(X(I), \underline{O}^{bi}_R) \subset \bigcap_{P \in X(I)} Q^{bi}_{R-P}(R) = S$.

Conversely, if $r \in S$ then the set $\Delta = \{r_P = r \in Q^{bi}_{R-P}(R),\ P \in X(I)\}$ defines a section: $X(I) \to Q^{bi}_{\_}(R) : P \mapsto r$. Since for each $t \in Q^{bi}_J(R) \subset Q^{bi}_o(R)$ either $t = r$ or $t \neq r$ it follows that $t^{-1}(\Delta) = X(I) \cap X(J)$ or $t^{-1}(\Delta) = \phi$.

<u>V.3.14. Corollary</u>. For any $M \in \underline{\underline{bi}}(R)$, finitely generated on the left, and any

X(I) we have $\Gamma(X(I), \underline{O}^{bi}_M) = \bigcap_{P \in X(I)} Q^{bi}_{R-P}(M)$.

V.3.15. Proposition. If X(I) is such that κ_I is a t-functor in $\underline{\underline{bi}}(R)$ then $\Gamma(X(I), \underline{O}^{bi}_M) = Q^{bi}_I(M)$, for every $M \in \underline{bi}(R)$ such that M is absolutely torsion free.

V.3.16. Corollary. If R is a left Noetherian prime ring then for any $f \in Z(R)$ we have $\Gamma(X(f), \underline{O}^{bi}_R) = R_f$.

Proof. Clearly κ_{Rf} is a t-functor in $\underline{\underline{bi}}(R)$ and $Q^{bi}_{Rf}(R) = Q_{Rf}(R) = R_f$ is central localization at $\{1, f, f^2, \ldots\}$.

V.3.17. Theorem. Let R be a left Noetherian prime ring such that Spec R has a basis of open sets which are geometric in the bimodule sense, then $Q^{bi}_{-}(R)$ is a sheaf of rings.

Proof. Consider kernel functors κ_I, κ_J for ideals I,J of R, such that κ_I and κ_J have property (T) for bimodules (then also $\kappa_{IJ} = \kappa_I \vee \kappa_J$ is a t-functor in the bimodule sense). For any $x \in Q^{bi}_{IJ}(R)$ there is an $n \in \mathbb{N}$ such that $(IJ)^n x \subset R$. But then $Q^{bi}_J(R)(IJ)^n x \subset Q^{bi}_J(R)$ implies $I^n x \subset Q^{bi}_J(R)$ (use the fact that $Q^{bi}_J(R)I$ is an ideal, whereas $Q^{bi}_J(R)J = Q^{bi}_J(R)$). Since $Q^{bi}_I(Q^{bi}_J(R))$ contains both $Q^{bi}_I(R)$ and $Q^{bi}_J(R)$ as subrings, we obtain $Q^{bi}_I(R)x = Q^{bi}_I(R)I^n x \subset Q^{bi}_I(Q^{bi}_J(R))$ i.e. $x \in Q^{bi}_I(Q^{bi}_J(R))$. Conversely, one easily shows that $Q^{bi}_I(Q^{bi}_J(R)) \subset Q^{bi}_{IJ}(R)$, thus we may identify $Q^{bi}_{IJ}(R)$ with $Q^{bi}_I(Q^{bi}_J(R))$. Let $\{X_\alpha = X(I_\alpha)\}$ be a basis of open sets which is geometric in the bimodule sense i.e. such that $\kappa_\alpha = \kappa_{I_\alpha}$ is a t-functor for bimodules. Let us introduce the following notation : assume that κ_I is a t-functor for R-bimodules, put $S = Q^{bi}_I(R)$ and put $M_\alpha = Q^{bi}_{I_\alpha}(M)$ for any R-bimodule M. We have to prove the following : if X(I) is covered by the basic open sets $X_1, \ldots, X_n$, then we have an exact sequence :

$$0 \to S \xrightarrow{\varphi} \prod_{1 \leq \alpha \leq n} Q^{bi}_{II_\alpha}(R) \underset{\nu}{\overset{\mu}{\rightrightarrows}} \prod_{1 \leq \alpha \leq n} \prod_{1 \leq \beta \leq n} Q^{bi}_{II_\alpha I_\beta}(R)$$

where φ maps $s \in S$ to the sequence of its images in $Q^{bi}_{II_1}(R),\ldots,Q^{bi}_{II_n}(R)$, and where $\mu((r_\alpha)_\alpha) = (r'_{\alpha\beta})_{\alpha,\beta}$ with $r'_{\alpha\beta}$ the image of r_α in $Q^{bi}_{II_\alpha I_\beta}(R)$, and similarly $\nu(r_\alpha)_\alpha) = (r''_{\alpha\beta})_{\alpha,\beta}$ with $r''_{\alpha\beta}$ the image of r_β in $Q^{bi}_{II_\alpha I_\beta}(R)$. If we agree to write T for the ring $\prod_\alpha Q^{bi}_{II_\alpha}(R)$ then we have to establish exactness of the sequence

$$0 \to S \to T \overset{i_1}{\underset{i_2}{\rightrightarrows}} T \underset{S}{\otimes} T$$

where φ is the canonical inclusion and i_1, i_2 are defined by $i_1(t) = 1 \otimes t$, $i_2(t) = t \otimes 1$. Actually this amounts to showing that the Amitsur complex associated to φ is exact upto the first degeneracy. (by Proposition 6.2. in [95]). It suffices to check that φ is a faithfully flat extension for bimodules. Obviously φ is flat for S-bimodules because all localizations involved are localizations at t-functors for bimodules. If $M \in \underline{\underline{bi}}(S)$ then $\prod_\alpha M \underset{S}{\otimes} T = M \underset{S}{\otimes} (\prod_\alpha S_\alpha) = \prod_\alpha (M \underset{S}{\otimes} S_\alpha) =$

$$= \prod_\alpha (M \underset{S}{\otimes} S \underset{R}{\otimes} R_\alpha) = \prod_\alpha (M \underset{R}{\otimes} R_\alpha) = \prod_\alpha M_\alpha .$$

The problem is now reduced to showing injectivity of the map $M \to \prod_{1 \leq \alpha \leq n} M_\alpha$: $m \mapsto (m_\alpha)_\alpha$, while this is an immediate consequence of the fact that $Q^{bi}_{-}(M)$ is a separated presheaf.∎

<u>V.3.18. Remarks</u>. 1. In the situation of V.3.17., $\underline{O}^{bi}_R$ is a sheaf of local rings (i.e. the stalks are "local" rings).

2. In V.3.17. we may replace R by any $M \in \underline{\underline{bi}}(R)$ which is finitely generated on the left and which is torsion free for every symmetric kernel functor on R-mod.

In view of Corollaries II.3.9. the foregoing results applied to an Azumaya algebra R with center $Z(R) = C$ simplify a lot. Indeed Spec(R)is homeomorphic to Spec(C)under restriction of ideals and if we identify these topological spaces then we may think of Spec(R)as being endowed with the structure sheaves $\underline{O}^{bi}_R$ and $\underline{O}_C$, the canonical structure sheaf on Spec(C). The above yields that $\underline{O}^{bi}_R \cong \underline{O}_C \underset{C}{\otimes} R$ and we deduce :

$$\mathbb{k}_R(P) = Q_{R-P}(R/P) = Q_{cl}(R/P) = \mathbb{k}_C(p) \underset{C}{\otimes} R ,$$

where $p = P \cap C$ and $\mathbb{k}_C(p) = Q_{C-p}(C/p)$.

V.3.19. Theorem. If R is a prime Azumaya algebra then we have
$(\mathrm{Spec}(R), \underline{O}_R^{bi}) \cong (\mathrm{Spec}(C), \underline{O}_C \underset{C}{\otimes} R)$

This theorem states that from the geometrical point of view Azumaya algebras form a trivial extension of commutative rings. Note also that V.3.19. has been stated without any Noetherian hypotheses; although one could use II.3.11. here, one may also use the results for Zariski central rings and radical localization which do not depend on finiteness conditions.

Now we return to the projective case. To each open set $X_+(I)$ of Proj(R) where R is a positively graded left Noetherian prime ring, we associate the ring $bi(Q_I^g(R))_o$. Thus we have defined a separated presheaf of rings on Proj(R) such that for each central element c of degree 1 we have that $bi(Q_c^g(R))_o = (Q_c^g(R))_o$, because of the following.

V.3.20. Lemma. Let R be any graded ring such that $C = Z(R)$ contains an invertible element c of degree 1. For any graded R-bimodule M we have $(bi\ M)_o = bi(M_o)$.

Proof. First note that bi M is graded. Take $m_o \in (bi\ M)_o$, then $m_o = \sum_{i=1}^k r_i z_i$, with $r_i \in R_{d_i}$, $z_i \in Z_R(M)_{-d_i}$. Therefore $m_o = \sum_{i=1}^k r_i c^{-d_i} \in R_o Z_{R_o}(M_o) = bi(M_o)$.
Conversely, if $x \in Z_{R_o}(M_o)$ and $r \in R_d$ then we have $c^d(rx - xr) = o$ i.e. $x \in Z_R(M)$; consequently $bi(M_o) \subset (bi\ M)_o$. ∎

Now define $\underline{O}_R^{+,bi}$ to be the sheaf associated to the presheaf $Q_{\_}^{g,bi}(R)$. By the foregoing, the restriction of $(Q_{\_}^{g,bi}(R))_o = ((Q_{\_}^g(R))_o)^{bi}$ to an open subset $X_+(Rc)$, where c is central of degree 1, is the structure sheaf $\underline{O}_S^{bi}$ over Spec S, where $S = (Q_c^g(R))_o$, the latter topological space identified with $X_+(Rc)$. Clearly, similar constructions allow to construct a presheaf $Q_{\_}^{g,bi}(M)$ and a sheaf $\underline{O}_M^{+,bi}$ for an arbitrary graded R-bimodule M.

V.3.21. Theorem. Let R be a positively graded prime left Noetherian ring satisfying the conditions PS1, 2 then Proj(R) may be endowed with a sheaf of rings $\underline{O}_R^{+,bi}$ such that for each homogeneous central element c of degree 1 we have :

$$(X_+(Rc), \underline{O}_R^{+,bi}|X_+(Rc)) = (\mathrm{Spec}(Q_c^g(R)_o), \underline{O}^{bi}_{Q_c^g(R)_o}) .$$

Let us now fulfill the promises made at the beginning of this section and show how these new constructions yield a satisfactory functorial behaviour. As each scheme or variety is built up from its irreducible components, there is, up to a certain point no real restriction in limiting ourselves to irreducible spaces. Nevertheless in what follows, it will sometimes be convenient to use reducible (affine) varieties. Unless otherwise mentioned topological spaces or related concepts will be irreducible. Recall that a topological space is irreducible, provided every finite intersection of nonempty open subsets is non empty. A generic point of an irreducible topological space X is a point, the closure of which is X itself.

V.3.22. Proposition. Let X_R be the topological space Spec(R) endowed with its Zariski topology, then for every $P \in X_R$ the set V(P) is irreducible, when endowed with the induced topology. Conversely, every irreducible closed subset $S \subset X$ is of the form V(P) for some $P \in X_R$ and P is the unique generic point for S.

Proof. Straightforward.

V.3.23. Corollary. Spec(R) is irreducible if and only if rad(o) is prime, i.e. if R has a unique minimal prime ideal.

If $\varphi : R \to S$ is a ring homomorphism, we do not necessarily obtain an associated map ${}^a\varphi : \mathrm{Spec}(S) \to \mathrm{Spec}(R)$, since a subring of a prime ring is not necessarily prime. However, things improve considerably, when we assume φ to be a ring extension in the sense explained before. Let us first recall some easy results, due to Procesi [136] .

V.3.24. Proposition. Let $\varphi : R \to S$ be an extension, then the following statements hold.

1. if $P \in \mathrm{Spec}(S)$, then $\varphi^{-1}(P) \in \mathrm{Spec}(R)$;
2. the map ${}^a\varphi : \mathrm{Spec}(S) \to \mathrm{Spec}(R)$ given by ${}^a\varphi(P) = \varphi^{-1}(P)$, is continuous when Spec(R) and Spec(S) are endowed with their Zariski topologies;
3. we thus obtain a contravariant functor from the category of rings and extensions to the category of topological spaces and continuous morphisms.

Let us illustrate the usefulness of this topological Spec-map in the study of ringextensions. Recall the following well known definitions for a ring inclusion $R \subset S$; cf. section II.4.

1°. If for any prime ideal P of R there is a prime ideal Q of S with $Q \cap R = P$, then $R \subset S$ is said to have lying over (LO).

2°. If for any pair $P \subset P_1$ of prime ideals of R and any prime ideal Q of S such that $Q \cap R = P$, there is a prime ideal Q_1 of S such that $Q \subset Q_1$ and $Q_1 \cap R = R_1$, then $R \subset S$ is said to have going up (GU).

3°. If for two prime ideals $Q \subset Q_1$ of the relation $Q \cap R = Q_1 \cap R$ forces $Q = Q_1$, then $R \subset S$ is said to have incomparability (INC).

4°. If for any pair $P \subset P_1$ of prime ideals of R and any prime ideal Q_1 of S with $Q_1 \cap R = P_1$ there is a prime ideal Q of S with $Q \subset Q_1$ and $Q \cap R = P$, then $R \subset S$ is said to have going down (GD).

Let us mention some results related to II.4.18.

V.3.25. Proposition. Let $R \hookrightarrow S$ be an extension of p.i. rings; if ${}^a\varphi : \mathrm{Spec}(S) \to \mathrm{Spec}(R)$ is an open map, then $R \subset S$ has GD.

Proof. Let $P_1 \subset P$ be two primes of R such that we may find a prime ideal Q of S lying over P. Let $W = \{Q_\alpha \mid Q_\alpha \in \mathrm{Spec}(S)$ and $Q_\alpha \cap R = P_1\}$, and set $I = \cap\{Q_\alpha ; Q_\alpha \in W\}$. If $W = \phi$, we set $I = R$. Then II.4.4 states that there is a prime ideal Q_1 of S with $Q_1 \subset Q$ and $Q_1 \cap R = P_1$ if and only if $I \subset Q$. So, let us show $I \subset Q$. Suppose first that $W = \phi$, i.e. there is no prime ideal Q_α of S with $Q_\alpha \cap R = P_1$, then, since

Spec(S) is open in Spec(S) and ${}^a\varphi$ is assumed to be an open map, we find that ${}^a\varphi$(Spec(S)) is an open set sitting in Spec(R), not containing P_1. So Spec(R) - ${}^a\varphi$(Spec(S)) is of the form V(K) for some $K \subset P_1$. Hence P is also in this closed subset, contradicting the fact that $P \in {}^a\varphi(\mathrm{Spec}(S))$, since $Q \cap R = P$. Therefore $W \neq \phi$. Now, suppose $I \not\subset Q$. Let $T = \mathrm{Spec}(S) - V(I)$, then T is open in Spec(S) and $Q \in T$. So ${}^a\varphi(T)$ is open in R and $P \in {}^a\varphi(T)$. This gives the same contradiction as above, since $P_1 \notin {}^a\varphi(T)$. Thus $I \subset Q$, and it follows by loc.cit.that we may find $Q_1 \in \mathrm{Spec}(S)$ such that $Q_1 \subset Q$ and $Q_1 \cap R = P_1$, yielding the conclusion. ∎

The following should also be compared to II.4.18.

<u>V.3.26. Theorem.</u> Let $R \subset S$ be an injective ringextension. Suppose that for any ideal I of R there exist only a finite number of prime ideals which are minimal over I. Assume moreover that either $R \subset S$ has GU and INC or that $R \subset S$ is a finitely generated extension of P.I. rings, then if $R \subset S$ has GD then the map ${}^a\varphi : \mathrm{Spec}(S) \to \mathrm{Spec}(R)$ is open.

<u>Proof</u>. Let us first show that if $\{P_\alpha; \alpha \in A\}$ is a set of nonzero prime ideals of R, such that $\bigcap_\alpha P_\alpha = o$, then we may find a set of primes $\{Q_\alpha; \alpha \in A\}$ in S, such that $\bigcap_\beta Q_\beta = o$ and $Q_\beta \cap R \in \{P_\alpha; \alpha \in A\}$ for every β. Indeed, let $\{P_\alpha; \alpha \in A\}$ be such as asserted, then by GU we may find for each $\alpha \in A$ a prime ideal Q_α of S with $Q_\alpha \cap R = P_\alpha$. Let $I = \bigcap_\alpha Q_\alpha$. Observe that $I \cap R = (\bigcap_\alpha Q_\alpha) \cap R = \bigcap_\alpha (Q_\alpha \cap R) = \bigcap_\alpha P_\alpha = o$. If $I \neq o$, then we may enlarge I to an ideal I' of S, which is maximal with respect to $I' \cap R = o$. But then I' is a prime ideal of S containing the prime ideal o of S. This contradicts INC in $R \subset S$, yielding $\bigcap_\alpha Q_\alpha = o$ as claimed. On the other hand, for S finitely generated over R and satisfying a polynomial identity, the result is known, cf.[107]. Now, let U be an open subset of Spec(S). We wish to show that ${}^a\varphi(U)$ is an open subset of Spec(R). It will suffice to show that $\{P_\alpha; \alpha \in A\} = \mathrm{Spec}(R) - {}^a\varphi(U)$ is closed. Let $I = \bigcap_\alpha P_\alpha$, then $\{P_\alpha; \alpha \in A\}$ is closed in Spec(R), if and only if $V(I) = \{P_\alpha; \alpha \in A\}$, i.e. exactly then when $I \subset P$ implies $P \in \{P_\alpha; \alpha \in A\}$ for any prime P of R. Let us show the validity of this assertion. Suppose that there is a prime ideal

P of R <u>not</u> in $\{P_\alpha;\alpha\in A\}$ but containing I. Let us assume at this point that P is minimal over I. Let $\{P=P_1,P_2,\dots,P_n\}$ be the collection of primes of R, which are minimal over I. For each $1\leq i\leq n$, we put

$$J_i=\cap\{P_\alpha;P_i\subset P_\alpha\}\quad .$$

Since each P_α contains one of the $P_i\in\{P_1,\dots,P_n\}$, it follows that $J_1\cap\dots\cap J_n\subset I\subset P$. Since P is prime we find $J_k\subset P$ for some k. If $k\neq 1$, then $P_k\subset J_k$ and $P_k\not\subset P_1$, hence $J_k\not\subset P_1$. Thus $J_1\subset P$, which implies that $J_1=P$ in view of $P\subset J_1$ and the definition of J_1. So P is an intersection of some subset $\{P_r\}\subset\{P_\alpha;\alpha\in A\}$. Furthermore, since $P\notin\{P_\alpha:\alpha\in A\}=\mathrm{Spec}(R)-{}^a\varphi(U)$, there is in Spec(S) a prime ideal $Q\in U$, such that $Q\cap R=P$.

We now consider the extension of prime rings $R/P=\overline{R}\subset\overline{S}=S/Q$. It is easy to see that, if $R\subset S$ has GU and INC, then so does $\overline{R}\subset\overline{S}$. Similarly, if $R\subset S$ is a finitely generated extension of PI rings, then so is $\overline{R}\subset\overline{S}$. We also have a collection of prime ideals $\{\overline{P}_r\}$, all of them nonzero, with $\overline{P}_r=P_r/P$ and $\bigcap_r\overline{P}_r=\overline{P}=\overline{o}$ in $\overline{R}$. If $\overline{R}\subset\overline{S}$ is finitely generated we will use [407] , in the other case the first part of this proof. In either case we find a set of prime ideals $\{\overline{Q}_\beta;\beta\in B\}$ of $\overline{S}$ so that $\overline{Q}_\beta\cap\overline{R}\in\{\overline{P}_r\}$ for each $\beta\in B$ and such that $\bigcap_\beta\overline{Q}_\beta=\overline{o}$. Pulling back, we have a set $\{Q_\beta;\beta\in B\}\subset\mathrm{Spec}(S)$ with $Q_\beta\cap R\in\{P_\alpha;\alpha\in A\}$ for each $\beta\in B$ and $\bigcap_\beta Q_\beta=o$.

Now, U is open in Spec(S) and $Q\in U$, so we derive that $Q_\beta\in U$ for some β. Indeed, otherwise $\{Q_\beta;\beta\in B\}\subset\mathrm{Spec}(S)-U$, a closed subset V(K) of Spec(S). Since $Q_\beta\notin U$ implies $K\subset Q_\beta$, it follows that $K\subset\bigcap_\beta Q_\beta=Q$, implying $Q\in V(K)$, a contradiction. We must have $Q_\beta\in U$ for some $\beta\in B$. But then $Q_\beta\cap R\in{}^a\varphi(U)$ and $Q_\beta\cap R\in\{P_\alpha;\alpha\in A\}$-contradiction. Hence any prime ideal of R minimal over I is in $\{P_\alpha,\alpha\in A\}$.

Finally, let P be any prime ideal of R containing I and take $I\subset P_1\subset P$, where $P_1\in\mathrm{Spec}(R)$ is minimal over I. If $P\notin\{P_\alpha;\alpha\in A\}$, there is $Q\in U$ such that $Q\cap R=P$. By GD in $R\subset S$ we may pick $Q_1\in\mathrm{Spec}(S)$ with $Q_1\subset Q$ and $Q_1\cap R=P_1$. Since $P_1\in\{P_\alpha;\alpha\in A\}$ by the minimality assumption, we obtain $Q_1\notin U$, i.e. $Q_1\in V(K)=\mathrm{Spec}(S)-U$, so $K\subset Q_1\subset Q$, which yields $Q\in V(K)$. This contradicts $Q\in U$, hence $P\in\{P_\alpha;\alpha\in A\}$ for any $P \in \mathrm{Spec}(R)$ which contains I. So we have obtained that ${}^a\varphi(U)$ is open in Spec(R). ∎

In what follows, II.2.10 and II.2.11 play an important though implicite rôle.

V.3.27. Lemma. Let $\varphi : R \to S$ be a ring extension and let Q be a prime ideal of S. If we write P for $\varphi^{-1}(Q)$, then

$$\varphi_* \kappa_{R-P} \leq \kappa_{S-Q} .$$

Proof. If $I \in \mathcal{L}^2(\varphi_* \kappa_{R-P})$, then $\varphi^{-1}(I) \in \mathcal{L}^2(\kappa_{R-P})$. As $\varphi^{-1}(I)$ is twosided too, there is $s \in R-Q$ such that $s \in \varphi^{-1}(I)$. But this implies that $I \in \mathcal{L}(\kappa_{S-Q}) = \mathcal{L}(S-Q)$, for $\varphi(s) \in Q$ implies that $s \in \varphi^{-1}(Q) = P$. ∎

V.3.28. Corollary. Let $\varphi : R \to S$ be a ring extension and Q a prime ideal of S. Write P for the prime ideal $\varphi^{-1}(Q)$ of R. If κ_{R-P} has property (T) for R-bimodules (resp. left R-modules), then there is a canonical ringextension

$$\varphi_Q : Q^{bi}_{R-P}(R) \to Q^{bi}_{S-Q}(S)$$

(resp. $\varphi_Q : Q_{R-P}(R) \to Q_{S-Q}(S)$.) which extends φ.

Proof. This follows easily (at least in the bimodule case) from the technique of torsion free extensions, but can be seen directly as follows : compose the canonical morphisms

$$Q^{bi}_{R-P}(R) \to Q^{bi}_{R-P}(S) \cong Q^{bi}_{\varphi_*\sigma_{R-\varphi^{-1}(Q)}}(S) \to Q^{bi}_{S-Q}(S),$$

where the first map is induced by φ, the second is deduced from IV.2.5. and the third from IV.1.37. That φ_Q extends φ follows from its construction. For the one sided case one proceeds similarly. ∎

This will prove to be useful in the sequel. If we know e.g. that all prime ideals of R induce a kernel functor which has property (T) for R-bimodules, and if

$\varphi: R\to S$ is an absolutely torsion free extension, or, in particular, if S is prime, then we may derive a morphism between the corresponding rings of quotients, extending φ. This allows, for example, for any two prime Zariski central rings over an algebraically closed field, to introduce a geometric product of the corresponding varieties, without knowing whether their tensorproduct is Zariski central too. Another application is to arbitrary semiprime P.I. rings. These rings are known to be birational extensions of their center, and hence possess on an open subset of their spectrum kernel functors which have property (T) for R-bimodules.

Recall some definitions. A <u>morphism of ringed spaces</u>

$$\Phi : X = (X,\underline{O}_X) \to Y = (Y,\underline{O}_Y)$$

is a couple (φ,θ), where $\varphi: X\to Y$ is a continuous map between the underlying topological spaces and where $\theta:\underline{O}_Y\to\varphi_*(\underline{O}_X)$ is a sheaf morphism between the corresponding sheaves of rings on Y. Recall that the induced sheaf $\varphi_*(\underline{O}_X)$ on Y is defined by

$$\Gamma(U,\varphi_*(\underline{O}_X)) = \Gamma(\varphi^{-1}(U),\underline{O}_X) ,$$

for each open subset U of Y.

<u>V.3.29. Lemma</u>. Let (φ,θ_1) and (φ,θ_2) be two morphisms between the ringed spaces $(X,\underline{O}_X)$ and $(Y,\underline{O}_Y)$. These morphisms induce for each $x\in X$, $y=\varphi(x)\in Y$ morphisms

$$\theta_{i,x}:\underline{O}_{Y,y}\to\underline{O}_{X,x} \quad (i=1,2)$$

between the corresponding stalks. Suppose that for each $x\in X$ these morphisms coincide, then $\theta_1=\theta_2$.

<u>Proof</u>. It suffices to prove that for each open subset U of X the morphisms $\theta_1(U)$ and $\theta_2(U):\underline{O}_Y(U)\to\underline{O}_X(\varphi^{-1}(U))$ coincide. Consider the following commutative diagram

$$\begin{array}{ccc}
\underline{O}_Y(U) & \overset{\theta_1(U)}{\underset{\theta_2(U)}{\rightrightarrows}} & \underline{O}_X(\varphi^{-1}(U)) \\
\Big\downarrow \rho^U_y & & \Big\downarrow \rho^{\varphi^{-1}(U)}_x \\
\underline{O}_{Y,y} & \xrightarrow[\theta_{1,x} = \theta_{2,x}]{} & \underline{O}_{X,x}
\end{array}$$

Take $s \in \underline{O}_Y(U)$, then

$$(\rho^{\varphi^{-1}(U)}_x \circ \theta_1(U))(s) = (\rho^{\varphi^{-1}(U)}_x \circ \theta_2(U))(s),$$

hence there exists an open neighborhood V_x of x contained in $\varphi^{-1}(U)$ such that

$$(\ast) \qquad \rho^{\varphi^{-1}(U)}_{V_x} (\theta_1(U)(s)) = \rho^{\varphi^{-1}(U)}_{V_x} (\theta_2(U)(s)) .$$

As this may be realized for each $x \in \varphi^{-1}(U)$, we may cover $\varphi^{-1}(U)$ by a family of open sets $\{V_x;\ x \in \varphi^{-1}(U)\}$, such that $(\ast)$ holds for each x; but as $\underline{O}_X$ is a sheaf, this yields $\theta_1(U)(s) = \theta_2(U)(s)$, which yields the assertion. ∎

<u>V.3.30. Proposition.</u> **Let $\varphi : R \to S$ be an extension and Q a prime ideal of S. If $P = \varphi^{-1}(Q)$, then there is at most one ringmorphism $\varphi_Q : Q^{bi}_{R-P}(R) \to Q^{bi}_{S-Q}(S)$ and at most one ringmorphism $\varphi_Q : Q_{R-P}(R) \to Q_{S-Q}(S)$ extending φ.**

<u>Proof</u>. Assume that φ_Q and φ'_Q both extend φ, i.e. in the following diagram

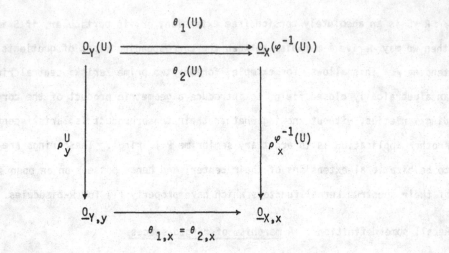

$$\begin{array}{ccc} R & \xrightarrow{\varphi} & S \\ {\scriptstyle j_P}\downarrow & & \downarrow{\scriptstyle j_Q} \\ Q^{bi}_{R-P}(R) & \overset{\varphi_Q}{\underset{\varphi'_Q}{\rightrightarrows}} & Q^{bi}_{S-Q}(S) \end{array}$$

the morphisms φ_Q and φ'_Q coincide on R. Hence $\psi = \varphi_Q - \varphi'_Q$, which is left and right R-linear, factorizes through $Q^{bi}_{R-P}(R)/R$:

$$\begin{array}{ccc} Q^{bi}_{R-P}(R) & \xrightarrow{\pi} & Q^{bi}_{R-P}(R)/R \\ & {\scriptstyle \psi}\searrow \quad \swarrow{\scriptstyle \overline{\psi}} & \\ & Q^{bi}_{S-Q}(S) & \end{array}$$

Now, since $Q^{bi}_{R-P}(R)/R$ is κ_{R-P}-torsion by construction, and since $Q^{bi}_{S-Q}(S)$ is κ_{S-Q}-torsion free, hence $\varphi_* \kappa_{R-P}$-torsion free, and thus κ_{R-P}-torsion free, we get that $\overline{\psi} = o$. But then $\psi = o$, i.e. $\varphi_Q = \varphi'_Q$. Note that φ_Q is then obviously a ringextension. For the one sided case, one proceeds similarly. ∎

To each extension $\varphi : R \to S$ we want to associate a morphism of ringed spaces

$$\Phi = ({}^a\varphi, \theta) : (\mathrm{Spec}(S), \underline{O}^{bi}_S) \to (\mathrm{Spec}(R), \underline{O}^{bi}_R) .$$

The morphism between the underlying topological spaces is the following :

$$\begin{array}{rl} {}^a\varphi : \mathrm{Spec}(S) & \to \mathrm{Spec}(R) \\ Q & \mapsto \varphi^{-1}(Q) \quad . \end{array}$$

It is continuous and if I is an ideal of R and $J = (\varphi(I))$, the ideal of S generated by $\varphi(I)$, then $({}^a\varphi)^{-1}(X(I)) = Y(J)$, where X = Spec(R), Y = Spec(S).

We still have to define the sheaf morphism

$$\theta : \underline{O}_R^{bi} \to ({}^a\varphi)_*(\underline{O}_S^{bi}).$$

Note that for each open subset X_I of Spec(R), we have

$$\Gamma(X_I, ({}^a\varphi)_*(O_S^{bi})) = \Gamma({}^a\varphi^{-1}(X(I)), \underline{O}_S^{bi}) = \Gamma(Y(J), \underline{O}_S^{bi}),$$

hence for each ideal I of R it suffices to give in a functorial way a ring morphism

$$\theta_I : Q_I^{bi}(R) \to Q_J^{bi}(S),$$

and it is natural to want this morphism to be such that $\theta_R = \varphi : R \to S$. Exactly as in V.3.30 , one may prove that this forces any such morphism Φ associated to φ to be unique. We will effectively construct such a morphism.

In the one sided case, things evolve somewhat differently. Indeed, we have not succeeded in constructing a morphism $\theta : \underline{O}_R \to ({}^a\varphi)_*(\underline{O}_S)$ for any extension $\varphi : R \to S$, and for reasons which we will explain below, it seems highly improbable that such a morphism actually may be exhibited. Nevertheless in some particular cases, it will be shown how to construct the desired morphism θ, e.g. when φ is injective or a flat epimorphism.

<u>V.3.31. Lemma</u>. Let $\varphi : R \to S$ be an extension and I an ideal of R. If we denote by $J = (\varphi(I))$ the ideal of S generated by its image, then $\varphi_*\sigma_I = \sigma_J$.

<u>Proof</u>. Let M be a left S-module, then $m \in (\varphi_*\sigma_I)M$ iff there exists an ideal K of R such that $\text{rad} K \subset I$ and $\varphi(K)m = o$. As this is equivalent to $(\varphi(K))m = o$, it suffices to verify that $\text{rad}(\varphi(K)) \supset J$, to finish the proof.

But, if for some prime ideal Q of S, we have $Q \supset (\varphi(K))$, then $Q \supset \varphi(K)$ and $\varphi^{-1}(Q) \supset K$. As φ is an extension, $\varphi^{-1}(Q)$ is a prime of R, hence $\varphi^{-1}(Q) \supset \text{rad}(K) \supset I$, so $Q \supset \varphi(I)$ and finally this implies $Q \supset (\varphi(I)) = J$. As this holds for any prime ideal of S containing $(\varphi(K))$, we indeed get that $\text{rad}(\varphi(K)) \supset J$. ∎

<u>V.3.32. Lemma</u>. Let $\varphi : R \to S$ be an injective extension of prime p.i. rings then φ

induces an extension $\tilde{\varphi} : Q(R) \to Q(S)$ and, $\tilde{\varphi}(Z(Q(R))) \subset Z_S(Q(S))$.

Proof. Here obviously $Q(R) = Q^{bi}_{R-0}(R)$, the classical ring of quotients of R. Now, the first part of the statement has already been pointed out, and since $\tilde{\varphi}(Z(Q(R))) \subset Z(Q(S))$, it suffices to prove that $Z(Q(S)) = Z_S(Q(S))$. Take $s \in Q(S)$ and suppose that s commutes with each element of S if $s' \in Q(S)$, then there exists an ideal I such that $Is' \subset S$, hence for all $i \in I$, we have $iss' = sis' = (is')s$, so $I(ss' - s's) = o$, hence $ss' = s's$ since S is prime. This finishes the proof. ∎

V.3.33. Lemma. Let $\pi : R \to U$ be a surjective morphism of prime P.I. rings and let I be an ideal of R. If we put $K = \pi(I)$, the image of I, then there exists a unique extension $Q^{bi}_I(R) \to Q^{bi}_K(U)$ making the following diagram commutative

$$\begin{array}{ccc} R & \xrightarrow{\ \pi\ } & U \\ \downarrow & & \downarrow \\ Q^{bi}_I(R) & \longrightarrow & Q^{bi}_K(U) \end{array}$$

Proof. In view of V.3.31., this is a direct application of IV.2.5. ∎

V.3.34. Lemma. Let $i : U \to S$ be an injective extension of prime P.I. algebras and let K be an ideal of U. If we call $J = (i(K))$ the S-ideal generated by K, then there exists a unique extension $Q_K(U) \to Q_J(S)$ making the following diagram commutative

$$\begin{array}{ccc} U & \xrightarrow{\ i\ } & S \\ \downarrow & & \downarrow \\ Q_K(U) & \longrightarrow & Q_J(S) \end{array}$$

Proof. The inclusion $i : U \to S$ yields an extension $\rho : Q(U) \to Q(S)$, and obviously $Q_K(U) \hookrightarrow Q(U)$ resp $Q_J(S) \hookrightarrow Q(S)$. In order to derive the desired morphism $Q_K(U) \to Q_J(S)$, it suffices to check that ρ maps $Q_K(U)$ into $Q_J(S)$. Now, if $q \in Q_K(U) \subset Q(U)$, then

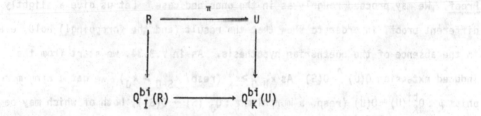

$K^n q \subset U$ for some $n \in \mathbb{N}$. But then

$$\rho(K)^n \rho(q) = SK^n q \subset S,$$

hence $J^n \rho(q) \subset S$, i.e. $\rho(q) \in Q_J(S)$. ∎

V.3.35. Lemma. Under the same assumptions, we may find a unique extension $Q_K^{bi}(U) \to Q_J^{bi}(S)$, making the following diagram commutative

$$\begin{array}{ccc} U & \xrightarrow{\ i\ } & S \\ \downarrow & & \downarrow \\ Q_K^{bi}(U) & \longrightarrow & Q_J^{bi}(S) \end{array}$$

Proof. We may proceed roughly as in the onesided case. Let us give a slightly different proof, in order to show that the result (and the foregoing!) hold, even in the absence of the noetherian hypothesis. As in V.3.34. we start from the induced extension $Q(U) \overset{\rho}{\to} Q(S)$. As $\kappa_{U-0} \geq \kappa_K$ (resp. $\kappa_{S-0} \geq \kappa_J$) we get a ring morphism $\mu : Q_K^{bi}(U) \to Q(U)$ (resp. a morphism $\nu : Q_J^{bi}(S) \to Q(S)$), both of which may be viewed as inclusions. Take $x \in Q^{bi}(U)$, then $x = \Sigma u_i z_i$, $u_i \in U$, $z_i \in Z(Q_K^{bi}(U)) \subset Z(Q(U))$, and there exists an ideal L such that $\mathrm{rad}(L) \supset K$ with $Lz_i \subset U$ for all i. Then

$$\rho\mu(Lz_i) = \rho\mu(L)\rho\mu(z_i) \subset S,$$

hence $(\rho\mu(L))\rho\mu(z_i) \subset S$. If we can prove that $\mathrm{rad}(\rho\mu(L)) \supset J$, then $\rho\mu(z_i) \in Z(Q(U))$ belongs to $Q_J^{bi}(S)$.

Furthermore, putting $i_K(x) = \Sigma i(u_i)(z_i)$ defines a map

$$i_K : Q_K^{bi}(U) \to Q_J^{bi}(S),$$

such that the following diagram is commutative :

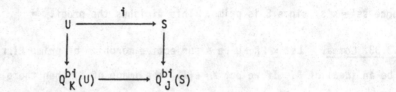

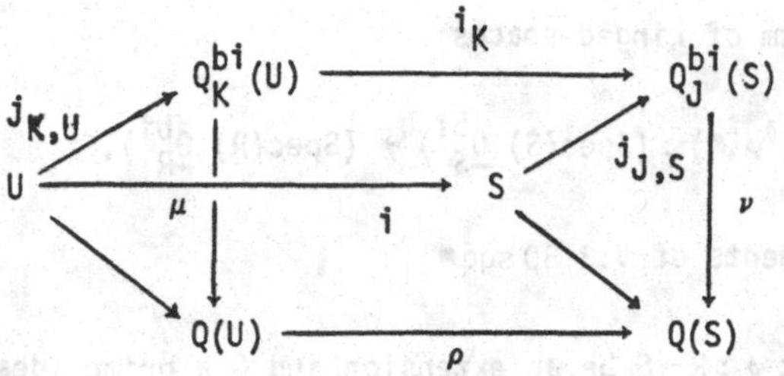

But the last statement is clear, for, if $Q \supset (\rho\mu(L))$ for some prime ideal $Q \in \mathrm{Spec}(S)$, then $\mu^{-1}\rho^{-1}(Q) \supset \mu^{-1}\rho^{-1}(\rho\mu(L)) \supset \mu^{-1}\mu(L) \supset L$, hence $\mu^{-1}\rho^{-1}(Q) \supset K$, so $Q \supset \rho\mu\mu^{-1}\rho^{-1}(Q) \supset$ $\supset \rho\mu(K)$. Thus, finally, $Q \supset (\rho\mu(K))$. The unicity of this morphism is checked as usually. ∎

Using the foregoing one now easily verifies that we have actually defined for each extension $\varphi : R \to S$ of prime p.i. algebras an essentially unique extension

$$\theta_I : Q_I^{bi}(R) \to Q_J^{bi}(S) ,$$

extending φ. Indeed, we first factorize φ through

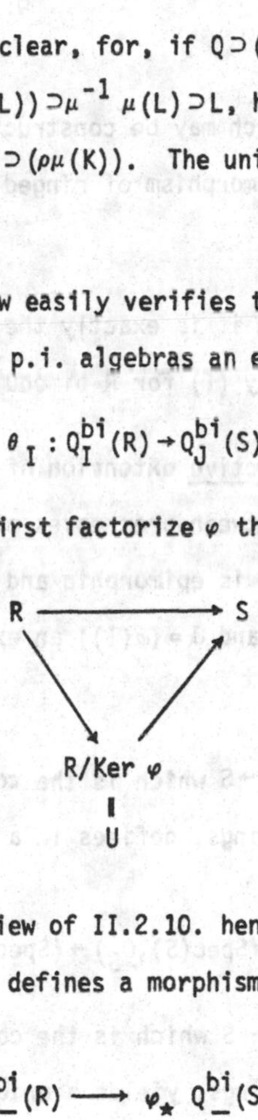

Note that U is prime, in view of II.2.10. hence we may apply V.3.33 to $R \to U = R/\mathrm{Ker}\,\varphi$ and V.3.35. to $U \hookrightarrow S$. This defines a morphism between the presheaves of rings on Spec(R):

$$\theta : Q_{-}^{bi}(R) \longrightarrow \varphi_{*}\, Q_{-}^{bi}(S)$$

hence by passing to the associated sheaves, we derive

<u>V.3.36. Theorem.</u> Each extension $\varphi : R \to S$ of prime P.I. rings defines in a

canonical way a morphism of ringed spaces

$$({}^a\varphi,\theta) : (\mathrm{Spec}(S),\underline{O}_S^{bi}) \to (\mathrm{Spec}(R),\underline{O}_R^{bi}),$$

satisfying the requirements of V.3.30 sqq ■

V.3.37. Corollary. Let $\varphi : R \to S$ be an extension and Q a prime ideal of S. If $P = \varphi^{-1}(Q)$, then there exists a unique ring morphism $\varphi_Q : Q_{R-P}^{bi}(R) \to Q_{S-Q}^{bi}(S)$ extending φ.

Proof. The desired ringmorphism (which may be constructed in a direct way), is the stalk map at Q obtained from the morphism of ringed spaces defined by the foregoing theorem. ■

Note that φ_Q is an extension and that it is exactly the map constructed in V.3.28 provided that κ_{R-P} has property (T) for R-bimodules.

The foregoing also shows that an injective extension of prime P.I. rings $\varphi : R \to S$ defines a morphism $\theta : \underline{O}_R^{\circ} \to \varphi_*(\underline{O}_S^{\circ})$ between presheaves of rings on Spec(R). Note that if $\varphi : R \to S$ is an extension which is epimorphic and flat in R-mod, then V.3.31. yields for each ideal I of R and $J = (\varphi(I))$ an extension : $\theta_I : Q_I^{bi}(R) \to Q_J^{bi}(S)$, which extends φ. Hence, as before

V.3.38. Theorem. Each extension $\varphi : R \to S$ which is the composition of inclusions and flat epimorphisms of prime P.I. rings, defines in a canonical way a morphism of ringed spaces

$$({}^a\varphi,\theta) : (\mathrm{Spec}(S),\underline{O}_S) \to (\mathrm{Spec}(R),\underline{O}_R),$$

V.3.39. Corollary. An extension $\varphi : R \to S$ which is the composition of inclusions and flat epimorphisms of prime P.I. rings, yields a unique ring morphism $\varphi_Q : Q_{R-P}(R) \to Q_{S-Q}(S)$ extending φ. ■

Let us now conclude this section by studying some functoriality aspects of Proj(R) and the sheaves on it. In the sequel of this section, R and S will be positively graded left noetherian prime P.I. rings, satisfying the usual conditions. Let $\varphi : R \to S$ be a graded extension (implying that the center of R is mapped into the center of S and that the inverse image $\varphi^{-1}(P)$ of a (graded prime ideal of S is a (graded) prime ideal of R). Put $\mathrm{Proj}(R) = Y_+(R)$, $\mathrm{Proj}(S) = X_+(S)$. Let U_φ be the open set $X_+(\varphi(R_+))$ of $X_+(S)$.

V.3.40. Lemma. φ induces a continuous morphism ${}^a\varphi : U_\varphi \to Y_+(R)$, given by $P \mapsto \varphi^{-1}(P)$.

Proof. Check that $({}^a\varphi)^{-1}(Y_+(I)) = U_\varphi \cap X_+(\varphi(I))$. ∎

V.3.41. Remark. Note that if $\varphi_d : R_d \to S_d$ is an isomorphism for every $d > d_o$ for some fixed $d_o \in \mathbb{N}$, then $U_\varphi = \mathrm{Proj}(S)$ and ${}^a\varphi$ is a homeomorphism, while φ need not be injective.

V.3.42. Lemma. If c is a homogeneous central element of R of degree 1, then the graded extension $\varphi : R \to S$ induces a graded extension $\varphi_c : Q^g_c(R) \to Q^g_{\varphi(c)}(S)$ and the ringmorphism $(\varphi_c)_o : Q^g_c(R)_o \to Q^g_{\varphi(c)}(S)_o$ is also a ringextension.

Proof. Since $\varphi(c)$ is central in S, the first statement is easily verified. Since $Q^g_{\varphi(c)}(S)$ is a $Q^+_c(R)$-bimodule, the second statement follows directly from V.3.20. ∎

V.3.43. Theorem. If we equip U_φ with the induced sheaf, then the morphism $\varphi : R \to S$ induces a morphism of ringed spaces

$$({}^a\varphi, \theta) : (U_\varphi, \underline{O}^{+,bi}_S | U_\varphi) \to (\mathrm{Proj}(R), \underline{O}^{+,bi}_R) .$$

Proof. Some details, easy enough to verify, will be left to the reader and we give a blue-print of the proof. Consider the structure sheaves $\underline{O}^{+,bi}_R$ and $\underline{O}^{+,bi}_S \big| U_\varphi$. We shall first establish that for each central homogeneous element c of degree 1 (afterwards : for any such c of positive degree) there is a sheaf morphism

$$\underline{O}^{+,bi}_R | Y_+(Rc) \to ({}^a\varphi)_*(\underline{O}^{+,bi}_S | U_\varphi) | Y_+(Rc).$$

Then it will be sufficient to glue these morphisms together in the obvious way, in order to obtain the desired morphism $\underline{O}_R^{+,bi} \to ({}^a\varphi)_*(\underline{O}_S^{+,bi}|U_\varphi)$.

Let $W = Y_+(I)$ be an open subset in $Y_+(Rc)$, then :

$$
\begin{aligned}
({}^a\varphi)_*(\underline{O}_S^{+,bi}|U_\varphi)(W) &= (\underline{O}_S^{+,bi}|U_\varphi)(({}^a\varphi)^{-1}(Y_+(I))) \\
&= (\underline{O}_S^{+,bi}|U_\rho)(U_\varphi \cap X_+(\varphi(I))) \\
&= \underline{O}_S^{+,bi}(U_\varphi \cap X_+(\varphi(I))) \\
&= (\underline{O}_S^{+,bi}|X_+(S\varphi(c)))(U_\varphi \cap X_+(\varphi(I))) \\
&= (\underline{O}_S^{+,bi}|X_+(S\varphi(c)))(U_\varphi \cap ({}^a\varphi)^{-1}(W)).
\end{aligned}
$$

Upon identifying $Y_+(Rc)$ with $\mathrm{Spec}(Q_c^g(R)_o)$ and $X_+(S\varphi(c))$ with $\mathrm{Spec}(Q_{\varphi(c)}^g(S)_o)$, we see that the induced sheaves $\underline{O}_R^{+,bi}|Y_+(Rc)$ resp. $\underline{O}_S^{+,bi}|X_+(S\varphi(c))$ correspond to the structure sheaves on $\mathrm{Spec}(Q_c^g(R)_o)$ resp. $\mathrm{Spec}(Q_{\varphi(c)}^g(S)_o)$. For simplicity's sake we will denote the latter by $\underline{O}_{R,c}$ resp. $\underline{O}_{S,\varphi(c)}$. Let τ be the ring extension $\varphi_{c,o} : Q_c^g(R)_o \to Q_{\varphi(c)}^g(S)_o$ obtained before. Hence the affine theory yields that τ induces a morphism of ringed spaces

$$(\mathrm{Spec}(Q_{\varphi(c)}^g(S)_o), \underline{O}_{S,\varphi(c)}) \to (\mathrm{Spec}(Q_c^g(R)_o), \underline{O}_{R,c}) \text{ ,}$$

and in particular a sheaf morphism

$$\theta : \underline{O}_{R,c} \to \tau_* \underline{O}_{S,\varphi(c)} \text{ .}$$

For each pair of open sets $V' \subset U'$ of $\mathrm{Spec}(Q_c^g(R)_o)$, this yields a commutative diagram :

$$
\begin{array}{ccccc}
\underline{O}_{R,c}(U') & \xrightarrow{\theta_{U'}} (\tau_* \underline{O}_{S,\varphi(c)})(U') = & \underline{O}_{S,\varphi(c)}(\tau^{-1}(U')) & \to & \underline{O}_{S,\varphi(c)}(\tau^{-1}(U')\varphi) \\
\downarrow & \downarrow & & & \downarrow \\
\underline{O}_{R,c}(V') & \xrightarrow[\theta_{V'}]{} (\tau_* \underline{O}_{S,\varphi(c)})(V') = & \underline{O}_{S,\varphi(c)}(\tau^{-1}(V')) & \to & \underline{O}_{S,\varphi(c)}(\tau^{-1}(V)\varphi)
\end{array}
$$

This yields the desired sheaf morphism :

$$\underline{O}_R^{+,bi}|Y_+(Rc) \to (a_\varphi)_*(\underline{O}_S^{+,bi}|U_\varphi)|Y_+(Rc)$$

and the rest is verification.

Let A be a commutative ring. We shall say that a graded ring R is a graded A-algebra if there is given a ring morphism $\xi : A \to R$ such that $\xi(A) \subset C_0$.

V.3.44. Proposition. If R is a graded A-algebra then $(\mathrm{Proj}(R), \underline{O}_R^{+,bi})$ is a ringed space over $(\mathrm{Spec}(A), \underline{O}_A)$, the commutative scheme associated with A in the canonical way.

Proof. It suffices to exhibit the underlying topological morphism and the sheaf morphism between the structure sheaves. Pick $c \in C_1$; then c induces a ring morphism $\xi_c : A \to Q_c^g(R)_0$ yielding a continuous map $\xi_c^a : \mathrm{Spec}(Q_c^g(R)_0) \to \mathrm{Spec}(A)$. Since ξ_c is an extension, it induces also a sheaf morphism between the structure sheaves, say $\tau_c : \underline{O}_A \to \xi_{c,*}^a \underline{O}_{R,c}$. In order to construct the desired sheaf morphism $\tau : \underline{O}_A \to (\xi^a)\underline{O}_R^{+,bi}$, it suffices to verify that the morphisms τ_c agree on intersections of $X_+(Rc)$ and $X_+(Rd)$ which are identified with $\mathrm{Spec}(Q_c^g(R)_0)$ and $\mathrm{Spec}(Q_d^g(R)_0)$ (d another element of C_1). The latter follows immediately from the commutativity of the following diagram of ringmorphisms :

$$\begin{array}{ccc} & & Q_c^g(R)_0 \\ & \overset{\xi_c}{\nearrow} & \downarrow \\ A & & \\ & \underset{\xi_{cd}}{\searrow} & \\ & & Q_{cd}^g(R)_0 \end{array}$$

where the morphisms are the obvious ones. ∎

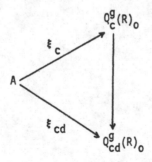

V.3.45. Remark. The foregoing results may be adapted such as to apply to the sheaves $\underline{O}_R^+$ on Proj(R), if we restrict to graded extensions which are composed of inclusions and graded flat epimorphisms of graded prime p.i. rings. Then we obtain results in the vain of V.3.38.

V.3.46. References. Some work on localization and bimodules has been done by J.P. Delale [46] and the authors [189], [197]. The sheaf theoretic application of these techniques was begun by A. Verschoren in his thesis, [194]. The graded theory is rooted in F. Van Oystaeyen's [181] and this is brought to bear on sheaves and schemes over Proj in [196]. In connection with the results on P.I. theory and Jacobson rings we may refer to C. Procesi [136] and M. Artin [18]. Of course the whole theory should be compared to the commutative algebraic geometry e.g. [51], [52], [118], [119] ... a.o., let us mention in particular some work of B. Iversen dealing with some geometrical interpretations of ring theoretical facts about graded rings, cf. [86]. Proceeding along the lines of development of V.2, V.3 one is lead soon enough to the investigation of coherent and quasi-coherent sheaves over Spec and Proj (Serre's theorems etc....). We shall return to these problems in VII.

VI. ALGEBRAIC VARIETIES.

VI.1. Affine Geometric Spaces.

A topological space X is said to be a Jacobson space if the canonical inclusion $|X| \hookrightarrow X$, where $|X|$ is the set of closed points of X equipped with the induced topology, is a quasihomeomorphism. This implies that each non-trivial locally closed subset $Z \subset X$ contains a closed point of X, or equivalently, each closed subset of X is the closure of the set of closed points contained in it. The following lemma is easily checked :

VI.1.1. Lemma. Let $\{U_\alpha ; \alpha \in A\}$ be an open covering of X. Then X is a Jacobson space if and only if each U_α is a Jacobson space.

VI.1.2. Proposition. The ring R is a Jacobson ring if and only if Spec(R) is a Jacobson space.

Proof. A closed subset of $X = \text{Spec}(R)$ is of the form $V(I) = X - X(I)$ for some semiprime ideal I of R. Here $|X|$ consists of the maximal ideals of R. The fact that $V(I) \cap |X|$ is dense in V(I) means that I is an intersection of maximal ideals. Thus any intersection of prime ideals of R can be written as an intersection of maximal ideals of R, hence R is a Jacobson ring. The converse is obvious. ■

VI.1.3. Theorem. [130] (Nullstellensatz). A finitely generated prime P.I. algebra R over a field K is a Hilbert algebra.

Proof. It suffices to verify that every prime ideal P of R is intersection of maximal ideals of finite codimension. Since R is a P.I. algebra we may embed $\overline{R} = R/P$ in a ring $M_n(L)$ for some field extension L of K, $n \in \mathbb{N}$. If $R \to M_n(\tau(R))$ is the universal map in matrices then $\tau(R)$ is a finitely generated algebra over K (since R is). The following diagram of ring homomorphisms is commutative :

$$\begin{array}{ccc} R & \longrightarrow & M_n(\tau(R)) \\ \downarrow & & \downarrow M_n(\lambda) \\ \overline{R} & \longrightarrow & M_n(L) \end{array}$$

for some suitable morphism $\lambda : \tau(R) \to L$. Putting $A = \lambda(\tau(R))$, we obtain an inclusion $\overline{R} \to M_n(A)$ and A is a finitely generated commutative domain. By the celebrated commutative Nullstellensatz we obtain :

$$o = \cap \{M,\ M \in \Omega(A) \text{ such that } \dim_K A/M < \infty\} .$$

Put $U_M = \mathrm{Ker}(\overline{R} \to M_n(A) \to M_n(A/M))$. Clearly, $\overline{R}/U_M$ is then a finite dimensional K-algebra and therefore the Jacobson radical $J(\overline{R}/U_M)$ of $\overline{R}/U_M$ is the intersection of maximal ideals of finite codimension, and $J(\overline{R}/U_M)^h = o$. Consequently, putting T equal to $\cap\{M \in \Omega(R),\ \mathrm{codim}_K M < \infty\}$, then $T^h \subset U_M$ for all M. However, the intersection $\bigcap_M U_M$ has to be zero hence $T^h = o$ i.e. $T = o$.

From now on k denotes an algebraically closed field. An extension $\varphi : R \to S$ is said to be a finitely generated extension if there exist $a_1, \dots, a_m \in Z_R(S)$, such that $S = \varphi(R)\{a_1, \dots, a_m\}$. If A and B are k-algebras then $[A,B]$ is the set of all k-algebra morphisms $A \to B$ which are central extensions, modulo k-automorphisms of B.

VI.1.4. Proposition. Let R be a (prime) finitely generated P.I. algebra over k, satisfying a polynomial identity of degree d, then there is a one-to-one correspondence between $\Omega(R)$ and $\bigcup_n [R, M_n(k)]$ where $n \in \{1, \dots, s\}$ for some $s \in \mathbb{N}$, $s < d/2$.

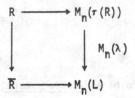

Proof. Take $M \in \Omega(R)$ and consider the morphism $\pi : k \to R \to R/M$ where $k \to R$ is the structural morphism of the k-algebra R. Since π is an extension we may use the results of Section II.2., in particular II.2.35, and conclude that R/M is finite dimensional over k. Hence $Z(R/M)$ is finite dimensional over k and also a field i.e. $Z(R/M) = k$. Since k is algebraically closed $R/M = M_t(k)$ for some $t \in \mathbb{N}$ and $R \to M_t(k)$ is obviously a central extension. Conversely, a central extension $\varphi : R \to M_s(k)$ with $P = \ker \varphi$ yields an extension of P.I. rings $R/P \to M_s(k)$. Thus R/P is prime and $Z(R/P) \hookrightarrow k$. Hence $Z(R/P) = k$. If $I \neq o$ were an ideal of R/P then from the fact that R/P is a prime P.I. ring it follows that $o \neq I \cap Z(R/P) = I \cap k$. However, this entails $I = R/P$ and it follows that R/P is simple i.e. $P \in \Omega(R)$. The statements about the bound $d/2$ are easy consequences of p.i. degree considerations or of Proposition II.4.8. ∎

With notations as in II.4.8. we have the following :

VI.1.5. Corollary. The subsets $\Sigma_n(R)$, with $2n \leq d$, form a closed filtration of Spec(R). More precisely. If $J_n = J(V_n)$ is the T-ideal corresponding to the variety of $n \times n$ matrices i.e. the minimal ideal of R such that R/J_n satisfies all the identities of $n \times n$ matrices, then $\Sigma_n(R) = \{P \in \mathrm{Spec}(R),\ P \supset J_n\} = V(J_n)$. ∎

In order to clarify the constructions we are about to introduce let us reconsider the commutative case.

Ring morphisms of commutative ring $C \to D$ correspond bijectively to morphisms of affine schemes. Basic in this is that Spec(C) is a locally ringed space, for any commutative ring C; i.e. the stalks of the structure sheaf $\underline{O}_C$ are local rings. In general, if X is a scheme and x a "point" of X, then we have an exact sequence :

$$o \longrightarrow \underline{m}_x \longrightarrow \underline{O}_{X,x} \xrightarrow{\pi_x} \mathbb{k}(x) \longrightarrow o$$

where $\underline{O}_{X,x}$ is the stalk at x of the structure sheaf on X, $\underline{m}_x$ is the unique maximal ideal of $\underline{O}_{X,x}$ and $\mathbb{k}(x) = \underline{O}_{X,x}/\underline{m}_x$ is the function field at x. If $X = \mathrm{Spec}(C)$ and $x = P \in \mathrm{Spec}(C)$ then $\underline{O}_{X,x} = C_P$, $\underline{M}_x = PC_P$ and $\mathbb{k}(x) = C_P/PC_P$. In this situation, knowledge of the restriction morphism

$$\varphi_x : \Gamma(X,\underline{O}_X) = C \to \underline{O}_{X,x} = C_P$$

permits us to recover the point $x = P$ by $P = \rho_x^{-1}(\mathrm{Ker}\ \pi_x)$.

Now in the noncommutative case the stalks of the structure sheaf need not be "local" rings, however we still may try to make use of the morphisms π_x.

VI.1.6. Lemma. (A. Grothendieck, [73]). Let $(X,\underline{O}_X)$ be a ringed space and let $\{F_x,\ x\in X\}$ be a family of abelian groups such that for each $x\in X$ we have that $F_x \in \underline{O}_{X,x}$-mod. There exists a sheaf of $\underline{O}_X$-modules $\underline{F}$ such that for any other sheaf of $\underline{O}_X$-modules $\underline{E}$ we have a bijection :

$$\mathrm{Hom}_{\underline{O}_X}(\underline{E},\underline{F}) \to \prod_{x\in X} \mathrm{Hom}_{\underline{O}_{X,x}}(\underline{E}_x, F_x) .$$

Proof. For each open subset U of X define :

$$\Gamma(U,\underline{F}) = \prod_{x\in U} F_x .$$

In this way, using the obvious restriction morphisms, we obtain a sheaf $\underline{F}$ of $\underline{O}_X$-modules satisfying the requirements of the statement, as is easily checked. Alternatively, define a sheaf of $\underline{O}_X$-modules $\underline{F}^x$ by putting for U open in X:

$$\Gamma(U,\underline{F}^x) = \begin{cases} F_x \text{ if } x\in U, \\ 0 \text{ if } x\notin U. \end{cases}$$

Then we obtain the sheaf $\underline{F}$ as $\underline{F} = \prod_{x\in X} \underline{F}^x$ (product of sheaves here). ∎

This construction enables us to "glue together" the morphisms $\pi_x : \underline{O}_{X,x} \to \mathbb{k}(x)$ by considering the sheaf $\underline{K}_X$ of $\underline{O}_X$-algebras defined by putting $F_x = \mathbb{k}(x)$ in the foregoing lemma, and associating to the family $(\pi_x)_{x\in X} \in \prod_{x\in X} \mathrm{Hom}_{\underline{O}_{X,x}}(\underline{O}_{X,x}, \mathbb{k}(x))$ the corresponding sheaf morphism $\pi_X \in \mathrm{Hom}_{\underline{O}_X}(\underline{O}_X, \underline{K}_X)$. This leads us to

VI.1.7. Definition. A pregeometric space is a system $(X, \underline{O}_X, \underline{K}_X, \pi_X)$ where X is a topological space endowed with two sheaves of rings $\underline{O}_X$ and $\underline{K}_X$ and a morphism of sheaves of rings $\pi_X : \underline{O}_X \to \underline{K}_X$.

VI.1.8. Lemma. With notations as in VI.1.6. we have that, for any $x\in X : \underline{F}_x = \varinjlim_{x\in U} \Gamma(U,\underline{F}) = \prod_{y\in \eta(x)} F_y$, where $\eta(x) = \cap\{U \text{ open in } X,\ x\in U\}$.

Proof. Obvious. ∎

The observation that for a T_1-space X we have that $\eta(x) = \{x\}$ motivates the following modification of VI.1.7.

<u>VI.1.9. Definition</u>. A pregeometric space over a T_1-space X is called a <u>geometric space</u>.

<u>VI.1.10. Corollary</u>. To a commutative Jacobson ring, there corresponds a geometric space $(\Omega(C),\underline{O}'_C,\underline{K}'_C,\pi'_C)$, where $\Omega(C)$ has the topology induced by the Zariski topology on Spec(C) and where $\underline{K}'_C$, and $\pi'_C : \underline{O}'_C \longrightarrow \underline{K}'_C$ are constructed in the obvious way.

It is now clear how to generalize the idea of a local morphism in terms of geometric spaces.

<u>VI.1.11. Definition</u>. A <u>morphism of geometric spaces</u>, $\Phi : (X, \underline{O}_X, \underline{K}_X, \pi_X) \to (Y,\underline{O}_Y,\underline{K}_Y,\pi_Y)$, is given by a triple (φ,θ,ψ) where $(\varphi,\theta) : (X,\underline{O}_X) \to (Y,\underline{O}_Y)$ and $(\varphi,\psi) : (X,\underline{K}_X) \to (Y,\underline{K}_Y)$ are morphisms of ringed spaces which make the following diagram commutative :

$$\begin{array}{ccc} \underline{O}_Y & \xrightarrow{\ \theta\ } & \varphi_*\underline{O}_X \\ \downarrow{\scriptstyle \pi_Y} & & \downarrow{\scriptstyle \varphi_*\pi_X} \\ \underline{K}_Y & \xrightarrow[\ \varphi\ ]{} & \varphi_*\underline{K}_X \end{array}$$

and such that ψ induces injective morphisms :

$$\psi_x : \underline{K}_{Y,y} \to (\varphi_*\underline{K}_X)_y \longrightarrow \underline{K}_{X,x}$$

for each $x \in X$ and $y = \varphi(x) \in Y$.

Note that although in general the image of a closed point is not necessarily closed, this will be the case if we assume that both X and Y are T_1-spaces.

Furthermore it is clear that a ring morphism between commutative Jacobson rings $R \to S$ induces a morphism of geometric spaces :

$$(\Omega(S), \underline{O}'_S, \underline{K}'_S, \pi'_S) \to (\Omega(R), \underline{O}'_R, \underline{K}'_R, \pi'_R)$$

if and only if the associated map $^a\varphi : \mathrm{Spec}(S) \to \mathrm{Spec}(R)$ maps $\Omega(S)$ into $\Omega(R)$.

<u>VI.1.12. Commutative Examples.</u>

1. Let S be an R-algebra of finite type where R is a (commutative) Jacobson ring. The structural morphism $\varphi : R \to S$ is such that for any $N \in \Omega(S)$, $\varphi^{-1}(N) = N \cap R \in \Omega(R)$.

2. If R and S are affine (commutative) k-algebras then any k-algebra morphism, $\varphi : R \to S$, induces a morphism $^a\varphi : \Omega(S) \to \Omega(R)$. This is an easy consequence of Hilbert's Nullstellensatz.

In this case the sheaf $\underline{K}'_R$ is given by $\Gamma(U, \underline{K}'_R) = k^U$ for each open U of $\Omega(R)$.

Let us now investigate the noncommutative case more closely. The results of chapter V learn that, to each <u>left Noetherian P.I. ring R</u> we may associate pre-geometric spaces $(\mathrm{Spec}(R), \underline{O}_R, \underline{K}_R, \pi_R)$ and $(\mathrm{Spec}(R), \underline{O}^{bi}_R, \underline{K}_R, \pi_R)$, where $\underline{O}_R$ and $\underline{O}^{bi}_R$ are the canonical sheaves on Spec(R), where $\underline{K}_R$ is the sheaf of $\underline{O}_R$-algebras (resp. of $\underline{O}^{bi}_R$-algebras) corresponding to the family of simple rings $\{Q^{bi}_P(R/P) = Q(R/P);\ P \in \mathrm{Spec}(R)\}$, and where π_R is the sheaf morphism corresponding to the family of ring extensions $\pi_P : Q_{R-P}(R) \to Q(R/P)$ (resp. $\pi_P : Q^{bi}_P(R) \to Q^{bi}_P(R/P) = Q(R/P)$, restriction of the foregoing) making the following diagrams commutative :

$$\begin{array}{ccc} R & \xrightarrow{\ \pi\ } & R/P \\ \downarrow & & \downarrow \\ Q_{R-P}(R) & \xrightarrow[\ \pi_P\ ]{} & Q(R/P) \end{array}$$

$$\begin{array}{ccc} R & \xrightarrow{\ \pi\ } & R/P \\ \downarrow & & \downarrow \\ Q^{bi}_P(R) & \xrightarrow[\ \pi_P\ ]{} & Q^{bi}_P(R/P) = Q(R/P) \end{array} .$$

Assume now that R is also a Jacobson ring, i.e. Spec(R) is a Jacobson space (see Proposition VI.1.2.), then the closed points of Spec(R), i.e. the maximal ideals of R are dense in every closed subset of Spec(R). Note also that in the induced topology on $\Omega(R)$, an open set $X(I) \cap \Omega(R)$ depends only on the prime radical of the ideal I (exactly because R is supposed to be a Jacobson ring). Now $\Omega(R)$ is quasihomeomorphic to Spec(R). Recall that a continuous morphism $\psi : X \to Y$ is a quasihomeomorphism if one of the following equivalent properties holds : 1°. the map $\text{Open}(Y) \to \text{Open}(X)$, $U \to \psi^{-1}(U)$ is bijective.
2°. the map $\text{Closed}(Y) \to \text{Closed}(X)$, $F \to \psi^{-1}(F)$ is bijective.
3°. the topology on X is the inverse image of the topology on Y and $\psi(X)$ is very dense in Y.

VI.1.13. Lemma. Let $\psi : X \to Y$ be a continuous map between topological spaces, let $\{U_\alpha, \alpha \in A\}$ be a covering for Y. If, for each $\alpha \in A$ the restriction $\psi^{-1}(U_\alpha) \to U_\alpha$ is a quasihomeomorphism then ψ is a quasihomeomorphism.

Proof. Straightforward. ∎

It follows that we may endow $\Omega(R)$ with sheaves $\underline{O}_{|R|}$ and $\underline{O}^{bi}_{|R|}$ by putting $\Gamma(U \cap \Omega(R), \underline{O}_{|R|}) = \Gamma(U, \underline{O}_R)$ and $\Gamma(U \cap \Omega(R), \underline{O}^{bi}_{|R|}) = \Gamma(U, \underline{O}^{bi}_R)$, for each open U of Spec(R). Similarly, we define a sheaf $\underline{K}_{|R|}$ on $\Omega(R)$ by putting $\Gamma(V, K_{|R|}) = \prod_{P \in V} Q^{bi}_P(R/P) = \prod_{P \in V} Q(R/P)$, for each open V in $\Omega(R)$.
We thus have defined geometric spaces $(\Omega(R), \underline{O}_{|R|}, \underline{K}_{|R|}, \pi_{|R|})$ and $(\Omega(R), \underline{O}^{bi}_{|R|}, \underline{K}_{|R|}, \pi_{|R|})$ where the morphisms $\pi_{|R|} : \underline{O}_{|R|} \to \underline{K}_{|R|}$ and $\pi_{|R|} : \underline{O}^{bi}_R \to \underline{K}_{|R|}$ are obtained by glueing the morphisms π_P together as before. All this is just an implicit application of the following lemma (cf. [ ] p. 83, (3.8.1)).

VI.1.14. Lemma. Let $f : X_o \to X$ be a continuous map of topological spaces. Then f is a quasihomeomorphism if and only if the functor $\underline{F} \to f^{*}(\underline{F})$ from the category of sheaves on X to the category of sheaves on X_o is an equivalence of categories. In particular, the canonical homomorphism :

$$\Gamma(U,\underline{F}) \to \Gamma(f^{-1}(U), f^{-1}(\underline{F}))$$

is an isomorphism, for any $U \in \mathrm{Open}(X)$, and this isomorphism is functorial in $\underline{F}$. Note that $f^{-1}(\underline{F})$ is the sheaf representing the functor $\underline{G} \to \mathrm{Hom}_f(\underline{G},\underline{F})$ where $\mathrm{Hom}_f(\underline{G},\underline{F})$ is the set of all f-morphisms $\underline{G} \to \underline{F}$ i.e. the set of all morphisms $u : \underline{G} \to f_*(\underline{F})$.

This states that, as far as sheaves are concerned, we did not loose information in restricting from Spec(R) to $\Omega(R)$.

VI.1.15. Notation. We write $\mathbb{k}_R(P)$ or simply $\mathbb{k}(P)$ for $Q_P^{bi}(R/P) = Q(R/P)$. We will write $\underline{O}_R$(resp. $\underline{O}_R^{bi}$, resp. $\underline{K}_R$) instead of $\underline{O}_{|R|}$ (resp. $\underline{O}_{|R|}^{bi}$, resp. $\underline{K}_{|R|}$) whenever it is perfectly clear from the context that we are working over $\Omega(R)$. Also we write $\Omega(R)$ ($\Omega^{bi}(R)$, Spec(R), $\mathrm{Spec}^{bi}(R)$) where we mean $(\Omega(R), \underline{O}_R, \underline{K}_R, \pi_R)$ (resp. $(\Omega(R), \underline{O}_R^{bi}, \underline{K}_R, \pi_R)$, $(\mathrm{Spec}(R), \underline{O}_R, \underline{K}_R, \pi_R)$, $(\mathrm{Spec}(R), \underline{O}_R^{bi}, \underline{K}_R, \pi_R)$).

VI.1.16. Proposition. Let R be a Jacobson ring and $R \to S$ a finitely generated extension, then $R \to S$ induces a morphism of geometric spaces $(\Omega(S), \underline{O}_S^{bi}, \underline{K}_S, \pi_S) \to (\Omega(R), \underline{O}_R^{bi}, \underline{K}_R, \pi_R)$.

Proof. Clearly, $R \to S$ yields a continuous map $\Omega(S) \to \Omega(R)$ and from Chapter V we retain that it also induces a morphism of ringed spaces $(\Omega(S), \underline{O}_S^{bi}) \to (\Omega(R), \underline{O}_R^{bi})$. For each $M \in \Omega(S)$, putting $N = R \cap M$, the canonical map $\tau : \mathbb{k}_R(N) \to \mathbb{k}_S(M)$ is injective, since $\mathbb{k}_R(N) = Q(R/N) = R/N$ and similarly for $\mathbb{k}_S(M)$ - note that automatically $N \in \Omega(R)$. The rest is verification.∎

If we work harder we can actually prove a stronger result.

VI.1.17. Proposition. Let $\varphi : R \to S$ be a finitely generated extension of left

Noetherian P.I. rings. Let $P \in \mathrm{Spec}(S)$ and put $Q = \varphi^{-1}(P)$. Then the canonical morphism : $Q^{bi}_{R-Q}(R)/Q^{bi}_{R-Q}(Q) \to Q^{bi}_{S-P}(S)/Q^{bi}_{S-P}(P)$, is injective.

Proof. Since $\varphi_* \kappa_{R-Q} < \kappa_{S-P}$ it follows that $\varphi_P : Q^{bi}_{R-Q}(R) \to Q^{bi}_{S-P}(S)$ maps $Q^{bi}_{R-Q}(Q)$ into $Q^{bi}_{S-P}(P)$. Now consider the following commutative diagrams :

$$\begin{array}{ccc} Q^{bi}_{R-Q}(R) & \xrightarrow{\varphi_P} & Q^{bi}_{S-P}(S) \\ \downarrow p_Q & & \downarrow p_P \\ Q^{bi}_{R-Q}(R)/Q^{bi}_{R-Q}(Q) & \xrightarrow{\gamma} & Q^{bi}_{S-P}(S)/Q^{bi}_{S-P}(P) \end{array}$$

$$\begin{array}{ccc} R & \longrightarrow & R/Q \\ \downarrow & & \downarrow \\ Q^{bi}_{R-Q}(R) & \xrightarrow{\pi_Q} & \mathbb{k}_R(Q) \\ & \searrow & \uparrow \alpha \\ & & Q^{bi}_{R-Q}(R)/Q^{bi}_{R-Q}(Q) \end{array} \qquad \begin{array}{ccc} S & \longrightarrow & S/P \\ \downarrow & & \downarrow \\ Q^{bi}_{S-P}(S) & \xrightarrow{\pi_P} & \mathbb{k}_S(P) \\ & \searrow & \uparrow \beta \\ & & Q^{bi}_{S-P}(S)/Q^{bi}_{S-P}(P) \end{array}$$

$$\begin{array}{ccc} Q^{bi}_{R-Q}(R) & \xrightarrow{\varphi_P} & Q^{bi}_{S-P}(S) \\ \downarrow \pi_Q & & \downarrow \pi_P \\ \mathbb{k}_R(Q) & \xrightarrow{\overline{\varphi}_P} & \mathbb{k}_S(P) \end{array}$$

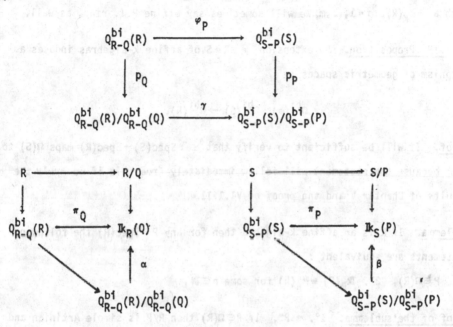

The maps α and β derive from the left exactness of Q^{bi}_{-}. Commutativity of the above diagrams yields the following equalities : $\beta \circ \gamma \circ p_Q = \beta \circ p_P \circ \varphi_P = \pi_P \circ \varphi_P = \overline{\varphi}_P \circ \pi_Q = \overline{\varphi}_P \circ \alpha \circ p_Q$. Since the map p_Q is obviously surjective it follows that $\beta \circ \gamma = \overline{\varphi}_p \circ \alpha$. Now α and β are injective, but also $\overline{\varphi}_p$ is injective ($\mathbb{k}_R(Q)$ is simple!), hence $\beta \circ \gamma = \overline{\varphi}_p \circ \alpha$ implies that γ is also injective. ∎

VI.1.18. Definition. Let K be any commutative field. An affine K-algebra is a ring R with a structural morphism $K \to R$ such that R is a left Noetherian prime P.I. ring which is a finitely generated extension of K, i.e. $R = K\{a_1, \ldots, a_m\}$, with $a_i \in Z_K(R)$, $i = 1, \ldots, m$, We will sometimes say affine P.I. ring, as well.

VI.1.19. Proposition. An extension $\varphi : R \to S$ of affine k-algebras induces a morphism of geometric spaces :

$$\Omega^{bi}(\varphi) : \Omega^{bi}(S) \longrightarrow \Omega^{bi}(R).$$

Proof. It will be sufficient to verify that ${}^a\varphi$: $\mathrm{Spec}(S) \to \mathrm{Spec}(R)$ maps $\Omega(S)$ to $\Omega(R)$ because the statement will follow immediately from this if we apply the results of Chapter V and the proof of VI.1.11.∎

Sublemma. If R is an affine k-algebra then for any $P \in \mathrm{Spec}(R)$ the following statements are equivalent :

1°. $P \in \Omega(R)$, 2°. $\mathbb{k}_R(P) \cong M_n(k)$ for some $n \in \mathbb{N}$.

Proof of the sublemma. 1°. ⇒ 2°. If $P \in \Omega(R)$ then R/P is simple Artinian and $R/P = Q_P^{bi}(R/P) = \mathbb{k}_R(P)$. Since $Z(R/P)$ is a field and finite dimensional over k, the hypothesis that k is algebraically closed entails $Z(R/P) = k$ and then $\mathbb{k}_R(P) = M_n(k)$ follows. Moreover, it is clear that $n = \text{p.i. deg } P$ and therefore $\Omega_n(R) = \Omega(R) \cap \mathrm{Spec}_n(R)$ may be characterized as being the set of all prime ideals P of R such that $\mathbb{k}_R(P) \cong M_n(k)$.

2°. ⇒ 1°. The canonical morphism $R \to R/P \to Q_P^{bi}(R/P) \cong M_n(k)$ is a central extension with kernel P, hence P is maximal (cf. II.2.30).

Returning to the proof of the proposition, if $M \in \Omega(S)$, then we have commutative diagrams of ring extensions :

$$\begin{array}{ccc} \mathbb{k}_R(\varphi^{-1}(M)) & \xrightarrow{\quad i \quad} & \mathbb{k}_S(M) \\ & \nwarrow \qquad \nearrow & \\ & k & \end{array}$$

$$\begin{array}{ccc} R/\varphi^{-1}(M) & \longrightarrow & S/M \\ \downarrow & & \downarrow \\ \mathbb{k}_R(\varphi^{-1}(M)) & \longrightarrow & \mathbb{k}_S(M) \end{array}$$

Commutativity of the latter diagram yields that i is an injective extension. The extension $\mathbb{k}_R(\varphi^{-1}(M)) \to \mathbb{k}_S(M)$ is an extension of central simple algebras and since the center of $\mathbb{k}_S(M)$ is k, it follows that the center of $\mathbb{k}_R(\varphi^{-1}(M))$ is k too. Hence $\mathbb{k}_R(\varphi^{-1}(M))$ is a full matrix ring over k and thus M is a maximal ideal of R by the sublemma. ∎

Now we turn to the main result of this section which gives a noncommutative analogue of the commutative fact that there is equivalence between the category of affine schemes and the category of rings.

<u>VI.1.20. Theorem</u>. Let $\underline{R}$ be the category of rings with ring extensions for the morphisms and let $\underline{G}$ be the category of geometric spaces and morphisms of geometric spaces. For each couple of affine k-algebras (R,S) there is a bijective correspondence between $\mathrm{Hom}_{\underline{R}}(S,R)$ and $\mathrm{Hom}_{\underline{G}}(\Omega^{bi}(R), \Omega^{bi}(S))$.

<u>Proof</u>. Because of Proposition VI.1.14. we have a map:

$$\mathrm{Hom}_{\underline{R}}(S,R) \longrightarrow \mathrm{Hom}_{\underline{G}}(\Omega^{bi}(R), \Omega^{bi}(S)).$$

Let us now establish an inverse map.

Let $(\varphi,\theta,\psi) : (\Omega(R), \underline{O}^{bi}_R, \underline{K}_R, \pi_R) \to (\Omega(S), \underline{O}^{bi}_S, \underline{K}_S, \pi_S)$ be a morphism of geometric spaces. For any $M \in \Omega(R)$ and $N = \varphi(M) \in \Omega(S)$ there is thus given a "local" morphism

$$\theta_M : Q^{bi}_{S-N}(S) \to Q^{bi}_{R-M}(R).$$

Furthermore, we have a morphism between the rings of global sections : $\gamma = \theta(\Omega(S))$: $S \to R$. These morphisms fit into the following commutative diagram :

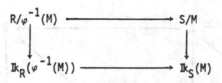

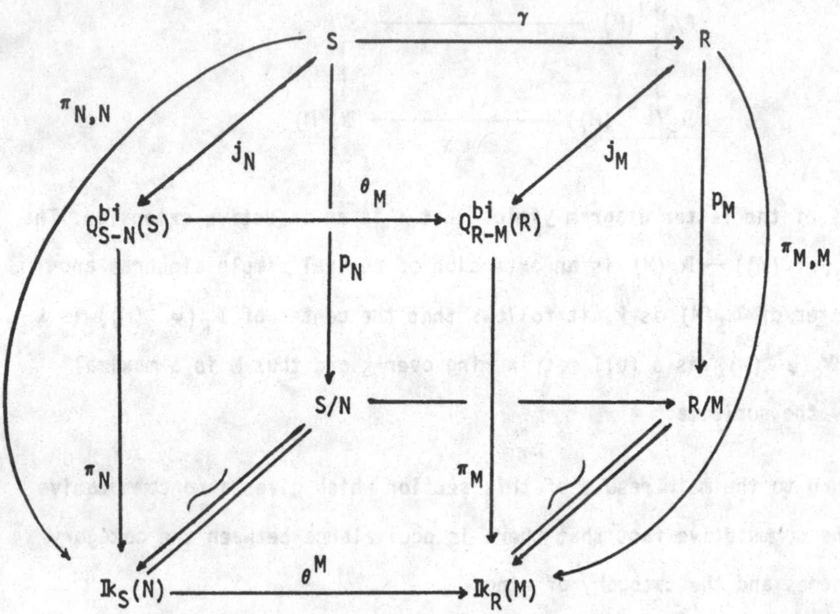

where θ^M is injective by assumption. Put $\pi_{M,M}$ (resp. $\pi_{N,N}$) for the morphism $R \to R/M \tilde{\to} Ik_R(M)$, (resp. $S \to S/N \tilde{\to} Ik_S(N)$).

Clearly, $a \in M$ if and only if $p_M(a) = o$ i.e. if and only if $\pi_{M,M}(a) = o$ (resp. $a \in N$ if and only if $\pi_{N,N}(a) = o$).

Claim : for any $M \in \Omega(R)$ we have $({}^a\gamma)(M) = \varphi(M)$ i.e. $\gamma^{-1}(M) = N$. Indeed, we have : $\theta^M \pi_{N,N} = \theta^M \pi_N j_N = \pi_M \theta_M j_N = \pi_M j_M \gamma = \pi_{M,M} \gamma$ and by hypothesis θ^M is injective, hence :

$$a \in \gamma^{-1}(M) \Leftrightarrow \gamma(a) \in M \Leftrightarrow \pi_{M,M}(\gamma(a)) = o \Leftrightarrow \theta^M \pi_{N,N}(a) = o \Leftrightarrow \pi_{N,N}(a) = o$$
$$\Leftrightarrow a \in N .$$

Finally γ induces a stalk map $\gamma_M : Q^{bi}_{S-N}(S) \to Q^{bi}_{R-M}(R)$ and there is a unique morphism $Q^{bi}_{S-N}(S) \to Q^{bi}_{R-M}(R)$ which does extend γ, therefore the morphisms γ_M and θ_M coincide. This states exactly that the morphism of ringed spaces associated with γ coincides with (φ, θ).

The reader will now easily convince himself that the correspondence $(\varphi, \theta, \psi) \leftrightarrow \gamma$ is a bijective correspondence as required. ▪

VI.1.21. Definition. A geometric space is called an affine geometric space if and only if there exists an affine k-algebra R such that it is isomorphic to $\Omega^{bi}(R) = (\Omega(R), \underline{O}^{bi}_R, \underline{K}_R, \pi_R)$.

Although we do not use the notion of a morphism of pregeometric spaces, we shall sometimes refer to an isomorphism of pregeometric spaces $(X, \underline{O}_X, \underline{K}_X, \pi_X)$ and $(Y, \underline{O}_Y, \underline{K}_Y, \pi_Y)$ meaning that we can find a homeomorphism $\varphi : X \to Y$ which "transforms" $\underline{O}_X$, $\underline{K}_X$ and π_X to $\underline{O}_Y$, $\underline{K}_Y$, π_Y.

VI.2. Algebraic Varieties.

In the commutative case $\underline{O}^{bi}_R$ and $\underline{O}_R$ coincide, in the noncommutative case both structure sheaves have different advantages (and disadvantages) creating shortcomings in an "algebraic geometry" based on the choice of either one of them. The answer to this dilemma is that we do not choose one of the possible structure sheaves but in complete neutrality we study both sheaves and their relations.

VI.2.1. Definition. A varietal space is a five-tuple $X = (X, \underline{O}_X, \underline{O}^{bi}_X, \underline{K}_X, \pi_X)$ such that $\underline{O}^{bi}_X \subset \underline{O}_X$ and such that $(X, \underline{O}_X, \underline{K}_X, \pi_X)$ and $(X, \underline{O}^{bi}_X, \underline{K}_X, \pi_X|_{\underline{O}^{bi}_X})$ are geometrical spaces.

To each prime Jacobson ring satisfying a polynomial identity we associate a varietal space $(\Omega(R), \underline{O}_R, \underline{O}^{bi}_R, \underline{K}_R, \pi_R)$ where all the ingredients are defined as in VI.1. Where no ambiguity arises we refer to this varietal space by $\Omega(R)$ it is called the affine scheme associated to R.

VI.2.2. Definitions. A morphism of varietal spaces, $\Phi : (X, \underline{O}_X, \underline{O}^{bi}_X, \underline{K}_X, \pi_X) \to (Y, \underline{O}_Y, \underline{O}^{bi}_Y, \underline{K}_Y, \pi_Y)$ is just a morphism of geometrical spaces $(\varphi, \theta, \psi) : (X, \underline{O}^{bi}_X, \underline{K}_X, \pi_X) \to (Y, \underline{O}^{bi}_Y, \underline{K}_Y, \pi_Y)$.

A <u>strong morphism of varietal spaces</u> is a morphism of geometric spaces, $\Phi = (\varphi, \theta, \psi) : (X, \underline{O}_X, \underline{K}_X, \pi_X) \to (Y, \underline{O}_Y, \underline{K}_Y, \pi_Y)$ which restricts to a morphism of geometric spaces :

$\Phi^{bi} = (\varphi, \theta|\underline{O}_X^{bi}, \psi) : (X, \underline{O}^{bi}_X, \underline{K}_X, \pi_X) \to (Y, \underline{O}^{bi}_Y, \underline{K}_Y, \pi_Y)$, such that the following diagram is commutative :

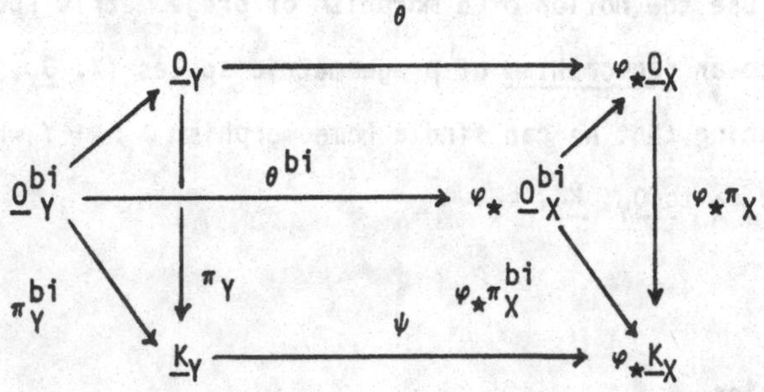

where $\pi_Y^{bi} = \pi_Y|\underline{O}^{bi}_Y$, $\pi_X^{bi} = \pi_X|\underline{O}^{bi}_X$ (we have always omitted the exponents bi and will continue to do so after this definition), and $\theta^{bi} = \theta|\underline{O}^{bi}_X$.

An <u>affine k-variety</u> (k is as always an algebraically closed field) is a varietal space which is strongly isomorphic to the varietal space $\Omega(R)$ associated to an affine k-algebra R. This is compatible with the following definition. An <u>algebraic k-variety</u> is a couple (X, σ) where X is a varietal space and $\sigma : X \to \Omega(k)$ is a strong morphism of varietal spaces, such that the following properties hold:

AV.1. $X = (X, \underline{O}_X, \underline{O}^{bi}_X, \underline{K}_X, \pi_X)$, where the underlying topological space X is covered by a finite number of open subsets U_α of X such that :

AV.2. The open subsets U_α are pairwise non disjoint,

AV.3. For each α, $(U_\alpha, \underline{O}_X|U_\alpha, \underline{O}^{bi}_X|U_\alpha, \underline{K}_X|U_\alpha, \pi_X|U_\alpha)$ is an (irreducible) affine k-variety isomorphic to some $\Omega(R_\alpha)$ where R_α is an affine k-algebra .

<u>VI.2.3. Remark</u>. An algebraic k-variety has an underlying topological space which is compact and irreducible.

Indeed, such an algebraic k-variety is covered by open subsets, pairwise non disjoint and all of them compact and irreducible; whence the statement above.

VI.2.4. Corollary. The structural map $\sigma : X \to \Omega(k)$ endows $\Gamma(X) = \Gamma(X, \underline{O}_X)$ and also $\Gamma^{bi}(X) = \Gamma(X, \underline{O}^{bi}_X)$ with a k-algebra structure.

Proof. σ induces a morphism :

$$k = \Gamma(\Omega(k), \underline{O}_k) \to \Gamma(\sigma^{-1}(\Omega(k)), \underline{O}_X) = \Gamma(X, \underline{O}_X) = \Gamma(X)$$

and on the other hand σ induces a morphism :

$$k = \Gamma(\Omega(k), \underline{O}^{bi}_k) \to \Gamma(\sigma^{-1}(\Omega(k)), \underline{O}^{bi}_X) = \Gamma(X, \underline{O}^{bi}_X) = \Gamma^{bi}(X),$$

and these may serve as structural morphisms.∎

There is some redundancy in the above because :

VI.2.5. Proposition. If $X = (X, \underline{O}_X, \underline{O}^{bi}_X, \underline{K}_X, \pi_X)$ is an algebraic k-variety then $\underline{O}_X$ is completely determined by $\underline{O}^{bi}_X$.

Proof. Let $\{U_\alpha, \alpha \in A\}$ be an open covering of X such that for each $\alpha \in A$ the induced varietal space $X|U_\alpha$ is affine i.e. for each $\alpha \in A$,

$X|U_\alpha = (\Omega(R_\alpha), \underline{O}_{R_\alpha}, \underline{O}^{bi}_{R_\alpha}, \underline{K}_{R_\alpha}, \pi_R)$ for some affine k-algebra R_α.

Therefore $\Gamma(\Omega(R_\alpha), \underline{O}^{bi}_{R_\alpha}) = \Gamma(U_\alpha, \underline{O}^{bi}_X) = R_\alpha$, thus the structure of $\underline{O}^{bi}_X$ determines $\underline{O}_{R_\alpha}$ on U_α. Since $\{U_\alpha; \alpha \in A\}$ covers X, the statement follows.

VI.2.6. Proposition. If $X = (X, \underline{O}_X, \underline{O}^{bi}_X, \underline{K}_X, \pi_X)$ is an algebraic variety, then

$$\Gamma(X) = \Gamma(X, \underline{O}_X) = \Gamma(X, \underline{O}^{bi}_X) = \Gamma^{bi}(X) .$$

Proof. Let $\{U_\alpha; \alpha \in A\}$ be an open covering of X with the property that $X|U$ is affine for each $\alpha \in A$, say

$$X|U_\alpha = (\Omega(R_\alpha), \underline{O}_{R_\alpha}, \underline{O}^{bi}_{R_\alpha}, \underline{K}_{R_\alpha}, \pi_{R_\alpha}).$$

Let $U_{\alpha\beta} = U_\alpha \cap U_\beta$ for each couple of indices $\alpha, \beta \in A$ and let us consider the following commutative exact diagram :

$$\begin{array}{ccccccc}
0 \longrightarrow & \Gamma(X, \underline{O}^{bi}_X) & \xrightarrow{\ i\ } & \prod_\alpha \Gamma(U_\alpha, \underline{O}^{bi}_X) & \xrightarrow{\ p\ } & \prod_{\alpha,\beta} \Gamma(U_{\alpha\beta}, \underline{O}^{bi}_X) \\
& \downarrow & & \downarrow g & & \downarrow \ell \\
0 \longrightarrow & (X, \underline{O}_X) & \xrightarrow[\ j\ ]{} & \prod_\alpha \Gamma(U_\alpha, \underline{O}_X) & \xrightarrow[\ q\ ]{} & \prod_{\alpha,\beta} \Gamma(U_{\alpha\beta}, \underline{O}_X)
\end{array}$$

Here, all arrows are the obvious ones, e.g. $p : \prod \Gamma(U_\alpha, \underline{O}^{bi}_X) \to \prod_{\alpha,\beta} \Gamma(U_{\alpha\beta}, \underline{O}^{bi}_X)$ is the difference of the structural morphisms

$$\mu, \nu : \prod_\alpha \Gamma(U_\alpha, \underline{O}^{bi}_X) \rightrightarrows \prod_{\alpha,\beta} \Gamma(U_{\alpha\beta}, \underline{O}^{bi}_X)$$

which characterize the fact that $\underline{O}^{bi}_X$ is a sheaf. Now

$$h : \prod_\alpha \Gamma(U_\alpha, \underline{O}^{bi}_X) = \prod_\alpha R_\alpha = \prod_\alpha \Gamma(U_{\alpha\beta}, \underline{O}^{bi}_X)$$

is an identification, since the $X|U_\alpha$ are affine for each $\alpha \in A$. Moreover, g and ℓ are obviously injective. Viewing i and j as inclusions and h as an identification, then any $\omega \in \Gamma(X, \underline{O}_X) \subset \prod_\alpha \Gamma(U_\alpha, \underline{O}_X) = \prod_\alpha \Gamma(U_\alpha, \underline{O}^{bi}_X)$ has the property that $0 = q(\omega) = \ell(p(\omega))$, hence $p(\omega) = 0$. However, then $\omega \in \mathrm{Ker}(p) = \Gamma(X, \underline{O}^{bi}_X)$, yielding the statement of the proposition.∎

VI.2.7. Note. The foregoing result indicates that the second statement in VI.2.4. reduces to the first one. Actually, in the definition of an algebraic k-variety it would have been sufficient to take σ to be a morphism i.e. not necessarily a strong morphism.

We will permit ourselves another "abus de langage". Instead of $(\mathcal{X}, \sigma)$ we will frequently say that $\mathcal{X}$ or X is an algebraic k-variety, when σ is determined by the context. Similarly, when no ambiguity concerning k can possibly arise, we will speak of an algebraic variety, without specifying k.

VI.2.8. Proposition. If X is an algebraic variety, then for any $x \in X$ we can find a positive $n \in \mathbb{N}$ such that $\underline{K}_{X,x}$ is isomorphic to $M_n(k)$.

Proof. Immediate from the fact that X may be covered by open subsets U_α with the property that for every index α, the geometric spaces induced on U_α are of the form $\Omega(R_\alpha)$ and $\Omega^{bi}(R_\alpha)$ for some affine k-algebra R_α. Now $x \in X$ is in some U_α and then the property is known to be true for $\Omega(R_\alpha)$. ■

The stalk $\underline{K}_{X,x}$ of $\underline{K}_X$ at $x \in X$ will be denoted throughout by $\mathbb{k}_X(x)$ or simply $\mathbb{k}(x)$ if no ambiguity arises. For any $n \in \mathbb{N}$ we may define $X_n = \{x \in X;\ \mathbb{k}_X(x) = M_n(k)\}$ and the varietal space induced on X_n will also be denoted by X_n. Similarly $\underline{O}_{X_n}$ stands for $\underline{O}_X | X_n$ etc.... If X is of the form $\Omega(R)$ then $\Omega(R)_n = \Omega_n(R)$ in our previous notations. If $f \in \Gamma(X, \underline{O}_X)$, $x \in X$ then $f(x)$ denotes the image of f under the composite map :

$$\begin{array}{ccc} & \Gamma(X, \underline{K}_X) & \\ \Gamma(X, \pi_X) \nearrow & & \searrow \\ \Gamma(X, \underline{O}_X) & \longrightarrow & \mathbb{k}_X(x) \\ f & \longmapsto & f(x) \end{array}$$

An open set U of an algebraic variety is called an <u>affine open set of X</u> if the varietal space structure induced on U is affine i.e. strongly isomorphic to $(\Omega(R), \underline{O}_R, \underline{O}^{bi}_R, \underline{K}_R, \pi_R)$ where R is an affine k-algebra. The intersection of affine open sets need not be affine and in general the affine open sets of X cannot form a basis for the topology of X. Special algebraic varieties where these phenomena do occur will be constructed further on in this section. The following theorem is fundamental in understanding where the obstruction lies.

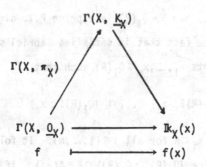

VI.2.9. Theorem. Let R be an affine k-algebra and let I be an ideal of R. If X(I) is an affine open subset of $\Omega(R)$ then κ_I is a t-functor.

Proof. Since X(I) is affine we may find an affine k-algebra S such that :

$$(X(I), \underline{O}_R|X(I), \underline{O}^{bi}_R|X(I), \underline{K}_R|X(I), \pi_R|X(I))$$

$$\cong (\Omega(S), \underline{O}_S, \underline{O}^{bi}_S, \underline{K}_S, \pi_S).$$

Consequently, $S = \Gamma(\Omega(S), \underline{O}_S) = \Gamma(X(I), \underline{O}_R) = Q_I(R)$, and the topological spaces $\Omega(S) = \Omega(Q_I(R))$ and X(I) are homeomorphic. In particular, it is easy to check that this homeomorphism is necessarily given by $M \in \Omega(Q_I(R)) \to M \cap R \in X(I)$. In order to show that κ_I has property T it is sufficient to establish that $Q_I(R)I = Q_I(R)$ because then $Q_I(R)J = Q_I(R)$ holds for every $J \in \mathcal{L}(\kappa_I)$.
The left ideal $Q_I(R)I$ is a left essential left ideal of $Q_I(R)$ (indeed, if $x \in L$, L a left ideal of $Q_I(R)$, then $I^n x \subset R$ yields $I^{n+1} x \subset I \cap L$ and $I^{n+1} x \neq o$ since otherwise $x \in \kappa_I(Q_I(R)) = o$). Now $S = Q_I(R)$ is a prime P.I. algebra and left Noetherian we may utilize the fact that it satisfies Gabriel's condition(H), cf. [33], and so we find elements $x_1, \dots, x_m \in Q_I(R)$ such that :

$$[Q_I(R)I : Q_I(R)] = [Q_I(R)I : x_1] \cap \dots \cap [Q_I(R)I : x_m] .$$

Select $N \in \mathbb{N}$ such that $I^{N-1} x_i \subset R$ for all $i \in \{1, \dots, m\}$. It follows that $I^N \subset \bigcap_{i=1}^m [Q_I(R)I : x_i]$ and therefore $[Q_I(R)I : Q_I(R)] \cap R \in \mathcal{L}(\kappa_I)$. If the latter ideal of R is proper then we may select $M \in \Omega(Q_I(R))$ such that $[Q_I(R)I : Q_I(R)] \cap R \subset \subset M \cap R$, i.e. $M \cap R \in \mathcal{L}(\kappa_I)$. However, $M \in \Omega(Q_I(R))$ corresponds to $M \cap R \in X(I)$ i.e. $M \cap R \notin \mathcal{L}(\kappa_I)$, yielding a contradiction.
It follows that $[Q_I(R)I : Q_I(R)] = Q_I(R)$, a fortiori $Q_I(R)I = Q_I(R)$ and the assertion of the theorem follows. ∎

In order to establish a converse to the above result we have to reconsider some technicalities, some of these having a proper interest for their own sake.

VI. 2.10. Lemma. Let R be a left Noetherian prime P.I. ring and let κ_0 be a symmetric kernel functor in R-mod which has property (T) for R-bimodules, then :

1. If J is a left ideal of $Q_\kappa^{bi}(R)$ then $J = Q_\kappa^{bi}(R)(R \cap J)$.
2. If I is a left ideal of R then $Q_\kappa^{bi}(R)I \cap R = I_\kappa$, where $I_\kappa = \{x \in R;\ \exists H \in \mathcal{L}^2(\kappa)$ such that $Hx \subset I\}$.
3. If $P \in Spec(R)$ and $P \notin \mathcal{L}^2(\kappa)$ then $Q_\kappa^{bi}(R)P \in Spec(Q_\kappa^{bi}(R))$.
4. There is a one-to-one correspondence between prime ideals of $Q_\kappa^{bi}(R)$ and prime ideals of R not in $\mathcal{L}^2(\kappa)$.

Proof. Easy exercise, using the results of Chapter IV 3, and IV.1.36, IV.2.6, IV.2.7, IV.2.11.

VI.2.11. Proposition. Let R be a left Noetherian prime P.I. ring and let I be an ideal of R such that κ_I has property (T) for R-bimodules. The following topological spaces are homeomorphic :
1. $Spec(Q_I(R))$, 2. $Spec(Q_I^{bi}(R))$, 3. $X(I) = \{P \subset Spec(R),\ I \not\subset P\}$.

Proof. That the spaces in 1. and 2. are homeomorphic follows from the fact that here $Q_I^{bi}(R) = Q_I(R)$. Indeed, we always have $Q_I^{bi}(R) = bi\ Q_I(R) \subset Q_I(R)$. Now if $x \in Q_I(R)$ then $I^n x \subset R$ for some n, hence $Q_I^{bi}(R)I^n\ x \subset Q_I^{bi}(R)$. Since property (T) for bimodules yields that $Q_I^{bi}(R)I^n = Q_I^{bi}(R)$ it follows that $x \in Q_I^{bi}(R)$ (identified with bi $Q_I(R)$).
The set bijection established in VI.2.10.4. is easily seen to be a homeomorphism as follows : an open subset of X(I) is of the form X(J) with $J \subset I$, clearly any prime ideal of $Q_I^{bi}(R)$ not containing J does not contain $Q_I^{bi}(R)J$. Hence X(J) corresponding to $X'(Q_I^{bi}(R)J) = \{P \in Spec(Q_I^{bi}(R));\ P \not\supset Q_I^{bi}(R)J\}$, i.e. the spaces in 2. and 3. are homeomorphic. ∎

VI.2.12. Proposition. Let R be a left Noetherian prime P.I. Jacobson ring and let I be an ideal of R such that κ_I has property (T) for R-bimodules, then $Q_I(R)$ (= $Q_I^{bi}(R)$) is a left Noetherian Jacobson ring.

Proof. That $Q_I(R)$ is a prime left Noetherian ring follows from general localization theory (cf. I.4.6.), and that $Q_I(R)$ is a P.I. ring too has been noted before (cf. II.2.).

Take $P \in \mathrm{Spec}(Q_I(R))$, then $P = Q_I(p) = Q_I(R)p$, $p = P \cap R$, $p \not\supset I$. and let $p = \bigcap_{m \supset p} m$, $m \in \Omega(R)$. If $m \not\supset I$ then $Q_I(m)$ is a maximal ideal of $Q_I(R)$, and $P \subset \bigcap_{m \supset p} Q_I(m) = Q$. On the other hand $q = Q \cap R = \bigcap_{m \supset p} (Q_I(m) \cap R) = \cap\{m, m \supset p,\ m \not\supset I\}$ hence $q \cap (\cap\{m \supset p+I\}) = p$, i.e. $q \cap \mathrm{rad}(p+I) = p$. However $q.\mathrm{rad}(p+I) \subset p$ and $I \not\subset p$ yields $q = p$ i.e. $Q = P$.

VI.2.13. Remark. Note that the trick used on the last lines of the foregoing proof may be modified to prove the following :

If R is any Jacobson ring then any $P \in X(J) \subset \mathrm{Spec}(R)$ i.e. $P \not\supset J$ for some ideal J of R, is the intersection of maximal ideals M of R which are also in X(J)!

VI.2.14. Theorem. Let R be an affine k-algebra and suppose that the ideal I of R is such that κ_I has property (T) for R-bimodules. Then X(I) is an affine open set of $\Omega(R)$.

Proof. We have to establish that $(X(I), \underline{O}_R|X(I), \underline{O}_R^{bi}|X(I), \underline{K}_R|X(I), \pi_R)$ is of the form $(\Omega(S), \underline{O}_S, \underline{O}_S^{bi}, \underline{\kappa}_S, \pi_S)$. The obvious candidate for S is $Q_I(R) = Q_I^{bi}(R)$ since we already established the homeomorphism $X(I) \cong \Omega(Q_I^{bi}(R))$ (restrict to maximal ideals in Proposition VI.2.11.). Moreover to $X(J) \subset X(I)$ we associate $Q_J(R)$ (resp. $Q_J^{bi}(R)$) in order to obtain the induced sheaf $\underline{O}_R|X(I)$ (resp. presheaf $\underline{Q}^{bi}(R)|X(I)$), whereas to $X(J^e) \subset \Omega(S)$ with $J^e = Q_I(R)J$, we associate $Q_{J^e}(Q_I(R))$ (resp. $Q_{J^e}^{bi}(Q_I^{bi}(R))$ in order to obtain the structure sheaf $\underline{O}_S$ (resp. presheaf $\underline{Q}^{bi}(S)$). Now it is known that $Q_J(R) = Q_{J^e}(Q_I(R))$ (e.g. see I.4.19.) but then also :

$$Q_J^{bi}(R) = bi(Q_J(R)) = bi(Q_{J^e}(Q_I(R))) = bi(Q_{J^e}(Q_I^{bi}(R))) = Q_{J^e}^{bi}(Q_I^{bi}(R)).$$

Finally $\underline{O}_R|X(I) = \underline{O}_S$, $\underline{O}_R^{bi}\big|X(I) = \underline{O}_S^{bi}$. The rest is mere verification (actually one easily proves that $Q_{R-P}(R/P)$ is ring isomorphic to $Q_{S-P^e}(S/P^e)$ for every $P \in X(I)$, with $P^e = Q_I(R)P$.).

VI.2.15. Conclusion. For an ideal I of an affine k-algebra the following implications hold : If κ_I has property (T) for R-bimodules, then X(I) is an affine open set of $\Omega(R)$, then κ_I is a t-functor in R-mod.

<u>VI.2.16. Example</u>. If c is a central element of an affine k-algebra R then κ_c has property (T) for R-bimodules and therefore $X(c) = \{P \in \Omega(R);\ c \notin P\}$ is an affine open set of $\Omega(R)$.

<u>VI.2.17. Example</u>. Let R be a positively graded k-algebra which is affine and left Noetherian and which satisfies the conditions PS 1, PS 2 . Any homogeneous central element c of R yields isomorphisms of ringed spaces :

$$(X_+(Rc)\ ,\ \underline{O}^+_R|X_+(Rc) = (\mathrm{Spec}(Q^g_c(R))_o,\ \underline{O}_{(Q^g_c(R))_o})$$

$$(X_+(Rc)\ ,\ \underline{O}^{+,bi}_R|X_+(Rc)) = (\mathrm{Spec}(Q^g_c(R))_o,\ \underline{O}^{bi}_{(Q^g_c(R))_o})$$

Originally $X_+(Rc)$ is viewed as a subset of Proj(R) but there is no harm in considering $X_+(Rc) \cap |\mathrm{Proj}(R)|$ as we did in the affine case i.e. restrict to ideals maximal in Proj(R), the latter space with the induced topology will be denoted P(R). Thus, identifying $\Omega(Q^g_c(R))_o)$ and $|X_+(Rc)|$, we see that the sheaves $\underline{K}_{(Q^g_c(R))_o}$ glue together to form a sheaf of simple rings $\underline{K}^+_R$ on the topological space P(R). The stalk at the "maximal" ideal P, which is graded, is $\underline{K}^+_{R,P} = (Q^g(R/P))_o$. It is then easily seen that $(P(R),\ \underline{O}^+_R,\ \underline{O}^{+,bi}_R\ ,\ \underline{K}^+_R,\ \pi^+_R)$ is an algebraic k-variety, where π^+_R is the sheaf morphism $\pi^+_R : \underline{O}^+_R \to \underline{K}^+_R$ associated to the family of sheaf morphisms :

$$\pi_{(Q^g_c(R))_o} : \underline{O}_{(Q^g_c(R))_o} \longrightarrow \underline{K}_{(Q^g_c(R))_o}\ ,$$

where c runs through $Z(R)_1$. We shall refer to this algebraic k-variety by $\underline{\mathcal{P}}(R)$, likewise we refer to the affine k-variety associated with an affine k-algebra R by <u>Spec</u>(R).

If $\varphi : R \to S$ is a graded extension of positively graded left Noetherian affine k-algebras which satisfy the conditions PS1-2 , then $X_+(\varphi(R_+))$ is denoted by U_φ.

Put $\underline{U}_\varphi = (U_\varphi, \underline{O}^+_R|U_\varphi, \underline{O}^{+,bi}_R|U_\varphi, \underline{K}^+_R|U, \pi_R|U_\varphi) = P(R)|U_\varphi$. It is not hard to convince ourselves that we have obtained a morphism of varietal spaces : $Q(\varphi) = \underline{U}_\varphi \to \underline{P}(R)$.

Let R be an affine k-algebra and consider the polynomial ring $R[X_1,\dots,X_n]$ in commuting variables X_i over R. It is easy to see that $R[X_1,\dots,X_n]$ is an affine k-algebra too. Moreover, putting deg $X_i = 1$, $i = 1,\dots,n$, yields a graded structure on $R[X_1,\dots,X_n]$ such that the conditions PS 1 - 2 are being fulfilled. This enables us to introduce the following basic algebraic k-varieties associated with R :

$$\mathbb{A}^n_R = \underline{\Omega}(R[X_1,\dots,X_n])$$

$$\mathbb{P}^n_R = \underline{\mathcal{P}}(R[X_o,\dots,X_n]).$$

$\mathbb{A}^n_R$ is <u>affine n-space over R</u>, $\mathbb{P}^n_R$ is <u>projective n-space over R</u>.

In [20], M. Artin and W. Schelter associate to a quasicoherent sheaf $\underline{A}$ of associative $\underline{O}_Z$-algebras over a commutative scheme $(Z, \underline{O}_Z)$ a topological space Spec($\underline{A}$) lying over Z as follows. If $U \subset Z$ is an affine open set then $\underline{A}(U)$ is an $\underline{O}_Z(U)$-algebra and Spec($\underline{A}$) may then be considered as the Spec($\underline{A}(U)$) glued together in the obvious way. Although Spec($\underline{A}$) is equipped with a sheaf of rings defined on the inverse images of the affine open subsets of Z, the extension of this sheaf to all open sets of Spec($\underline{A}$) has not been studied in [20]. The main application of this construction is the introduction of projective space over affine P.I. rings over a field k. Let us assume, as always here, that R is a prime (left) Noetherian affine k-algebra. Let $Z = \mathbb{P}^n_k = \mathbb{P}^n$ be projective n-space over k, covered by the n+1 affine spaces $U_j = \Omega(k[u_{oj}, u_{1j},\dots,u_{nj}])$ where $u_{ij} = \frac{u_i}{u_j}$ and where u_j, $j = o,\dots,n$, are the canonical homogeneous coordinates. Write T_j for the ring $k[u_{oj},\dots,u_{nj}]$. Then $\mathbb{P}^n$ is the union of these spectra obtained from the diagram of rings and localizations :

$$(*) \qquad \begin{matrix} T_i & \searrow & \\ & & T_{ij} \\ T_j & \nearrow & \end{matrix} \qquad\qquad i,j = o,\dots,n$$

where $T_{ij} = T_i[u_{ji}^{-1}] = T_j[u_{ij}^{-1}] = T_{ji}$.

If $\underline{A}$ is a quasi coherent sheaf of algebras over $\mathbb{P}^n$, then $\underline{A}(U_i) = A_i$ is a T_i-algebra, and we obtain diagrams :

$$\begin{array}{ccc} A_i & \searrow & \\ & & A_{ij} \qquad i,j = o,\dots,n\,, \\ A_j & \nearrow & \end{array}$$

where the arrows represent central localizations, induced by the arrows of (*) by taking tensor products. The space Spec($\underline{A}$) is the union of the schemes Spec(A_i), identified along their open subsets Spec(A_{ij}). Projective n-space over R is then defined to be $\mathbb{P}^n_R = \mathrm{Spec}(R \underset{k}{\otimes} \underline{O}_{\mathbb{P}^n})$ and the rings A_i defining it are

$$R \underset{k}{\otimes} T_i = R[u_{oi},\dots,u_{ni}].$$

Let $\gamma : \mathbb{P}^n_R \to \mathbb{P}^n$ be the canonical structure morphism.

<u>VI.2.18. Proposition.</u> Let R be a prime noetherian affine k-algebra; put S = $R[y_o,\dots,y_n]$ for some variables $y_o,\dots,y_n$ and put

$$\underline{A} = R \underset{k}{\otimes} \underline{O}_{\mathbb{P}^n}$$

Then we have $$\mathbb{P}^n_R = P(S)\ .$$

<u>Proof</u>. The ring S is graded by putting deg $y_i = 1$, $i = o,\dots,n$. Then the graded ring S satisfies PS 1 - 2, so the statements of the theorem make sense. Clearly

$$\underline{A}(U_j) = (R \underset{k}{\otimes} \underline{O}_{\mathbb{P}^n})(U_j) = R \underset{k}{\otimes} O_{\mathbb{P}^n}(U_j)$$

$$= R \underset{k}{\otimes} k[u_{oj},\dots,u_{nj}]$$

$$= R[u_{oj},\dots,u_{nj}]$$

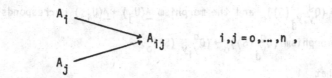

$$= (Q^g_{y_j}(R[y_o,\dots,y_n]))_o$$

$$= Q^g_{y_j}(S)_o \ ,$$

while $\underline{A}(U_{ij}) = (Q^g_{y_iy_j}(S))_o$ and the morphism $\underline{A}(U_j) \to \underline{A}(U_{ij})$ corresponds to the localization morphism $(Q^g_{y_j}(S))_o \to (Q^g_{y_iy_j}(S))_o$.

We obtained that

$$\mathbb{P}(S) = \bigcup_{c\in Z(S)_1} \Omega(Q^g_c(S)_o)$$

$$= \bigcup_{j=o}^{n} \Omega(Q^g_{y_j}(S)_o)$$

$$= \bigcup_{j=o}^{n} \Omega(\underline{A}(U_j))$$

$$= \Omega(\underline{A}) \ .$$

<u>VI.2.19. Remark.</u> The reader may verify that for <u>all</u> open affine subsets U of $\mathbb{P}^n$ we have that $\underline{A}(U) = (\gamma_* \underline{O}^{+,bi}_S)(U)$, thus showing that $\underline{O}^{+,bi}_S = \underline{O}^{+,bi}_{R[y_o,\dots,y_n]}$ is an extension of $\underline{A}$ to <u>all</u> open subsets of $\mathbb{P}^n_R = \mathrm{Proj}(R[y_o,\dots,y_n])$.

An algebraic k-variety X is said to be <u>pre-cellular</u> if it may be covered by affine open subsets $\{U_\alpha,\ \alpha\in A\}$ of X such that for each couple (α,β), $U_{\alpha\beta} = U_\alpha \cap U_\beta$ may again be covered by affine open subsets of X. Of course commutative algebraic varieties are examples of pre-cellular varieties.

An algebraic k-variety X is <u>cellular</u> if it is pre-cellular and every $U_{\alpha\beta}$ is itself affine. Any separated commutative algebraic k-variety is cellular. Other examples of cellular varieties are given by varieties $\mathbb{P}^n_R$ constructed before.

<u>VI.2.20. Theorem.</u> Let Y be an affine k-variety with coordinate ring $R = \Gamma(Y, \underline{O}_Y)$ and let X be an arbitrary pre-cellular algebraic k-variety. There is a bijective correspondence between weak morphisms $u : X \to Y$ and k-algebra extensions $\varphi : R \to \Gamma(X, \underline{O}_X)$.

<u>Proof</u>. (Modification of the proof of a similar statement in the commutative case). Choose a pre-cellular covering $\{U_\alpha, \alpha \in A\}$ for X. Any weak morphism $u : X \to Y$ yields a ring morphism :

$$f : R = \Gamma(Y, \underline{O}_Y) = \Gamma(Y, \underline{O}_Y^{bi}) \to \Gamma(X, \underline{O}_X) = \Gamma(X, \underline{O}_X^{bi}).$$

If $u, u' : X \to Y$ are different weak morphisms then there is an $\alpha \in A$ such that the induced maps u_α, $u'_\alpha : U_\alpha \to Y$ are different. Since U_α and Y are affine, there is a bijective correspondence between weak morphisms $U_\alpha \to Y$ and k-algebra extensions $f_\alpha, f'_\alpha : \Gamma(Y, \underline{O}_Y^{bi}) \to \Gamma(U_\alpha, \underline{O}_X^{bi} \mid U_\alpha)$, cf. VI.1.15, therefore f_α and f'_α are different. Let the extensions f and f' correspond to u resp u', then we have a commutative diagram :

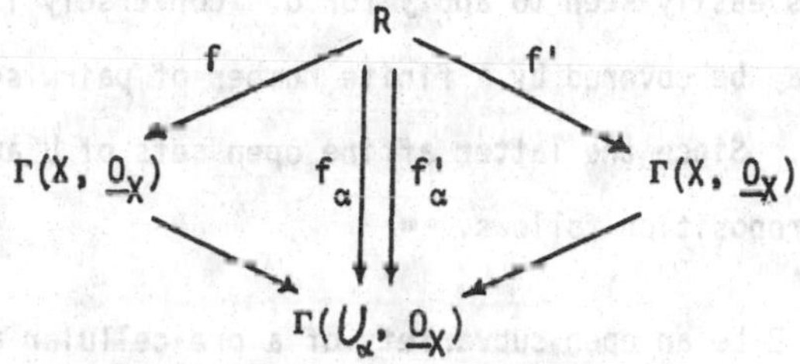

Hence it follows that f and f' are also different.

Conversely, for any $\alpha \in A$, the composed morphism

$$R = \Gamma(Y, \underline{O}_Y) \xrightarrow{\varphi} \Gamma(X, \underline{O}_X) \to \Gamma(U_\alpha, \underline{O}_X)$$

determines a morphism of affine k-varieties $u_\alpha : U_\alpha \to Y$ (again by VI.1.15.). In order to be able to glue these morphisms u_α together we have to check whether u_α and u_β agree on $U_{\alpha\beta}$, for all $\alpha, \beta \in A$. Since $U_{\alpha\beta}$ may be covered by affine open subsets of X the problem reduces to checking whether the restrictions of u_α and u_β to any open affine subset V of $U_{\alpha\beta}$ agree on V. Now $u_\alpha|V$ and $u_\beta|V$ both correspond to the ring extension :

$$R = \Gamma(Y, \underline{O}_Y) \xrightarrow{\varphi} \Gamma(X, \underline{O}_X) \xrightarrow{\rho} \Gamma(V, \underline{O}_X)$$

where ρ denotes the restriction morphism ρ_V^X of $\underline{O}_X$.

Again by VI.1.15, $u_\alpha|V = u_\beta|V$. ■

An open subset U of an algebraic k-variety X is called an <u>open subvariety</u> of X if the structure of X induces an algebraic k-variety structure on U. Clearly, since affine open sets of X need not form a basis for the Zariski topology of X, not every open subset of X yields an open subvariety.

<u>VI.2.21. Proposition.</u> An open subset U of an algebraic k-variety X is an open subvariety of X if and only if U may be covered by a finite number of affine open sets of X.

<u>Proof.</u> If $U = U_1 \cup \ldots \cup U_m$, U_i affine open subsets of X, then the definition of an algebraic k-variety is easily seen to apply for U. Conversely if U is an open subvariety of X then U may be covered by a finite number of pairwise non-disjoint affine open subsets of U. Since the latter affine open sets of U are also affine open subsets of X, the proposition follows. ■

<u>VI.2.22. Corollary.</u> Let U be an open subvariety of a pre-cellular algebraic k-variety X then the following properties are equivalent :

1. U is affine.
2. $\Gamma(U, \underline{O}_X)$ is an affine k-algebra and the identity on $\Gamma(U, \underline{O}_X)$ yields a weak morphism $U \to \Omega(\Gamma(U, \underline{O}_X))$.

<u>Proof.</u> Immediately from VI.2.21. and VI.2.20. ■

<u>VI.2.23. Note.</u> Implicitly we have used the fact that an algebraic k-variety which is weakly isomorphic to an affine k-variety is itself an affine k-variety. This is an easy consequence of the fact that for any algebraic k-variety the sheaf $\underline{O}_X$ is completely determined by $\underline{O}_X^{bi}$.

Let X be a fixed algebraic k-variety. In order to obtain a criterion for X to be affine, we present some easy lemmas.

<u>VI.2.24. Lemma.</u> Let $U = \Omega(B)$ be an affine open subset of X and let $\overline{F}$ be the

image of $f \in A = \Gamma(X, \underline{O}_X)$ in $B = \Gamma(U, \underline{O}_X)$. Put $X_f = \{x \in X; \pi_{X,x}(f) \neq o\}$, $U_{\overline{f}} = \{P \in \Omega(B); \overline{f} \notin P\}$. Then we have : $U \cap X_f = U_{\overline{f}}$.

Proof. The following diagram is commutative because $\Omega(B)$ is an affine open subset of X :

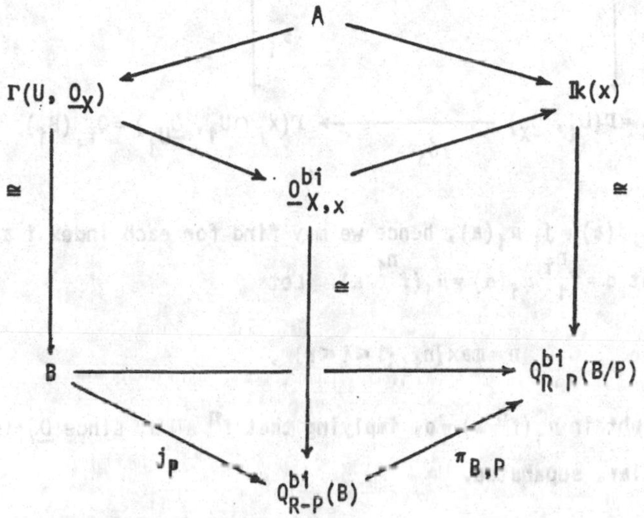

Here, vertical arrows are natural isomorphisms, and P is the maximal ideal of B corresponding to $x \in X$. Therefore, $\pi_{X,x}(f) = o$ if and only if $\pi_{B,P}\, j_P(\overline{f}) = o$ i.e. $\overline{f} \notin P$ for all $x \in U$. ∎

VI.2.25. Corollary. For each $f \in Z(\Gamma(X, \underline{O}_X))$, X_f is open in X.

Proof. Cover X by affine open subsets U_α, $\alpha \in A$. Since $U_\alpha \cap X_f$ is open for each $\alpha \in A$, the statement follows. ∎

VI.2.26. Lemma. Assume that under the canonical map $A = \Gamma(X, \underline{O}_X) \to \Gamma(X_f, \underline{O}_X)$ induced by a central $f \in A$, an element $a \in A$ maps to o, then we may find $n \in \mathbb{N}$ such that $f^n a = o$.

Proof. By assumption X may be covered by a finite number of open affines $U_i = \mathrm{Spec}(B_i)$, $1 \leq i \leq r$. The foregoing lemma yields that $U_i \cap X_f = U_{i,f_i}$, where f_i is the canonical image of f in B_i. Moreover, since the morphisms $A \to B_i = \Gamma(U_i, \underline{O}_X)$ are extensions, obviously f_i is central in B_i for each $1 \leq i \leq r$. Consider for each index i the following commutative diagram :

$$
\begin{array}{ccc}
A = \Gamma(X, \underline{O}_X) & \xrightarrow{\ \ j\ \ } & \Gamma(X_f, \underline{O}_X) \\
\mu_i \downarrow & & \downarrow \nu_i \\
B_i = \Gamma(U_i, \underline{O}_X) & \xrightarrow{\ \ j_i\ \ } & \Gamma(X_f \cap U_i, \underline{O}_{U_i}) = Q_{f_i}(B_i)
\end{array}
$$

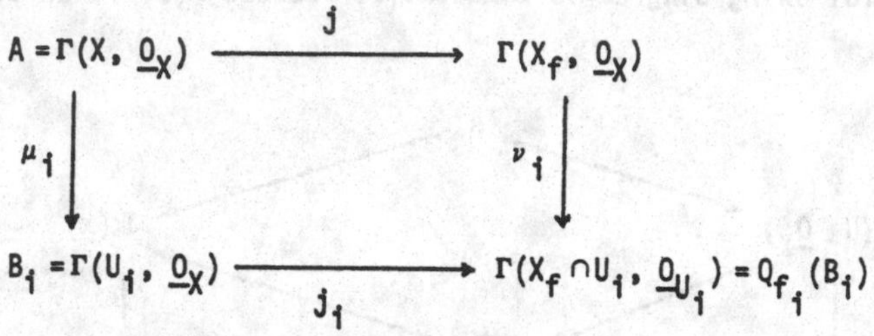

By definition $o = \nu_i\, j(a) = j_i\, \mu_i(a)$, hence we may find for each index i a positive integer n_i such that $o = f_i^{n_i}\, \mu_i(a) = \mu_i(f^{n_i} a)$. Let

$$n = \max\{n_{ij}; 1 \leq i \leq r\},$$

then for each i we obtain $\mu_i(f^n a) = o$, implying that $f^n a = o$, since $\underline{O}_X$ is a sheaf and thus in particular, separated. ∎

<u>VI.2.27. Lemma.</u> With assumptions as in VI.2.26. we may find for every $b \in \Gamma(X_f, \underline{O}_X)$ a positive integer n, such that $f^n b$ lifts to A.

<u>Proof.</u> Consider the following diagram :

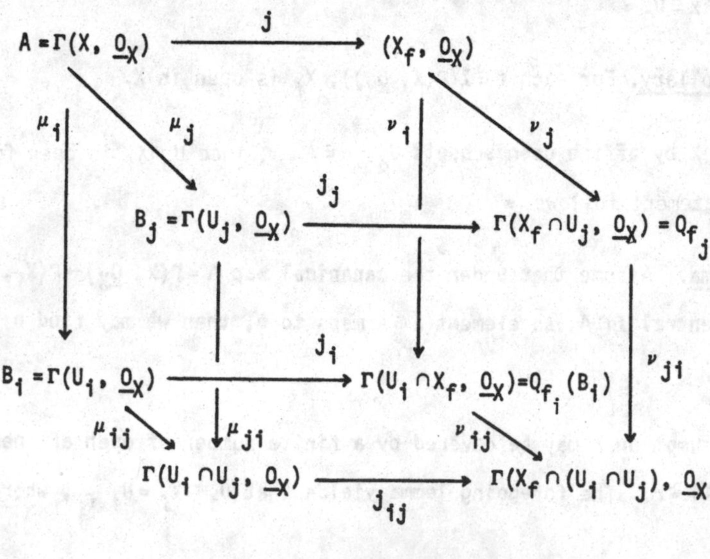

Now, $\nu_i(b) \in \Gamma(X_f \cap U_i, \underline{O}_X) = Q_{f_i}(B_i)$, hence we may find $n_i \in \mathbb{N}$, such that $f_i^{n_i}\, \nu_i(b) = \nu_i(f^{n_i}\, b) \in B_i$. Take $m = \max\{n_i;\ 1 \leq i \leq r\}$, then for each index i we have $\nu_i(f^m b) \in B_i$. Let $\alpha_{ij} = \mu_{ij}(\nu_i(f^m b))$, then

$$j_{ij}(\alpha_{ij}) = j_{ij}(\mu_{ij}(\nu_i(f^m b))) = \nu_{ij}(j_i(\nu_i(f^m b))) = \nu_{ij}\nu_i(f^m b)$$

and

$$j_{ij}(\alpha_{ji}) = j_{ij}(\mu_{ji}(\nu_j(f^m b))) = \nu_{ji}(\nu_j(f^m b))) = \nu_{ji}\nu_j(f^m b).$$

Hence $j_{ij}(\alpha_{ij} - \alpha_{ji}) = o$. Using the foregoing lemma, applied to $U_i \cap U_j$, we find for each pair of indices (i,j) a positive integer $n_{ij} \in \mathbb{N}$ such that $f_{ij}^{n_{ij}}(\alpha_{ji} - \alpha_{ij}) = o$, where f_{ij} is the canonical image of f in $\Gamma(U_i \cap U_j, \underline{O}_X)$.
Let $p = \max\{n_{ij};\ 1 \leq i,j \leq r\}$, then for each pair of indices (i,j) we find that

$$\mu_{ij}(\nu_i(f^{p+m}\, b)) = f_{ij}^p\, \mu_{ij}(\nu_i(f^m\, b)) = f_{ij}^p\, \mu_{ji}(\nu_j(f^m b)) = \mu_{ji}(\nu_i(f^{p+m}\, b)),$$

hence $\nu_i(f^{p+m}\, b)$ and $\nu_j(f^{p+m}\, b)$ agree on $U_i \cap U_j$ for each $1 \leq i,j \leq n$. But then, since $\underline{O}_X$ is a sheaf, $f^{p+m}\, b$ lifts to $A = \Gamma(X, \underline{O}_X)$. If we choose $n = p+m$, this finishes the proof. ∎

<u>VI.2.28. Proposition.</u> With the same notations and assumptions we have :
For each central $f \in A = \Gamma(X, \underline{O}_X)$:

$$\Gamma(X_f, \underline{O}_X) = A_f \quad .$$

<u>Proof</u>. Since f is central, we have $A_f = Q_{Rf}(A) = Q_{Rf}^{bi}(A)$ and the result follows from the foregoing lemma. ∎

<u>VI.2.29. Lemma</u>. Let $\Phi : \mathcal{X} \to \mathcal{Y}$ be a morphism of varietal spaces, with underlying continuous map $f : X \to Y$. Assume that Y may be covered by open subsets U_i such that for each index i the induced map

$$\phi|_{U_i} : \mathcal{X}|_{f^{-1}(U_i)} \to \mathcal{Y}|_{U_i}$$

is a weak (resp. a strong) isomorphism of varietal spaces, then so is Φ.

Proof. The obvious demonstration runs along the lines of the proof of a similar statement about locally ringed spaces. Details are left to the reader, cf. [194]. ∎

VI.2.30. Theorem. Let X be a pre-cellular algebraic k-variety, then the following statements are equivalent :

1. X is affine;
2. $A = \Gamma(X, \underline{O}_X)$ is an affine k-algebra with the property that there exist central elements $f_1, \dots, f_n \in A$ such that for each index i the subset X_{f_i} is affine and such that the f_i generate the unit of A.

Proof. The condition is obviously necessary, for indeed, if $X = \Omega(A)$, then the sets X_f with f central in A are affine and cover $\Omega(A)$, yielding that we may find $f_1, \dots, f_n$ central in A such that the conditions of 2 are met. Conversely, these conditions are also sufficient. Since the identity morphism 1_A on A corresponds via the canonical bijection

$$\mathrm{Hom}_{\underline{R}}(A,A) = \mathrm{Hom}_{\underline{G}}(X, \Omega(A))$$

to a morphism of algebraic k-varieties $\psi : X \to \Omega(A)$, all we have to do is to show that ψ is an isomorphism. Let $f : X \to \Omega(A)$ be the underlying topological morphism of ψ. By assumption $\Omega(A)$ is covered by the open affines $U_i = \{P \in \Omega(A);\ f_i \notin P\}$, while each of the X_{f_i} is affine, say $X_{f_i} = \Omega(A_{f_i})$, by the foregoing. Now, one easily sees that the induced map

$$f^{-1}(U_{f_i}) \to U_{f_i} = \Omega(A_{f_i})$$

corresponds to the identity morphism on A_{f_i}, yielding that ψ induces an isomorphism

$$f^{-1}(U_{f_i}) = X_{f_i} = \Omega(A_{f_i}).$$

But then we are in the situation of VI.2.29. and the result follows. ∎

VI.2.31. Note. Actually, in VI.2.30.,2 the assumption that $\Gamma(X, \underline{O}_X) = \Gamma(X, \underline{O}_X^{bi})$ should be an affine k-algebra is superfluous. Indeed, from VI.2.28. it follows that for each f_i we have $\Gamma(X_{f_i}, \underline{O}_X) = \Gamma(X_{f_i}, \underline{O}_X^{bi}) = A_{f_i}$. Now, each of these is an affine k-algebra, and the f_i generate the unit ideal of A. Since the canonical injection $A \to \prod_{i=1}^n A_{f_i}$ is an injective extension, A is a P.I. ring. Injectivity of each $A \to A_{f_i}$ entails that A is prime. That A is left Noetherian follows easily. If A_{f_i} is generated as a k-algebra by $\{a_{i,\alpha}\ f_i^{-n_{i,\alpha}}\ ;\ a_{i,\alpha} \in A,\ \alpha \in A_i\}$ then $\{a_{i,\alpha}\ ;\ 1 \le i \le n,\ \alpha \in A_i\}$ generates A as a k-algebra. Now each of these generator sets may be chosen to be finite, i.e. the assertion follows. ∎

An algebraic k-variety which may be covered by affine open sets of the form $\Omega(R_i)$ where R_i is a Zariski central affine k-algebra is called a Zariski central algebraic k-variety.

VI.2.32. Proposition. If X is a Zariski central algebraic k-variety then the following properties hold :

1. X is a pre-cellular variety.
2. We may identify $\underline{O}_X$ and $\underline{O}_X^{bi}$.

Proof. 1. Let X be covered by affine open subsets U_α, $\alpha \in A$, such that $\Gamma(U_\alpha, \underline{O}_X) = R_\alpha$ is a Zariski central affine k-algebra. Then any open V_α in U_α may be covered by open subsets of U_α of the form $X_{f_i}^{(\alpha)} = \{P \in \Omega(R_\alpha),\ f_i \notin P\}$, where $f_i \in Z(R_\alpha)$. Obviously, the open sets $X_{f_i}^{(\alpha)}$ are affine open subsets of U_α and of X. In particular, choosing $V_\alpha = U_\alpha \cap U_\beta$ we find that X is a pre-cellular variety.

2. Let $\{U_\alpha,\ \alpha \in A\}$ be as in 1, then $(\underline{O}_X | U_\alpha,\ \underline{O}_X^{bi} | U_\alpha) = (\underline{O}_{R_\alpha},\ \underline{O}_{R_\alpha}^{bi})$.

We have for each maximal ideal P of R_α :

$$\underline{O}^{bi}_{R_\alpha,P} = Q^{bi}_{R_\alpha - P}(R_\alpha) \overset{(*)}{=\!=} Q_{R_\alpha - P}(R_\alpha) = \underline{O}_{R_\alpha,P}$$

since $Q_{R-P_\alpha}(-)$ is a central localization (see Section II.1.).

The statement of 2. now follows from the fact that

we may "glue together" the isomorphisms $\underline{O}^{bi}_{R_{\alpha,P}} \cong \underline{Q}_{R_{\alpha,P}}$ on the covering $\{U_\alpha, \alpha \in A\}$. ∎

VI.2.33. Corollary. A Zariski central algebraic k-variety possesses a basis of Zariski central affine open subsets.

Proof. $\text{Spec}(R_\alpha)$ has a basis $X(R_\alpha f)$, $f \in Z(R_\alpha)$. The Corollary follows now from VI.2.32. and the results of II.1 on Zariski central rings. ∎

VI.3. Examples

In the concrete calculations included in this section, the next lemma plays a fundamental role.

VI.3.1. Lemma. Let R be a subring of S and suppose that $I \subset R$ is an ideal of R and of S, then :

1. For every nonzero prime ideal P of R there exist a prime ideal P' of S such that $o \neq P' \cap R \subset P$.
2. If $I \not\subset P$ then we may choose P' such that $P' \cap R = P$.

Proof. Taking P' to be an ideal of S maximal such that $P' \cap R \subset P$, one easily verifies 1. In order to check 2., consider the ideal IPI of S. Evidently $IPI \cap R \subset P$, hence $IPI \subset P'$ for some prime ideal P' as in 1. But then $ISPSI \subset P'$ with $I \not\subset P'$ yields $SPS \subset P'$, i.e. $P \subset P' \cap R$ and $P = P' \cap R$. Note that we have actually proved that any prime ideal P' with $P' \not\supset I$ intersects R in a prime ideal $P' \cap R$. ∎

Let C be a commutative domain, consider subrings A and B of C and ideals I and J of C such that $IJ \subset A \cap B$. If I and J are nonzero, then $R = \begin{pmatrix} A & I \\ J & B \end{pmatrix}$ is a prime P.I. ring of p.i. degree 2. Note that R is Noetherian if C is finite over A and B. Clearly $Z(R) \cong A \cap B$ and $T = \begin{pmatrix} IJ & IJ \\ IJ & IJ \end{pmatrix}$ is an ideal of R and of $M_2(C)$. The prime ideals of R not containing T are of the form $\langle P \rangle = P \cap R$ for some $P \in \text{Spec}\, M_2(C)$. Clearly, the prime ideals of R not containing T are in bijective correspondence with the prime ideals of C not containing IJ. Put $\hat{I} = \begin{pmatrix} A\cap I & I \\ IJ & B\cap I \end{pmatrix}$, $\hat{J} = \begin{pmatrix} A\cap J & IJ \\ J & B\cap J \end{pmatrix}$ and note that $\hat{I}\hat{J} \subset \begin{pmatrix} IJ & IJ \\ IJ & IJ \end{pmatrix}$. If a prime ideal Q of R contains T then it must contain either $\hat{I}$ or $\hat{J}$ and so it is easily verified that the following holds :

<u>VI.3.2. Proposition.</u> The prime ideals of Spec R are of the form :

1. $\langle P\rangle = P\cap R$ with $P\in \mathrm{Spec}(M_2(C))$ and $P\not\supset T$.
2. $Q^u = \begin{pmatrix} Q & I \\ J & B \end{pmatrix}$ with $Q\in\mathrm{Spec}(A)$ and $Q\not\supset (I\cap A)(J\cap A)$.
3. $Q^\ell = \begin{pmatrix} A & I \\ J & Q \end{pmatrix}$ with $Q\in\mathrm{Spec}(B)$ and $Q\not\supset (I\cap B)(J\cap B)$.

Let us now calculate the localizations at these ideals.

1. Since $\langle P\rangle$ does not contain T, R/P has p.i. degree 2 and localization at $\langle P\rangle$ is central localization at $\underline{p} = \langle P\rangle \cap Z(R) = P\cap Z(R)$, where Z(R) is $A\cap B$ embedded diagonally. Thus we find :

$$Q^{bi}_{\langle P\rangle}(R) = Q_{R-\langle P\rangle}(R) = \begin{pmatrix} A_{\underline{p}} & I_{\underline{p}} \\ J_{\underline{p}} & B_{\underline{p}} \end{pmatrix}$$

2. Let us write K for the field of fractions of Z(R) and let Q be the total ring of quotients of R, then :

$$Q(R) = \begin{pmatrix} AK & IK \\ JK & BK \end{pmatrix}.$$

We introduce the following notation : if U, $V\subset W$ then $W(U:V) = \{w\in W,\ Vw\subset U\}$. Straightforward calculation shows that

$$Q_{R-Q^u}(R) = \begin{pmatrix} A_Q\cap AK & I_Q\cap I_K \\ JK(A_Q:I) & BK(I_Q:I) \end{pmatrix}$$

and

$$Q^{bi}_{R-Q^u}(R) = (A_Q\cap K).R,$$

where $Q\in\mathrm{Spec}(A)$ is such that $Q\not\supset (I\cap A)(J\cap A)$.

3. Similar as in 2. one obtains :

$$Q_{R-Q^{\ell}}(R) = \begin{pmatrix} AK(J_Q : J) & IK(B_Q : J) \\ J_Q \cap JK & B_Q \cap BK \end{pmatrix}$$

and $Q^{bi}_{R-Q^{\ell}}(R) = (B_Q \cap K).R$,

where $Q \in \mathrm{Spec}(B)$ is such that $Q \not\supset (I \cap B)(J \cap B)$.

Let us apply the foregoing to the case $A = B = k[X,Y]$, $I = J = Xk[X,Y]$, where k is an algebraically closed field, i.e. consider $R = \begin{pmatrix} k[X,Y] & xk[X,Y] \\ xk[X,Y] & k[X,Y] \end{pmatrix}$.
Now $\mathrm{Spec}(k[X,Y])$ consists of the prime ideals : o, $(X-a, Y-b)$ with $a,b \in k$, $(f(X,Y))$ where $f(X,Y)$ is an irreducible polynomial in X and Y over k. Writing I for $Xk[X,Y]$ we may describe $\mathrm{Spec}(R)$ as being the set of the following prime ideals:

1. $\langle a,b \rangle = \begin{pmatrix} (X-a,\ Y-b) & (X-a,\ Y-b) \cap I \\ (X-a,\ Y-b) \cap I & (X-a,\ Y-b) \end{pmatrix}$, with $(a,b) \in k^* \times k$.

2. $\langle f \rangle = \begin{pmatrix} (f) & (f) \cap I \\ (f) \cap I & (f) \end{pmatrix}$, with f irreducible in $k[X,Y]$.

3. $\langle o,b \rangle^* = \begin{pmatrix} (X, Y-b) & I \\ I & k[X,Y] \end{pmatrix}$

$\langle o,b \rangle_* = \begin{pmatrix} k[X,Y] & I \\ I & (X, Y-b) \end{pmatrix}$

4. $I^* = \begin{pmatrix} I & I \\ I & k[X,Y] \end{pmatrix}$, $\quad I_* = \begin{pmatrix} k[X,Y] & I \\ I & I \end{pmatrix}$

The topological space $\mathrm{Spec}(R)$ consists of the (commutative) affine plane $\mathbb{A}^2(k)$ where the "Y-axis" is replaced by a split copy of itself, which may be visualised as follows :

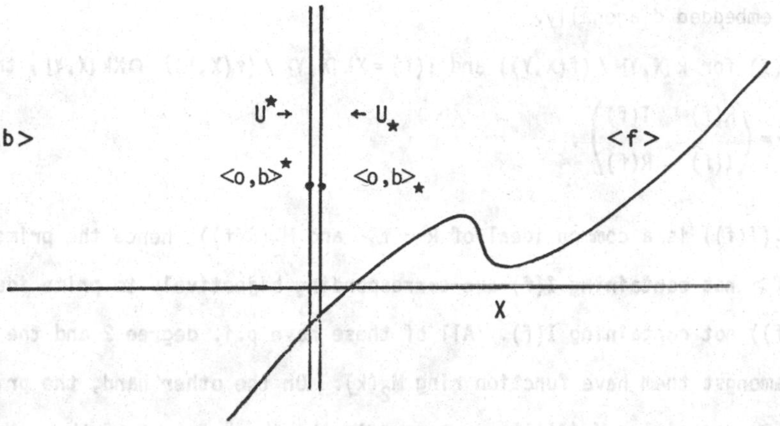

The above picture is what one would obtain by glueing together two copies of $\mathbb{A}^2(k)$ along open sets U and V, both being copies of $\mathbb{A}^2(k)$ - (Y-axis). The obtained prescheme is not separated and should therefore not be called a variety. Now Spec(R) and the prescheme constructed above are homeomorphic as topological spaces but the sheaves are very different; this makes it possible that Spec(R) is an affine and separated variety (in the non-commutative sense). The set of closed points $\Omega(R)$ in Spec(R) is given by :

$\Omega(R) = \{ \langle a,b \rangle, \langle o,b \rangle^{\star}, \langle o,b \rangle_{\star}, \text{ for } (a,b) \in k^{\star} \times k \}$.

It is easy enough to establish that $\mathbb{k}_R(\langle a,b \rangle) = M_2(k)$ whereas $\mathbb{k}_R(\langle o,b \rangle^{\star}) = \mathbb{k}_R(\langle o,b \rangle_{\star}) = k$. The surjective ring homomorphism :

$$R = \begin{pmatrix} k[X,Y] & Xk[X,Y] \\ Xk[X,Y] & k[X,Y] \end{pmatrix} \longrightarrow \begin{pmatrix} k[X] & Xk[X] \\ Xk[X] & k[X] \end{pmatrix} = S$$

defined by specializing Y to o in each entry, yields a morphism Spec(S) → Spec(R) which is just the embedding of the "affine k-line with split origin" as a closed subvariety of the "k-plane with split Y-axis" (referring to non-commutative (separated) varieties by giving it the name of a commutative object which looks like it but certainly is not a variety may be called a poetic liberty rather than "abus de langage".)

Further examples may be constructed from the foregoing as follows. Let f be an irreducible polynomial in k [X,Y], and take $f \neq X$. Since $\langle f \rangle \not\supset I$ we know that $R/\langle f \rangle$ has p.i. degree 2; as a matter of fact the Formanek center of R is just I^2 embedded diagonally.

Write $R(f)$ for $k[X,Y]/(f(X,Y))$ and $I(f) = Xk[X,Y]/(f(X,Y)) \cap Xk[X,Y]$, then

$$R/\langle f\rangle = \begin{pmatrix} R(f) & I(f) \\ I(f) & R(f) \end{pmatrix}.$$

Again $M_2(I(f))$ is a common ideal of $R/\langle f\rangle$ and $M_2(R(f))$, hence the prime ideals of $R/\langle f\rangle$ not containing $I(f)$ are corresponding bijectively to prime ideals of $\mathrm{Spec}(R(f))$ not containing $I(f)$. All of these have p.i. degree 2 and the maximal ideals amongst them have function ring $M_2(k)$. On the other hand, the prime ideals of $R/\langle f\rangle$ containing $M_2(I(f))$ correspond to the "bad" primes of the "Y-axis", this may be visualized as follows :

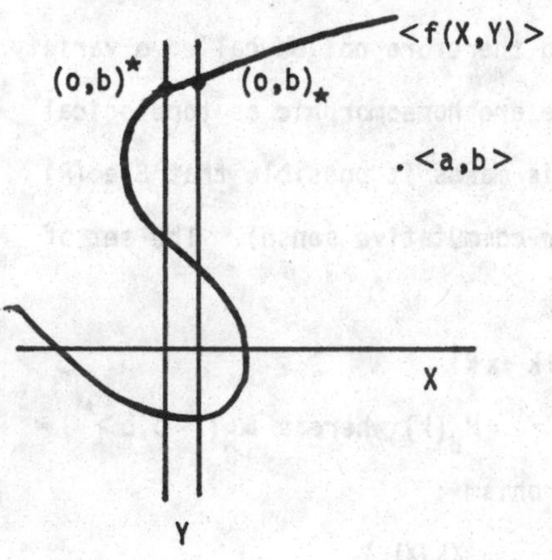

Spec(R)

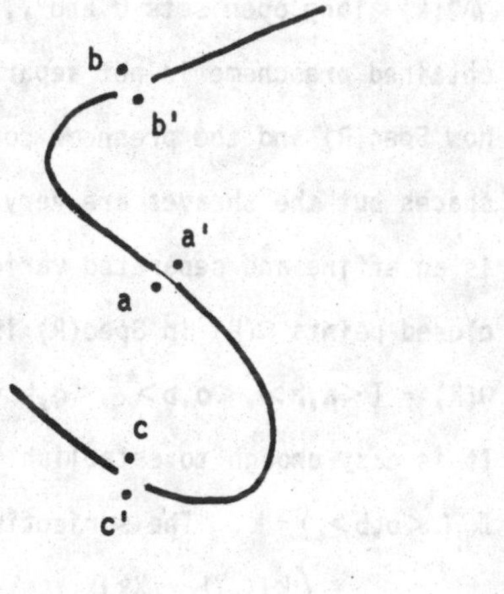

Spec(R/<f>)

The points $\{a,a',b,b',c,c'\}$ correspond to the intersection points of the prime ideal $\langle f\rangle$ of R and the prime ideals I^*, I_* i.e. $b = (\pi(o,b)^*)$ etc.... where $\pi : k[X,Y] \to R(f)$ is the canonical projection.

With notations as before, Proj(R) consists of the graded prime ideals $0, I^*$, I_* and Rf where f is a homogeneous polynomial of $k[X,Y]$ different from X. Its closed points are I^*, I_* and Rf with $f = aX + bY$ and $(a,b) \in k \times k^*$. The dense set of closed points of Proj(R) may be regarded as a projective line over k with one point, say the point at infinity, split.

Note that R satisfies all conditions necessary to obtain a decent scheme structure on Proj(R).

We may cover Proj(R) by the open sets $U = X_+(X)$ and $V = X_+(Y)$ where $U = \mathrm{Spec}((Q_X^g(R))_o)$ and $Q_X^g(R) = M_2(k[X, X^{-1}, Y])$, i.e. U corresponds to the Zariski open set of $\mathbb{P}_k^1 = \mathrm{Proj}(k[X,Y])$ consisting of all graded prime ideals not containing X. Obviously, localization at $P \in U$ is just central localization at $P \cap k[X,Y]$, and the function rings at the closed points of U are isomorphic to $M_2(k)$.

Let us now consider V. It is clear that :

$$(Q_Y^g(R))_o = \begin{pmatrix} k[X,Y,Y^{-1}]_o & Xk[X,Y,Y^{-1}]_{-1} \\ Xk[X,Y,Y^{-1}]_{-1} & k[X,Y,Y^{-1}]_o \end{pmatrix}$$

Thus, $\mathrm{Spec}((Q_Y^g(R))_o)$ consists of the ideals $q(Q_Y^g(R))_o$ with $q \in \mathrm{Spec}((k[X,Y,Y^{-1}])_o)$ not containing X, plus the ideals q^* and q_*,

$$q^* = \begin{pmatrix} q & X\,k[X,Y,Y^{-1}]_{-1} \\ X\,k[X,Y,Y^{-1}]_{-1} & k[X,Y,Y^{-1}]_o \end{pmatrix}$$

$$q_* = \begin{pmatrix} k[X,Y,Y^{-1}]_o & X\,k[X,Y,Y^{-1}]_{-1} \\ X\,k[X,Y,Y^{-1}]_{-1} & q \end{pmatrix},$$

where q does not contain X.

The first class of prime ideals mentioned consists of prime ideals which are uniquely lying over their central parts and localization at these ideals is central i.e. easy. The latter class of prime ideals consists of prime ideals which ly pairwise over their central parts and these have p.i. degree equal to 1.

In case $q = X\,k[X,Y,Y^{-1}]$, the function ring at q^*, q_* is nothing but the field k. The rings $Q_{q^*}((Q_y^g(R))_o)$ and $Q_{q_*}((Q_y^g(R))_o)$ are easily calculated and the stalks of the structure sheaf are seen to be :

$$\underline{O}^g_{R,q^*} = \underline{O}^g_{R,q_*} = Q^g_{Xk[X,Y]}(R),$$

i.e. central localization at the prime ideal $Xk[X,Y]$ of $k[X,Y]$.
The canonical inclusion $k[X,Y] \to R$ yields a morphism of schemes $\mathrm{Proj}(R) \to \mathrm{Proj}(k[X,Y]) = \mathbb{P}^1_k$ which may be visualized in :

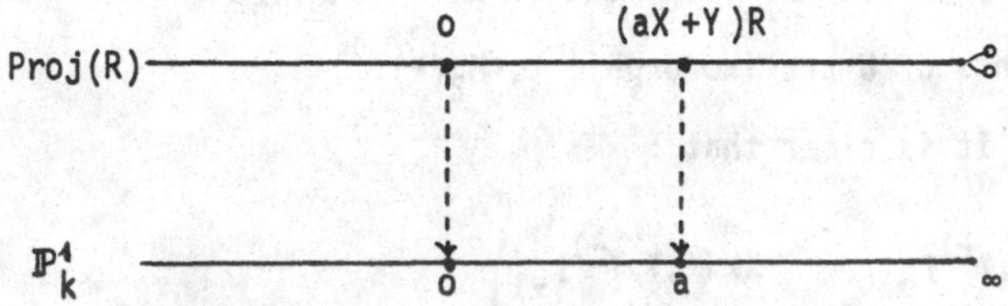

Examples of affine k-varieties with only closed points (except o) arise when considering HNP-rings, cf. [149]. J.C. Robson and L. Small have shown that a prime P.I. ring R such that all ideals are left projective is left and right hereditary and left and right noetherian. Moreover it follows that the center of R is a Dedekind domain and that R is finitely generated as a Z(R)-module. Probably the easiest non-trivial example of an HNP P.I. ring is the following :

$$R = \begin{pmatrix} k[X] & k[X] \\ X\,k[X] & k[X] \end{pmatrix}.$$

It is possible to make Spec(R) into an algebraic transformation space under the action of a commutative algebraic group, isomorphic to k^* in this case. The fix-points in Spec(R) for the action of k^* turn out to be the "origin" o and the degenerate points $\begin{pmatrix} X\,k[X] & k[X] \\ X\,k[X] & k[X] \end{pmatrix}$, and $\begin{pmatrix} k[X] & k[X] \\ X\,k[X] & Xk[X] \end{pmatrix}$, cf. [194] for details. This is then also an example of the effect of a general theory developed by B. Mueller, which links the occurrence of degenerate prime ideals of P.I. rings which are finite modules over their center, to certain invariants under the action of suitable algebraic groups.

VI.3.3. An example with p.i. deg R = 3.

Let D be a commutative domain, I a nonzero ideal of D. Consider the P.I.

ring

$$R = \begin{pmatrix} D & I & I \\ D & D & I \\ D & D & D \end{pmatrix}$$

Clearly $M_3(I)$ is an ideal of R and $M_3(D)$. Consequently Spec(R) consists of "central" prime ideals R_p, where $p \in \mathrm{Spec}(D)$ does not contain I, and of prime ideals containing $M_3(I)$. For each ideal J of D containing I we define ideals $J^{(1)}, J^{(2)}, J^{(3)}$ of R as follows :

$$J^{(1)} = \begin{pmatrix} J & I & I \\ D & D & I \\ D & D & D \end{pmatrix}, \quad J^{(2)} = \begin{pmatrix} D & I & I \\ D & J & I \\ D & D & D \end{pmatrix}, \quad J^{(3)} = \begin{pmatrix} D & I & I \\ D & D & I \\ D & D & J \end{pmatrix} .$$

If $Q \in \mathrm{Spec}(R)$ contains $M_3(I) \supset I^{(1)}\ I^{(2)}\ I^{(3)}$ then Q contains $I^{(\alpha)}$ for at least one $\alpha \in \{1,2,3\}$. It is easy enough to verify that $\mathrm{Spec}_2(R) = \phi$, $\mathrm{Spec}_1(R) = \{P^{(\alpha)}$, $\alpha = 1,2,3$, P prime ideal of D containing I}. Let us take $D = k[X,Y]$ and $I = (f(X,Y))$ where f(X,Y) is irreducible in $k[X,Y]$. Any prime ideal containing I is either maximal or equal to I and in the former case it is of the form $\langle a,b \rangle = (X-a, Y-b)$ where $f(a,b) = o$. Spec(R) may therefore be visualized as follows

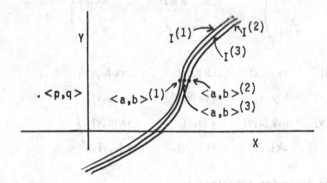

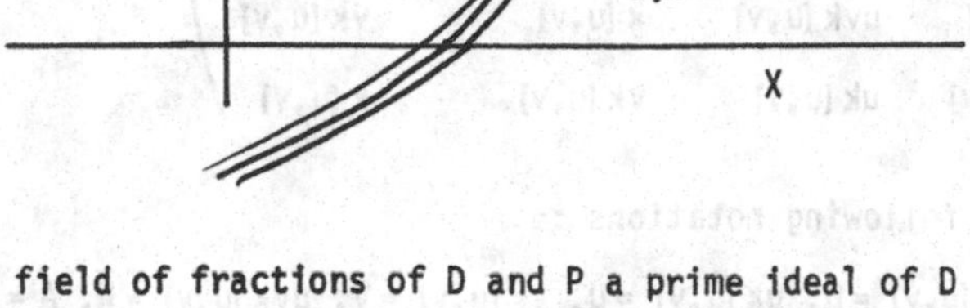

If K is the field of fractions of D and P a prime ideal of D containing I (the central localizations are easily calculated so we do not mention these here) then:

$$Q_{R-P^{(1)}}(R) = \begin{pmatrix} D_p & I_p & I_p \\ K(D_p : I) & K(I_p : I) & K(I_p : I) \\ K(D_p : I) & K(I_p : I) & K(I_p : I) \end{pmatrix}$$

$$Q_{R-P^{(2)}}(R) = \begin{pmatrix} D_p & D_p & I_p \\ D_p & D_p & I_p \\ K(D_p : I) & K(D_p : I) & K(I_p : I) \end{pmatrix}$$

$$Q_{R-P^{(3)}}(R) = \begin{pmatrix} D_p & D_p & D_p \\ D_p & D_p & D_p \\ D_p & D_p & D_p \end{pmatrix}$$

Hence $\qquad Q^{bi}_{R-P^{(\alpha)}}(R) = RD_p$.

VI.3.4. An example of the form $R \underset{k}{\otimes} R$ and p.i. degree 4.

Since the k-algebra $k[X] \underset{k}{\otimes} k[X]$ is isomorphic to $k[u,v]$ we may identify the k-algebra

$$\begin{pmatrix} k[X] & Xk[X] \\ Xk[X] & k[X] \end{pmatrix} \underset{k}{\otimes} \begin{pmatrix} k[X] & Xk[X] \\ Xk[X] & k[X] \end{pmatrix}$$

with

$$\begin{pmatrix} k[u,v] & vk[u,v] & uk[u,v] & uvk[u,v] \\ vk[u,v] & k[u,v] & uvk[u,v] & uk[u,v] \\ uk[u,v] & uvk[u,v] & k[u,v] & vk[u,v] \\ uvk[u,v] & uk[u,v] & vk[u,v] & k[u,v] \end{pmatrix}$$

Let us introduce the following notations :

$k[X] = R$, $Xk[X] = T$, $k[u,v] = D$, $uk[u,v] = U$, $vk[u,v] = V$, $uvk[u,v] = W$, $A = \begin{pmatrix} R & T \\ T & R \end{pmatrix}$ and then

$$A \underset{k}{\otimes} A = \begin{pmatrix} D & V & U & W \\ V & D & W & U \\ U & W & D & V \\ W & U & V & D \end{pmatrix} = B$$

First let us search for the prime ideals of $\Gamma = \begin{pmatrix} D & V \\ V & D \end{pmatrix}$. Since $M_2(V)$ is an ideal of Γ and $M_2(D)$ two cases have to be distinguished : 1. Prime ideals of p.i. degree 2; these are o, $<a,b> = (u-a,\ v-b)M_2(D) \cap \Gamma$ where $b \neq o$, and $<f> = (f)M_2(D) \cap \Gamma$ where $f \neq U$ is an irreducible polynomial in u and v over k.

2. Prime ideals of p.i. degree 1; these are

$$<a,o>^* = \begin{pmatrix} (u-a,\ v)D & V \\ V & D \end{pmatrix}, \quad <V>^* = \begin{pmatrix} V & V \\ V & D \end{pmatrix}$$

$$<a,o>_* = \begin{pmatrix} D & V \\ V & (u-a,\ v)D \end{pmatrix}, \quad <V>_* = \begin{pmatrix} D & V \\ V & V \end{pmatrix}$$

If Λ is the Γ-ideal $\begin{pmatrix} U & W \\ W & U \end{pmatrix}$, then $M_2(\Lambda)$ is a common ideal for B and $M_2(\Gamma)$, what enables us to unrafle the structure of Spec (B) as follows :

(α) Prime ideals not containing $M_2(\Lambda)$. These may be classified as before.

(α_1) Prime ideals of p.i. degree 4. These are : o, $\ll a,b \gg = M_2(<a,b>) \cap B$ for all $ab \neq o$, and $\ll f \gg = M_2(<f>) \cap B$ for $f \neq u$ and $f \neq v$.

(α_2) Prime ideals of p.i. degree 2. These are :
$\ll a,o>^* > = M_2(<a,o>^*) \cap B$, $\ll a,o>_* > = M_2(<a,o>_*) \cap B$ for $a \neq o$, and also $\ll v>^* > = M_2(<v>^*) \cap B$, $\ll v>_* > = M_2(<v>_*) \cap B$

(β) Prime ideals containing $M_2(\Lambda)$. These are of the form $<P>^* = \begin{pmatrix} P & \Lambda \\ \Lambda & \Gamma \end{pmatrix}$ or $<P>_* = \begin{pmatrix} \Gamma & \Lambda \\ \Lambda & P \end{pmatrix}$ where $P \in \mathrm{Spec}(\Gamma)$. Then we obtain the subdivision :

(β_1) Prime ideals of p.i. degree 2. These are : $\ll o,b>^* >$, $\ll o,b>_* >$ for $b \neq o$, and also $\ll u>^* >$, $\ll u>_* >$.

(β_2) Prime ideals of p.i. degree 1. These are : $\ll o,o>^* >^*$, $\ll o,o>_* >^*$, $\ll o,o>^* >_*$ and $\ll o,o>_* >_*$.

The set of maximal ideals of B is thus given by :

$\Omega(B) = \{\ll a,b \gg, \ll a,o>^* >, \ll a,o>_* >, \ll o,b>^* >, \ll o,b>_* >, \ll o,o>^* >^*, \ll o,o>^* >_*, \ll o,o>_* >^*, \ll o,o>_* >_*$ with $a,b \in k$, $ab \neq o\}$. On the other hand we have $\Omega(A) = \{\tilde{a}, o^+, o^-$, with $a \in k^*\}$ where $\tilde{a} = (X-a)A$ has p.i. degree 2, and $o^- = \begin{pmatrix} R & T \\ T & T \end{pmatrix}$, $o^+ = \begin{pmatrix} T & T \\ R & R \end{pmatrix}$.

To each couple of maximal ideals M and M' of A there corresponds an ideal

$M \boxtimes M' \in \Omega(B)$ as follows :

(M, M')	$M \boxtimes M'$
(a, b)	$\ll a,b \gg$
(a, o^+)	$\ll a,o >^{\star}>$
(a, o^-)	$\ll a,o >_{\star}>$
(o^+, b)	$\ll o,b >^{\star}>$
(o^-, b)	$\ll o,b >_{\star}>$
(o^+, o^+)	$\ll o,o >^{\star}>^{\star}$
(o^-, o^+)	$\ll o,o >_{\star}>^{\star}$
(o^+, o^-)	$\ll o,o >^{\star}>_{\star}$
(o^-, o^-)	$\ll o,o >_{\star}>_{\star}$

From p.i. deg M + p.i. deg M' = p.i. deg $M \boxtimes M'$ it follows that $\mathbb{k}_B(M \boxtimes M') = k_A(M) \underset{k}{\otimes} k_A(M')$. Moreover the above table corresponds with the isomorphism of the varieties $\Omega(A) \times \Omega(A)$ and $\Omega(A \underset{k}{\otimes} A)$.

VI.3.5. An analytically flavored example : quasi-Azumaya algebras.

A semiprime P.I. algebra R, of p.i. degree n say, over an algebraically closed field k is said to be a quasi-Azumaya algebra if $\cap\{M \in \Omega_n(R)\} = o$ i.e. the maximal ideals of maximal p.i. degree intersect in o. These algebras turn out to be useful in case one considers $k = \mathbb{C}$ and Banach algebras R satisfying the identities of $n \times n$ matrices which are quasi-Azumaya algebras. Although such a Banach P.I. algebra need not be affine, any Banach quasi-Azumaya algebra R has the property that, for any maximal ideal M of p.i. degree j we have that $R/M = M_j(\mathbb{C})$. This property makes these rings admissible for a generalized form of Gelfand-duality and it is fundamental in representing these quasi-Azumaya algebras as rings of continuous functions on a suitably defined spectrum. Because of the fact that a quasi-Azumaya Banach algebra need not be affine, this example may not be of direct interest as far as the non-commutative geometry is concerned. We have included this example because it may be used as an interesting test case for some ring theoretical methods involved in the geometry and also because it gives a hint that next to non-commutative algebraic geometry, non-commutative (complex) analysis (whatever that may be) seems to benefit from the theory of P.I. rings.

Put $R' = \{f : [o,2] \to M_2(\mathbb{C}),\ f$ continuous and triangular on $[1,2]$. Obviously R' is a Banach algebra satisfying the polynomial identities of 2×2-matrices. It is easily seen that $R'' = R'/J(R')$, $J(R')$ being the Jacobson radical of R', is of the following form : $R'' = \{f : [o,2] \to M_2(\mathbb{C}),\ f$ continuous and diagonal on $[1,2]\}$ Let I_2 be the intersection of the maximal ideals of R'' which are of p.i. degree 2, let I_1 be the intersection of maximal ideals of p.i. degree 1. Then we have : $R''/I_2 = \{f : [o,1] \to M_2(\mathbb{C})$ continuous and $f(1)$ diagonal$\}$ because $I_2 = \{f : [o,2] \to M_2(\mathbb{C})$ continuous and $f|[o,1] = o.\}$ $R''/I_1 = \{f : [1,2] \to \mathbb{C} \oplus \mathbb{C}$ continuous$\}$, because $I_1 = \{f : [o,2] \to M_2(\mathbb{C})$ continuous and $f|[1,2] = o\}$ $R = R''/I_2$ is obviously (by definition) a quasi-Azumaya algebra which embeds into $C_\infty([o,1[, M_2(\mathbb{C}))$.

VI.3.6. References.

For VI.1 and VI.2, let us just mention A. Verschoren's thesis [194] and F. Van Oystaeyen, A. Verschoren [188] as main references. For details (and more examples) on HNP P.I. rings we may refer to J.C. Robson, L. Small [149] and some related results of E. Nauwelaerts, F. Van Oystaeyen [125] or E. Nauwelaerts [123]. The analytical example in VI.3.5 was brought to our attention by D. Luminet (V.U.B., Brussels).

VII. COHERENT AND QUASICOHERENT SHEAVES OF MODULES OVER AN ALGEBRAIC k-VARIETY.

One motivation for studying coherent sheaves of modules is that these are very useful for studying closed subvarieties of an irreducible algebraic variety. Since the algebraic varieties and subvarieties considered are usually taken to be irreducible, it follows that for most properties we may restrict to torsion-free sheaves of modules in the sense that restriction morphisms are injective. For general (i.e. not necessarily torsion-free) sheaves of modules it turns out that, under acceptable conditions on the underlying variety, we still obtain some results generalizing the commutative theory. It is well-known that in the commutative case the equivalent properties characterizing (quasi-) coherent sheaves of modules over a scheme, cf. [159], [52], follow from the fundamental fact that the schemes considered possess a basis of affine subsets. Since the latter fact is not true in the non-commutative case, the well-known characterizations of (quasi-) coherent sheaves of modules over an algebraic variety which are equivalent in the commutative case, are no longer equivalent here. On the other hand, when a basis of open affines is available e.g. in the Zariski central case, then again, the characterizations used in the commutative case, remain valid.

Let $M \in R$-mod. To M we have associated a presheaf $\underline{Q}_M$ of left $\underline{O}_R$-modules over Spec(R) which was seen to be a separated presheaf, the associated sheaf has been denoted by $\underline{O}_M$. If M is absolutely torsion-free then $\underline{Q}_M = \underline{\mathcal{O}}_M$. More generally, a <u>sheaf</u> will be said to be <u>torsion-free</u> whenever its restriction homomorphisms are injective.

<u>VII.1. Lemma</u>. Let R be a left Noetherian ring, then $\underline{O}_M$ is torsion-free if and only if M is absolutely torsion-free.

<u>Proof</u>. One implication is obvious. Conversely, if $\underline{O}_M$ is torsion-free, choose $s \in M$ and $o \neq I \subset R$ and assume that $Is = o$. Consider the following commutative diagram, where X = Spec(R):

$$\begin{array}{ccc}
\underline{O}_M(X) & \rightarrowtail & \underline{O}_M(X(I)) \\
\downarrow & & \downarrow \\
M = \underline{Q}_M(X) & \xrightarrow[i]{\qquad} & \underline{Q}_M(X(I)) = Q_I(M)
\end{array}$$

The upper horizontal arrow is injective by assumption, both vertical arrows are injective, since $\underline{Q}_M$ is separated. Hence $i : M \to Q_I(M)$ is injective, i.e. $o = \text{Ker}\, i = \kappa_I M$ implying that $s = o$, since $Is = o$ implies $s \in \kappa_I M$. This finishes the proof. ∎

A sheaf of left $\underline{O}_X$-modules $\underline{M}$ on an algebraic k-variety X is said to be <u>arithmetically quasicoherent</u> if for all affine open subsets $U = \Omega(R)$ of X we may find a left R-module M such that $\underline{M}|_U = \underline{O}_M$ on $\Omega(R)$. It is said to be <u>arithmetically prequasicoherent</u> if it may be covered by a collection of open affine sub- $\{U_\alpha = \Omega(R_\alpha);\ \alpha \in A\}$, with the property that for each $\alpha \in A$ there may be found a left R_α-module M_α such that $\underline{M}|U_\alpha = \underline{O}_{M_\alpha}$ on $\Omega(R_\alpha)$. If in these definitions the left R-module M (resp. the left R_α-modules M_α) may be chosen to be finitely generated, hence left noetherian, then we speak of an <u>arithmetically coherent</u> (resp. an <u>arithmetically precoherent</u>) sheaf of left $\underline{O}_X$-modules.

<u>VII.2. Proposition.</u> Let R be an affine k-algebra and M an absolutely torsion free left R-module, then $\underline{O}_M$ is an arithmetically quasi coherent sheaf of left $\underline{O}_R$-modules. If M is finitely generated as a left R-module, then $\underline{O}_M$ is arithmetically coherent. In particular, $\underline{O}_R$ is an arithmetically coherent sheaf of left $\underline{O}_R$-modules.

<u>Proof.</u> Since M is assumed to be absolutely torsion free, we know that for any open subset X(I) of $\Omega(R)$, the sections of $\underline{O}_M$ over X(I) are given by

$$\Gamma(X(I), \underline{O}_M) = Q_I(M).$$

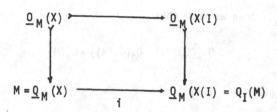

Now, if $X(I) \subset \Omega(R)$ is affine, then in particular κ_I induces a perfect localization in R-mod. So, if $X(J) \subset X(I)$ and if Q_{J^e} denotes the localization functor in $Q_I(R)$-mod induced by κ_{J^e}, then we know that

$$Q_{J^e}(Q_I(M)) = Q_J(Q_I(M)) = Q_J(M) .$$

But this implies that

$$\Gamma(X(J), \underline{O}_{Q_I(M)}) = Q_{J^e}(Q_I(M)) = Q_J(M) = \Gamma(X(J)\underline{O}_M) ,$$

where $X(I) \subset \Omega(Q_I(R))$ is the open subset corresponding homeomorphically to $X(J) \subset \Omega(R)$ under the identification $X(I) = \Omega(Q_I(R))$. So we have proved the first assertion because

$$\underline{O}_M | X(I) = \underline{O}_{Q_I(M)}$$

on $\Omega(Q_I(M)) = X(I)$. The other statements are now obvious. If M is finitely generated as a left R-module, then there exists an epimorphism in R-mod of the form

$$R^n \to M \to o$$

for some $n \in \mathbb{N}$. Hence, with notations as above we obtain an epimorphism in $Q_I(R)$-mod

$$Q_I(R)^n = Q_I(R^n) \to Q_I(M) \to o,$$

since Q_I is a perfect localization. This yields the conclusion. ∎

Note that we have implicitly used the following observation :

<u>VII.3. Lemma</u>. If κ is an arbitrary idempotent kernel functor in R-mod, then for any left R-module M we have that $Q_\kappa(M)$ is an absolutely torsion free $Q_\kappa(R)$-module whenever M is an absolutely torsion free R-module.

<u>Proof</u>. Indeed, let $x \in Q_\kappa(M)$ and J an ideal of $Q_\kappa(R)$. If $Jx = o$, then $J = o$ or $x = o$, for assume that $J \neq o$, then $J \cap R = J_1 \neq o$. Choose an ideal $I \in \mathcal{L}(\kappa)$ such that $Ix \subset M$, then

$$J_1 Ix \subset JIx \subset Jx = o ,$$

hence $Ix = o$, since M is absolutely torsion free. Now this implies $x \in \kappa(Q_\kappa(M)) = o$, which proves our assertion. ∎

We define the category Atf(R) to be the full subcategory of R-mod consisting of all absolutely torsion free left R-modules. When we speak of an exact sequence in Atf(R), we mean a sequence of absolutely torsion free left R-modules, which is exact in R-mod. Note that the category Atf(R) is closed under taking injective hulls in R-mod. Indeed, if M is an absolutely torsion free left R-module and $E = E(M)$ is an injective hull for M, then for any ideal I of R we have

$$\kappa_I E \cap M = \kappa_I M = o \ ,$$

hence $\kappa_I E = o$, since $\kappa_I E \subset E$ is essential.

Let Aqc(R) be the category of arithmetically quasi coherent sheaves of left $\underline{O}_R$-modules on $\Omega(R)$ viewed as a full subcategory of $\sigma(\Omega(R), \underline{O}_R)$, the category of all sheaves of left $\underline{O}_R$-modules on $\Omega(R)$. Then associating the sheaf $\underline{O}_M$ to the left R-module $M \in Ob(\underline{Atf}(R))$ may be viewed as a functor

$$\underline{O} : \mathrm{Atf}(R) \longrightarrow \mathrm{Aqc}(R)$$

$$M \longmapsto \underline{O}_M$$

On the other hand consider the functor

$$Q_{R-P} : R\text{-mod} \longrightarrow R\text{-mod}$$

$$M \longmapsto Q_{R-P}(M)$$

We know that for any $P \in \Omega(R)$ the stalk of $\underline{O}_M$ at P is given by

$$\underline{O}_{M,P} = Q_{R-P}(M) \quad .$$

It is now clear that in general the functor $\underline{O}$ is far from being exact, since if it were, we would obtain from each exact sequence in Atf(R)

$$o \to M' \to M \to M'' \to o$$

an exact sequence in R-mod (actually in Aqc(R)!)

$$o \to Q_{R-P}(M') \to Q_{R-P}(M) \to Q_{R-P}(M'') \to o$$

which we know not to be true.

Let us denote by $R^i\underline{O}$ the i-th derived functor of $\underline{O}$ and by R^iQ_{R-P} the i-th derived functor of Q_{R-P}. For any $M \in$ Atf(R) we write $R^i\underline{O}_M$ for $(R^i\underline{O})(M)$ and $R^i\underline{O}_{M,P}$ for $(R^i\underline{O}_M)_P = [(R^i\underline{O})(M)]_P$. Similarly we write $R^iQ_{R-P}(M)$ where $(R^iQ_{R-P})(M)$ is meant. We thus obtain for any exact sequence

(*)
$$o \to M' \to M \to M'' \to o$$

in R-mod a long exact sequence

(**)
$$o \to \underline{O}_{M'} \to \underline{O}_M \to \underline{O}_{M''} \to R^1\underline{O}_{M'} \to R^1\underline{O}_M \to R^1\underline{O}_{M''} \to R^2\underline{O}_{M'} \to \dots$$

Note that in most applications the sequence (*) is considered in Atf(R), but in order to construct derived functors we have to work in the Grothendieck category R-mod! In general, it is not clear how $R^i\underline{O}_M$ and the different $R^iQ_{R-P}(M)$ are related. Things improve considerably, however, when classical varieties are considered. Let us develop some of the machinery first. Let R be an arbitrary ring and let $\mathcal{L}$ be a filter of left R-ideals. Recall that if M is a left R-module, then a fundamental system of neighborhoods of o for the $\mathcal{L}$-topology on M is given by

$$\mathcal{L}(M) = \{L < M;\ \forall x \in M\ (L : x) \in \mathcal{L}\}$$
$$= \{L < M;\ M/L \text{ is } \mathcal{L}\text{-torsion}\}$$

If M' is a submodule of M then the $\mathcal{L}$-topology on M induces on M' a subspace topology, which in general is weaker than the $\mathcal{L}$-topology on M'. The following result is due to P. Gabriel [61].

VII.4. Proposition. The following properties of a filter $\mathcal{L}$ on R are equivalent :

1. for every module M and submodule M' the $\mathcal{L}$-topology on M' coincides with the

subspace topology induced from the $\mathcal{L}$-topology on M;

2. the class of $\mathcal{L}$-torsion modules is closed under injective hulls;

3. for every module M, the $\mathcal{L}$-torsion submodule of M is essentially closed in M;

4. every injective left R-module is split by its $\mathcal{L}$-torsion submodule. ∎

An idempotent kernel functor κ is called <u>stable</u> if the class of its torsion modules T_κ is closed under injective hulls, i.e. the equivalent properties of VII.4. are valid. As an example the torsion theory for commutative integral domains is stable. More generally, we know that the Goldie torsion theory is stable.

A finitely generated left R-module M (where R is left noetherian!) is said to be an <u>Artin-Rees module</u> if for every ideal I of R, every left submodule L of M and every positive integer n we may find $h(n) \in \mathbb{N}$ such that

$$I^{h(n)}M \cap L \subset I^n L .$$

In particular, if every κ_I is stable, then <u>every</u> finitely generated left R-module is Artin-Rees. A ring R is said to be <u>left classical</u> if it is left noetherian and an Artin-Rees module over itself.

<u>VII.5. Proposition</u>. The following properties of a left noetherian ring R are equivalent :

1. R is left classical;
2. every finitely generated left R-module is Artin-Rees;
3. every idempotent symmetric kernel functor is stable.

<u>VII.6. Proposition</u>. (Krull's Intersection Theorem) If R is a left classical ring and I a twosided R-ideal, then for any finitely generated left R-module M we have

$$\bigcap_{n=0}^{\infty} I^n M = \{x \in M;\ \exists a \in I,\ (1-a)x = 0\}$$

<u>VII.7. Theorem</u>. If κ is a stable kernel functor, then for every left R-module M

we have

$$Q_\kappa(M) = \varinjlim \mathrm{Hom}_R(I,M),$$

where I runs through $\mathcal{L}(\kappa)$.

Proof. The exact sequence

$$o \longrightarrow \kappa M \longrightarrow M \longrightarrow M/\kappa M \longrightarrow o$$

induces an exact sequence

$$o \to \varinjlim \mathrm{Hom}(I,\kappa M) \to \varinjlim \mathrm{Hom}(I,M) \to \varinjlim \mathrm{Hom}(I,M/\kappa M) \to \varinjlim \mathrm{Ext}^1(I,\kappa M)$$

where the first term is known to be zero. If E is an injective hull of κM, then E is a κ-torsion module by hypothesis. The sequence

$$o \longrightarrow \kappa M \longrightarrow E \longrightarrow E/\kappa M \longrightarrow o$$

induces an exact sequence

$$\varinjlim \mathrm{Hom}(I,E/\kappa M) \to \varinjlim \mathrm{Ext}^1(I,\kappa M) \to o$$

Here the first term is again zero, hence so is the last term of the long exact sequence, thus yielding an isomorphism

$$\varinjlim \mathrm{Hom}(I,M) = \varinjlim \mathrm{Hom}(I,M/\kappa M) = Q_\kappa(M) \quad \blacksquare$$

Let M be an arbitrary left R-module. In general there are few things we can say about $\underline{Q}_M$ or about $\underline{O}_M$. However, we have :

VII.8. Proposition. If R is left classical and $M \in R\text{-mod}$, then

$$\underline{O}_{M,P} = Q_{R-P}(M).$$

Proof. $\underline{O}_{M,P} = \varinjlim_{P \in X(I)} Q_I(M) = \varinjlim_{I \not\subset P} Q_I(M) = \varinjlim_{I \not\subset P} \varinjlim_{J \in \mathcal{L}(I)} \mathrm{Hom}(J,M)$

$$= \varinjlim_{I \not\subset P} \varinjlim_{n \in \mathbb{N}} \mathrm{Hom}(I^n, M) = \varinjlim_{n \in \mathbb{N}} \varinjlim_{I \not\subset P} \mathrm{Hom}(I^n, M) =$$

$$= \varinjlim_{n \in \mathbb{N}} Q_{R-P}(M) = Q_{R-P}(M). \qquad \blacksquare$$

<u>VII.9. Theorem.</u> If R is left classical, then for any $M \in \underline{Atf}(R)$ we have

$$R^i \underline{O}_{M,P} = R^i Q_{R-P}(M) .$$

<u>Proof</u>. Take $M \in \underline{Atf}(R)$ and let

$$0 \to M \to E^0 \to E^1 \to E^2 \to \ldots$$

be an injective resolution of M in R-mod. Note that E^0 (= E(M), for example) may be chosen to be absolutely torsion free, but that the other E^i's may not be of this type in general. Viewing $\underline{O}$ as a functor R-mod $\to \sigma(\mathrm{Spec}(R), \underline{O}_R)$, we then have a complex

$$0 \to \underline{O}_M \xrightarrow{\eta} \underline{O}_{E^0} \xrightarrow{\varphi_0} \underline{O}_{E^1} \xrightarrow{\varphi_1} \underline{O}_{E^2} \xrightarrow{\varphi_2} \ldots$$

obtained by applying the functor $\underline{O}$. By well-known properties of derived functors we then have for each $i \geq 0$

$$R^i \underline{O}_M = \mathrm{Ker}\, \varphi_{i+1} / \mathrm{Im}\, \varphi_i \; .$$

On the other hand, by applying the functor Q_{R-P} for any $P \in \mathrm{Spec}(R)$ we obtain a complex

$$0 \to Q_{R-P}(M) \xrightarrow{\varepsilon} Q_{R-P}(E^0) \xrightarrow{\psi_0} Q_{R-P}(E^1) \xrightarrow{\psi_1} Q_{R-P}(E^2) \xrightarrow{\psi_2} \ldots$$

and for any $i \geq 0$ we thus have

$$R^i Q_{R-P}(M) = \mathrm{Ker}\, \psi_{i+1} / \mathrm{Im}\, \psi_i$$

Now $\mathrm{Ker}(\varphi_{i+1})_P = \mathrm{Ker}(\varphi_{i+1,P}) = \mathrm{Ker}\, \psi_{i+1}$ and $\mathrm{Im}(\varphi_i)_P = \mathrm{Im}(\varphi_{i,P}) = \mathrm{Im}\, \psi_i$ by the foregoing, while by exactness of taking stalks we get :

$$R^i \underline{O}_{M,P} = (\mathrm{Ker}\, \varphi_{i+1}/\mathrm{Im}\, \varphi_i)_P = \mathrm{Ker}(\varphi_{i+1,P})/\mathrm{Im}(\varphi_{i,P})$$

$$= \mathrm{Ker}\, \psi_{i+1}/\mathrm{Im}\, \psi_i = R^i\, Q_{R-P}(M). \quad \blacksquare$$

Before proceeding any further, let us recollect some examples of left classical rings

1. If R is a commutative noetherian ring, then all idempotent kernel functors in R-mod are stable and all finitely generated modules are Artin-Rees. In particular, R is classical.
2. If A is a commutative noetherian ring and R is an Azumaya A-algebra, then R is classical.
3. Fully bounded Zariski central rings are classical, cf. [180]; this applies in particular to Zariski central left Noetherian P.I. rings.
4. If R is finite over its center and $\rho : \mathrm{Spec}(R) \to \mathrm{Spec}(C)$ is bijective (hence homeomorphic!), then R is classical.

VII.10. Proposition. Let R be left classical, then the functor $\underline{O} : R\text{-mod} \to \sigma(\Omega(R), \underline{O}_R)$ is exact if and only if κ_{R-P} has property (T) for any $P \in \Omega(R)$ - in the affine case; otherwise we have to consider $\underline{O}_R$ on Spec(R).

Proof. First, let us note that a functor $F : \underline{C} \to \underline{D}$ is exact if and only if R^1F vanishes identically. Indeed, one direction of this equivalence being obvious, take an object $C \in \mathrm{Ob}\, \underline{C}$ with injective hull E and consider the following exact sequence in $\underline{C}$:

$$o \to C \to E \to E/C \to o$$

applying F we obtain two exact sequences

$$o \to FC \to FE \overset{p}{\to} F(E/C) \to o$$

and

$$o \to FC \to FE \overset{p}{\to} F(E/C) \overset{\partial}{\to} (R^1F)C \overset{q}{\to} (R^1F)E = o$$

Here the first sequence is exact in $\underline{\mathcal{D}}$, since F is assumed to be an exact functor, while $(R^1F)E = o$ by the injectivity of E. From the exactness of the first sequence we derive $\mathrm{Im}\, p = F(E/C)$, hence $\partial : F(E/C) \to (R^1F)C$ reduces to the zero map. But then $\mathrm{Ker}\, q = \mathrm{Im}\,\partial = o$, so q is injective and $(R^1F)C = o$, since it injects into $(R^1F)E = o$ through q.

Now, to conclude, the functor $\underline{O}$ is exact if and only if for each exact sequence

$$o \to M' \to M \to M' \to o$$

in R-mod, the following sequence in $\sigma(X_R, \underline{O}_R)$ is exact

$$o \to \underline{O}_{M'} \to \underline{O}_M \to \underline{O}_{M''} \to o .$$

But this is true exactly when for each $P \in X_R$ the next sequence is exact

$$o \to \underline{O}_{M',P} \to \underline{O}_{M,P} \to \underline{O}_{M'',P} \to o$$

By VII.9 this reduces to the exactness of

$$o \to Q_{R-P}(M') \to Q_{R-P}(M) \to Q_{R-P}(M'') \to o$$

i.e. Q_{R-P} is an exact functor in R-mod for each $P \in X_R$. Moreover, in the presence of the noetherian hypothesis on R, it is well-known that Q_{R-P} obviously commutes with direct sums, i.e. κ_{R-P} is a t-functor as asserted.

<u>VII.11. Corollary</u>. If X_R has a basis of open T-sets, i.e. open subsets $X(I)$ such that κ_I has property (T), then $\underline{O}$ is an exact functor. This applies in particular if X_R has a basis of open affines.

<u>Proof</u>. In the "classical" case this follows from the foregoing. In the general case, this is easily seen by taking for each $P \in X_R$ a filterbasis of open subsets $\{X_\alpha, \alpha \in A\}$ where $X_\alpha = X_R - V(I_\alpha)$ and where each κ_{I_α} is a t-functor. We then know

that κ_{R-P} is a t-functor, while

$$\underline{O}_{M,P} = \varinjlim_{X(I)\ni P} Q_I(M) = \varinjlim_{\alpha\in A} Q_I(M) = \varinjlim_{\alpha\in A} [Q_I(R) \underset{R}{\otimes} M]$$

$$= \left[\varinjlim_{\alpha\in A} Q_I(R)\right] \underset{R}{\otimes} M = Q_{R-P}(R) \underset{R}{\otimes} M = Q_{R-P}(M) ,$$

which, in view of the foregoing proves the assertion. ∎

Let us now restrict to Zariski central rings. Let $\underline{M}$ be an arithmetically prequasicoherent sheaf of left $\underline{O}_R$-modules on X_R, then X_R may be covered by a finite number of open sets $U_\alpha = X(Rg_\alpha)$ where $g_\alpha \in C$, the center of R, and such that the restriction of $\underline{M}$ to U_α is of the form $\underline{O}(M_\alpha, Q_{g_\alpha}(R))$. Here we denote by $\underline{O}(M;A)$ the sheaf of $\underline{O}_A$-modules on Spec(A) associated with the left A-module M in the usual way. Indeed, since $\underline{M}$ is arithmetically prequasicoherent, we may find open affine subsets U = Spec(S) covering X_R and such that $\underline{M}|U = \underline{O}(M,S)$ for some left S-module M (this is so, because R is assumed to be Zariski central). We know that the X(Rc), with $c\in C$ form a basis for the Zariski topology on X_R, so

$$U = \bigcup X(Rg_\alpha)$$

for certain $g_\alpha \in C$. To an inclusion $X(Rg_\alpha) \subset V$ there corresponds a ringmorphism $S \to R_{g_\alpha} = Q_{g_\alpha}(R)$, so

$$\underline{M}|X(Rg) = \underline{O}(R_g \underset{S}{\otimes} M, R_g) .$$

Since X_R is quasi compact because each of its open subsets is, this yields the finiteness argument that finishes the proof.

VII.12. Proposition. Let R be a Zariski central ring, let U = X(I) be an open subset of X_R and let $\underline{M}$ be arithmetically prequasicoherent on $X_R = X$, then

1. $\mathrm{Ker}\, \underline{M}^X_U = \kappa_I\, \underline{M}(X)$
2. If $s\in\Gamma(U,\underline{M})$ then for some $n>o$ we have $I^n s \subset \mathrm{Im}\, \underline{M}^X_U$.

Proof. 1) By the foregoing remarks, X may be covered by a finite number of open sets of the form $U_i = X(Rg_i)$, with $g_i \in C$, where for each i

$$\underline{M}|X(Rg_i) = \underline{O}(M_i, Q_{Rg_i}(R) = R_{g_i}) .$$

Assume that $s \in \Gamma(X,\underline{M})$ maps to o under $\underline{M}^X_U$, let $s_i = \underline{M}^X_{U_i}(s) \in \Gamma(U_i, \underline{O}(M_i, R_{g_i})) = M_i$. Since $X_I \cap X_{Rg_i} = X_{I_i}$ where $I_i =: Ig_i$ is twosided, since $g_i \in C$, we have

$$\underline{M}|X(I_i) = \underline{O}(Q_I(M_i)),$$

thus the image of s_i in $Q_I(M_i)$ is zero, hence for some $n_i \in \mathbb{N}$ we find $I^{n_i} s_i = o$. Since the g_i are finite in number, we may find $n \in \mathbb{N}$ such that $I^n s_i = o$ for all i. But the $X(Rg_i)$ cover X, hence $I^n s = o$, so $s \in \kappa_I \underline{M}(X) = \kappa_I \Gamma(X,\underline{M})$. The converse being obvious, this proves our assertion.

2) Take $s \in \Gamma(U,\underline{M})$ and restrict it for each g_i to yield elements $s_i \in \underline{M}^U_{X(Ig_i)}$ $s \in Q_I(M_i) = \Gamma(X(Ig_i),\underline{M})$. Since $M_i = \Gamma(X(g_i),\underline{M})$, we may find a positive integer n_i such that $I^{n_i} s_i \in M_i$, and since the $X(g_i)$ are finite in number there exists an $n \in \mathbb{N}$ such that $I^n s_i \in M_i$ for each i. Choose $\alpha \in I^n$, then this means that for some $t_i \in M_i$ we have

$$\underline{M}^{X(g_i)}_{X(Ig_i)}(t_i) = \alpha s_i .$$

Now, if $r_{ij} = \underline{M}^{X(g_i)}_{X(g_ig_j)}(t_i)$, then

$$\underline{M}^{X(g_ig_j)}_{X(Ig_ig_j)}(r_{ij}) = \underline{M}^{X(g_ig_j)}_{X(Ig_ig)} \underline{M}^{X(g_i)}_{X(g_ig_j)}(t_i) = \underline{M}^{X(g_i)}_{X(T_{g_ig_j})}(t) =$$

$$= \underline{M}^{X(g_iI)}_{X(Ig_ig_j)} \underline{M}^{X(g_i)}_{X(g_iI)}(t_i) = \underline{M}^{X(g_iI)}_{X(Ig_ig_j)}(\alpha s_i) =$$

$$= \alpha \underline{M}^{X(g_iI)}_{X(Ig_ig_j)} \underline{M}^{U}_{X(g_iI)}(s) = \alpha\, \underline{M}^{U}_{X(Ig_ig_j)}(s) ,$$

hence r_{ij} and r_{ji} agree on $X(Ig_ig_j)$. The first part of the lemma then provides us a positive integer $m(i,j,\alpha)$ such that

$$I^{m(i,j,\alpha)}(r_{ij}-r_{ji}) = o \quad .$$

Since the $X(g_i)$ are finite in number and I is finitely generated we may find $m \in \mathbb{N}$ such that $I^m(r_{ij}-r_{ji}) = o$ for all i,j. This means that for all $\beta \in I^m$ we have

$$\underline{M}^{X(g_i)}_{X(g_ig_j)}(\beta t_i) = \underline{M}^{X(g_j)}_{X(g_ig_j)}(\beta t_j)$$

hence there exists $t \in \Gamma(X,\underline{M})$ depending on β, such that for all i

$$\underline{M}^{X}_{X(g_i)}(t) = \beta t_i$$

thus

$$\underline{M}^{X}_{X(g_iI)}(t) = \alpha\beta s_i \quad .$$

So we have proved that for each $\gamma \in I^{n+m}$ we may find $t \in \Gamma(X,\underline{M})$ such that $\underline{M}^X_{X(g_iI)}(t) = \gamma s_i = \underline{M}^X_{X(g_iI)}(\gamma s)$, hence

$$\underline{M}^U_{X(g_iI)}(\underline{M}^X_U(t) - \gamma s) = o$$

for all i, so $\underline{M}^X_U(t) = \gamma s$, i.e. $I^{n+m}s \subset \mathrm{Im}\ \underline{M}^X_U$. ■

<u>VII. 13. Corollary</u>. A sheaf of left $\underline{O}_X$-modules $\underline{M}$ on a Zariski central variety X is arithmetically prequasi coherent if and only if for every open affine subset $U = \mathrm{Spec}(R)$ of X there is a left R-module M such that $\underline{M}|_U = \underline{O}(M,R)$. Since X is noetherian, $\underline{M}$ is arithmetically precoherent iff the same holds with M finitely generated over R.

<u>Proof</u>. If $\underline{M}$ is quasicoherent and U is an open affine subset of X, then $\underline{M}|_U$ is arithmetically prequasicoherent, since we may find a basis for the Zariski topology consisting of open affines for which the restriction of $\underline{M}$ is associated to a module, according to our previous remarks. In other words, we may assume X

to be affine, i.e. $X = \mathrm{Spec}(R)$. Moreover, as we have seen R is then a Zariski central ring itself. If $M = \Gamma(X,\underline{M})$, then we obtain a canonical map

$$\alpha : \underline{O}(M,R) \to \underline{M} .$$

Since $\underline{M}$ is arithmetically prequasicoherent, we know that X may be covered by open subsets $X(Rg_i)$ such that $\underline{M}|_{X(Rg_i)} = \underline{O}(M_i, R_{g_i})$ for some $M_i \in R_{g_i}$-mod. The foregoing result applied to $X(Rg_i)$ yields that

$$\Gamma(X(Rg_i),\underline{M}) = M_{g_i}$$

hence $M_i = M_{g_i}$, implying that $\alpha | X(Rg_i)$ is an isomorphism. But then α is an isomorphism too, since the $X(Rg_i)$ cover X.
The arithmetically precoherent case may be treated similarly. ∎

<u>VII.14. Corollary</u>. Let R be a Zariski central ring and $X_R = \mathrm{Spec}(R)$ (or $\Omega(R)$), then the functor

$$\begin{aligned} \underline{O} : R\text{-mod} &\to \sigma(\underline{O}_R, X_R) \\ M &\to \underline{O}_M \end{aligned}$$

yields an equivalence of categories between R-mod and <u>Aqc</u>(R).

Let us recall that for any ringed space $(X,\underline{O}_X)$ and any sheaf of left $\underline{O}_X$-modules $\underline{M}_X$ on X we say that $\underline{M}_X$ is <u>(geometrically) quasicoherent</u> if and only if for any point $x \in X$ we may find an open neighborhood U of x and an exact sequence of left $\underline{O}_X$-modules

$$\underline{O}_X|U^{(I)} \to \underline{O}_X|U^{(J)} \to \underline{M}_X|U \to o$$

As a nontrivial example, let M be an arbitrary left R-module, and let $\underline{O}_R \otimes_R M$ be the sheaf associated to the presheaf which to each open subset U of X_R associates $(U,\underline{O}_R) \otimes_R M$, then $\underline{O}_R \otimes_R M$ is geometrically quasi coherent. Indeed, we may find an exact sequence in R-mod of the form

$$R^{(I)} \longrightarrow R^{(J)} \longrightarrow M \longrightarrow o$$

yielding an exact sequence

$$\underline{O}_X(V) \underset{R}{\otimes} R^{(I)} \longrightarrow \underline{O}_X(V) \underset{R}{\otimes} R^{(J)} \longrightarrow \underline{O}_X(V) \underset{R}{\otimes} M \longrightarrow o$$

hence an exact diagram of sheaves

$$\underline{O}_R^{(I)} \longrightarrow \underline{O}_R^{(J)} \longrightarrow \underline{O}_R \underset{R}{\otimes} M \longrightarrow o$$

$$(\underline{O}_X|U)^{(I)} \longrightarrow (\underline{O}_X|U)^{(J)} \longrightarrow \underline{M}_X|U \longrightarrow o$$

It follows that $\underline{M}_X|U = \underline{O}_R \underset{R}{\otimes} M$, which proves the assertion. ∎

Let us define a sheaf of left $\underline{O}_X$-modules $\underline{M}_X$ to be <u>geometrically affine quasicoherent</u> if for any point $x \in X$ we may find an open neighborhood U of x which is <u>affine</u> and an exact sequence of left $\underline{O}_X$-modules

$$(\underline{O}_X|U)^{(I)} \longrightarrow (\underline{O}_X|U)^{(J)} \longrightarrow \underline{M}_X|U \quad \to o \quad .$$

It is said to be <u>arithmetically, almost prequasicoherent</u> if for each point $x \in X$ we may find an open affine neighborhood U of x with the property that there exists a left $R = (U, \underline{O}_X)$-module M such that

$$\underline{M}_X|U = \underline{O}_R \underset{R}{\otimes} M \quad .$$

<u>VII.15. Proposition</u>. Let $\underline{M}_X$ be a sheaf of left $\underline{O}_X$-modules on an algebraic variety X, then the following statements are equivalent :

1. $\underline{M}_X$ is geometrically affine quasicoherent;
2. $\underline{M}_X$ is arithmetically almost prequasicoherent.

<u>Proof</u>. One direction of this equivalence follows from the previous remarks; on

the other hand, if we choose U affine such that the sequence

$$(\underline{O}_X|U)^{(I)} \longrightarrow (\underline{O}_X|U)^{(J)} \longrightarrow \underline{M}|U \longrightarrow o$$

is exact, then we know that on $U = X_R$ we have $\underline{O}_X|U = \underline{O}_R$, hence this sequence reduces to

$$\underline{O}_X^{(I)} \xrightarrow{\varphi} \underline{O}_R^{(J)} \longrightarrow \underline{M}|U \longrightarrow o \ .$$

Let $M = \text{Coker}\ \Gamma(U,\varphi)$, i.e. we have an exact sequence of left R-modules

$$R^{(I)} \longrightarrow R^{(J)} \longrightarrow M \longrightarrow o$$

yielding for each $V \subset U$ an exact sequence

$$(\Gamma(U,\underline{O}_R) \underset{R}{\otimes} M)^{(I)} \longrightarrow (\Gamma(U,\underline{O}_R) \underset{R}{\otimes} M)^{(J)} \longrightarrow \Gamma(U,\underline{O}_R) \underset{R}{\otimes} M \longrightarrow o$$

hence by sheafification an exact sequence of sheaves of left $\underline{O}_R$-modules

$$\underline{O}_R^{(I)} \longrightarrow \underline{O}_R^{(J)} \longrightarrow \underline{O}_R \underset{R}{\otimes} M \longrightarrow o \ .$$

In particular, if X_R possesses a basis of open T-sets, then $\underline{O}_R \underset{R}{\otimes} M = \underline{O}_M$, which yields that $\underline{O}_M$ is then geometrically quasicoherent.

<u>VII. 16. Theorem</u>. For any Zariski central variety X and any sheaf of left $\underline{O}_X$-modules $\underline{M}_X$ the following statements are equivalent :

1. $\underline{M}_X$ is arithmetically prequasicoherent;
2. $\underline{M}_X$ is arithmetically quasicoherent;
3. $\underline{M}_X$ is geometrically quasicoherent.

The proof of this result is an easy adaptation of the commutative analogue, we will not go into its (boring) details. Note however

<u>VII. 17. Lemma</u>. Assume that the algebraic variety X admits a basis of open affines then a sheaf of left $\underline{O}_X$-modules $\underline{M}_X$ is arithmetically quasicoherent if and only if it is geometrically quasicoherent.

Proof. 1. Let $\underline{M}_X$ be geometrically quasicoherent then we may find an open covering $\{U_\alpha;\ \alpha\in A\}$ of X such that we obtain exact sequences

$$(\underline{O}_X|U_\alpha)^{(I_\alpha)} \longrightarrow (\underline{O}_X|U_\alpha)^{(J_\alpha)} \longrightarrow \underline{M}_X|U_\alpha \longrightarrow o$$

and it is clear that we may choose the U_α to be affine, say $(U_\alpha, \underline{O}_X|U_\alpha) = (X_{R_\alpha}, \underline{O}_{R_\alpha})$. Then

$$(\underline{O}_X|U_\alpha)^{(I_\alpha)} = \underline{O}(R_\alpha^{(I_\alpha)}, X_{R_\alpha})\ ; (\underline{O}_X|U_\alpha)^{(J_\alpha)} = \underline{O}(R_\alpha^{(J_\alpha)}, X_{R_\alpha})\ .$$

So we obtain exact sequences

$$\underline{O}_{R_\alpha}^{(I_\alpha)} \xrightarrow{\varphi_\alpha} \underline{O}_R^{(J_\alpha)} \longrightarrow M_X|U_\alpha \longrightarrow o$$

On the other hand, if $\psi_\alpha : R_\alpha^{(I_\alpha)} \to R_\alpha^{(J_\alpha)}$ is the R_α-linear morphism induced by taking global sections of φ_α, then for $M_\alpha = \mathrm{Coker}\ \psi_\alpha$ we obtain an exact sequence

$$\underline{O}_{R_\alpha}^{(I_\alpha)} \xrightarrow{\varphi_\alpha} \underline{O}_{R_\alpha}^{(J_\alpha)} \longrightarrow \underline{O}_{M_\alpha} \longrightarrow o$$

i.e. $\underline{M}_X|U_\alpha = \underline{O}_{M_\alpha} = \underline{O}(M_\alpha, U_\alpha = X_{R_\alpha})$ and $\underline{M}_X$ is arithmetically quasicoherent.

2. The converse is obvious. ∎

If one defines geometrically coherent sheaves of left modules as usually, then it is clear that geometrically and arithmetically quasicoherent sheaves coincide.

VII.18. Lemma. Let R be a left classical ring and E an injective R-module, then the presheaf $\underline{Q}_E$ on X_R is flasque (= flabby).

Proof. This is almost obvious. Take $X(J)\subset X(I)$ then we have to show that the canonical map $Q_I(E)\to Q_J(E)$ is surjective; but $Q_K(E) = \varinjlim \mathrm{Hom}_R(K^n,E)$ for any ideal K of R, so if $q\in Q_J(E)$ is represented by $\varphi : J^n\to E$, then the injectivity of E yields that we may find $\psi : I^n\to E$ extending φ , i.e.

$$\begin{array}{ccc} I^n & \longrightarrow & J^n \\ & \psi \searrow \quad \swarrow \varphi & \\ & E & \end{array}$$

is commutative. It is then clear that the element $t \in Q_I(E)$ represented by ψ is mapped onto q under the canonical map, proving its surjectivity. ▪

<u>VII.19. Corollary</u>. If Spec(R) possesses a basis of open T-sets (e.g. if R is Zariski central) or if E is the injective hull of an absolutely torsion free left R-module, then $\underline{O}_E$ is a flasque sheaf of $\underline{O}_R$-modules on X_R. ▪

<u>VII.20. Corollary</u>. Assume one of the following properties

1. R is Zariski central;
2. R is (left) classical and possesses a basis of open T-sets,

then for any quasi coherent sheaf $\underline{M}$ on X_R and any $n \in \mathbb{N}_o$ we have

$$H^n(X_R, \underline{M}) = o \ .$$

<u>Proof</u>. Let $\underline{M} = \underline{O}_M$, and take an injective resolution

$$(20,*) \qquad o \to M \to E_o \to E_1 \to \ldots$$

of M in R-mod, then we derive an exact sequence of sheaves of left $\underline{O}_R$-modules

$$o \longrightarrow \underline{O}_M \longrightarrow \underline{O}_{E_o} \longrightarrow \underline{O}_{E_1} \longrightarrow \ldots$$

Since each $\underline{O}_E$ is flabby, we may use this to calculate cohomology : applying the functor $\Gamma(X_R,-)$ we recover the original exact sequence (20,*), i.e. $H^o(X_R, \underline{M}) = M$ and $H^n(X_R, \underline{M}) = o$ for $i > o$. ▪

Using (VI.2.30.) we may prove the following non-commutative version of Serre's Theorem.

VII.21. Theorem. Let X be a Zariski central variety, then the following conditions are equivalent

1. X is affine;
2. $H^i(X,\underline{M})=o$ for every (geometrically) quasicoherent sheaf $\underline{M}$ and every $i>o$;
3. $H^1(X,\underline{I})=o$ for all coherent sheaves of ideals $\underline{I}$ of $\underline{O}_X$. ∎

VII.22. Corollary. Let R be a Zariski central ring and consider the following exact sequence of sheaves of left $\underline{O}_R$-modules on X_R

$$o \to \underline{M}' \to \underline{M} \to \underline{M}'' \to o\,.$$

If F' is (geometrically) quasicoherent, then the following, induced sequence of left R-modules is exact too

$$o \to \Gamma(X_R,\underline{M}') \to \Gamma(X_R,\underline{M}) \to \Gamma(X_R,\underline{M}'') \to o$$

Proof. This follows immediately from the long exact cohomology sequence

$$o \to \Gamma(X_R,\underline{M}') \to \Gamma(X_R,\underline{M}) \to \Gamma(X_R,\underline{M}'') \to H^1(X_R,\underline{M}') \to \ldots$$

where $H^1(X_R,\underline{M}')=o$ in view of Theorem VII.21. ∎

VII.23. Proposition. Let $\varphi : R\to S$ be a morphism of Zariski central rings and M an S-module (on the left!), then, if we denote by $f:\Omega(S)\to\Omega(R)$ the associated continuous morphism, we have

$$f_*(\underline{O}_M)=\underline{O}_{{}_RM}\ .$$

Proof. Let X(I)be an open subset of $\Omega(R)$, then we have

$$f_*(\underline{Q}_M)(X(I)) = \underline{Q}_M(f^{-1}(X(I)) = \underline{Q}_M(X(\varphi(I))) = Q_{(\varphi(I))}(M),$$

while we also have

$$\underline{Q}_{R}M(X(I)) = Q_I({}_RM).$$

If we choose I such that κ_I induces a perfect localization in S-mod, then we know that both terms coincide. Now, since we have assumed that R is Zariski central, we may find a basis {X(I)} with this property, hence the presheaves $f_*(\underline{O}_M)$ and $\underline{O}_M$ coincide on a basis of open subsets of $\Omega(R)$, so their associated sheaves $f_*(\underline{O}_M)$ and $\underline{O}_{_RM}$ are identical. This proves the assertion. ∎

<u>VII.24. Proposition</u>. Let R be a Zariski central ring then the kernel, cokernel and image of any morphism of quasicoherent sheaves on $\Omega(R)$ is quasicoherent.

<u>Proof</u>. This is an obvious consequence of the fact that $M \to \underline{O}_M$ is an exact functor. ∎

<u>VII.25. Corollary</u>. If X is a Zariski central variety, then the kernel, cokernel and image of any morphism of quasicoherent sheaves on X is quasicoherent.

<u>Proof</u>. This is an easy consequence of VII.24. since the notion of quasicoherence is local. ∎

<u>VII.26. Remarks</u>.

1. The same statements remain valid for <u>coherent</u> sheaves of modules;
2. Any extension of (quasi) coherent sheaves on a Zariski variety is (quasi) coherent (same proof as in the commutative case).

<u>VII.27. Corollary</u>. Let $X \xrightarrow{f} Y$ be a morphism of Zariski central varieties and $\underline{M}$ a quasicoherent sheaf of $\underline{O}_X$-modules, then $f_*\underline{M}$ is a quasicoherent sheaf of $\underline{O}_Y$ modules.

<u>Proof</u>. We may again assume Y to be affine, i.e. $Y = \Omega(R)$ with R Zariski central and cover X with a finite number of open affines U_i, with each intersection $U_i \cap U_j$ covered by open affine subsets U_{ijk}. Now, for any open subset V of Y a section $s \in \Gamma(V, f_*\underline{M}) = \Gamma(f^{-1}(V), \underline{M})$ is given by a collection of sections $s_i \in \Gamma(f^{-1}(V) \cap U_i, \underline{M})$, whose restrictions to the open sets $f^{-1}(V) \cap U_{ijk}$ are equal. So there is an exact sequence of sheaves on Y

$$o \longrightarrow f_*\underline{M} \longrightarrow \prod_i f_*(\underline{M}|U_i) \longrightarrow \prod_{i,j,k} f_*(\underline{M}|U_{ijk}) .$$

Here $f_*(\underline{M}|U_i)$ and $f_*(\underline{M}|U_{ijk})$ are quasicoherent by the foregoing, therefore $f_*\underline{M}$ is quasicoherent too. ∎

Let us now consider the projective case. In the sequel of this section R is assumed to be a positively graded, prime left Noetherian ring satisfying the conditions PS 1-2.

Recall that $Q^g_I(M)$, for a graded ideal I of R and arbitrary $M \in R\text{-gr}$ is the graded ring of quotients associated to κ^+_I which is given by the graded filter consisting of all left ideals of R containing a graded ideal J of R such that $\text{rad}(J) \supset I_+$. Assigning $Q^g_I(M)$ to the Zariski open set $X_+(I)$ of Proj(R) yields a separated presheaf $\underline{Q}^+_M$ over Proj(R). The associated sheaf of $\underline{Q}^+_M$, denoted by $L\underline{Q}^+_M$, yields a sheaf $\underline{O}^+_M$ defined by putting $\underline{O}^+_M(U) = (Q^g_I(M))_o$ for all $U = X_+(I)$ open in Proj(R). One easily checks the following properties :

<u>VII.28. Proposition</u>. 1. If either one of the following conditions holds, a. R is left classical and bounded, b. M is absolutely torsion free, c. R is Zariski central, then the stalk of $\underline{Q}^+_M$ at $P \in \text{Proj}(R)$ is exactly $\underline{Q}^+_{M,P} = Q^g_{R-P}(M)$, thus we also have that $\underline{O}^+_{M,P} = (Q^g_{R-P}(M))_o$.

2. For any homogeneous central element $c \in C_1$, we have ; $\underline{Q}^+_M|X_+(c) \cong \underline{Q}_{Q^g_c(M)}$. Furthermore we have that $\underline{O}^+_M\ X_+(c) \cong \underline{O}_{(Q^g_c(M))_o}$.

<u>Proof</u>. 1. Starting from a. or c. one easily sees that κ^+_{R-P} is a graded t-functor for every $P \in \text{Proj}(R)$ i.e. κ^+_{R-P} is of finite type, hence the stalk theorem follows by classical methods. Starting from b. one may mimic the original proof given in case $M = R$.

2. Since Q^g_c is a graded localization having the graded property (T), it follows that $X_+(c)$ is an open T-set in Proj(R) i.e. the first statement follows from the fact that κ^+_c is a graded geometric kernel functor, whereas the second statement is a direct consequence of sheafification methods (on a geometric open set!). ∎

VII.29. Corollary. For any $M \in R$-gr the sheaf $\underline{O}^+_M$ is arithmetically prequasicoherent. If M is finitely generated then $\underline{O}^+_M$ is arithmetically precoherent.

VII.30. Remarks. 1. The assumptions b., or c., of VII.28.1., imply that $\underline{O}^+_M$ is a sheaf.

2. For any $n \in \mathbb{N}$ we define $\underline{O}^+_R(n) = \underline{O}^+_{R(n)}$, where R(n) is defined by $R(n)_m = R_{n+m}$.

We call $\underline{O}^+_R(1)$ the twisting sheaf (Serre's) just as in the commutative case. If $\underline{M}$ is any sheaf of left $\underline{O}^+_R$-modules, put $\underline{M}(n) = \underline{O}^+_R(n) \underset{\underline{O}^+_R}{\otimes} \underline{M}$.

A sheaf of twosided $\underline{O}_X$-modules $\underline{\mathcal{L}}$ on an algebraic variety $X = (X, \underline{O}_X, \underline{O}^{bi}_X, \underline{K}_X, \pi_X)$ is said to be invertible if X may be covered by open affine subsets U with the property that

$$\underline{\mathcal{L}}|_U \cong \underline{O}_X|_U = \underline{O}_U .$$

VII.31. Proposition. 1. For any $n \in \mathbb{N}$, $\underline{O}^+_R(n)$ is an invertible sheaf over Proj(R).
2. For any $M \in R$-gr : $\underline{O}^+_M(n) = \underline{O}^+_{M(n)}$ for any $n \in \mathbb{N}$, in particular, for each couple $(n,m) \in \mathbb{N}^2$ we have :

$$\underline{O}^+_R(n) \otimes \underline{O}^+_R(m) \cong \underline{O}^+(m+n).$$

Proof. 1. Our assumptions on R yield that the open sets $X_+(c)$ with $c \in Z(R)_1$ cover Proj(R). By VII.28. it follows that $\underline{O}^+_R(n)|X_+(c) = \underline{O}^+_{R(n)}|X_+(c) \cong \underline{O}^+_{(Q^g_c(R(n)))_o}$, with $X_+(c) \cong \mathrm{Spec}((Q^g_c(R))_o)$. We claim that $(Q^g_c(R(n)))_o$ is a free $(Q^g_c(R))_o$-module of rank one. This claim is easily established if one notes that $(Q^g_c(R(n)))_o$ is nothing but $(Q^g_c(R))_n$ which is a free $(Q^g_c(R))_o$-module on the basis $\{c^n\}$ Therefore $\underline{O}^+_R(n)$ is locally free of rank one i.e. invertible.
2. Follows easily from the fact that Q^g_c is just central localization (graded!) at the set $\{1,c,c^2,\dots\}$.

Let B be a graded ring and let $\underline{M}$ be a sheaf of left $\underline{O}^+_R$-modules on $\mathrm{Proj}(R) \simeq X_+(R)$ then we may associate to it

$$\Gamma_*(\underline{M}) = \bigoplus_{n\in\mathbb{Z}} \Gamma(X_+(R), \underline{M}(n)) \quad .$$

The set $\Gamma_*(\underline{M})$ may be endowed with the structure of a graded left R-module in the following way. If $r\in R_d$, then we may view r as an element of

$$(X_+(R), \underline{O}^+_R(d)) ,$$

so if $m\in\Gamma(X_+(R), \underline{M}(n)) = \Gamma_*(\underline{M})_n$, then r.m is defined to be the image of $r\otimes m\in\Gamma(X_+(R), \underline{O}^+_R(d)) \otimes\Gamma(X_+(R), \underline{M}(n))$ under the morphism which identifies $\underline{M}(n+d)$ and $\underline{O}^+_R(d) \otimes\underline{M}(n)$.

<u>VII.32. Proposition</u>. For any affine left noetherian prime PI algebra R over k and any $n\in\mathbb{N}$ we have

$$\Gamma_*(\underline{O}_{\mathbb{P}^n_R}) = R[X_o,\dots,X_n] \quad .$$

<u>Proof</u>. This is exactly the same proof as in the commutative case. Note also that the ring $R[X_o,\dots,X_n]$ satisfies the conditions put forward at the beginning of the "graded-part" of this section.

From here on we assume that R is a graded ring satisfying the conditions mentioned before but with the property of being Zariski central. In this case the structure sheaf $\underline{O}^+_R$ is a sheaf of local rings. What is special in this case is that the associated variety Proj(R) (actually P(R)!) posseses a <u>basis</u> of open geometric subsets, i.e. open subsets $X_+(I)$ with the property that κ^+_I induces a geometric localization.

<u>VII.33. Proposition</u>. Let $\underline{\mathcal{L}}$ be an invertible sheaf on X = Proj(R) and let

$$Z(\Gamma(X,\underline{\mathcal{L}})) = \{f\in\Gamma(X,\underline{\mathcal{L}}); \forall r\in\Gamma(X,\underline{O}_X), rf = fr\} .$$

For any $f\in Z(\Gamma(X,\underline{\mathcal{L}}))$ we denote by X_f the set of all $x\in X$ with the property that $\underline{\mathcal{L}}^X_x(f) = f(x) \notin \underline{M}_x\underline{\mathcal{L}}_x$. Let $\underline{F}$ be a (arithmetically pre-) quasicoherent sheaf of left $\underline{O}_X$-modules on X, then the following statements hold.

1. if $s \in \Gamma(X,\underline{F})$ has the property that $\underline{F}^X_{X_f}(s) = o$, then we may find $n \in \mathbb{N}$ such that $f^n s = o$, considered as a global section of $\underline{\mathcal{L}}^{\otimes n} \underset{\underline{O}_X}{\otimes} \underline{F}$, i.e.

$$\mathrm{Ker}\, \underline{F}^X_{X_f} \subset \kappa_f \underline{F}(X)$$

2. if $t \in \Gamma(X_f,\underline{F})$ then for some $n > o$ we have $f^n t \in \mathrm{Im}(\underline{F} \otimes \underline{\mathcal{L}}^{\otimes n})^X_X$, i.e. $f^n t \in \Gamma(X_f, F \otimes \mathcal{L}^{\otimes n})$ extends to a global section of $F \otimes \mathcal{L}^{\otimes n}$).

Proof. We cover $X = \mathrm{Proj}(R)$ with a finite number of open affine subsets $U = \mathrm{Spec}(A)$ such that $\Gamma(X,\underline{O}_X) \to \Gamma(U,\underline{O}_X) = A$ is an extension and such that $\underline{\mathcal{L}}|_U$ is free, i.e. there is an isomorphism of two-sided $\underline{O}_U$-modules $\psi : \underline{\mathcal{L}}|_U = \underline{O}_U$. Since $\underline{F}$ is quasi-coherent we may find a left A-module M with $\underline{F}|_U = \underline{O}_M$. The section $s \in \Gamma(X,\underline{F})$ in the first statement restricts to an element of M, which by abuse of notation will also be denoted by s. On the other hand, the section $f \in \Gamma(X,\underline{\mathcal{L}})$ restricts to give a section of $\underline{\mathcal{L}}|_U$, which yields an element $g = \psi(f) \in A$. Since $\Gamma(X,\underline{O}_X) \to \Gamma(U,\underline{O}_X)$ is an extension and ψ is an isomorphism of twosided modules, obviously $g \in Z(A)$, the center of A Moreover $X_f \cap U = X(g)$, whence obviously X_f is open in X. Now $\underline{F}^X_{X_f}(s) = o$, so it follows from (VII.12) that we may find a positive integer n such that $g^n s = o$ in M. But then the isomorphism

$$\underline{F} \otimes \psi^{\otimes n} : \underline{F} \otimes \underline{\mathcal{L}}^n|_U = \underline{F}|_U$$

yields that $f^n s \in \Gamma(U, \underline{F} \otimes \underline{\mathcal{L}}^{\otimes n})$ vanishes. Since one easily verifies that this statement is actually independent of ψ, we may do this for each open set U in the covering, choose a single positive integer n, large enough, and we find $f^n s = o$ on X. This proves the first statement.

The second statement may be proved in roughly the same way as VII.12.2. ∎

VII.34. Theorem. Let $\underline{F}$ be a quasicoherent sheaf of left $\underline{O}^+_R$-modules on $X = \mathrm{Proj}(R)$, where R is Zariski-central. Then there is a natural isomorphism

$$\beta : O^+_{\Gamma_*(\underline{F})} \xrightarrow{\sim} \underline{F}\,.$$

Proof. Let us first define β for any $\underline{O}^+_R$-module $\underline{F}$. Take f central, homogeneous of degree 1 in R. Since $\underline{O}^+_{\Gamma_*(\underline{F})}$ is quasicoherent in any case, to define $\beta : \underline{O}^+_{\Gamma_*(\underline{F})} \to \underline{F}$ it suffices to give the image of a section of $\underline{O}^+_{\Gamma_*(\underline{F})}$ over $X_+(f)$. Such a section is represented by a fraction mf^{-d}, where $m \in \Gamma(X, \underline{F}(d))$ for some $d \in \mathbb{N}$. We may think of f^{-d} as a section of $\underline{O}^+_R(-d)$ defined over $X_+(f)$, while then $m \otimes f^{-d}$ may be viewed as a section of $\underline{F}$ over $X_+(f)$. This defines β. If $\underline{F}$ is quasicoherent, then we would like to identify $(Q^g_f(\Gamma_*(\underline{F}))_o)$ with $\Gamma(X_+(f), \underline{F})$ in order to prove that β is an isomorphism. Consider f as a global section of the invertible sheaf $\underline{\mathcal{L}} = \underline{O}^+_R(1)$. Since f is central in R_1, obviously $f \in Z(\Gamma(X, \underline{\mathcal{L}}))$. Moreover, by assumption we may find finitely many elements $f_o, \dots, f_r$ which are central in R_1 such that $X = \cup X_+(f_i)$. The intersections $X_+(f_i) \cap X_+(f_j) = X_+(f_i f_j)$ are also affine and for each index i the restriction $\underline{\mathcal{L}}|_{X_+(f_i)}$ is free. We thus may apply the foregoing lemma to show that $\underline{F}(X_+(f_i)) = Q_{f_i}(\Gamma_*(\underline{F}))_o$ which finishes the proof (using a straightforward local global argument). ∎

VIII. PRODUCTS, SUBVARIETIES etc...

First let us make precise the notion of a separated variety, already hinted at in foregoing sections. An algebraic k-variety X is <u>separated</u> if for all algebraic k-varieties Y and for all morphisms $f,g \in \mathrm{Hom}(Y,X)$ the set $\{y \in Y;\ f(y) = g(y)\}$ is a closed subset of Y.

VIII.1. Lemma. Let $\Omega(R)$ be an affine k-variety. If $x,y \in \Omega(R)$ are such that $\mathbb{k}(x) = \mathbb{k}(y)$ and $s(x) = s(y)$ for all $s \in R$, then $x = y$.

Proof. Let M,N be the maximal ideals of R representing x,y. Let e_x, e_y be the canonical morphisms given by

$$e_x : R \to Q^{bi}_{R-M}(R) \longrightarrow Q^{bi}_{R-M}(R/M) = \mathbb{k}_R(M)$$

$$e_y : R \to Q^{bi}_{R-N}(R) \longrightarrow Q^{bi}_{R-N}(R/N) = \mathbb{k}_R(N)$$

Then : $s(x) = e_x(s)$ and $s(y) = e_y(s)$.

If $x \neq y$ i.e. there is an $s \in M-N$, then $e_x(s) = o$ and $e_y(s) \neq o$ (indeed $e_y(s) = o$ would entail $s \in N$). Therefore $x \neq y$ yields the existence of an $s \in R$ such that $s(x) = e_x(s) \neq e_y(s) = s(y)$. ∎

VIII.2. Corollary. Let X be an algebraic k-variety. If $x,y \in X$ are such that $\mathbb{k}(x) = \mathbb{k}(y)$ and for all $s \in \Gamma(X, \underline{O}_X)$, $s(x) = s(y)$, then $x = y$.

Proof. Reduces locally to Lemma VIII.1. ∎

VIII.3. Theorem. Each affine k-variety is separated.

Proof. Put $X_R = \Omega(R)$ and suppose $f,g \in \mathrm{Hom}(Y,X_R)$, Y an arbitrary k-variety. Clearly $\{y \in Y;\ f(y) = g(y)\} = \{y \in Y;\ sf(y) = sg(y)$ for all $s \in \Gamma(X, \underline{O}_X)\}$. Since $M_n(k)$ is endowed with the discrete topology (and is therefore separated as a topological space), the maps $s : X_R \to M_n(k)$ are continuous for all $n \in \mathbb{N}$. Thus $\{y \in Y;\ sf(y) = sg(y)$ for all $s \in \Gamma(X, \underline{O}_X)\}$ is closed in Y. ∎

VIII.4. Proposition. If for each couple x_1, $x_2 \in X$, an algebraic k-variety, there exists an open affine subvariety of X containing x_1 and x_2 then X is separated.

Proof. As in the commutative case.■

VIII.5. Corollary. If the graded ring R satisfies the usual conditions (making Proj(R) into a k-variety) then Proj(R) is separated.

From now on in this section, all algebraic varieties are assumed to be precellular, unless explicitely mentioned otherwise. Mostly we will restrict further to consider only irreducible affine varieties but usually this will be no real restriction.

A product of algebraic k-varieties X and Y should be an object $X \times Y$ such that for each k-variety Z there is a bijection : $\mathrm{Hom}(Z,X) \times \mathrm{Hom}(Z,Y) = \mathrm{Hom}(Z,X \times Y)$. Now such an object $X \times Y$ need not exist in general, even when X and Y are affine. Indeed, if $X = \Omega(R)$, $Y = \Omega(S)$ then $\Omega(R \otimes_k S)$ is the only plausible candidate for $X \times Y$. Note first the following :

VIII.6. Lemma. The varietal space $\Omega(R \otimes_k S)$ is a well-defined affine k-variety.

Proof. A result of G. Bergman [28], yields that the product of two prime affine k-algebras over an algebraically closed field k is prime too. A result of A.Regev, [143], yields that the tensor product of P.I. algebras over a field is a P.I. algebra. Now $R \otimes_k S$ is affine (resp. left noetherian) if R and S are - the proof of the noetherian case is due to L. Small, cf. APPENDIX - proving the assertion.

If $\Omega(R \otimes_k S)$ were a product in the category of (cellular) algebraic varieties over k, then for an arbitrary k-variety $\Omega(T)$ we would find that :
$\mathrm{Hom}(R,T) \times \mathrm{Hom}(S,T) = \mathrm{Hom}(\Omega(T),\Omega(R)) \times \mathrm{Hom}(\Omega(T),\Omega(S)) = \mathrm{Hom}(\Omega(T),\Omega(R) \times \Omega(S)) = \mathrm{Hom}(\Omega(T),\Omega(R \otimes_k S)) = \mathrm{Hom}(R \otimes_k S,T)$ i.e. then $R \otimes_k S$ is a coproduct of R and S in the category of P.I. k-algebras. However if $u \in \mathrm{Hom}(R,T)$, $v \in \mathrm{Hom}(S,T)$ have noncommuting images in T then there cannot be a k-algebra morphism $R \otimes_k S \to T$ extending u and v

in the sense that the following diagram is commutative :

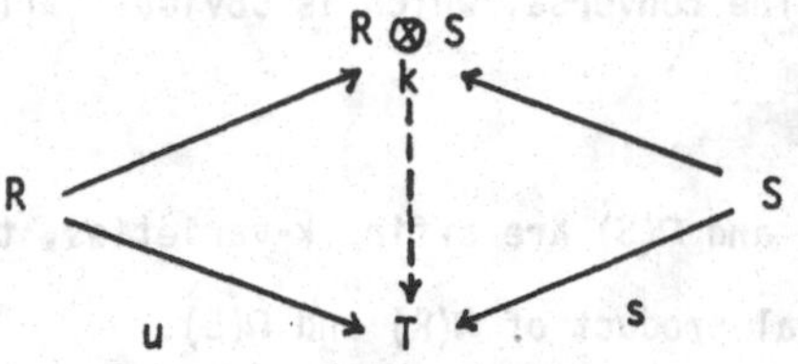

This failure forces us to introduce the following :

<u>VIII.7. Definition</u>. If X and Y are algebraic k-varieties then a <u>geometric product</u> of X and Y is an algebraic k-variety $X \hat{\times} Y$ such that the following property holds : if Z is an algebraic k-variety and if Hom(Z;X,Y) denotes the set of all couples (φ,ψ) in $\mathrm{Hom}(Z,X) \times \mathrm{Hom}(Z,Y)$ such that the maps :

$$\Gamma(\varphi) \in \mathrm{Hom}_k(\Gamma(X),\Gamma(Z)) \quad \text{and} \quad \Gamma(\psi) = \mathrm{Hom}_k(\Gamma(Y),\Gamma(Z))$$

commute in the following sense : for all $(x,y) \in \Gamma(X) \times \Gamma(Y)$, $\Gamma(\varphi)(x)\Gamma(\psi)(y) = \Gamma(\psi)(y)\Gamma(\varphi)(x)$, then :

$$\mathrm{Hom}(Z,X \hat{\times} Y) = \mathrm{Hom}(Z;X,Y) .$$

Uniqueness of the geometric product (if it exists!) up to isomorphism follows from the universal property; as to the existence of such a geometric product :

<u>VIII.8. Proposition</u>. If $\Omega(R)$ and $\Omega(S)$ are algebraic k-varieties then $\Omega(R \otimes_k S)$ is a geometric product of $\Omega(R)$ and $\Omega(S)$ <u>in the category of affine k-varieties</u>.

<u>Proof</u>. Let Z be an arbitrary affine k-variety and let $(\varphi,\psi) \in \mathrm{Hom}(Z;\Omega(R),\Omega(S))$, then $(\Gamma(\varphi),\Gamma(\psi)) \in \mathrm{Hom}_k(R,S;\Gamma(Z))$, where $\mathrm{Hom}_k(R,S;\Gamma(Z))$ is the set of all couples $(u,v) \in \mathrm{Hom}_k(R,\Gamma(Z)) \times \mathrm{Hom}_k(S,\Gamma(Z))$ such that for all $(r,s) \in R \times S$ we have

$$u(r)v(s) = v(s)u(r) .$$

But then, by the universal property of the tensorproduct of algebras, i.e.

$$\mathrm{Hom}_k(R,S;\Gamma(Z)) = \mathrm{Hom}_k(R \otimes_k S,\Gamma(Z))$$

we get that $(\Gamma(\varphi),\Gamma(\psi))$ corresponds to $\theta \in \mathrm{Hom}_k(R \otimes_k S, \Gamma(Z))$, hence we find a corresponding $\Theta \in \mathrm{Hom}(Z, \Omega(R \otimes_k S))$. The converse, which is obvious (!), finishes the proof. ■

<u>VIII.9. Proposition</u>. If $\Omega(R)$ and $\Omega(S)$ are affine k-varieties, then the <u>set</u> $\Omega(R \otimes_k S)$ is the set theoretical product of $\Omega(R)$ and $\Omega(S)$.

<u>Proof</u>. The structural extension $i_R : R \to R \otimes_k S$ and $i_S : S \to R \otimes_k S$ induce morphisms of k-varieties $\Omega(R \otimes_k S) \to \Omega(R)$ and $\Omega(R \otimes_k S) \to \Omega(S)$, hence set-theoretical maps $\Omega(R \otimes_k S) \to \Omega(R) \times \Omega(S)$. Conversely, to each couple $(N_1, N_2) \in \Omega(R) \times \Omega(s)$ we associate $N_1 \otimes_k S + R \otimes_k N_2 \in \Omega(R \otimes_k S)$ and one proves as in the commutative case that these maps are inverse to each other. ■

<u>VIII.10. Proposition</u>. If $(M,N) \in \Omega(R) \times \Omega(S) = \Omega(R \otimes_k S)$, then the function ring at (M,N) is

$$\mathbb{k}_{R \otimes S}(M,N) = \mathbb{k}_R(M) \otimes_k \mathbb{k}_S(N)$$

<u>Proof</u>. If $\underline{M} = R \otimes N + M \otimes S$ is the maximal ideal of $R \otimes_k S$ corresponding to $(M,N) \in \Omega(R) \times \Omega(S)$, then clearly

$$(R \otimes_k S)/\underline{M} = R/M \otimes_k S/N$$

Now M and N being maximal ideals, it is obvious that R/M, S/N and $(R \otimes_k S)/\underline{M}$ are prime PI rings which are simple, hence also artinian. Now this implies that R/M, S/N and $(R \otimes_k S)/\underline{M}$ equal their total rings of quotients and that $\mathbb{k}_R(M) \otimes_k \mathbb{k}_S(N) = \mathbb{k}_{R \otimes S}(\underline{M})$. ■

This result is in correspondence with the fact that $\mathbb{k}_{R \otimes S}(\underline{M})$ should again be a matrix ring. ■

<u>VIII.11. Proposition</u>. Assume that $(M,N) \in \Omega(R \otimes_k S) = \Omega(R) \times \Omega(S)$ is such that the localization at M (resp. at N) has property (T) (= induces a perfect localization) for R-bimodules (resp. S-bimodules), then the stalk at (M,N) of the structure

sheaf of $\Omega(R \underset{k}{\otimes} S)$ is given by the localization of $Q^{bi}_{R-M}(R) \otimes Q^{bi}_{S-N}(S)$ at $Q^{bi}_{R-M}(M) \underset{k}{\otimes} Q^{bi}_{S-N}(S) + Q^{bi}_{R-M}(R) \underset{k}{\otimes} Q^{bi}_{S-N}(N)$, i.e.

$$\underline{O}^{bi}_{(x,y)} = Q^{bi}_{\underline{M}}(\underline{O}^{bi}_x \otimes \underline{O}^{bi}_y) ,$$

where $\underline{M} = \underline{O}_x \otimes \underline{M}_y + \underline{M}_x \otimes \underline{O}_y$.

<u>Proof.</u> Consider the following inclusions

$$R \otimes S \to \underline{O}^{bi}_x \otimes S \to \underline{O}^{bi}_x \otimes \underline{O}^{bi}_y .$$

This yields an exact sequence

$$o \longrightarrow \underline{O}^{bi}_x \otimes S/R \otimes S \longrightarrow \underline{O}^{bi}_x \otimes \underline{O}^{bi}_y/R \otimes S \longrightarrow \underline{O}^{bi}_x \otimes \underline{O}^{bi}_y/\underline{O}^{bi}_x \otimes S \longrightarrow o$$

$$o \longrightarrow (\underline{O}^{bi}_x/R) \otimes S \longrightarrow \underline{O}^{bi}_x \otimes \underline{O}^{bi}_y/R \otimes S \longrightarrow \underline{O}^{bi}_x \otimes (\underline{O}^{bi}_y/S) \longrightarrow o$$

Let us identify the couple (M,N) and the maximal ideal $R \otimes N + M \otimes S$ of $R \otimes S$ corresponding to it and let us show that the first and the last term in this sequence are torsion at (M,N). Take $q \in \underline{O}^{bi}_x/R$ then there exists an ideal $I \subset M$ such that $Iq = o$. Let $z = \Sigma q_\alpha \otimes s_\alpha \in (\underline{O}^{bi}_x/R) \otimes S$ with $q_\alpha \in \underline{O}^{bi}_x/R$ and $s_\alpha \in S$, then the foregoing remark yields the existence of an ideal $J \not\subset M$ such that $Jq_\alpha = o$ for each α, hence $(J \otimes S)z = o$ and $J \otimes S$ is not contained in (M,N). This proves that $(\underline{O}^{bi}_x/R) \otimes S$ is torsion at (M,N). Similar reasoning yields that $\underline{O}^{bi}_x \otimes (\underline{O}^{bi}_y/S)$ is torsion at (M,N), hence so is $\underline{O}^{bi}_x \otimes \underline{O}^{bi}_y/R \otimes S$.

Clearly it follows that :

$$Q^{bi}_{(M,N)}(R \underset{k}{\otimes} S) = Q^{bi}_{(M,N)}(\underline{O}^{bi}_x \underset{k}{\otimes} \underline{O}^{bi}_y)$$

and it remains to show that $Q^{bi}_{(M,N)}(\underline{O}^{bi}_x \otimes \underline{O}^{bi}_y) = Q^{bi}_{\underline{M}}(\underline{O}^{bi}_x \otimes \underline{O}^{bi}_y)$.

Let us first note that $\underline{O}^{bi}_x \otimes \underline{O}^{bi}_y$ has maximal ideal $\underline{M} = \underline{O}^{bi}_x \otimes \underline{M}_y + \underline{M}_x \otimes \underline{O}^{bi}_y$. Indeed,

$$\underline{O}^{bi}_x \underset{k}{\otimes} \underline{O}^{bi}_y/\underline{M} = \underline{O}^{bi}_x/\underline{M}_x \underset{k}{\otimes} \underline{O}^{bi}_y/\underline{M}_y$$

$$= M_r(k) \underset{k}{\otimes} M_s(k) \quad \text{for some } r,s \in \mathbb{N}$$

$$= M_{rs}(k)$$

and $\underline{M}$ is maximal.

Now obviously $\mathcal{L}^2(\kappa_{\underline{M}}) = \{I \subset \underline{O}^{bi}_x \otimes \underline{O}^{bi}_y;\ I \not\subset \underline{M}\}$. On the other hand

$$\mathcal{L}^2(i_* \kappa_{(M,N)}) = \{I \subset \underline{O}^{bi}_x \otimes \underline{O}^{bi}_y;\ I \cap (R \otimes S) \not\subset (M,N)\}$$

where $i : R \underset{k}{\otimes} S \to \underline{O}_x \underset{k}{\otimes} \underline{O}_y$ is the canonical inclusion. Since $\underline{M} \cap (R \otimes S) = (M,N)$, clearly $i_* \kappa_{(M,N)} \leq \kappa_{M}$. Moreover, the converse inequality is valid too. Indeed, $I \not\subset M$ implies $I \cap (R \otimes S) \not\subset (M,N)$, as may be seen as follows. Take $x = \Sigma\, r_\alpha \otimes s_\alpha \in I - \underline{M}$ where $r_\alpha \in \underline{O}_x$ and $s_\alpha \in \underline{O}_y$, then there exists an ideal $J \not\subset M$ and an ideal $K \not\subset N$ such that $(J \otimes K)x \subset R \otimes S$. Assume that $I \cap (R \otimes S) \subset (M,N)$, then $(J \otimes K)x \subset I \cap (R \otimes S) \subset (M,N)$, hence $(\underline{O}^{bi}_x \otimes \underline{O}^{bi}_y)(J \otimes K)x \subset (\underline{O}^{bi}_x \otimes \underline{O}^{bi}_y)(M,N) = (\underline{O}^{bi}_x \otimes \underline{O}^{bi}_y)(R \otimes N + M \otimes S) = \underline{O}^{bi}_x \otimes \underline{M}_y + \underline{M}_x \otimes \underline{O}^{bi}_y = \underline{M}$. But, since $(\underline{O}^{bi}_x \otimes \underline{O}^{bi}_y)(J \otimes K) = \underline{O}^{bi}_x J \otimes \underline{O}^{bi}_y K = \underline{O}^{bi}_x \otimes \underline{O}^{bi}_y$ by property (T), this yields that $x \in \underline{M}$, contradiction.

We thus proved that $i_* \kappa_{(M,N)} = \kappa_{\underline{M}}$, hence $Q^{bi}_{\underline{M}}(\underline{O}^{bi}_x \otimes \underline{O}^{bi}_y) = Q^{bi}_{i_* \kappa_{(M,N)}}(\underline{O}^{bi}_x \otimes \underline{O}^{bi}_y)$ and the latter equals $Q^{bi}_{(M,N)}(\underline{O}^{bi}_x \otimes \underline{O}^{bi}_y)$, since the inclusion

$$i : R \otimes S \hookrightarrow \underline{O}^{bi}_x \otimes \underline{O}^{bi}_y$$

is obviously an extension. This finishes the proof. ∎

At this point it may be useful to reconsider the example given in ,
i.e. $R = \begin{pmatrix} k[X] & Xk[X] \\ Xk[X] & k[X] \end{pmatrix}$. There we have calculated that a couple $(M,M') \in \Omega(R) \times \Omega(R)$ corresponded to some $M \boxtimes M' \in \Omega(R \underset{k}{\otimes} R)$ such that :

p.i.deg(M) + p.i.deg(M') = p.i.deg($M \boxtimes M'$) and $\mathbb{k}(M \boxtimes M') \cong \mathbb{k}(M) \underset{k}{\otimes} \mathbb{k}(M')$.

<u>VIII.12. Theorem</u>. Let $X = \Omega(R)$ and $Y = \Omega(S)$ be affine k-varieties, then the

geometric product of X and Y exists and it is given by $X \hat{\times} Y = \Omega(R \underset{k}{\otimes} S)$. (Note that this time we mean the product in the category of all cellular varieties!).

<u>Proof</u>. Consider the following diagram of precellular algebraic k-varieties with $(\varphi,\psi) \in \mathrm{Hom}(T;X,Y)$

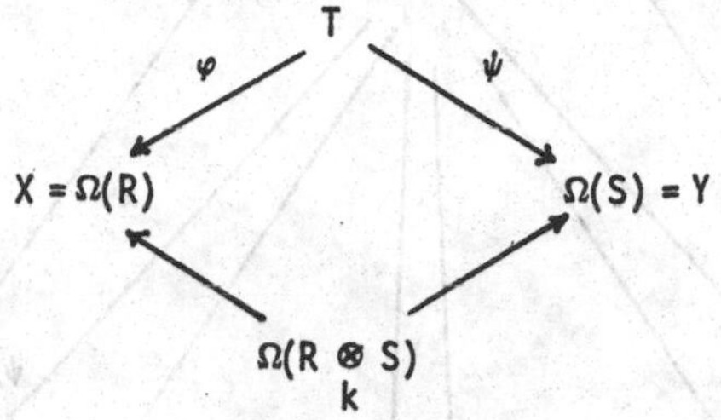

Here T is covered by affines $\{T_\alpha;\ \alpha \in A\}$ with the property that for each pair of indices $(\alpha,\beta) \in A^2$ the intersection $T_\alpha \cap T_\beta$ is a union of affines too.
Since the T_α are affine, we obtain for each α :

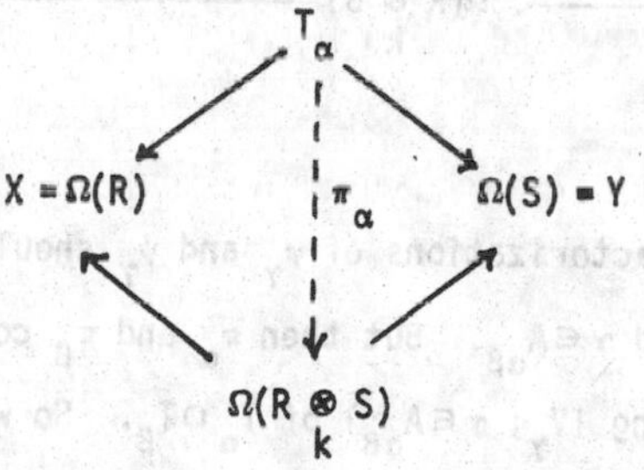

It now suffices to glue the π_α together to obtain a morphism $\pi : T \to \Omega(R \underset{k}{\otimes} S)$. To this purpose, consider $T_\alpha \cap T_\beta$ covered by the affines $\{V_\gamma;\ \gamma \in A_{\alpha\beta}\}$, then we find for each index $\gamma \in A_{\alpha\beta}$, morphisms :

$V_\gamma \to T_\alpha \xrightarrow{\varphi} T$ and $V_\gamma \to T_\beta \xrightarrow{\psi} T$, yielding the following commutative diagram

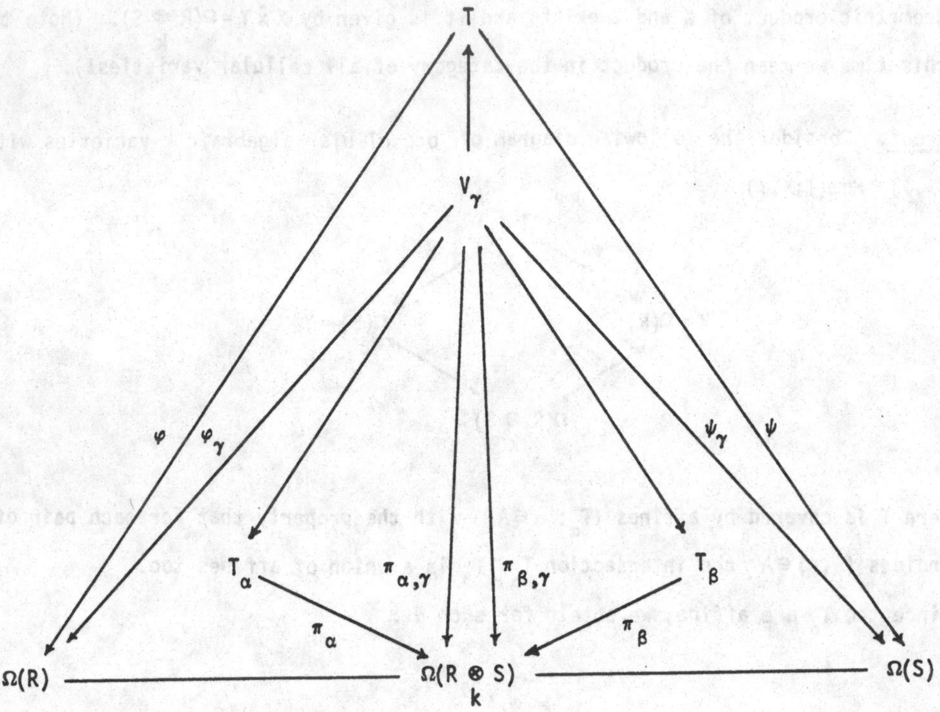

Since the V_γ are affine, the factorizations of φ_γ and ψ_γ should be unique, hence $\pi_{\alpha,\gamma} = \pi_{\beta,\gamma}$, for each $\alpha,\beta \in A$ and $\gamma \in A_{\alpha\beta}$. But then π_α and π_β coincide on each open affine subset V_γ of the covering $\{V_\gamma;\ \gamma \in A_{\alpha\beta}\}$ of $T_\alpha \cap T_\beta$. So we obtain a unique global morphism $\pi : T \to \Omega(R \otimes_k S)$, with the required properties. ∎

<u>VIII.13. Corollary</u>. Let $\Omega(R)$ be an affine k-variety and $\Omega(C)$ a commutative affine k-variety then the geometric product $\Omega(R)\hat{\times}\,\Omega(C)$ is actually a product of $\Omega(R)$ and $\Omega(C)$.

<u>Proof</u>. It suffices to note in the foregoing that the tensorproduct of an algebra and a central algebra over the same ring S is an amalgamated sum in the category of S-algebras. If in particular S is commutative,say the field k itself, then a central extension of it is commutative too, and conversely. Apply this to the k-algebras R and C. ∎

VIII.14. Corollary. Let X and Y be affine k-varieties then for any commutative k-variety Z we have a bijection

$$\mathrm{Hom}(Z, X \hat{\times} Y) = \mathrm{Hom}(Z,X) \times \mathrm{Hom}(Z,Y) .$$

Proof. This is obvious, as $\Gamma(Z)$ is commutative, hence

$$\mathrm{Hom}_k(\Gamma(X) \otimes \Gamma(Y), \Gamma(Z)) = \mathrm{Hom}_k(\Gamma(X), \Gamma(Z)) \times \mathrm{Hom}_k(\Gamma(Y), \Gamma(Z)) . \blacksquare$$

VIII.15. Theorem. If X and Y are precellular varieties, then there exists a geometric product $X \hat{\times} Y$ in the category of precellular varieties.

Proof. A globalization of local constructions, formally similar to the construction of the product of algebraic varieties in the commutative case.■

We shall return to the existence of products after a short disgression dealing with separatedness and products.

It is not hard to see that a cellular k-variety X is separated if and only if the diagonal, $\Delta(X) = \{Z \in X \hat{\times} X, \pi_1(Z) = \pi_2(Z)\}$ is closed in $X \hat{\times} X$. Moreover if Y is another cellular k-variety then $X \hat{\times} Y$ is separated if X and Y are separated. Indeed, let $f,g : Z \to X \hat{\times} Y$ be morphisms of cellular k-varieties then $f(z) = g(z)$ if and only if $\pi_1 f(z) = \pi_1 g(z)$ and $\pi_2 f(z) = \pi_2 g(z)$. Thus, $\{z \in Z, f(z) = g(z)\} = \{z \in Z, \pi_1 f(z) = \pi_1 g(z)\} \cap \{z \in Z, \pi_2 f(z) = \pi_2 g(z)\}$. The latter sets are closed because X and Y are separated, therefore it follows that $X \hat{\times} Y$ is separated.

VIII.16. Proposition. If X and Y are cellular k-varieties then $(X \hat{\times} Y)_n = \cup X_r \hat{\times} Y_s$ as topological spaces, the union being taken over all r,s with $rs = n$.

Proof. Pick $(x,y) \in X_r \hat{\times} Y_s$; then $\mathbb{k}(x) = M_r(k)$, $\mathbb{k}(y) = M_s(k)$ and $\mathbb{k}(x,y) = \mathbb{k}(x) \otimes_k \mathbb{k}(y) \cong M_{rs}(k)$, i.e. $(x,y) \in (X \hat{\times} Y)_{rs}$. Also, conversely, if $(x,y) \in (X \hat{\times} Y)_n$ then $rs = n$ follows. Note that for the "commutative" part of X i.e. X_1 we find $(X \hat{\times} Y)_1 = X_1 \hat{\times} Y_1 = X_1 \times Y_1$.■

<u>VIII.17. Proposition.</u> Let X,Y and Z be precellular algebraic k-varieties.

1. The projections $X \hat{x} Y \to Y$ and $X \hat{x} Y \to X$ are surjective.

2. The geometric products $(X \hat{x} Y) \hat{x} Z$ and $X \hat{x} (Y \hat{x} Z)$ are canonically isomorphic.

<u>Proof.</u> 1. Pick $x \in X$, $y \in Y$. Then $\mathbb{k}(x)$, $\mathbb{k}(y)$ and $K = \mathbb{k}(x) \otimes_k \mathbb{k}(y)$ are matrix rings. Let $\alpha : \Omega(K) \to X$, resp. $\beta : \Omega(K) \to Y$ be the map induced by the extension :

$$\Gamma(X) \to \underline{O}_{X,x} \to \mathbb{k}(x) \longrightarrow K ,$$

resp.

$$\Gamma(Y) \to \underline{O}_{Y,y} \to \mathbb{k}(y) \longrightarrow K .$$

Then $\alpha(z) = x$ and $\beta(z) = y$, where z is the unique point of $\Omega(K)$. Now the images of $\Gamma(X)$ and $\Gamma(Y)$ in K commute, therefore we may define a morphism $\gamma : \Omega(K) \to X \hat{x} Y$ such that $\pi_1\gamma = \alpha$ and $\pi_2\gamma = \beta$, where π_1, resp. π_2, is the canonical projection $X \hat{x} Y \to X$, resp. $X \hat{x} Y \to Y$. Putting $t = \mathrm{Im}\gamma$, then $\pi_1(t) = x$ and $\pi_2(t) = y$, hence the projections are surjective.

2. It is sufficient to remark that $(\Gamma(X) \otimes_k \Gamma(Y)) \otimes_k \Gamma(Z)$ and $\Gamma(X) \otimes_k (\Gamma(Y) \otimes_k \Gamma(Z))$ are isomorphic k-algebras. ∎

Although for our purposes the restriction to cellular varieties is not restrictive, it may be worthwhile to consider products of algebraic k-varieties whenever they exist (within the category of algebraic k-varieties).
In this set-up we have :

<u>VIII.18. Proposition.</u> Let X and Y be algebraic k-varieties where X is a commutative variety. Assume that $X \times Y$ exists. If U is open in X then there is a structure of an algebraic k-variety on $p^{-1}(U)$, where $p : X \times Y \to X$ is the canonical projection. The k-variety $p^{-1}(U) = W$ may be viewed as a product of U and Y.

<u>Proof.</u> The following diagram is commutative

$$\begin{array}{ccc} W & \longrightarrow & U \\ \downarrow & & \downarrow \\ Y & \longrightarrow & k \end{array}$$

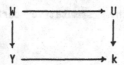

Consider a pair of morphisms $\varphi,\psi : W' \to U,Y$, then this induces a unique π given by

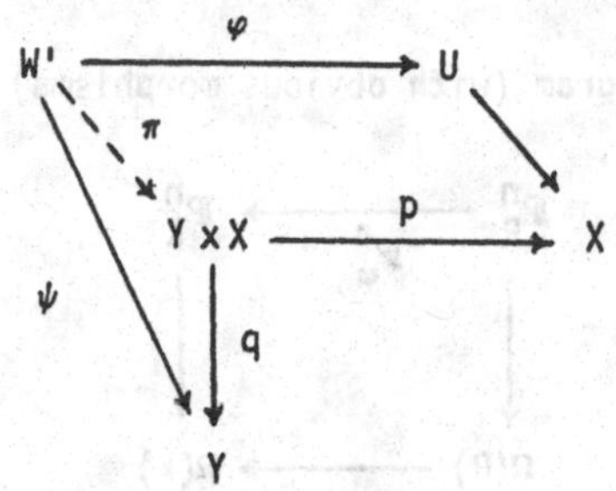

Moreover $p \circ \pi(W') = \varphi W' \subset V$, hence π maps W' into W (while this morphism is unique). This makes the above diagram commutative and establishes the assertion. ∎

VIII.19. Proposition. Let X and Y be algebraic k-varieties, assume that X is commutative and let $\{X_\alpha;\alpha\in A\}$ and $\{Y_\beta;\beta\in B\}$ be open coverings of X resp. Y. If a product of X_α and Y_β exists for each couple of indices $(\alpha,\beta)\in A\times B$, then there exists a product of X and Y.

Proof. Using the foregoing result this is an easy exercise in cutting and glueing only utilizing category-theoretical universal properties of products. We leave details to the reader. ∎

VIII.20. Corollary. If X and Y are algebraic k-varieties and X is commutative, then the product $X\times Y$ exists - and is unique.

Proof. Cover X and Y by open affines and apply VIII.13. ∎

VIII.21. A special case. Let R be an affine graded PI ring satisfying the conditions PS1-2 and let $\mathbb{P}^n_k$ be projective n-space over k, then

$$\mathbb{P}^n_R = \mathbb{P}^n_k \times \Omega(R)$$

Proof. First, since $\mathbb{P}^n_k$ is a commutative variety we know by the foregoing that $\mathbb{P}^n_k \times\Omega(R)$ does exist.

The canonical morphism $\psi : k\to k[X_0,\dots,X_n]$ induces a morphism of graded algebras

$$\varphi : k[X_o,\dots,X_n] \to R[X_o,\dots,X_n] = R \underset{k}{\otimes} k[X_o,\dots,X_n] = k[X_o,\dots,X_n] \underset{k}{\otimes} R \ .$$

Consider the following diagram (with obvious morphisms).

$$\begin{array}{ccc} \mathbb{P}^n_R & \xrightarrow{\ \mathbb{P}^n_\varphi\ } & \mathbb{P}^n_k \\ \downarrow & & \downarrow \\ \Omega(R) & \longrightarrow & \Omega(k) \end{array}$$

That this diagram is a commutative one is an easy consequence of the fact that for each central $r \in k[X_o,\dots,X_n]$ which is homogeneous of degree 1 we have commutative diagrams

$$\begin{array}{ccc} X_+(\varphi(r)) & \longrightarrow & X_+(r) \\ \downarrow & & \downarrow \\ \Omega(R) & \longrightarrow & \Omega(k) \end{array}$$

where $X_+(\varphi(r)) \to X_+(r)$ is obtained by localization of φ. Now it is sufficient to note that $X_+(\varphi(R)) = X_+(r) \times \Omega(R)$. This follows from $R[X_o,\dots,X_n] = k[X_o,\dots,X_n] \underset{R}{\otimes} R$ i.e. $Q^g_{\varphi(r)}(R[X_o,\dots,X_n]) = Q^g_r(k[X_o,\dots,X_n]) \underset{k}{\otimes} R = k[X_o,\dots,X_n]_{(r)} \underset{k}{\otimes} R$.∎

Since products of precellular algebraic k-varieties seem to work reasonably well one may want to apply these techniques to the study of "non-commutative" algebraic group varieties. It will turn out that the latter varieties have to be defined over Azumaya algebras i.e. they are actually commutative varieties. The sequel of this section deals with algebraic k-monoids and certain transformation spaces. An <u>algebraic k-monoid</u> is an algebraic k-variety G with an $e \in G$ and a given morphism $\mu : G \hat{\times} G \to G : (x,y) \longrightarrow xy$ such that the set of points of G is a monoid.

If T is a terminal object in the category of algebraic k-varieties V, i.e. $\Omega(k)$ up to isomorphism, then an algebraic k-monoid may be viewed as a system

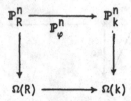

$(G,\mu,\eta,\varphi,\lambda,\rho)$ where $\mu \in \mathrm{Hom}(G \hat{\times} G, G)$, $\eta \in \mathrm{Hom}(T,G)$, φ the canonical isomorphism $G \hat{\times} (G \hat{\times} G) \cong (G \hat{\times} G) \hat{\times} G$, λ an isomorphism $T \hat{\times} G \cong G$, ρ an isomorphism $G \hat{\times} T \cong G$, such that the following diagrams are commutative :

(g.1)

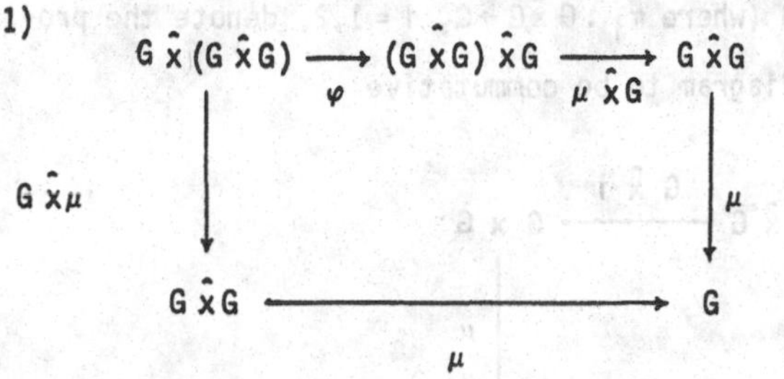

(g.2)

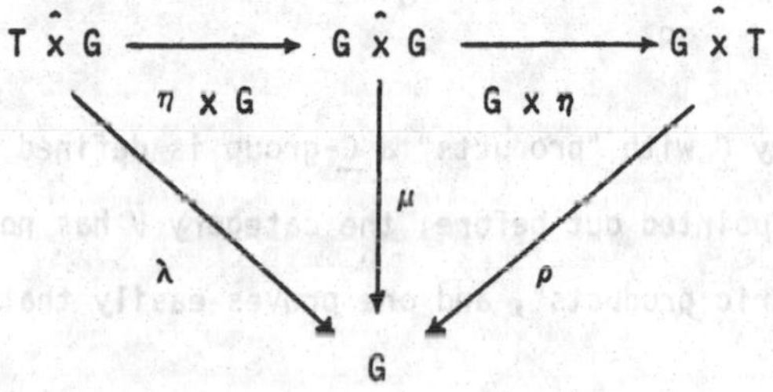

<u>A morphism of algebraic k-monoids</u> is a morphism of algebraic k-varieties which is also a monoid homomorphism.
In other words, a morphism $\psi \in \mathrm{Hom}(G_1, G_2)$, where G_1 and G_2 are algebraic k-monoids is a morphism of algebraic k-monoids, provided the following diagrams commute.

(mg.1)

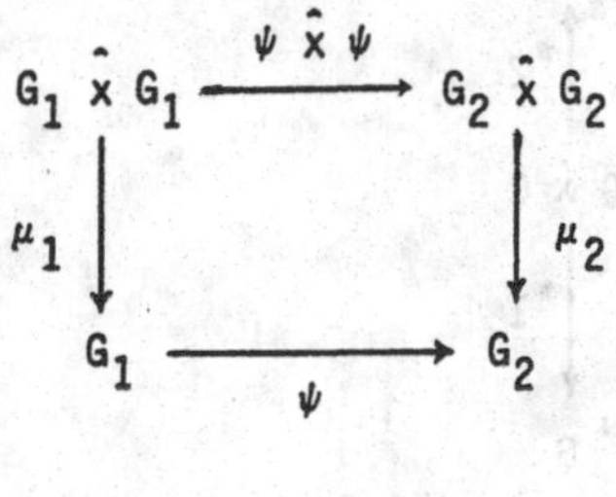

(mg.2)

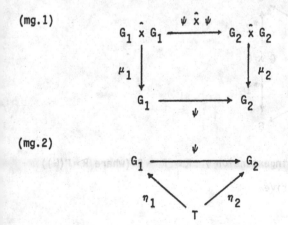

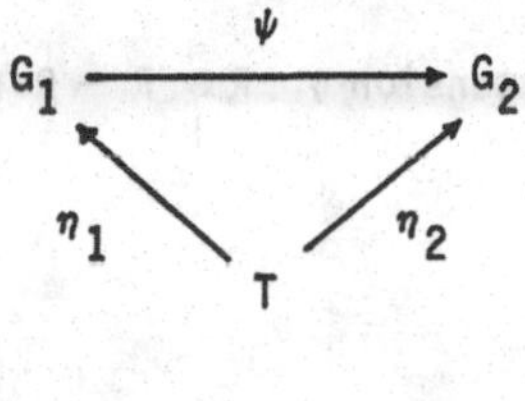

An <u>algebraic k-group</u> is an algebraic k-monoid G together with a morphism $i : G \to G : x \to x^{-1}$, such that the data (e, μ, i) equip G with a group structure. In the commutative case, denoting by $c : G \to T$ the structural map and by δ_G the diagonal map defined by $\pi_i \circ \delta_G = G$ (where $\pi_i : G \hat{x} G \to G$, $i = 1,2$, denote the projections) we want the following diagram to be commutative

(g.3)

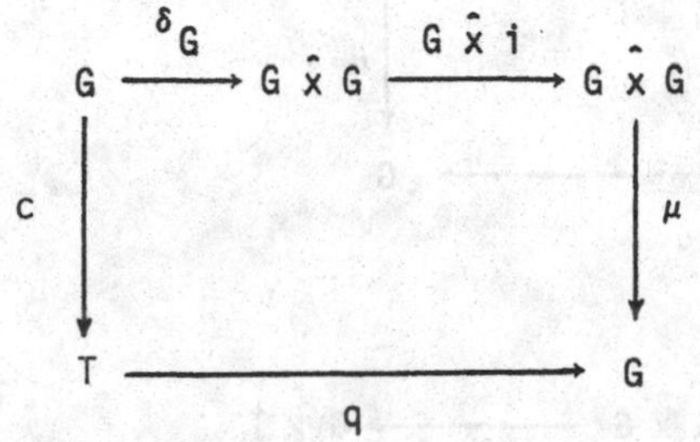

More generally, in any category <u>C</u> with "products" a <u>C</u>-group is defined by the diagrams (g.1-3). But, as we pointed out before, the category V has no categorical products, but only "geometric products", and one proves easily that a "diagonal map"

$$\delta_G : G \to G \hat{x} G$$

exists if and only if $\Gamma(G)$ is commutative. Indeed δ_G is the unique morphism (if it exists) making the following diagram commutative :

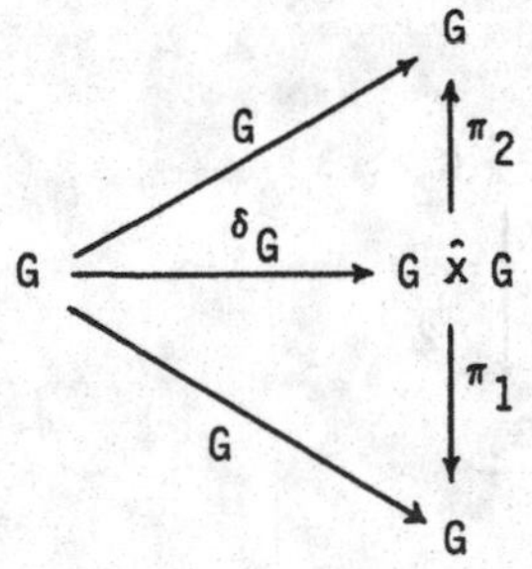

To $\delta_G : G \to G \hat{x} G$ there corresponds a ringextension $\gamma : R \otimes_k R \to R$ (where $R = \Gamma(G)$) making the following diagrams commutative

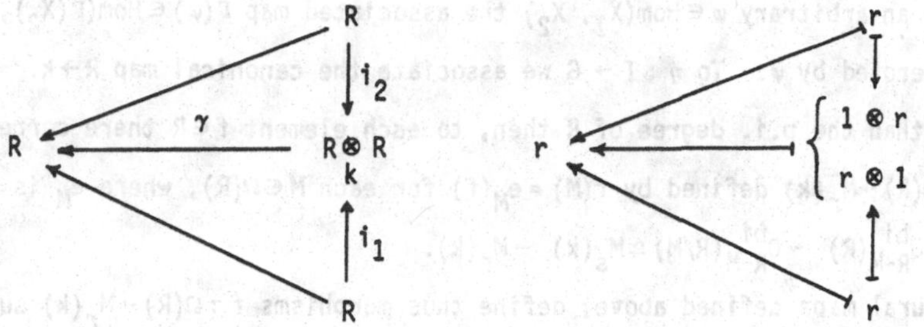

The ringmorphism γ is uniquely determined by these data, since for any $r,s \in R$ we have

$$\gamma(r \otimes s) = \gamma((r \otimes 1)(1 \otimes s)) = \gamma(r \otimes 1)\gamma(1 \otimes s) = \gamma i_1(r)\gamma i_2(s) = rs$$

Furthermore, if γ is a ringmorphism, then

$$rs = \gamma(r \otimes s) = \gamma((1 \otimes s)(r \otimes 1)) = \gamma(1 \otimes s)\gamma(r \otimes 1) = sr$$

which implies that R is commutative.

As a consequence, we find that the only algebraic k-groups in V are the usual ones arising in commutative algebraic geometry!

We now consider affine algebraic k-monoids only, i.e. of the form $\Omega(R)$. The diagrams g.1. and g.2. give rise to diagrams h.1. and h.2. below

(h.1)

$$
\begin{array}{ccc}
R \underset{k}{\otimes} R \underset{k}{\otimes} R & \xrightarrow{\quad\quad\quad} & R \underset{k}{\otimes} R \\[2mm]
& \mu_0 \otimes R & \\[2mm]
R \otimes \mu_0 \Big\uparrow & & \Big\uparrow \mu_0 \\[2mm]
R \underset{k}{\otimes} R & \xleftarrow{\quad\mu_0\quad} & R
\end{array}
$$

(h.2)

$$
\begin{array}{ccc}
R & \xleftarrow{\quad\quad\quad} & R \underset{k}{\otimes} R \\[2mm]
& (\eta_0, 1_R) & \\[2mm]
(1_R, \eta_0) \Big\uparrow & & \Big\uparrow \mu_0 \\[2mm]
R \underset{k}{\otimes} R & \xleftarrow{\quad\mu_0\quad} & R
\end{array}
$$

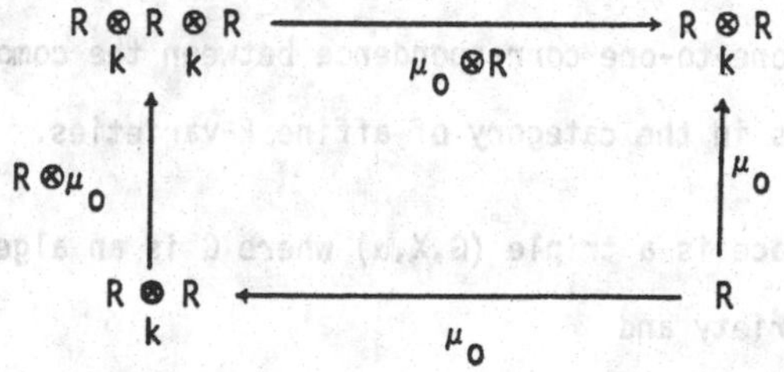

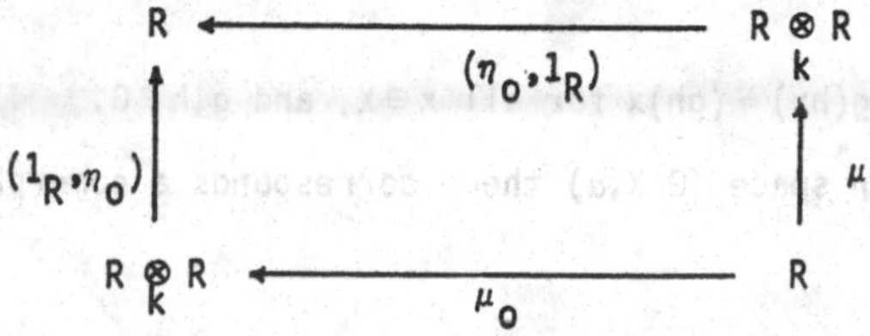

where, for an arbitrary $\varphi \in \mathrm{Hom}(X_1, X_2)$ the associated map $\Gamma(\varphi) \in \mathrm{Hom}(\Gamma(X_2), \Gamma(X_1))$ has been denoted by φ. To $\eta : T \to G$ we associate the canonical map $R \to k$. If $r \in \mathbb{N}$ is larger than the p.i. degree of R then, to each element $f \in R$ there corresponds a map $f : \Omega(R) \to M_r(k)$ defined by $f(M) = e_M(f)$ for each $M \in \Omega(R)$, where e_M is given by $e_M : R \to Q^{bi}_{R-M}(R) \to Q^{bi}_{R-M}(R/M) \cong M_s(k) \to M_r(k)$.

The structural maps defined above, define thus morphisms $f : \Omega(R) \to M_r(k)$ such that, if $\mu_o f = \sum_i g_i \otimes h_i$ then $f(xy) = \sum_i g_i(x) \otimes h_i(y)$. This follows from the commutativity of the following diagram :

$$\begin{array}{ccc} R & \xrightarrow{\mu_o} & R \underset{k}{\otimes} R \\ \downarrow & & \downarrow \\ \mathbb{k}(xy) & \longrightarrow & \mathbb{k}(x) \underset{k}{\otimes} \mathbb{k}(y) \end{array} .$$

The map $\varphi : G \hat{x} (G \hat{x} G) \to (G \hat{x} G) \hat{x} G$ expresses the associativity of the tensor product whereas, finally $\lambda : t \hat{x} G \to G$ resp. $\rho : G \hat{x} t \to G$ represent the canonical isomorphism $\lambda_o : R \to k \underset{k}{\otimes} R$ resp. $\rho_o : R \to R \underset{k}{\otimes} k$.

As in the commutative case one calls a k-algebra R endowed with the structure defined above an associative coassociative augmented and coaugmented coalgebra, let us say a comonoidal k-algebra for simplicity's sake.

<u>VIII.22. Theorem</u>. There is a one-to-one correspondence between the comonoidal k-algebras in R and the monoids in the category of affine k-varieties. ∎

An algebraic transformation space is a triple (G,X,α) where G is an algebraic k-monoid, X is an algebraic k-variety and

$$\alpha : G \hat{x} X \to X : (g,x) \longmapsto gx = \alpha(g,x)$$

is a morphism with $ex = x$ and $g(hx) = (gh)x$ for all $x \in X$, and $g,h \in G$.

To an algebraic transformation space (G,X,α) there corresponds a comorphism

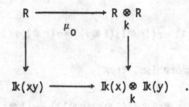

$$\alpha_0 : \Gamma(X) \to \Gamma(G) \underset{k}{\otimes} \Gamma(X)$$

of finitely generated k-algebras, inducing for each $x \in X$, $y \in G$ an inclusion $\mathbb{k}(yx) \to \mathbb{k}(y) \underset{k}{\otimes} \mathbb{k}(x)$ on the corresponding function rings.

Take $g \in G$, then we construct a morphism $\Omega(\mathbb{k}(g)) \to G$ with image g as follows. To g there corresponds a maximal ideal M of $\Gamma(G)$ with $\Gamma(G)/M = \mathbb{k}(g)$ and $\pi_g : \Gamma(G) \to \Gamma(G)/M = \mathbb{k}(g)$ induces the desired morphism. Note that we made use of the fact that $\Omega(\mathbb{k}(g))$ is well-defined, $\mathbb{k}(g)$ being a finitely generated simple P.I. ring over k by Posner's theorem and its bimodule-interpretation. Let us define a morphism $\lambda_g : \Gamma(X) \to \mathbb{k}(g) \underset{k}{\otimes} \Gamma(X)$ by the following commutative diagram

$$\begin{array}{ccc} \Gamma(X) & \xrightarrow{\ \lambda_g\ } & \mathbb{k}(g) \underset{k}{\otimes} \Gamma(X) \\ & {\scriptstyle \alpha_0} \searrow \quad \nearrow {\scriptstyle \pi_g \otimes \Gamma(X)} & \\ & \Gamma(G) \underset{k}{\otimes} \Gamma(X) & \end{array}$$

The corresponding morphism $\Lambda_g : \Omega(\mathbb{k}(g)) \hat{\times} X \to X$ will be called the left translation by $g \in G$. Note that $\Omega(\mathbb{k}(y)) \hat{\times} X$ and X are homeomorphic topological spaces. Indeed, g being closed in $\mathrm{Spec}(\Gamma(G))$, we know that $\mathbb{k}(g)$ is isomorphic to some matrix ring $M_r(k)$ over k, hence $\mathbb{k}(g) \underset{k}{\otimes} \Gamma(X) = M_r(k) \underset{k}{\otimes} \Gamma(X) = M_r(\Gamma(X))$, i.e. as affine schemes we have

$$\Omega(\mathbb{k}(y)) \hat{\times} X = \Omega(M_r(\Gamma(X)))$$

But $\Omega(M_r(\Gamma(X))) \cong \Omega(\Gamma(X))$ because of the correspondence $P \to P \cap \Gamma(X)$. Let us identify $\Omega(M_r(\Gamma(X)))$ and $\Omega(\Gamma(X))$ under this homeomorphism, and assume that P is a maximal ideal of $\Gamma(X)$ with $\kappa_{\Gamma X-P}$ having property (T) for $\Gamma(X)$-bimodules, then the stalk of the structure sheaf on $\Omega(M_r(\Gamma(X)))$ at this point is just the stalk of X at this point tensored with $M_r(k) = \mathbb{k}(g)$. Indeed, let $\Gamma(X)$ be denoted by R and assume that P induces a perfect localization, then we have to show that

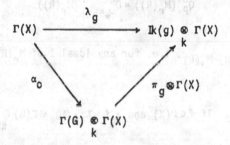

$$Q^{bi}_{M_r(P)}(M_r(R)) = Q^{bi}_P(R) \underset{R}{\otimes} M_r(k) = M_r(Q^{bi}_P(R)) .$$

First, the fact that κ_{R-P} has property (T) for R-bimodules yields that

$$M_r(Q^{bi}_P(R)) = Q^{bi}_P(R) \underset{R}{\otimes} M_r(R) = Q^{bi}_P(M_r(R)) ,$$

so it remains to prove that $Q^{bi}_P(M_r(R)) = Q^{bi}_{M_r(P)}(M_r(R))$. But the canonical morphism

$$\omega : R \to M_r(R)$$

is an extension, hence, since κ_{R-P} has property (T),

$$Q^{bi}_P(M_r(R)) = Q^{bi}_{\omega_\star \kappa_{R-P}}(M_r(R)).$$

Finally $\omega_\star \kappa_{R-P} = \kappa_{M_r(R)-M_r(P)}$, i.e. for any ideal I of $M_r(R)$, $I \not\subset M_r(P)$ iff $R \cap I = \omega^{-1}(I) \not\subset P$. ∎

<u>VIII.23. Proposition</u>. If $f \in \Gamma(X)$ and $\alpha_o f = \Sigma f_i \otimes g_i \in \Gamma(G) \underset{k}{\otimes} \Gamma(X)$, then $\lambda_g f = \Sigma f_i(g) \otimes g_i$.

<u>Proof</u>. Let q be the unique closed point in $\Omega(\mathbb{k}(g))$ and let $x \in X$. Because

$$\lambda_g f(q,x) = \Sigma \pi_g (f_i)(q) \otimes g_i(x)$$

and since the map $\Gamma(G) \to \Gamma(G)/M = \mathbb{k}(g)$ is just evaluation at g, we find that $\pi_g(f_i)(q) = f_i(g)$. Hence $\lambda_g f(q,x) = (\Sigma f_i(g) \otimes g_i)(q,x)$, i.e. $\lambda_g f = \Sigma f_i(g) \otimes g_i$. ∎

A subset S of $\Gamma(X)$ is said to be stable under left translation by G if for all $g \in G$ we have $\lambda_g S \subset \Gamma(G) \underset{k}{\otimes} S$, where $\Gamma(G) \underset{k}{\otimes} S$ stands for the set of all $\Sigma g_i \otimes s_i$, with $g_i \in \Gamma(G)$, $s_i \in S$.

<u>VIII.24. Proposition</u>. (cf. [30]). Let F be a finite dimensional subvectorspace of $\Gamma(X)$ over k. Then there exists a finite dimensional subspace E which contains F and which is stable under left translation by G. A necessary and sufficient condition that F be invariant under left translation is that

$$\alpha_o F \subset \Gamma(G) \underset{k}{\otimes} F$$

Proof. We may assume that F is spanned by a single $f \in \Gamma(X)$, since the general case follows then by taking sums of the E's obtained for each element of a k-basis for F. If we write $\alpha_o f = \sum_{i=1}^{n} f_i \otimes g_i$, where n is minimal as such, then $\lambda_g f = \Sigma f_i(g) \otimes g_i$. The k-vectorspace spanned by $\{f_i, g_i\}$ satisfies our requirements. The second part of the proof parallels the commutative case: let $\{f_i\} \cup \{h_j\}$ be a basis for $\Gamma(X)$ such that $\{f_i\}$ spans F. For any $f \in F$, $g \in G$ we then find $\lambda_g f = \Sigma r_i(g) \otimes f_i + \Sigma s_j(g) \otimes h_j$, if $\alpha_o f = \Sigma r_i \otimes f_i + \Sigma s_j \otimes h_j$. Hence $\lambda_g f \in \mathbb{k}(g) \otimes F$ if and only if $s_j(g) = o$ for all j. ∎

Consider the algebraic transformation space (G, G, μ). By the foregoing we obtain two actions :

left translation : $\lambda_g : \Gamma(G) \to \mathbb{k}(g) \underset{k}{\otimes} \Gamma(G)$

right translation: $\rho_g : \Gamma(G) \to \Gamma(G) \underset{k}{\otimes} \mathbb{k}(g)$

where (denoting $\mu_o f$ by $\Sigma f_i \otimes g_i$) we have $\lambda_g f = \Sigma f_i(g) \otimes g_i$ and $\rho_g f = \Sigma f_i \otimes g_i(g)$.

VIII.25. Proposition. For $g, h \in G$ we have

1. $\rho_{hg} = (\rho_h \otimes 1)\rho_g$
2. $\lambda_{hg} = (1 \otimes \lambda_g)\lambda_h$
3. $(1 \otimes \rho_g)\lambda_h = (\lambda_h \otimes 1)\rho_g$.

Proof. The identities stated all make sense because : $(\rho_h \otimes 1) \circ \rho_g : \Gamma(G) \to \Gamma(G) \otimes \mathbb{k}(g) \to \Gamma(G) \otimes \mathbb{k}(h) \otimes \mathbb{k}(g)$, and $\rho_{hg} : \Gamma(G) \to \Gamma(G) \otimes \mathbb{k}(h_g) = \Gamma(G) \underset{k}{\otimes} \mathbb{k}(h) \underset{k}{\otimes} \mathbb{k}(g)$ (and similarly for the left translations).

1. Take $f \in \Gamma(G)$ and write $\mu_o f = \Sigma f_i \otimes g_i$, $\mu_o f_i = \Sigma h_{ij} \otimes k_{ij}$, $\mu_o g_i = \Sigma p_{ik} \otimes q_{ik}$, then the associativity of the group law on G yields

$$(1 \otimes \mu_o)\mu_o = (\mu_o \otimes 1)\mu_o$$

hence

$$\Sigma f_i \otimes p_{ik} \otimes q_{ik} = \Sigma f_i \otimes \mu_o g_i = (1 \otimes \mu_o)(\Sigma f_i \otimes g_i) = (1 \otimes \mu_o)\mu_o f = (\mu_o \otimes 1)\mu_o f =$$

$$= (\mu_o \otimes 1)(\Sigma f_i \otimes g_i) = \Sigma \mu_o f_i \otimes g_i = \Sigma h_{ij} \otimes k_{ij} \otimes g_i \quad .$$

Now $(\rho_h \otimes 1)\rho_g f = (\rho_h \otimes 1)(\Sigma f_i \otimes g_i(g)) = \Sigma \rho_h(f_i) \otimes g_i(g) = \Sigma h_{ij} \otimes k_{ij}(h) \otimes g_i(g)$

$$= \Sigma f_i \otimes p_{ik}(h) \otimes q_{ik}(g) = (\Sigma f_i \otimes \mu_o g_i)(h,g) = (\Sigma f_i \otimes g_i)(hg) = \rho_{hg} f$$

which proves 1. The proof of 2. is similar.

3. $(1 \otimes \rho_g)\lambda_h f = (1 \otimes \rho_g)(\Sigma f_i(h) \otimes g_i) = \Sigma f_i(h) \otimes \rho_g(g_i) = \Sigma f_i(h) \otimes p_{ik} \otimes q_{ik}(g) =$

$$= \Sigma h_{ij}(h) \otimes k_{ij} \otimes \rho_i(g) = \Sigma \lambda_h(f_i) \otimes g_i(g) = (\lambda_h \otimes 1)(\Sigma f_i \otimes g_i(g)) =$$

$$= (\lambda_h \otimes 1)\rho_g f. \quad \blacksquare$$

<u>VIII.26. Corollary</u>. Every finite dimensional subspace of $\Gamma(G)$ is contained in a finite dimensional subspace which is stable under both left and right translation by G.

If G is an affine algebraic k-monoid then $\Gamma(G)$ is of the form $k\{f_1,\dots,f_n\}$ where the f_i may be chosen such that the k-subspace E of $\Gamma(G)$ generated by them is stable under right translation i.e. $\mu_o E \subset E \underset{k}{\otimes} \Gamma(G)$. Hence, for $i \in \{1,\dots,n\}$, $\mu_o f_i = \sum_j f_j \otimes m_{ji}$ with $m_{ji} \in \Gamma(G)$. If $g \in G$ then $\rho_g f_i = \sum_j f_j \otimes m_{ji}(g)$. Since $\rho_{hg} = (\rho_h \otimes 1)\rho_g$ we obtain :

$$\Sigma f_j \otimes m_{ji}(hg) = \rho_{hg} f_i = (\rho_h \otimes 1)\rho_g f_i$$

$$= (\rho_h \otimes 1)(\sum_k f_k \otimes m_{ki}(g)) = \sum_k \rho_h f_k \otimes m_{ki}(g)$$

$$= \Sigma f_j \otimes m_{jk}(h) \otimes m_{ki}(g).$$

Now since $\{f_1,\dots,f_n\}$ is a k-basis for E we obtain :

$$(*)\quad m_{ji}(hg) = \sum_k m_{jk}(h) \otimes m_{ik}(g) \in \mathbb{k}(h) \underset{k}{\otimes} \mathbb{k}(g) = \mathbb{k}(h,g)$$

and this is compatible with the embedding $\mathbb{k}(hg) \hookrightarrow \mathbb{k}(h,g)$ induced by $\mu: G \hat{\times} G \to G$. The relation (*) may be viewed as the non-commutative counterpart of the commutative fact that G may be embedded in $GL_n(k)$ for some n. The latter statement fails in the non-commutative case because, although $\mathbb{k}(g)$ may be embedded into some $M_{n(g)}(k)$ these embeddings cannot be constructed such that everything is compatible with the group law of G.

<u>VIII.27. Example</u>. Put $R = \begin{pmatrix} k[Y] & k[Y] \\ Yk[Y] & K[Y] \end{pmatrix}$. Let $A = k[X,X^{-1}]$ be the coordinate ring of the algebraic group $G_m(k)$ and identify $S = A \underset{k}{\otimes} R$ with the ring :

$$\begin{pmatrix} k[X,X^{-1},Y] & k[X,X^{-1},Y] \\ Yk[X,X^{-1},Y] & k[X,X^{-1},Y] \end{pmatrix}.$$

The ring morphism $\varphi : R \to A \underset{k}{\otimes} R = S$ given by $\varphi_{ij}(X) = X$ and $\varphi_{ij}(Y) = XY$, $i,j = 1,2$, is a central extension and one easily checks that the following diagram commutes:

$$\begin{array}{ccccc}
R & \xrightarrow{\varphi} & A \underset{k}{\otimes} R & \xrightarrow{A \otimes \varphi} & A \underset{k}{\otimes} (A \underset{k}{\otimes} R) \\
\downarrow{\scriptstyle\varphi} & & & & \downarrow{\scriptstyle\wr} \\
A \underset{k}{\otimes} R & & \xrightarrow[\psi \otimes R]{\hspace{4em}} & & (A \underset{k}{\otimes} A) \underset{k}{\otimes} R
\end{array}$$

where $\psi : A \to A \underset{k}{\otimes} A : X \to X \otimes X$.

Hence (R,A,φ) defines an algebraic transformation space $(G_m(k),\Omega(R),\alpha)$. To exhibit its structure, let us first calculate $\Omega(S)$.

The ideals of $k[X,X^{-1},Y]$ are $\{o,(X-a,Y-b),(f(X,Y))\}$ where $a \neq o$ and $f \neq X$ is irreducible. Next, let $M_2(Yk[X,X^{-1},Y]) = I$ be a common ideal of S and $M_2(k[X,X^{-1},Y])$, then,

$$\mathrm{Spec}(S) - V(I) = \{o, \langle a,b\rangle, \langle f(X,Y)\rangle;\ ab \neq o,\ f \neq X,\ f \neq Y\}.$$

Outside of V(I) the following possibilities remain

$$Y_\star = \begin{pmatrix} k[X,X^{-1},Y] & k[X,X^{-1},Y] \\ Yk[X,X^{-1},Y] & Yk[X,X^{-1},Y] \end{pmatrix} \qquad Y^\star = \begin{pmatrix} Yk[X,X^{-1},Y] & k[X,X^{-1},Y] \\ Yk[X,X^{-1},Y] & k[X,X^{-1},Y] \end{pmatrix}$$

$$\langle a,o\rangle_\star = \begin{pmatrix} k[X,X^{-1},Y] & k[X,X^{-1},Y] \\ Yk[X,X^{-1},Y] & (X-a,Y) \end{pmatrix} \qquad \langle a,o\rangle^\star = \begin{pmatrix} (X-a,Y) & k[X,X^{-1},Y] \\ Yk[X,X^{-1},Y] & k[X,X^{-1},Y] \end{pmatrix}$$

The maximal ideals are $\langle a,b\rangle$, $\langle a,o\rangle_\star$ and $\langle a,o\rangle^\star$.
Obviously to the closed joint $((X-a),(Y-b)) \in \Omega(k[X,X^{-1}]) \times \Omega(R)$ (resp. $((X-a),M_\star)$, resp. $((X-a),M^\star)$) there corresponds the maximal ideal $\langle a,b\rangle$ (resp. $\langle a,o\rangle_\star$, resp $\langle a,o\rangle^\star$), where

$$M_\star = \begin{pmatrix} k[Y] & k[Y] \\ Yk[Y] & Yk[Y] \end{pmatrix} \quad \text{and} \quad M^\star = \begin{pmatrix} Yk[Y] & k[Y] \\ Yk[Y] & k[Y] \end{pmatrix}$$

Furthermore we find :

$$\varphi^{-1}(\langle a,b\rangle) = \begin{pmatrix} (Y-ab) & (Y-ab) \\ (Y-ab) \cap Yk[X,X^{-1},Y] & (Y-ab) \end{pmatrix}$$

$$\varphi^{-1}(\langle a,o\rangle_\star) = M_\star \quad \text{and} \quad \varphi^{-1}(\langle a,o\rangle^\star) = M^\star .$$

Let us check this for $\varphi^{-1}\begin{pmatrix} k[X,X^{-1},Y] & k[X,X^{-1},Y] \\ Yk[X,X^{-1},Y] & (X-a,Y) \end{pmatrix}$. Clearly $\varphi^{-1}(\langle a,o\rangle_\star)$ is of the form $\begin{pmatrix} k[Y] & k[Y] \\ Yk[Y] & I \end{pmatrix}$, where I is an ideal of $k[Y]$ and it is prime and contains $\varphi_{22}^{-1}((X-a)Y + aY) = Y$. Since I is maximal, we find that $\varphi^{-1}(\langle a,o\rangle_\star) = \begin{pmatrix} k[Y] & k[Y] \\ Yk[Y] & k[Y] \end{pmatrix} = M_\star$, as desired. If we identify $\mathbb{G}_m(k)$ and $k^\star$ and if we use the notation $[a] = (Y-a)$ for the corresponding maximal ideal of $k[Y]$, then the action of $\mathbb{G}_m(k)$ on $\Omega(R)$ (denoted multiplicatively), becomes :

$$\alpha[a] = [\alpha a]$$

$$\alpha M^* = M^*, \ \alpha M_* = M_* .$$

Note that the fixed points of Spec(R) under the action of $\mathbb{G}_m(k)$ are exactly the "origin" [o] and the degenerate points M_* and M^*. This is thus a special case of some theory developed by B. Mueller, which links the occurrence of degenerate prime ideals (i.e. the non-Azumaya-ness) of P.I. algebras to certain invariants of the corresponding prime spectra under the action of algebraic groups.

In the sequel of this chapter we intend to investigate subvarieties of algebraic k-varieties.

Let $\mathcal{X} = (X, \underline{O}_X)$ be a ringed space. Classically a <u>closed subspace</u> of X is defined to be a couple $(\mathcal{Y}, \pi)$, where $\mathcal{Y} = (Y, \underline{O}_Y)$ is another ringed space with underlying topological space Y closed in X, where $\pi : \underline{O}_X \to i_* \underline{O}_Y$ is a surjective sheaf morphism and where $i : Y \to X$ denotes the canonical inclusion. To a closed subspace Y of X one associates a sheaf of ideals of $\underline{O}_X$ by setting $\underline{J}(X,Y) = \mathrm{Ker}\,\pi$. Clearly $\underline{J}(X,Y)$ determines Y upto isomorphism. Conversely, however, each sheaf of ideals $\underline{J}$ of $\underline{O}_X$ does not necessarily determine a closed subvariety; in the commutative case, it is well-known that some coherence condition on $\underline{J}$ has to be imposed.

A morphism of ringed spaces $(\varphi, \theta) : (X, \underline{O}_X) \to (Y, \underline{O}_Y)$ is said to be a <u>closed immersion</u> if and only if the following conditions are not :

1. φ is closed and injective on the underlying topological spaces;
2. for any $x \in X$ and $y = \varphi(x) \in Y$ the induced map

$$\theta_x : \underline{O}_{Y,y} \to \underline{O}_{X,x}$$

is surjective.

This means that (φ, θ) induces an isomorphim of Y with a closed subspace of X.

<u>In the commutative case</u> one then says that (φ,θ) is a closed immersion of varieties resp. that Y is a closed subvariety of X if X and Y are actually algebraic k-varieties.

Furthermore in [79] one proves that for a (prime) ideal P of R the projection $\pi : R \to R/P$ induces a closed immersion $\pi : \mathrm{Spec}(R/P) \to \mathrm{Spec}(R)$ and conversely that each closed immersion of varieties $Y \to \mathrm{Spec}(R)$ is essentially of this form. Using this it is easy to see that for an arbitrary scheme X and a closed subset Z of X there exists a unique reduced structure sheaf $\underline{O}_Z$ on Z making $(Z,\underline{O}_Z)$ into a closed subscheme of X. Similarly if $f : X \to Y$ is a morphism of schemes, then f is a closed immersion if and only if for all affine open sets $U \subset X$ the inverse image $f^{-1}(U)$ is affine and the map

$$\Gamma(U,\underline{O}_X) \to \Gamma(f^{-1}(U), \underline{O}_Y)$$

is surjective.

<u>In the non-commutative case</u>, definitions similar to the foregoing do not lead to similar nice results. Indeed the morphism ${}^a\pi : \mathrm{Spec}(R/P) \to \mathrm{Spec}(R)$ associated to the canonical $\pi : R \to R/P$ is not necessarily a closed immersion for this would imply the localized morphisms $Q_Q^{bi}(\pi) : Q_Q^{bi}(R) \to Q_Q^{bi}(R/P)$ to be surjective. Now if $Q \in \mathrm{Spec}(R)$ does not induce a kernel functor κ_{R-Q} satisfying property (T) for bimodules, then $Q_Q^{bi}(\pi)$ need not be surjective. We avoid restricting to rings having the property that κ_{R-Q} has property (T) for bimodules for all $Q \in \mathrm{Spec}(R)$ and proceed as follows. Let $X = (X, \underline{O}_X, \underline{O}_X^{bi}, \underline{K}_X, \pi_X)$ be a varietal space (cf. Section VI.1.). A <u>closed varietal pre-subspace</u> of X is a triple $(\mathcal{Y},\mu,\nu)$ where $\mathcal{Y}$ is a varietal space $(Y, \underline{O}_Y, \underline{O}_Y^{bi}, \underline{K}_Y, \pi_Y)$ such that topologically Y is a closed subspace of X and where $\mu : \underline{O}_X^{bi} \to i_* \underline{O}_Y^{bi}$ and $\nu : \underline{K}_X \to i_* \underline{K}_Y$ are morphisms in $\sigma(X, \underline{O}_X^{bi})$, such that (denoting by $i : Y \to X$ the canonical inclusion of topological spaces) :

CS1. The following diagram is commutative :

$$\begin{array}{ccc} \underline{O}^{bi}_X & \xrightarrow{\ \mu\ } & i_* \underline{O}^{bi}_Y \\ \downarrow{\scriptstyle \pi_X} & & \downarrow{\scriptstyle i_* \pi_Y} \\ \underline{K}_X & \xrightarrow[\ \nu\ ]{} & i_* \underline{K}_Y \end{array}$$

CS2. The morphism $\nu : \underline{K}_X \rightarrow i_* \underline{K}_Y$ is an isomorphism.

CS3. For any $x \in X$ we have that x is in Y if and only if $\underline{(\mathrm{Ker}\ \mu)}_x \subsetneq \underline{O}^{bi}_{X,x}$.

A <u>closed preimmersion of varietal spaces</u> is now defined in the obvious way. If X and Y are algebraic k-varieties then we speak of <u>closed presubvarieties</u> and closed preimmersions of algebraic k-varieties.

<u>VIII.28. Proposition</u>. Let R be an affine left noetherian prime P.I. ring and let P be a prime ideal of R. The canonical projection $p : R \rightarrow R/P$ induces a closed preimmersion of varieties

$$^a p : \Omega(R/P) \rightarrow \Omega(R)$$

<u>Proof</u>. The map $^a p : \Omega(R/P) \rightarrow \Omega(R) : Q \rightarrow p^{-1}(Q)$ induces a bijection between $\Omega(R/P)$ and V(P), the set of all maximal ideals containing P. More generally, if $J \supset P$ is an ideal of R and $\bar{J} = J/P$ then $^a p$ maps $V(\bar{J})$ onto V(J), i.e. $^a p$ is closed. Furthermore one derives from (V.3.36) a morphism

$$\mu(P) : \underline{O}^{bi}_R \rightarrow (^a p)_* \, \underline{O}^{bi}_{R/P}$$

associated to $p : R \rightarrow R/P$ in the usual way. Next, recall that if $\varphi : R \rightarrow S$ is a central extension of P.I. rings and Q a prime ideal of S, then p.i. $\deg(R/\varphi^{-1}(Q))$ = p.i. deg(S/Q). Applying this to the central extension $R \rightarrow R/P$ one obtains a morphism of sheaves

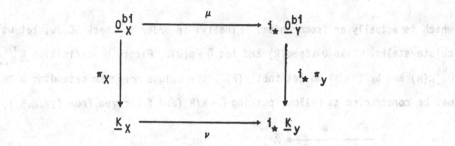

$$\nu(P) : \underline{K}_R \to ({}^a p)_* \underline{K}_{R/P}$$

which is actually an isomorphism. Finally, in order to check SC.3., let us calculate stalks. Take $Q \in \mathrm{Spec}(R)$ and let $\overline{Q} = p(Q)$. First, by definition $\underline{Q}^{bi}_{R,Q} = Q^{bi}_{R-Q}(R)$ and by (V.3.30) we get that $\mu(P)_Q$, the unique morphism extending p to $Q^{bi}_{R-Q}(R)$ may be constructed as follows putting $\overline{R} = R/P$ (and f derived from (IV.2.5.)) :

$$\begin{array}{ccccc}
R & \xrightarrow{\quad p \quad} & \overline{R} & & \\
\downarrow{\scriptstyle j_{Q,R}} & & \downarrow{\scriptstyle j_{Q,\overline{R}}} & \searrow{\scriptstyle j_{\overline{Q},\overline{R}}} & \\
Q^{bi}_{R-Q}(R) & \xrightarrow[Q^{bi}_{R-Q}(p)]{} & Q^{bi}_{R-Q}(\overline{R}) = Q^{bi}_{p_*\sigma_{R-Q}}(\overline{R}) & \xrightarrow[f]{} & Q^{bi}_{\overline{R}-\overline{Q}}(\overline{R})
\end{array}$$

Surjectivity of p yields : $\mathcal{L}^2(p_* \kappa_{R-Q}) = \mathcal{L}^2(\kappa_{\overline{R}-\overline{Q}})$, hence $p_* \kappa_{R-Q} = \kappa_{\overline{R}-\overline{Q}}$ or f reduces to the identity. Hence $(\mathrm{Ker}\, \mu(P))_Q = \mathrm{Ker}\, Q^{bi}_{R-Q}(p)$.
If $Q \notin V(P)$, then $P \in \mathcal{L}^2(\kappa_{R-Q})$ and $Q^{bi}_{R-Q}(R/P) = o$ hence $Q^{bi}_{R-Q}(R) = \mathrm{Ker}\, Q^{bi}_{R-Q}(p)$. Conversely, if $Q^{bi}_{R-Q}(R) = \mathrm{Ker}\, Q^{bi}_{R-Q}(p)$, then $Q^{bi}_{R-Q}(P)$ is the zero morphism, whence $P \in \mathcal{L}^2(\kappa_{R-Q})$. ∎

VIII.29. The degree-filtration of an affine k-variety.

Consider $\mathrm{Spec}_1(R) \to \mathrm{Spec}(R)$, the subset of all prime ideals P of R corresponding to an absolutely irreducible representation of dimension 1, i.e. such that R/P may be embedded centrally into some commutative field K. Since $\mathrm{Spec}_1(R)$ may be identified topologically with $\mathrm{Spec}(R/J_1)$, where J_1 is the "T-ideal", cf. [136] corresponding to the variety of commutative algebras, we find a closed immersion of of topological spaces (since the foregoing arguments are still valid for semiprime quotients of R)

$$\mathrm{Spec}(R/\mathrm{rad}(J_1)) \to \mathrm{Spec}(R),$$

Note that in [136] one shows that J_1 is actually semiprime, hence it is not necessary to replace J_1 by $\mathrm{rad}(J_1)$. The same arguments hold for the T-ideals J_m

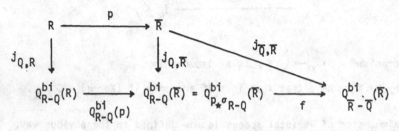

of R corresponding to the variety of m by m matrices : more generally, defining $\Sigma_m(R) = V(\mathrm{rad}(J_m)) = \mathrm{Spec}(R/\mathrm{rad}(J_m))$ we obtain

$$\Sigma_1(R) \hookrightarrow \Sigma_2(\ ) \hookrightarrow \Sigma_3(R) \hookrightarrow \dots \hookrightarrow \Sigma_n(R) = \mathrm{Spec}(R)$$

for some n such that $2n \leq$ p.i.deg(R). This yields a closed filtration of the affine k-variety $\Omega(R)$ by restriction to maximal ideals. Unfortunately, it is in general not a filtration by closed subsets with a structure of algebraic variety (since all of our varieties were assumed to be irreducible). Note however that reducible algebraic varieties may be considered as well, cf.[194]. It is obvious how a similar filtration of an arbitrary algebraic k-variety may be exhibited : locally it is an affine problem that may be globalized afterwards. As a matter of fact for any algebraic variety X we have

$$X_n = \{x \in X;\ \underline{K}_{X,x} = M_n(k)\}.$$

Let us now return to closed subvarieties of algebraic k-varieties.
A <u>closed subvariety of an affine k-variety</u> $\Omega(R)$ is an algebraic k-variety Z isomorphic to $\Omega(R/P)$ for some $P \in \mathrm{Spec}(R)$ One easily verifies that for any open affine subvariety Y of $\Omega(R)$ the induced varietal space $Y \cap Z$ is a closed subvariety of the affine k-variety Y. This is mainly due to the fact that if $Y = \Omega(S)$, where $S = Q_I(R)$ for some ideal I of R, then κ_I has property (T) in R-mod and $Q_I(R) = Q_I^{bi}(R)$, implying that ideals localize to ideals under κ_I. This motivates the definition: a <u>closed subvariety</u> of an arbitrary algebraic k-variety X is a varietal space Z such that for each open affine subspace Y of X the induced varietal space $Y \cap Z$ is a closed subvariety of the affine subvariety Y. Note that this agrees with the commutative case and the following :

<u>VIII.30. Proposition.</u> (Classical situation) Let Y be a cellular, closed subvariety of an affine k-variety $\Omega(R)$; if $\underline{J}$ is the subsheaf of ideals defining Y and $P = \Gamma(\Omega(R), \underline{J})$, then $Y = \Omega(R/P)$, i.e. we obtain a commutative diagram of algebraic

k-varieties

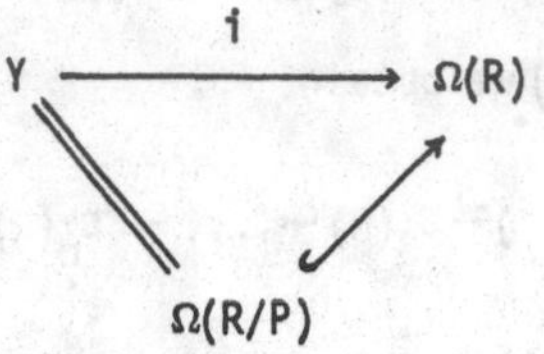

Proof. It is clear that we may assume that $P = o$, i.e. the induced map $j = \Gamma(i) : R \to \Gamma(Y)$ is injective. Indeed, from (VI.2.20) and the fact that Y is cellular one deduces that $\Gamma(i)$ factorizes through R/P Since the space $\Omega(R)$ is quasicompact, Y may be covered by a finite number of affine open subsets $\{\Omega(S_i)\}$, whose intersections may be assumed to be affine too. We have to check first that i is a homeomorphism. As i is closed, it suffices to check surjectivity. Since $i(Y)$ is a closed subset of $\Omega(R)$, we find that $Y = V(P)$ for some ideal P of R, which is prime by the irreducibility of Y. Clearly $\Gamma(i)s = o$ in each point of Y, so, in particular, we get restrictions

$$\mathrm{res}_{Y,\Omega(S_i)}(\Gamma(i)s) = \sigma_i \in S_i$$

such that for each $Q \in \Omega(S_i)$ the composed morphism

$$S_i \to S_i/Q \to \mathbb{K}_{S_i}(Q)$$

annihilates σ_i, i.e. $\sigma_i \in \mathrm{rad}\, o = o$. But then $s = o$, because Y is covered by the affine pieces $\Omega(S_i)$. This yields $P = o$, hence $Y = V(P) = V(o) = \Omega(R)$. We are now in the situation of a single topological space endowed with two sheaves of rings $\underline{O}_X$ and $\underline{O}_Y$ and a surjective homomorphism.

(NOTE. Actually we should include $\underline{O}_X^{bi}$ and $\underline{O}_Y^{bi}$ as well, but this does not yield any new information).

To round off, let us argue as in the commutative case to check that this homomorphism is injective.

Take $y \in Y$, $i(y) = P \in \Omega(R)$. We want the map

$$i_y : \underline{O}_{X,P} = Q^{bi}_{R-P}(R) \to \underline{O}_{Y,y}$$

to be injective. Suppose on the contrary that there exists an element $r \in R$ in the kernel of i_y. Consider an open covering $\{\Omega(R_\alpha) = U_\alpha\}$ of X and extensions $\varphi_\alpha : R \to R_\alpha$ such that the following diagram is commutative :

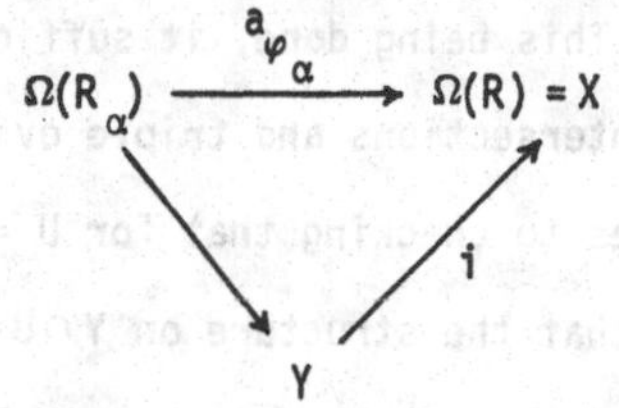

We know that for each $y \in Y$ the restrictions $res_{Y,y}\, i(r)$ vanish, hence there exists an open subset X(I) with $I \not\subset P$ such that for each $y \in U = i^{-1}(X(I))$ we have $res_{Y,y}\, j(r) = o$. If we denote by I_α the ideal generated by $\varphi_\alpha(I)$ in R_α, this gives

$$U \cap U_\alpha = X_\alpha = \{Q \in \Omega(R_\alpha);\ I_\alpha \not\subset Q\}$$

and $\Gamma(X_\alpha, \underline{O}_Y) = Q_\alpha(R_\alpha)$, where $Q_\alpha^{bi}(-)$ is the localization at κ_{I_α}. For each index α, we may choose $r_\alpha \in R_\alpha$ representing $j(r_\alpha)$ on $\Omega(R_\alpha)$ and an ideal L_α such that there exists L with $(\varphi_\alpha(L))r_\alpha = o$ for all α and rad $L \supset I$. But then for each index the restrictions $res_{R,R_\alpha}(j(Lr))$ of $j(Lr)$ to $\Omega(R_\alpha)$ vanish, whence $j(Lr) = o$. Since j is injective we get $Lr = o$, i.e. $r \in \kappa_{R-P}R = o$, which proves the assertion, i.e. the injectivity of i_y. ∎

In the Zariski central case we see that the definitions we have given coincide with the classical ones. Moreover, it is easy to see that every irreducible closed subset of a Zariski central k-variety is in an essentially unique way endowed with the structure of a closed subvariety. Indeed, if $X = \Omega(R)$ is affine and $Y = V(P)$ is an irreducible closed subset, then Y is homeomorphic with $\Omega(R/P)$, and it is this structure that does the job. In general, if Y is a closed, irreducible subset of an arbitrary Zariski central variety X, then for each open affine subset U_α of X we consider the closed subset $Y_\alpha = Y \cap U_\alpha$ of $U_\alpha = \Omega(R_\alpha)$. If Y_α is irreducible, then we give it the structure just defined. Otherwise we give it the reduced structure defined by the homeomorphism $Y_\alpha \cong \Omega(R_\alpha/I_\alpha)$, where $I_\alpha = \cap\{P \in Y_\alpha\}$. Note that this ideal I_α is semiprime, but not prime in the reducible case. Since R_α is Zariski central, however, there is no problem in constructing the appropriate sheaves in this case, even for nonprime R_α. This being done, it suffices to check that these structures behave well on intersections and triple overlaps. As in the commutative case one easily reduces to checking that for $U = \Omega(R)$ open affine and $c \in Z(R)$ we have for $V = X(c) = \Omega(R_c)$ that the structure on $Y \cap U$ obtained from R gives by restriction to $Y \cap V$ the same as that obtained from R . This is an easy consequence of the fact that the occuring localizations are central.

Note that it actually suffices to consider varieties X which posses a basis of open affines. Let X be a Zariski central variety, and let $i : Y \subset X$ be a closed subvariety with structural morphism $\mu : \underline{O}_X \to i_* \underline{O}_Y$. Note that we may write $\underline{O}_X$ and $\underline{O}_Y$ for $\underline{O}^{bi}_X$ and $\underline{O}^{bi}_Y$, since these coincide by the fact that X (and Y!) is Zariski central. Then $\underline{J}_Y = \mathrm{Ker}\,\mu$ is a coherent sheaf of ideals of $\underline{O}_X$. Indeed, $\underline{O}_X$ is quasi coherent, while we know from (VII.27) that $i_*\underline{O}_Y$ is quasicoherent too. But then so is $\underline{\mathrm{Ker}}\,\mu = \underline{J}_Y$ by (VII.25). Now, if we take $U = \Omega(R) \subset X$ to be affine, then R is left noetherian, so $I = \Gamma(U, \underline{J}_Y)$ is finitely generated, proving the assertion. Moreover, it is clear that Y is completely determined by $\underline{J}_Y$.
More concretely : any quasicoherent sheaf of ideals of $\underline{O}_X$, say $\underline{J}$, determines a unique closed varietal pre-subspace of X. Indeed, let Y be the support of the quotient sheaf $\underline{O}_X/\underline{J}_Y$, then $(Y, \underline{O}_X/\underline{J}_Y)$ with obvious supplementary structure, does the trick. The unicity is easily checked and so it suffices to check that $(Y, \underline{O}_X/\underline{J}_Y)$ is a closed varietal presubspace. This is only a local verification, so we choose X affine, say $X = \Omega(R)$, with R Zariski central. Then $\underline{J}_Y = \underline{O}_I$ for some ideal I of R and we see that $(Y, \underline{O}_X/\underline{J}_Y) = (\Omega(R/I), \underline{O}_{R/I})$, which finishes the proof of the claim, up to some straightforward verification left to the reader.

<u>VIII.31. Corollary</u>. For any Zariski central ring R the prime ideals of R and the closed subvarieties of $\Omega(R)$ correspond bijectively. ■

VIII.32. Corollary. Every closed subvariety of the spectrum of a Zariski central ring is affine. ▪

VIII.33. Corollary. Every irreducible closed subset of the spectrum of a Zariski central ring determines uniquely a corresponding affine closed subvariety.

APPENDIX.

In this section we have made use several times of the following result, due to L.W. Small (unpublished) :

Proposition. If R is a left noetherian ring containing a field k and if S is an affine left noetherian P.I. algebra over k, then $R \otimes_k S$ is left noetherian too.

Proof. It is clear that S may be assumed to be prime. As indicated in the remarks following II.3.21. the trace ring, denoted by S[T] here, cf. [12], yields a finite central extension $S \to S[T]$, hence a finite central extension $R \otimes_k S \to R \otimes_k S[T]$. So, using a well-known result due to Eisenbud and Eakin, it suffices to show that $R \otimes_k S[T]$ is noetherian, in order to obtain the result. But, since S[T] is well known to be affine and finite over a polynomial ring, say $k[Y] = k[Y_1, \dots, Y_n]$, cf. [12], we obtain another finite central extension $R \otimes_k k[Y] \to R \otimes_k S[T]$ in which the first term is just R[Y], a noetherian ring by the Hilbert Basis Theorem. Hence so is $R \otimes_k S[T]$, proving the assertion. ■

IX. REPRESENTATION THEORY REVISITED.

The central problem in representation theory is to construct a left adjoint V_n of the functor M_n i.e. find a functor in the category of k-algebras such that for each pair of k-algebras R and S we have a functorial bijection $\mathrm{Hom}_k(R,M_n(S)) = \mathrm{Hom}_k(V_n(R),S)$. In order to obtain a workable object we only seek to represent the functor $S \to \mathrm{Hom}_k(R,M_n(S))/\sim$ where morphisms $\varphi_1 : R \to M_n(S)$, $\varphi_2 : R \to M_n(S)$ are said to be equivalent if there exists a k-automorphism ψ of $M_n(S)$ such that $\psi\varphi_1 = \varphi_2$. However, even in the commutative case this functor need not be representable, and we are forced quite naturally into the following situation.

IX.1. Definition. If C is a commutative k-algebra and A an Azumaya algebra of rank n^2 over C then a morphism $\varphi : R \to A$ is said to be an *irreducible representation of degree n over C* if and only if $\varphi(R)C = A$. Irreducible representations $\varphi_1 : R \to A_1$, $\varphi_2 : R \to A_2$ will be called *equivalent representations over C* if there exists a C-algebra isomorphism $\psi : A_1 \to A_2$ such that $\psi\varphi_1 = \varphi_2$.

Let $Q_n(R,C)$ be the *set* of all equivalence classes of irreducible representations of degree n over C. Obviously, Q_n is a covariant functor in C and contravariant in R if one restricts ring morphisms to be central extensions. A result of C. Procesi, [136], states that, for fixed R, Q_n is representable by an open set of an affine scheme which is a quotient of a suitable open subset U_R of $\mathrm{Spec}(V_n(R))$ under the canonical action of the algebraic group A_n defined by $A_n(S) = \mathrm{Aut}_S(M_n(S))$. More specifically, we want U_R to be an open subset of $\mathrm{Spec}(V_n(R))$ endowed with a sheaf of Azumaya algebras, such that morphisms of ringed spaces $\mathrm{Spec}(C) \to U_R$ correspond bijectively to elements of the set $Q_n(R,C)$.

Let us point out an alternative construction in the case where we consider k-algebras satisfying the identities of $n \times n$ matrices (also due to C. Procesi). Let R be such a k-algebra and write $\Gamma(R)$ for $k + F(R)$ where $F(R)$ is the Formanek center of R (cf. Chapter II).

IX.2. Theorem. 1. If $\alpha \in \Gamma(R)$ then $S = R \underset{\Gamma(R)}{\otimes} \Gamma(R)[\alpha^{-1}]$ is an Azumaya algebra.
2. The open subset $\tilde{U}_R = \mathrm{Spec}(\Gamma(R)) - V(F(R))$ of $\mathrm{Spec}(\Gamma(R))$ is endowed with a sheaf of Azumaya algebras and as a ringed space $\tilde{U}_R$ is essentially the same as U_R.

Proof. 1. Follows directly from the results in Chapter II. because $SF(S) = S$.
2. Let $\varphi : R \to A$ be an absolutely irreducible representation of degree n over C. If $M \in \Omega(C)$ then A/MA is a central simple algebra of rank n^2 over C/M and the composition $R \overset{\varphi}{\to} A \overset{\pi}{\to} A/MA$ is absolutely irreducible, therefore $\pi\varphi(F(R)) \neq o$ i.e. $\varphi(F(R)) \not\subset M$. The latter holds for any $M \in \Omega(C)$, hence $\varphi(F(R))C = C$. Therefore, the morphism $\varphi : \Gamma(R) \to C$ yields a map $\varphi^* : \mathrm{Spec}(C) \to \tilde{U}_R$. Conversely, to a scheme morphism $\mathrm{Spec}(B) \to \tilde{U}_R$ there corresponds a ring morphism $\psi : \Gamma(R) \to B$ (for a commutative ring B). By 1., $R \underset{\Gamma(R)}{\otimes} B$ is an Azumaya algebra (utilizing the well-known local-global properties of Azumaya algebras). Since $\tilde{U}_R$ and U_R an classified by the same functor, these ringed spaces are to be isomorphic.

IX.3. Corollary. If a k-algebra R satisfies the identities of n x n matrices then $\tilde{U}_R$ is homeomorphic to $\mathrm{Spec}_n(R)$. The stalk at $P \in \tilde{U}_R$ is isomorphic to $Q_{C-p}(R)$ where $C = Z(R)$, $p = P \cap C$.

Proof. The canonical inclusion $\Gamma(R) \to R$ yields a continuous bijection $j : \mathrm{Spec}_n(R) \to \tilde{U}_R$ and the only thing left to check is that j is a closed map. If I is an ideal of R then $j(V(I)) = V(I \cap \Gamma(R)) \cap \tilde{U}_R$. Indeed, if $P \in \mathrm{Spec}_n(R) - V(I)$ then $\bar{I} = I \bmod P$ is nonzero in R/P. So $\bar{I}$ is a prime ring of degree n and $F(\bar{I}) \neq o$. Since F(I) maps to $F(\bar{I})$ we may select an $\alpha \in F(I) - P$. On the other hand $\alpha \in F(I) \subset F(R)$ yields $\alpha \notin P \cap \Gamma(R)$, $P \cap \Gamma(R) \in \tilde{U}_R$ and $P \cap \Gamma(R) \notin V(I \cap \Gamma(R))$. The converse inclusion, i.e. $j(V(I)) \subset V(I \cap \Gamma(R)) \cap \tilde{U}_R$, being obvious, this proves the fact that j is closed, hence a homeomorphism. The stalk at P is $R \underset{\Gamma(R)}{\otimes} \Gamma(R)_p$, by construction. The fact that central localization has property (T) (together with the assumptions on $\tilde{U}_R$) yields that this is nothing but $Q_{C-p}(R)$.

IX.4. Corollary. If R is left Noetherian and satisfies the identities of n x n matrices then the following ringed spaces are isomorphic :

$\tilde{U}_R$, $(\mathrm{Spec}_n(R), \underline{O}_R|\mathrm{Spec}_n(R))$, $(\mathrm{Spec}_n(R), \underline{O}_R^{bi}|\mathrm{Spec}_n(R))$

The left Noetherian hypothesis is needed only in the global construction of sheaves $\underline{O}_R^{bi}$, $\underline{O}_R$. However if R is not necessarily left Noetherian then still $\underline{O}_R^{bi}|\mathrm{Spec}_n(R)$ is a sheaf.

Recall from Rowen [152] that an ideal I of R is identity faithful if there exists a common regular central polynomial for R and R/I. A prime ideal of a semiprime k-algebra satisfying the $n \times n$ identities is exactly then identity faithful if it does not contain F(R).

IX.5. Memo. (cf. Chapter II) 1. Let R be a semiprime k-algebra satisfying the identities of $n \times n$ matrices, then R is a Zariski algebra (over its center) with respect to the central n-kernel J_n and RJ_n (in the sense of II.2.15.).

2. For any $P \in \mathrm{Spec}(R)$ such that $F(R) \not\subset P$ the kernel functor κ_{R-P} is a T-functor for R-bimodules and $Q_{R-P}(R) = Q^{bi}_{R-P}(R) = Q_{Z(R)-p}(R)$ is a local Azumaya algebra of rank n^2 over its center $Q_{Z(R)-p}(Z(R))$, where $p = P \cap Z(R)$. In particular $Q(R/P) = R_p/P_p$.

Let us now show how the foregoing may be applied to the study of dimension questions. First, let X be an algebraic k-variety with structure sheaf $\underline{O}_X$, then we define the function ring of X by

$$R(X) = \varinjlim_{\phi \neq U \in \mathrm{Open}(X)} \Gamma(U, \underline{O}_X)$$

Clearly if V is an arbitrary non empty open subset of X, then the set of all open subsets of V is cofinal in Open(X). In particular, if U is an open affine subvariety of X, then $R(U) = R(X)$.

IX.6. Theorem. For any algebraic k-variety the function ring R(X) is simple.

Proof. As X is covered by affine pieces, we may choose an affine open subvariety $U = \Omega(R)$ of X and we thenknow that $R(X) = R(U)$. As $\Omega(R)$ is irreducible, R is prime, hence by Posner's theorem $Q(R) = Q_{R-0}(R)$ is simple and finite dimensional over its center. But

$$\varinjlim_{\phi \neq V \in \mathrm{Open}(U)} \Gamma(V, \underline{O}_U) = \varinjlim Q_I(R) = Q_{R-0}(R) = Q(R),$$

hence $R(U) = R(X)$ is simple. ∎

Note that the foregoing proof also shows that

$$R(X) = \varinjlim_{\phi \neq U \in \mathrm{Open}(X)} \Gamma(U, \underline{O}_X^{bi})$$

Indeed, for an affine piece $U = \Omega(R)$, we see that

$$R(U) = \varinjlim_{\phi \neq V \in \mathrm{Open}(U)} \Gamma(V, \underline{O}_R) = \varinjlim Q_I(R) = Q_{R-0}(R) = Q(R) = Q_{R-0}^{bi}(R) = \varinjlim Q_I^{bi}(R) =$$

$$= \varinjlim_{\phi \neq V \in \mathrm{Open}(U)} \Gamma(V, Q_R^{bi}) = \varinjlim_{\phi \neq V \in \mathrm{Open}(U)} \Gamma(V, \underline{O}_R^{bi})$$

<u>IX.7. Definition.</u> The <u>dimension</u> of an algebraic k-variety X is

$$\dim X = \mathrm{tr.deg}_k\, Z(R(X)) .$$

Let us motivate this definition. The <u>topological dimension</u> of an irreducible topological space X is by definition the maximal number of elements in a strictly increasing chain of irreducible subspace

$$\phi \neq Z_1 \subset Z_2 \subset \dots \subset Z_r .$$

If $X = \Omega(R)$ is affine, then an irreducible subspace of X is necessarily of the form $V(P)$, where $P \in \mathrm{Spec}(R)$. We thus see that the topological dimension of an affine k-variety is the Krull-dimension of the ring of global sections of its structure sheaf. Recall that the Krull dimension of a ring R is defined as $\dim R = \sup \mathrm{rk}\, P$, where for a prime ideal P of R the <u>rank</u> rk P of P is the supremum of all lengths of chains $P = P_0 \supset P_1 \supset P_2 \supset \dots \supset P_n$ of prime ideals. Since it is reasonable to expect the dimension of an algebraic k-variety to coincide with its topological dimension, we could like to know whether for any finitely generated prime P.I. algebra R :

$$\dim R = \text{tr.deg}_k Z(Q(R)) .$$

This is proved by Procesi in [136]. Let us recall the proof here :

<u>IX.8. Theorem.</u> Let $R = k\{a_1, \dots, a_m\}$ be an affine (prime) P.I. algebra, let Q be its total ring of fractions and let Z be the center of Q; then $\dim R = \text{tr.deg}_k Z$.

<u>Proof.</u> In II.2.37. we have established that $\dim R \leq \text{tr.deg}_k Z$.
The foregoing remarks learn that $\text{Spec}_n(R)$ is homeomorphic to U_R, i.e. homeomorphic to an open subscheme of $\text{Spec}(\Gamma(R))$, with $\Gamma(R) = k + F(R)$. The latter open set may be given as a union of spaces $\text{Spec}(\Gamma(R)[\alpha^{-1}])$, where $\Gamma(R)[\alpha^{-1}]$ is affine over k and :

$$\text{tr.deg}_k \Gamma(R)[\alpha^{-1}] = \text{tr.deg}_k Z.$$

This obviously finishes the proof. ∎

<u>IX.9. Proposition.</u> Let Y be a proper closed k-subvariety of the algebraic k-variety X, then $\dim Y < \dim X$.

<u>Proof.</u> 1°. <u>Affine case.</u> Put $X = \Omega(R)$, $Y = \Omega(R/P)$ for some prime ideal P of R and let π be the canonical map $R \to R/P$. A chain of prime ideals $P_1 \supset \dots \supset P_n$ in R/P yields a chain of prime ideals $\pi^{-1}(P_1) \supset \dots \supset \pi^{-1}(P_n)$ in R, hence for each $Q \in \text{Spec}(R/P)$, $\text{rk}\, Q \leq \text{rk}\, \pi^{-1}(Q)$ and $\dim R/P \leq \dim R$. In calculating the supremum of ranks of prime ideals we may restrict to considering maximal ideals.
Therefore, if $M \supset P_1 \supset \dots \supset P_n$ is a saturated chain for M in R/P then the corresponding chain in R, $\pi^{-1}(M) \supset \pi^{-1}(P_1) \supset \dots \pi^{-1}(P_n) = P \supset o$ is a chain of strictly larger length. Since this holds for each maximal ideal of R we find that $\dim R/P < \dim R$ unless $\dim R/P = \infty$ or $P = o$, and both these cases are excluded.
2°. <u>The general case.</u> Choose $U \subset X$ affine with $U \cap Y \neq \phi$. Let R be the coordinate

ring of U and let P be the prime ideal corresponding to the closed subset $U \cap Y$ of U, then $\dim U \cap Y < \dim U = \dim X$, and as $\dim Y = \dim U \cap Y$, this finishes the proof. ∎

If Y is a closed subvariety then the codimension of Y in X is denoted by $\text{codim}_X Y$ and is defined to be the difference $\dim X - \dim Y \geq o$. The foregoing proposition may then be restated as

IX.10. Corollary. If Y is a closed subvariety of X, then $X = Y$ if and only if $\text{codim}_X Y = o$. ∎

Note that the foregoing follows even easier from the fact that under our assumptions

$$\dim R = \dim R/P + \text{rk } P$$

for any prime ideal P of R. This property will also be used in

IX.11. Theorem. Let X be a Zariski central k-variety, $U \subset X$ an open subset, $g \in Z(\Gamma(U, \underline{O}_X))$ and Y an irreducible component of

$$\{x \in U;\ g(x) = o\}.$$

Assume that $g \neq o$ then

$$\dim Y = \dim X - 1 \quad .$$

Proof. Choose an open affine k-sub variety V of X such that $V \subset U$ and $V \cap Y \neq \phi$ and assume that $V = \Omega(R)$ for some prime noetherian affine PI ring R. The restriction h of g to V lies in Z(R) and the closed subset $Y \cap V$ corresponds to a prime ideal P of R. Since Y is a maximal irreducible subset of (g=o), clearly $Y \cap V$ is a maximal irreducible subset of $(h = o)$, i.e. P is an isolated prime of Rh. From a well-known result of Jategaonkar's [90] it then follows that P contains no nonzero, strictly smaller prime ideal of R, i.e. ht $P = 1$. Now recall Schelter's result [156] stating that under these assumptions :

$$\dim R = \dim R/P + \text{ht } P$$

for any prime ideal P of R, thus yielding the statement. ∎

This may be generalized to :

<u>IX.12. Proposition.</u> If X possesses a basis of open affines and if $U \subset X$ is an open subset, $g \in Z(\Gamma(U,\underline{O}_X))$ and Y is an irreducible component of $\{x \in U; g(x) = o\}$, then $g \neq o$ implies $\dim Y = \dim X-1$.

<u>IX.13. Proposition.</u> If Y is a closed subvariety of an algebraic k-variety X which is an irreducible component of the zero-set of $g \in Z(\Gamma(U,\underline{O}_X))$ for some U open in X, then $\dim Y = \dim X-1$.

<u>IX.14. Definition.</u> Let X be as before. If Z is a closed subset then we say that Z has <u>pure dimension (codimension) n</u> if each of its irreducible components has dimension (codimension) n. From the foregoing theorem, it then follows that for any central, nonzero element g of $\Gamma(X,\underline{O}_X)$ the closed subset

$$V(g) = \{x \in X : g(x) = o\}$$

has pure codimension 1. Conversely, if Z is an arbitrary irreducible closed subset of X of codimension 1, let U be an open affine subset such that $U \cap Z \neq \phi$. If $o \neq f \in Z(\Gamma(U,\underline{O}_U))$ has the property that $Z \subset V(f)$, then $Z \cap U$ is an irreducible component of V(f). Indeed, should W be an irreducible component of V(f) containing $Z \cap U$ (which is obviously irreducible!), then we get

$$\dim X - 1 = \dim(Z \cap U) \leq \dim W < \dim X,$$

hence $\dim W = \dim(Z \cap U)$, yielding $W = Z \cap U$ by IX.10.

This obviously yields that a maximal closed irreducible subset of an algebraic k-variety strictly contained in it has codimension 1, hence by induction :

<u>IX.15. Corollary</u>. The dimension of a Zariski central k-variety is equal to its topological dimension. ∎

IX.16. Proposition. For any pair of precellular algebraic k-varieties X and Y we have

$$\dim(X \hat{\times} Y) = \dim X + \dim Y \ .$$

Proof." It clearly suffices to prove this for affine k-varieties. Since the function ring of affine k-varieties are central simple, one easily checks that their tensorproduct is central simple too. Now recall from [136] and [58] that Posner's theorem may be stated as follows: if R is a prime P.I. ring then R has a central simple classical ring of quotients Q(R) which is generated by R and Z(Q(R)). Moreover Z(Q(R)) is the commutative ring of quotients Q(Z(R)) of the center Z(R) of R. We then find

$$\begin{aligned}
\dim \Omega(R \underset{k}{\otimes} S) &= \mathrm{tr.deg}_k\, Z(Q(R \underset{k}{\otimes} S)) \\
&= \mathrm{tr.deg}_k\, Q(Z(R \underset{k}{\otimes} S)) \\
&= \mathrm{tr.deg}_k\, Q(Z(R) \underset{k}{\otimes} Z(S)) \\
&= \mathrm{tr.deg}_k\, Q(Z(R)) + \mathrm{tr.deg}\, Q(Z(S)) \\
&= \mathrm{tr.deg}_k\, (Z(Q(R))) + \mathrm{tr.deg}_k Z(Q(S)) \\
&= \dim \Omega(R) + \dim \Omega(S) \ . \ \blacksquare
\end{aligned}$$

X. BIRATIONALITY AND QUASIVARIETIES.

It is well-known that a commutative affine k-variety is embeddable in some affine k-space k^n which may be given as $\Omega(k[X_1,\dots,X_n])$. In the non-commutative case affine n-space over k is not that easy to define and as a matter of fact we will have to weaken the concept of an algebraic k-variety and use quasivarieties in order to obtain something like affine n-space.

In the sequel the terminology (and results) of Chapter II will be used freely.

Let {R} denote the variety (of algebras) generated by a given algebra R. Let $k\{X_s, s\in S\}$ be the free associative k-algebra over S. We have defined the ideal $I(\{R\}, k\{X_s\})$ to be the ideal of polynomial identities of R (cf. II.1.), we shall denote it by I(R). Let $k\{\xi_s^n\}$ be the free algebra in card(S) variables in the variety V_n generated by the n by n matrices over commutative fields. Since $k\{\xi_s^n\}$ has no zero-divisors it makes sense to define the space $\Omega(k\{\xi_s^n\})$, the closed points of $\mathrm{Spec}(k\{\xi_s^n\})$, as those points corresponding to central extensions $\varphi : k\{\xi_s^n\}\to M_d(k)$ where $d\leq\frac{1}{2}$. p.i. $\deg(k\{\xi_s^n\})$, i.e., the set of $(m_s, s\in S)\in M_d(k)^S$ such that $\xi_s^n\to m_s$ defines a central extension $k\{\xi_o^n\}\to M_d(k)$, or again, such that $\{m_s, s\in S\}$ generates $M_d(k)$ as a k-algebra. Note that $\Omega_d(k\{\xi_s^n\})$ is <u>not</u> a closed subvariety of $M_d(k)^S=\Omega(k[X_s^{ij}, s\in S, i,j\in\{1,\dots,d\}])$.

$$\text{Put } \mathbb{A}_n^m(k)=\Omega(k\{\xi_s^n, s=1,\dots,m\})$$
$$=\Omega(k\{X_s, 1\leq s\leq m\}/I_{m,n}),$$

where $I_{m,n}=I(V_n, k\{X_s, 1\leq s\leq m\})$.

We call $\mathbb{A}_n^m(k)$ the <u>affine k-space of size m</u> and <u>degree n</u>. Obviously all this amounts to say that $\mathbb{A}_n^m(k)_d=\{[a_1,\dots,a_m]; a_i\in M_d(k)$ such that $k\{a_1,\dots,a_m\}=M_d(k)\}$ for any $1\leq d\leq n$, where $[a_1,\dots,a_m]$ is the class of $(a_1,\dots,a_m)$ modulo automorphisms of $PGL_n(k)$. Putting $\mathbb{A}_{n,p}^m(k)=\bigcup_{d\leq p}\mathbb{A}_n^m(k)_d$, we obtain a filtration by closed subsets $\mathbb{A}_{n,1}^m(k)\to\mathbb{A}_{n,2}^m(k)\to\dots\to\mathbb{A}_{n,n}^m(k)=\mathbb{A}_n^m(k)$. On the other hand, the chain of semiprime ideals $I_{m,1}\supset\dots\supset I_{m,n}$ induces a filtration by closed subsets $\mathbb{A}_1^m(k) \hookrightarrow \mathbb{A}_2^m(k) \hookrightarrow \ldots \hookrightarrow \mathbb{A}_n^m(k)$.

To prove that these filtrations are essentially the same it is sufficient to prove that for any $p<n$, $\mathbb{A}_{n,p}^m(k)$ and $\mathbb{A}_{p,p}^m(k)$ coincide, where $\mathbb{A}_{n,p}^m(k)$ is the space of points of degree p in $\mathbb{A}_n^m(k)$. Since $p<n$ we may define a k-algebra morphism $\tau : k\{\xi_1^n,\ldots,\xi_m^n\} \to k\{\xi_1^p,\ldots,\xi_m^p\}$ by $\xi_i^n \to \xi_i^p$. Note that τ corresponds to the inclusion $\mathbb{A}_p^m(k) \hookrightarrow \mathbb{A}_n^m(k)$ in the usual way. Now to $P \in \mathbb{A}_{p,p}^m(k)$ there corresponds a central extension $\alpha : k\{\xi_1^p,\ldots,\xi_m^p\} \to M_p(k)$ such that $P = \operatorname{Ker} \alpha$, hence $Q = \tau^{-1}(P) = \operatorname{Ker}(\alpha \circ \tau) \in \mathbb{A}_{n,p}^m(k)$. Conversely if $Q \in \mathbb{A}_{n,p}^m(k)$ then $Q = \operatorname{Ker} \beta$ for some central extension $\beta : k\{\xi_1^n,\ldots,\xi_m^n\} \to M_p(k)$. Since $k\{\xi_1^n,\ldots,\xi_m^n\}/Q$ then satisfies the identities of $p \times p$ matrices, there exists a surjective k-algebra morphism :

$$\psi : k\{\xi_1^p,\ldots,\xi_m^p\} \to k\{\xi_1^n,\ldots,\xi_m^n\}/Q \quad .$$

Clearly $P = \operatorname{Ker} \psi$ is in $\mathbb{A}_{p,p}^m(k)$ and it is easily verified that $\mathbb{A}_{p,p}^m(k)$ is homeomorphic to $\mathbb{A}_{n,p}^m(k)$ under the correspondence just defined. Moreover in corresponding points of $\mathbb{A}_{p,p}^m(k)$ and $\mathbb{A}_{n,p}^m$ we find isomorphic functions rings (we only have to check this for closed points!) because if M and N are corresponding closed points, then we have :

$$\mathbb{k}(N) = k\{\xi_1^p,\ldots,\xi_m^p\}/N \cong k\{\xi_1^n,\ldots,\xi_m^n\}/M \cong \mathbb{k}(M) \quad (\cong M_p(k)) \quad .$$

For any finitely generated prime P.I. algebra over k of p.i. degree n we obtain a surjective morphism $\overline{\varphi} : k\{\xi_1^n,\ldots,\xi_m^n\} \to R$. Since $\overline{\varphi}$ is also an extension it yields a closed immersion :

$$\operatorname{Spec}(R) \to \operatorname{Spec}(k\{\xi_1^n,\ldots,\xi_m^n\}) = \mathbb{A}_n^m(k)$$

and a closed immersion

$$\Omega(R) \to \Omega(k\{\xi_1^n,\ldots,\xi_m^n\}) = |\mathbb{A}_n^m(k)|$$

of $\Omega(R)$ into affine k-space of size m and degree n. One easily verifies that in this embedding the subspace $\Sigma_p(R)$ embeds into $\mathbb{A}_{n,p}^n(k) = \mathbb{A}_{n,p}^m(k) = \mathbb{A}_p^m(k)$ for $1 \leq p < n$. This should be viewed as a generalization of the embedding of a commu-

commutative affine k-variety into affine m-space (i.e. size m, degree 1 in our terminology).

More explicitly let $m,n,p \in \mathbb{N}$, where $m \geq 2$, and $p \leq m$, then we define for any $\underline{a} = (a_1,\dots,a_m) \in M_p(k)^m$

$$\psi^n_{\underline{a}} : k\{\xi^n_1,\dots,\xi^n_m\} \to M_p(k)$$

by $\psi^n_{\underline{a}}(\xi^n_i) = a_i$. Because $n \geq p$ this yields a well-defined k-algebra morphism. Let F be a subset of $k\{\xi^n_1,\dots,\xi^n_m\}$, then we put (note the abuse of notation)

$$V(F,p) = \{[\underline{a}] \text{"}\in\text{"} M_p(k)^m;\ k\{a_1,\dots,a_m\} = M_p(k)\ \&\ \forall f \in F,\ \psi_{\underline{a}}(f) = o\}.$$

Obviously V(F,p) only depends on the ideal of $k\{\xi^n_1,\dots,\xi^n_m\}$ generated by F. Moreover $V(\phi,p)$ may be identified with $|\mathbb{A}^m_{p;p}(k)| = \Omega_p(k\{\xi^n_s;\ 1 \leq s \leq m\})$, so for any set F we find an embedding

$$V(F,p) \to |\mathbb{A}^m_{n;p}(k)| \quad .$$

Let $V_d(F) = \bigcup_{p=1}^{d} V(F,p)$, then the local embeddings of degree p globalize to

$$V_d(F) \to |\mathbb{A}^m_{n,d}(k)| \ ,$$

since $|\mathbb{A}^m_{n,d}(k)| = \bigcup_p |\mathbb{A}^m_{n;p}(k)|$.

Assume now that $F \subset k\{\xi^n_1,\dots,\xi^n_n\}$ is a (semi) prime ideal and let $R = k\{\xi^n_1,\dots,\xi^n_m\}/F$. Then $\mathrm{Spec}_p(R) \cap \Omega(R) = \Omega_p(R)$ may be identified with V(F,p) and the morphisms

$$\Omega_p(R) \to |\mathbb{A}^m_{n;p}(k)|$$

and

$$|\Sigma_d(R)| \to |\mathbb{A}^m_{n,p}(k)|$$

deduced from the projection $k\{\xi^n_1,\dots,\xi^n_m\} \to R$ are exactly those defined above, by the noncommutative version of Hilbert's Nullstellensatz.

An algebraic k-variety may thus be considered as being obtained by patching together pieces cut out in some generalized affine k-space by polynomials whose "variables" are generic matrices of a certain dimension.

In the foregoing we have seen that it is very useful to treat spectra of certain nonnoetherian rings (such as the rings or algebras of generic matrices) as if they were algebraic varieties. It is obvious that it is impossible to endow the spectrum of a non-noetherian ring with a structure sheaf by mere localization techniques as we did before, since the idempotency of the occuring kernel functors depends heavily on certain finiteness assumptions. Nevertheless in the section dedicated to birational extensions, we have seen that in many circumstances localization behaves well on certain dense open sets. It is from this point of view that we may treat the nonnoetherian situation.

Let R be an affine prime P.I. algebra over k which is not necessarily left noetherian. We have seen that its prime spectrum splits according to the p.i. degree n of R in the following way

$$\mathrm{Spec}(R) = \mathrm{Spec}_n(R) + \Sigma_{n-1}(R)$$

where $\mathrm{Spec}_n(R)$ consists of the prime ideals of degree n of R and where $\Sigma_{n-1}(R) = \bigcup_{p<n} \mathrm{Spec}_p(R)$ is homeomorphic to $\mathrm{Spec}(R/J_{n-1})$, where J_{n-1} is the ideal of identities for n-1 by n-1 matrices, i.e. obtained by substituting elements of R as entries in the polynomials which yield identities for all n-1 by n-1 matrix algebras. Let us first construct a presheaf of rings on Spec(R), which behaves nicely on the open dense subset $\mathrm{Spec}_n(R)$ of Spec(R). Let us start by defining

$$\Gamma(X(I), \underline{Q}_R) = \{x \in Q(R);\ Lx \subset R \text{ for some } L \in \mathcal{L}(I)\}$$

Note that R being a prime P.I. algebra, we know that the total ring of fractions Q(R) of R does exist by Posner's theorem. Furthermore, in this definition $\mathcal{L}(I)$ is consisting of all left ideals L of R with the property that we may find a two-sided R-ideal K such that $I \subset \mathrm{rad}(K)$ and such that $K \subset L$.

<u>X.1. Proposition.</u> For each ideal I of R the set $\Gamma(X(I), \underline{Q}_R)$ is an R-subalgebra of Q(R).

<u>Proof</u>. First note that $\mathcal{L}(I)$ possesses the set $\mathcal{L}^2(I)$ of all twosided ideals K of R for which $I \subset \mathrm{rad}(K)$ as a filterbasis. Moreover, if $K_1, K_2 \in \mathcal{L}^2(I)$ then $K_1K_2 \in \in \mathcal{L}^2(I)$. Indeed,

$$\mathrm{rad}(K_1K_2) = \mathrm{rad}(K_1) \cap \mathrm{rad}(K_2) \supset I.$$

Now, if x and y are elements of $\Gamma(X(I), \underline{Q}_R)$, then by assumption we may find $K_1, K_2 \in \mathcal{L}^2(I)$ such that $K_1x \subset R$ and $K_2y \subset R$. But then

$$K_2K_1xy \subset K_2Ry = K_2y \subset R,$$

hence $xy \in \Gamma(X(I), \underline{Q}_R)$ in view of the foregoing. The fact that $\Gamma(X(I), \underline{Q}_R)$ is additively closed is even more obvious, as well as the fact that these operations define a ring-structure on $\Gamma(X(I), \underline{Q}_R)$. Finally, if $x \in \Gamma(X(I), \underline{Q}_R)$ then we may find $K \in \mathcal{L}^2(I)$ with $Kx \subset R$. But then $KR\,xR = K\,xR \subset R$ implies that $RxR \subset \Gamma(X(I), \underline{Q}_R)$, proving the R-algebra structure of $\Gamma(X(I), \underline{Q}_R)$. ∎

Since $\Gamma(X(I), \underline{Q}_R) = Q_I(R)$ when κ_I is idempotent, there is no ambiguity in writing $Q_I(R)$ for $\Gamma(X(I), \underline{Q}_R)$, even in the absence of the noetherian hypothesis. One proves exactly as in the noetherian case that $\underline{Q}_R$ is a separated presheaf of rings on Spec(R). Let us just point out how the restriction morphisms work. If $X(I) \subset X(J)$ then if I and J are chosen to be radical, i.e. when $J \supset I$, hence we construct

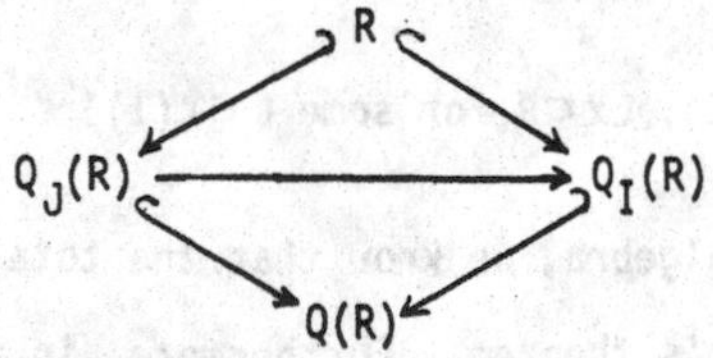

in the obvious way : if $x \in Q_J(R)$, then we may find $K \in \mathcal{L}^2(J)$ i.e. $J \subset \mathrm{rad}(K)$ such that $Kx \subset R$. But since $I \subset J$, clearly $K \in \mathcal{L}(I)$ too, showing that $x \in Q_I(R)$ as well. Let us note that even in the absence of any noetherian hypothesis, we know that Hilbert's nullstellensatz still holds, offering the possibility to induce $\underline{Q}_R$ on $\Omega(R)$. Let us now calculate the stalks of this presheaf.

Define

$$Q_{R-P}(R) = \{x \in Q(R); \exists L \in \mathcal{L}(R-P), Lx \subset R\}$$

where $\mathcal{L}(R-P)$ consists of all left ideals of R containing an ideal I of R with $I \not\subset P$. There is no ambiguity in the notation "$Q_{R-P}(R)$", again since in the left noetherian case $\mathcal{L}(R-P)$ defines the idempotent kernel functor κ_{R-P}.

X.2. Proposition. Under the above assumptions, we have for any prime ideal P of R

$$\underline{Q}_{R,P} = \varinjlim_{X(I) \ni P} \Gamma(X(I), \underline{Q}_R) = Q_{R-P}(R) .$$

Proof. We have to show that

$$Q_{R-P}(R) = \varinjlim_{I \not\subset P} Q_I(R)$$

Now, since all $Q_I(R)$ (and $Q_{R-P}(R)$) inject canonically into Q(R) it is easy to see that $Q_{R-P}(R) \subset \bigcup_{I \not\subset P} Q_I(R) = \varinjlim_{I \not\subset P} Q_I(R)$. Indeed, if $x \in Q_{R-P}(R)$, then for some $I \not\subset P$ we have $I \subset R$, hence by definition we obtain $x \in Q_I(R)$.
Conversely, if $x \in Q_I(R)$, for some $I \not\subset P$, then we may find $K \not\subset R$ with $I \subset \mathrm{rad}(K)$ and such that $Kx \subset R$. Let us show that $K \not\subset P$, then we are done. Indeed, if $K \subset P$, then $\mathrm{rad}(K) \subset P$ and hence $I \subset P$, contradiction. This proves the assertion. ∎

X.3. Proposition. Under the same assumptions, we have

$$\Gamma(X_R, \underline{Q}_R) = R \quad .$$

Proof. By definition $\Gamma(X_R, \underline{Q}_R) = \{x \in Q(R); \exists L \in \mathcal{L}(R), Lx \subset R\}$. Now $\mathcal{L}(R) = \{R\}$, hence $Rx \subset R$, implying $x \in R$! ∎

X.4. Proposition. With assumptions as before, we have

$$Q_I(R) = \bigcap_{I \not\subset P} Q_{R-P}(R) .$$

Proof. First, let us note that $Q_I(R) \subset Q_{R-P}(R)$ for any $I \not\subset P$. Indeed, if $x \in Q_I(R)$ then we may find a twosided ideal K such that $\mathrm{rad}(K) \supset I$ and with $Kx \subset R$. But $K \not\subset P$, for otherwise $\mathrm{rad}(K) \subset P$ and $I \subset P$, contradiction.

Next, $\cap Q_{R-P}(R) \subset Q_I(R)$. For, if $x \in \cap Q_{R-P}(R)$, then for each $P \not\supset I$ we may find an ideal $I_P \not\subset P$ such that $I_P x \subset R$. Hence, if $K = \sum_{P \in X_I} I_P$, then

$$Kx \subset R .$$

Now, $K \in \mathcal{L}(I)$ for otherwise $\mathrm{rad}(\Sigma I_P) \not\supset I$, which leads to a contradiction as follows: if $Q \supset \Sigma I_P$, then $Q \notin X(I)$, for if $Q \in X(I)$, then $I_Q \not\subset Q$ implies $\Sigma I_P \not\subset Q$, so $Q \supset I$, hence

$$\mathrm{rad} \sum_{P \in X(I)} I_P = \bigcap_{Q \supset \Sigma I_P} Q \supset I ,$$

a contradiction. This proves the result. ∎

X.5. Corollary. The presheaf $\underline{Q}_R$ on Spec(R) is a sheaf. ∎

We will write $\underline{O}_R$ for this sheaf independent of the fact whether it is considered on $X_R = \mathrm{Spec}(R)$ or $X_R = \Omega(R)$.

X.6. Proposition. If R is an affine prime P.I. algebra over k, then

$$(\Omega_n(R), \underline{O}_R|_{\Omega_n(R)})$$

is a locally ringed space with global sections

$$\Gamma(\Omega_n(R), \underline{O}_R) = Q_{J_{n-1}}(R),$$

where $J_n \subset R$ is the ideal which is the radical in R of the variety of $n \times n$ matrices.

Proof. This follows immediately from Rowen[152], E. Nauwelaerts [124]. ∎

Like in the left noetherian case we know that $Q_I(R)$ is a twosided R-module, so we may construct

$$Q_I^{bi}(R) = bi(Q_I(R)).$$

If we assign $Q_I^{bi}(R)$ to the open subset $X(I)$ of X_R, then we obtain, with obvious restriction morphisms a separated presheaf of rings on X_R, denoted by $\underline{Q}_R^{bi}$. It is in general not a sheaf; its sheafification will be denoted by $\underline{O}_R^{bi}$, with these assumptions we have :

<u>X.6'. Proposition.</u> Under these assumption we have

$$\Gamma(X_R, \underline{O}_R^{bi}) = R.$$

<u>Proof.</u> Similar to the left Noetherian case.∎

<u>X.7. Proposition.</u> For each $P \in X_R$ we have

$$\underline{O}_{R,P}^{bi} = Q_{R-P}^{bi}(R) \quad .$$

<u>Proof.</u> Exactly as in the left noetherian case. ∎

<u>X.8. Proposition.</u> If R is an arbitrary PI ring, then $Q_{R-P}(R/P) = R/P = Q(R/P)$ for any $P \in \Omega(R)$.

<u>Proof.</u> Easy modification of the Noetherian case.

In the same vein, but perhaps technically more complicated, let us include a proof for the following

<u>X.9. Proposition.</u> If $P \in \Omega(R)$, there is a canonical morphism

$$\pi_{R,P} : Q_{R-P}(R) \to R/P,$$

uniquely extending $\pi : R \to R/P$.

<u>Proof.</u> Consider the following diagram of $\overline{R}$-modules, where $\overline{R} = R/P$:

$$\begin{array}{ccc} R/P = \overline{R} & \xrightarrow{\overline{j}_P} & Q_{R-P} \\ & {\scriptstyle j} \searrow \quad \swarrow {\scriptstyle \varphi_P} & \\ & E(R/P) & \end{array}$$

Here $E(R/P)$ is an injective hull of R/P in $\overline{R}$-mod, while φ is an extension of j to $Q_{R-P}(R)/P$. Now, if $x \in Q_{R-P}(h)/P$, then we may find $I \not\subset P$ such that $Ix \subset \overline{R}$, hence $\overline{I}x \subset \overline{R}$ implying that $\overline{I}\varphi_P(x) \subset \overline{R}$. But, since P is maximal $\overline{I} = I + P/P = R/P = \overline{R}$, hence $\overline{R}\,\varphi_P(x) \subset \overline{R}$, implying $\varphi_P(x) \in R/P$, i.e. actually this factorization yields an $\overline{R}$-linear morphism $Q_{R-P}(R)/P \to R/P$, which is easily seen to be surjective. Composing with the canonical map $Q_{R-P}(R) \to Q_{R-P}(R)/P$, we obtain a morphism $Q_{R-P}(R) \to R/P$, which is R-linear. Let us show that it is unique in the commutative diagram

$$\begin{array}{ccc} R & \xrightarrow{\;j_P\;} & Q_{R-P}(R) \\ & {\scriptstyle\pi}\searrow \quad \swarrow{\scriptstyle \pi_{R,P}} & \\ & R/P & \end{array}$$

Assume that $\varphi_1, \varphi_2 : Q_{R-P}(R) \to R/P$ both have the property that $\varphi_1 j_P = \pi = \varphi_2 j_P$, then for $\psi = \varphi_1 - \varphi_2$ we have $\psi j_P = o$. If $x \in Q_{R-P}(R)$ we may find $I \not\subset P$ such that $Ix \subset R$, hence $o = \psi j_P(Ix) = I\psi(x) = o!$ So we find $\psi = o$, or $\varphi_1 = \varphi_2$, implying that $\pi_{R,P}$: $Q_{R-P}(R) \to R/P$ is the unique map extending π to $Q_{R-P}(R)$. To show that $\pi_{R,P}$ is a ring morphism, let $x,y \in Q_{R-P}(R)$ and choose $I \not\subset P$ such that $Ix \subset R$, then for each $i \in I$ we have by R-linearity

$$i\pi_{R,P}(xy) = \pi_{R,P}(i(xy)) = \pi_{R,P}((ix)y) = \pi_{R,P}(ix)\pi_{R,P}(y) = i\pi_{R,P}(x)\pi_{R,P}(y),$$

i.e. $I(\pi_{R,P}(xy) - \pi_{R,P}(x)\pi_{R,P}(y)) = o$, which yields $\pi_{R,P}(xy) = \pi_{R,P}(x)\pi_{R,P}(y)$, since $I \not\subset P$ yields that $Iz = o$ implies $z = o$ for any $z \in \overline{R}$. This proves our assertions. ∎

We now define an <u>affine</u> <u>k-quasivariety</u> to be a varietal space which is strongly isomorphic to a varietal space

$$\Omega(R) = (X_R, \underline{O}_R, \underline{O}_R^{bi}, \underline{K}_R, \pi_R) ,$$

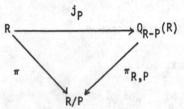

where R is an arbitrary affine prime P.I. algebra over k, where $\underline{O}_R$ and $\underline{O}_R^{bi}$ are the structure sheaves on $X_R = \Omega(R)$ we have constructed above, where $\underline{K}_R$ is the sheaf of simple rings on X_R obtained from the family $\{Q(R/P);\ P\in\Omega(R)\}$ in the usual way, and where finally $\pi_R : \underline{O}_R \to \underline{K}_R$ is obtained from the family

$$\{\pi_{R,P} : Q_{R-P}(R) \to Q(R/P)\ ;\ P\in X_R\}$$

à la Grothendieck [73]. An <u>algebraic</u> k-<u>quasivariety</u> is then a couple (X,σ) where $X=(X,\ \underline{O}_X,\underline{O}_X^{bi},\ \underline{K}_X,\pi_X)$ is a varietal space, whose underlying topological space X is covered by a finite number of open subsets U_α of X with the property that 1. the open subsets U_α are pairwise non-disjoint;

2. the induced geometric space $X_\alpha = X|U_\alpha$, where

$$X|U_\alpha = (U_\alpha,\ \underline{O}_X|U_\alpha,\ \underline{O}_X^{bi}|U_\alpha,\ \underline{K}_X|U_\alpha,\ \pi_X|U_\alpha)$$

is for each index α an (irreducible) affine algebraic k-variety, isomorphic to some $\Omega(R_\alpha)$, where R_α is a prime affine P.I. algebra, and where $\sigma : X\to\Omega(k)$ is a (strong) morphism of varietal spaces.

<u>X.10. Examples</u>. Very useful examples are of course given by

$$\mathbb{A}_R^n = \Omega(R[X_1,\dots,X_n])$$

and

$$\mathbb{P}_R^n = \mathrm{Proj}(R[X_0,\dots,X_n])$$

for any affine prime P.I. algebra R. Other examples are the affine quasivarieties

$$\mathbb{A}_n^m(k) = \Omega(k\{\xi_1^{(n)},\dots,\xi_m^{(n)}\})$$

associated to algebras of generic matrices.

<u>X.11. Proposition</u>. The underlying topological space of an algebraic k-quasivariety is compact and irreducible.

<u>Proof</u>. It is obvious that only the fact that the space X_R is compact should be verified. Now it clearly suffices to show that open sets of X_R satisfy the ascending chain condition. Let

$$X(I_1) \subset X(I_2) \subset \ldots \subset X(I_n) \subset \ldots$$

be such a chain, yielding a chain of ideals of R

$$\mathrm{rad}(I_1) \subset \mathrm{rad}(I_2) \subset \ldots \subset \mathrm{rad}(I_n) \subset \ldots$$

Recall now that it has been proved by Procesi ([136]) that any (semi) prime affine P.I. ring satisfies the ascending chain condition on semiprime ideals, thus proving the assertion. ∎

Other properties mimicing these obtained in the noetherian case may be derived similarly, their verification is left to the reader. A weak (strong) morphism of algebraic k-quasivariety is just a weak (strong) morphism of the underlying varietal spaces, which respects the k-structure on these quasivarieties. We then have :

X.12. Proposition. An injective extension $\varphi : R \to S$ of affine prime PI algebras induces a strong morphism $\Omega(S) \to \Omega(R)$ of affine k-quasivarieties.

Proof. Most of the proof is only routine. Let us just indicate that for any ideal I of R with image $J = (\varphi(I)) \subset S$ we obtain a commutative diagram

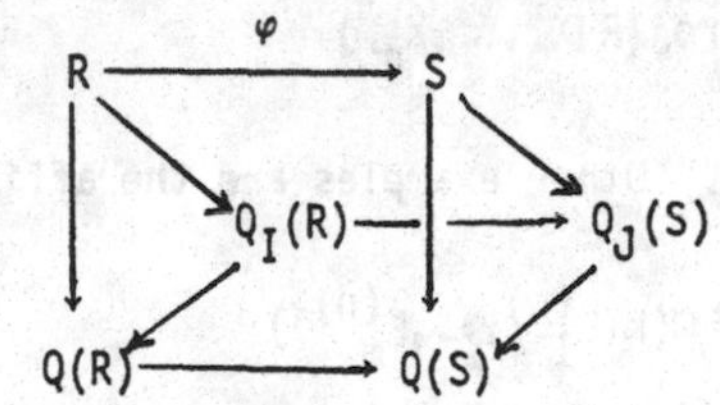

The morphism $Q(R) \to Q(S)$ extending φ is well-known to exist, in view of (II.2.11). Now, if $x \in Q_I(R)$ then for some twosided ideal K of R with $I \subset \mathrm{rad}(K)$ we have $Ix \subset R$. If we view φ as an inclusion, we have $(\varphi(K)) = SK$, hence $(\varphi(K))x = SKx \subset S$, so it suffices to check that $J \subset \mathrm{rad}(SK)$ If $Q \in \mathrm{Spec}(S)$ with $Q \supset SK$, then $Q \cap R \supset K$, hence $Q \cap R \supset I$ since $Q \cap R$ is prime, φ being an extension. But $Q \supset I$ implies $Q \supset SI = J$, proving the assertion. ∎

On the other hand, an arbitrary extension $\psi : R \to S$ does not in general induce a morphism of quasivarieties $\Omega(S) \longrightarrow \Omega(R)$, due to the absence of idempotency of certain occuring kernel functors. This leads us to define a rational map between varietal spaces to be an equivalence class of pairs (U, φ_U), where U is a nonempty open subset of X and $\varphi_U : U \to Y$ a morphism between the corresponding varietal spaces, two such pairs (U,φ_U) and (V,φ_V) being equivalent if there is an open nonempty $W \subset U \cap V$ such that φ_U and φ_V coincide on W. Note that this is actually an equivalence relation. We will write

$$\varphi : X \hookrightarrow Y$$

when $\varphi = (U,\varphi_V)$ is a rational map. We say that $\varphi : X \hookrightarrow Y$ is dominant when for some (U, φ_U) the image $\varphi_U(U)$ is dense Y.

X.13. Lemma. If $\varphi : X \to Y$ is dominant, then for all (U, φ_U) representing φ, the image $\varphi_U(U)$ is dense in Y.

Proof. Indeed, it is sufficient to show that if $\overline{\varphi_U(U)} = Y$, then for all $\phi \neq W \subset V$ we have $\overline{\varphi_U(W)} = Y$ as well. Now, $\overline{\varphi_U(U)} = Y$ means that for all $\phi \neq V \subset Y$ we have $\varphi_U(U) \cap V \neq \phi$, hence $U \cap \varphi_U^{-1}(V) \neq \phi$, so $W \cap \varphi_U^{-1}(V) \neq \phi$ implying $\varphi_U(W) \cap V \neq \phi$ i.e. $\overline{\varphi_U(W)} = Y$. ∎

It is now clear that dominant rational maps may be composed. Let $\varphi : X \hookrightarrow Y$ and $\psi : Y \hookrightarrow Z$ be dominant rational maps, represented by $\varphi_U : U \to Y$ and $\psi_V : V \to Z$. Let $W = U \cap \varphi_U^{-1}(V)$, which is nonempty since φ is dominant and let $\theta_W = \psi_V \circ \varphi_U|_W : W \to Z$, then θ_W represents a rational map, which is itself dominant

$$\theta : X \hookrightarrow Z ,$$

θ is clearly independent of the particular choice of φ_U and ψ_V as one easily verifies.

The class of all quasivarieties and dominant rational maps defines a category $\underset{\sim}{Q}$. An isomorphism in $\underset{\sim}{Q}$ is called a birational map. Hence a birational map $\varphi : X \to Y$

is a rational map which admits an inverse $\psi : Y \hookrightarrow X$ in $\mathcal{Q}$.

X.14. Proposition. Any birational extension $R \to S$ between affine prime PI algebras defines a birational map

$$\Omega(S) \hookrightarrow \Omega(R) .$$

Proof. This is an immediate consequence of II.1 . ∎

Define the function ring of a *varietal space* X to be

$$R(X) = \varinjlim_{\phi \neq U \in \mathrm{Open}(X)} \Gamma(U, \underline{O}_X)$$

If $\phi \neq U \subset X$ is open, then we have $R(U) = R(X)$ if U is given the induced structure of a varietal space. If R is an affine prime P.I. algebra then for X_R we have

$$R(X_R) = \varinjlim_{\phi \neq U \in \mathrm{Open}(X_R)} \Gamma(U, \underline{O}_R) = \varinjlim Q_I(R) = Q_{R-o}(R) = Q(R).$$

This shows in particular that for any algebraic k-quasivariety X the function ring R(X) is simple, which should be viewed as the noncommutative counterpart of the existence of a field of functions (= function field) in the commutative case. Using Procesi's results on the Krull dimension of prime PI rings, it is now possible to build a dimension theory for quasivarieties by putting

$$\dim X = \mathrm{tr.deg}_k \, Z(R(X))$$

Details are left to the reader.

X.15. Proposition. Any dominant rational map $X \to Y$ induces in a canonical way a k-algebra extension $R(Y) \to R(X)$.

Proof. It is obvious that we may restrict to the affine case. So, let

$$\varphi : \Omega(S) \hookrightarrow \Omega(R)$$

be a dominant rational map defined on $Y(J) \subset \Omega(S)$ by

$$\varphi_J : Y(J) \to \Omega(R) \quad .$$

Take $f \in Q(R)$, then $f = rc^{-1}$ for some $r \in R$ and $c \in C = Z(R)$; hence $cRf \subset R$, yielding $f \in Q_c(R) = \Gamma(X(c), \underline{O}_R) = \Gamma(X(c), \underline{O}_R^{bi})$. Now, since φ is dominant

$$\varphi_J(Y(J)) \cap X(c) \neq \phi,$$

so, let $\varphi_J^{-1}(X(c)) \cap Y(J)$ be the nonempty open set $Y(I)$ of $\Omega(S)$. Then $\varphi_J(Y(I)) \subset X(c)$, hence the induced map

$$\varphi_I =: \varphi_J | Y(I) : Y(I) \to \Omega(R)$$

induces over $X(c)$ a ringextension

$$\tau_c : R_c = \Gamma(X(c), \underline{O}_R) = \Gamma(X(c), \underline{O}_R^{bi}) \to \Gamma(Y(I), \underline{O}_S^{bi}) \subset Q(S)$$

and we let $\tau_c(f)$ be the element of $Q(S)$ corresponding to $f \in Q(R)$. It is easy to see that $\tau_c(f)$ is actually independent of c, for if $f \in Q_c(R)$ and $f \in Q_{c'}(R)$, then and we find $\tau_c(f) = \tau_{cc'}(f) = \tau_{c'}(f)$. Moreover, the map $i : Q(R) \to Q(S)$ thus obtained is a ring morphism. Indeed, choose f and g in $Q(R)$, then we may find a common subring $Q_c(R)$ containing both f and g : if $f = r_1c_1^{-1}$ and $g = r_2c_2^{-1}$, then we may take $c = c_1c_2$. Now the image $i(fg)$ of fg is just $\tau_c(fg) = \tau_c(f)\tau_c(g) = i(f)i(g)$. Finally that i is an extension is obvious from the fact that the morphisms τ_c are extensions for each $c \in C$. ∎

X.16. Remark. It is obvious that $i : Q(R) \to Q(S)$ is injective. Indeed, $\mathrm{Ker}(i)$ is a twosided ideal of $Q(R)$, hence $\mathrm{Ker}\ i = o$ or $\mathrm{Ker}\ i = Q(R)$, the latter being excluded for pertinent reasons.

The foregoing may be applied as follows. Let R be an affine prime P.I.algebra over k, say of p.i.degree n. Then we know that we may find a positive integer m such that there is a surjective morphism

$$\pi : k\{\xi_1^{(n)},\dots,\xi_m^{(n)}\} \to R .$$

This map does not yield a morphism of algebraic k-varieties since even if R is chosen to be left noetherian, this is certainly not true for $k\{\xi_1^{(n)},\dots,\xi_m^{(n)}\}$. Yet, since π is surjective, it certainly is a central extension. So, by [29] we know that if $P \in \mathrm{Spec}_p(R)$, then $\pi^{-1}(P) \in \mathrm{Spec}_p(k\{\xi_1^{(n)},\dots,\xi_m^{(n)}\})$. In particular this is also true when restricted to Ω_p, so that we obtain for $p = n$ a continuous map

$$\Omega_n(R) \to \Omega_n(k\{\xi_1^{(n)},\dots,\xi_m^{(n)}\}) .$$

Now note that these spaces are open subsets of $\Omega(R)$ and $\Omega(k\{\xi_1^{(n)},\dots,\xi_m^{(n)}\})$ and that we may extend this map to a morphism of the induced varietal spaces, since Ω_n is just the open set of birationality with the center for these affine k-algebras. It follows that we obtain a rational map

$$\Omega(R) \to |\mathbb{A}_n^m(k)| ,$$

which may be represented topologically by an injective continuous map. We will call this a rational injection. Hence

X.17. Proposition. Each affine k-quasivariety injects rationally into some $|\mathbb{A}_n^m(k)|$ for suitably chosen positive integers m,n. ∎

Note that this rational injection does obviously not yield a map between the corresponding function rings. Indeed, being a closed embedding, this injection is never dominant, unless it is the identity of course!

Note also that we have actually proved

X.18. Proposition. Any central extension $\varphi : R \to S$ between affine prime P.I. algebras of the same pi degree induces a rational map

$$\Omega(S) \to \Omega(R). \blacksquare$$

X.19. Proposition. Any (injective) k-algebra extension $R(Y) \to R(X)$ defines a dominant rational map

$$X \hookrightarrow Y \quad .$$

Proof. Again it is clear that we may restrict to the affine case. So, let $\theta : Q(R) \to Q(S)$ be a k-algebra extension. Let $R = k\{x_1, \dots, x_n\} \subset Q(R)$ and put $f_i = \theta(x_i) \in Q(S)$. Clearly we may find c_i such that $f_i \in Q_{c_i}(S)$, since f_i is of the form $s_i\, c_i^{-1}$ for some $s_i \in S$ and $c_i \in D = Z(S)$. If we put $c = c_1 \dots c_n$, then $f_i \in Q_c(S)$ for any $1 \leq i \leq n$. We thus get by restriction a map $R \to Q_c(S)$ which makes the following diagram commutative

$$\begin{array}{ccc} Q(R) & \xrightarrow{\ \theta\ } & Q(S) \\ \uparrow & & \uparrow \\ R & \longrightarrow & Q_c(S) \end{array}$$

This map is not necessarily an extension! However, we may proceed as follows, to enlarge $Q_c(S)$ a little in order to make the induced map an extension. Since S is an affine k-algebra, i.e. $S = k\{g_1, \dots, g_m\}$ for a finite number of k-commuting elements $g_1, \dots, g_m \in S$ and since θ is an extension, we know that each of these generators g_p may be written as

$$g_p = \sum_q r_{pq} \frac{s_{pq}}{d_{pq}} ,$$

where $r_{pq} \in R$ and $s_{pq}\, d_{pq}^{-1} \in Z_R(Q(S))$ with $s_{pq} \in S$ and $d_{pq} \in D$. Let $d = \prod_{p,q} d_{pq} \in D$ and consider the induced morphism

$$R \longrightarrow Q_c(S) \longrightarrow Q_{cd}(S) \quad .$$

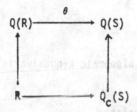

We claim that it is an extension. Indeed, first note that $S \subset RZ_R(Q_{cd}(S))$, for any element of S may be written as a polynomial in the g_p and each of these is a linear combination of elements of R (the $r_{pq}!$) and of $Z_R(Q_{cd}(S))$ (the $s_{pq}\ d_{pq}^{-1}!$). Next, we easily see that $(cd)^{-1} \in Z_R(Q_{cd}(S))$, since $cd \in D$. So we find that $RZ_R(Q_{cd}(S)) = Q_{cd}(S)$, which shows that $R \to Q_{cd}(S)$ is an extension. Now this extension corresponds to a morphism of affine quasivarieties

$$\Omega(Q_{cd}(S)) \longrightarrow \Omega(R)$$

and it suffices to remark that $\Omega(Q_{cd}(S))$ is an (affine) open subset of $\Omega(S)$ to show that we thus obtain a rational map

$$\Omega(S) \longmapsto \Omega(R) \ ,$$

which proves the assertion. ∎

X.20. Corollary. For any pair of algebraic k-quasivarieties X and Y the following assertions are equivalent :

1. X and Y are birationally equivalent;
2. there are open subsets $U \subset X$ and $V \subset Y$ such that the varietal spaces U and V are isomorphic;
3. the k-algebras R(X) and R(Y) are isomorphic. ∎

Note that the foregoing gives an equivalence between the category of algebraic k-quasivarieties with dominant rational maps and the category of central simple algebras finite dimensional over their center which is a finitely generated field extension of k with extensions. This follows from a well-known result of Procesi's [136] and the foregoing.

X.21. Comments for Chapters VIII, IX, X.

Whereas the theory of coherent sheaves over algebraic k-varieties does not seem to be fully satisfactory unless the varieties considered are Zariski central varieties, the theory of geometric products, closed subvarieties and birationality works rather well and provides a geometrical setting worthy of that name. Of course the theory is by far not complete at this stage e.g. what about coherent sheaves over Proj and coherent ideals etc.... Indeed much of the projective theory remains to be settled.

On the other hand, our sheaf theoretic methods did give us some new insights. In particular the calculation of $\underline{O}^{bi}_{R,P}$ at a point P where the localization is non central contains new information relating then non-Azumayaness of $\underline{O}^{bi}_{R,P}$ to the defect of κ_{R-P} being a non-central kernel functor. Any possible value of the non-commutative geometry should perhaps be weighted in a very intuitive case i.e. the case of "curves". In the next chapter we shall try to show how the non-commutative Riemann-Roch theorem for curves is linked to the arithmetical theory of certain central simple algebras and in an addendum we point out some applications which are the subject of recent research.

References for the somewhat technical chapters VIII, IX and X are A. Verschoren's thesis [194] and the papers by F. Van Oystaeyen and A. Verschoren [183], [189], [195]. We have heavily used C. Procesi's results [130], [136] and also some of M. Artin's and W. Schelter's, [20], [21].

XI. A NON-COMMUTATIVE VERSION OF THE RIEMANN-ROCH THEOREM FOR CURVES

XI.1. Rational Points.

As always k is an algebraically closed field. Let k_o be an arbitrary subfield of k. The theory developed th far deals with algebraic k-(quasi) varieties, but we will now also consider algebraic (quasi) varieties over k_o.

An affine k-variety $\Omega(R)$ is defined over k_o if there is an affine k_o-algebra R_o (i.e. a finitely generated prime P.I. k_o-algebra) such that the varietal spaces $\Omega(R)$ and $\Omega(R_o) \underset{k_o}{x} k$ are isomorphic, where the latter stands for the product of $\Omega(R_o)$ and $\Omega(k)$ as $\Omega(k_o)$-spaces (since k and k_o are commutative this is indeed a genuine fibre product). Reformulating this definition : there is an affine k_o-algebra such that $R \cong R_o \underset{k_o}{\otimes} k$, as k-algebras. For $\Omega(R_o)$ to be a varietal space with the usual properties it is necessary that R_o is a Hilbert algebra but since R_o is an affine k_o-algebra this presents no problem.

An algebraic k-(quasi) variety X is defined over k_o if there is an "algebraic k_o-(quasi) variety" X_o and a k-isomorphism of varietal spaces $X \cong X_o \underset{k_o}{x} k$. The meaning of "algebraic k_o- (quasi) variety" is clear, i.e. a varietal space obtained by glueing together affine k_o-varieties. As a matter of fact if X_o is defined using the affine covering $\{U_\alpha = \Omega(R_\alpha), \alpha \in A\}$ where for each $\alpha \in A$, R_α is an affine k_o-algebra, then glueing together the opens $U_\alpha \underset{k_o}{x} k = \Omega(R_\alpha \underset{k_o}{\otimes} k)$, $\alpha \in A$, yields $X = X_o \underset{k_o}{x} k$.

For simplicity's sake all our varieties will be assumed to be precellular.

A point $x \in X_o$ of an algebraic k_o-(quasi) variety X_o is rational over k_o if $\mathbb{k}(x)$ is a central simple k_o-algebra. Let $X_o(k_o)$ be the set of k_o-rational points of X_o. If R is an affine k_o-algebra we write $\Omega_{k_o}(R)$ for the set of all k_o-rational points of $\Omega(R)$.

XI.1.1. Proposition. Any extension $\varphi : R \to S$ of affine k_o-algebras defines a morphism $\Omega_{k_o}(S) \to \Omega_{k_o}(R) : M \mapsto \varphi^{-1}(M)$.

Proof. If $M \in \Omega_{k_o}(S)$ then $S/M = Q(S/M)$ is a central simple k_o-algebra i.e. $k_o = Z(S/M)$. If $P = \varphi^{-1}(M)$ then we obtain inclusions of k_o-algebras : $k_o \to R/P \to S/M$. Since these inclusions are extensions we obtain inclusions : $k_o \to Z(R/P) \to Z(S/M) = k_o$, hence $Z(R/P) = k_o$. It follows that R/P is simple, P is maximal and $P \in \Omega_{k_o}(R)$. ■

In the sequel X is defined over k_o, $X = X_o \times_{k_o} k$. Any k_o-automorphism σ of k defines a conjugation map $\sigma_X : X \to X$, which restricts to the identity on X_o. The action of σ_X on $\Omega(k)$ is the map $\Omega(k) \to \Omega(k)$ derived from $\sigma^{-1} : k \to k$. Equivalently, for any affine U_α in a covering defining X, say $U_\alpha = \Omega(R_\alpha \otimes_{k_o} k)$ for an affine k_o-algebra R_α, we define $\sigma_\alpha : \Omega(R_\alpha \otimes_{k_o} k) \to \Omega(R_\alpha \otimes_{k_o} k)$, as being the variety map associated to the k_o-algebra morphism, $R_\alpha \otimes \sigma^{-1} : R_\alpha \otimes_{k_o} k \to R_\alpha \otimes_{k_o} k$ (which is an extension!). It is an easy exercise to verify that the local components σ_α globalize to σ_X while σ_X is independent of the selected affine covering of X. Since $(\sigma\tau)_X = \sigma_X\tau_X$, it follows that we thus obtain an action of $Gal(k/k_o)$ on X.

XI.1.2. Proposition. σ_X is given by coordinate-wise conjugation.

Proof. We may assume $X_o = \Omega(R_o)$ to be affine. So $X_o = \Omega(R_o)$ with $R_o = k_o\{\xi_1^{(m)}, \dots, \xi_n^{(m)}\}/I_o$ and $X = X_o \times_{k_o} k = \Omega(R)$. Elements of $\Omega(R)$ correspond bijectively to equivalence classes of central extensions $R \to M_p(k)$ where $1 < p < \frac{d}{2}$ with $d = $ p.i. deg R. Since $R \cong k\{\xi_1^{(m)}, \dots, \xi_n^{(m)}\}/I$ with $I = I_o k$, a central extension as before will be completely determined by specifying the images of the $\xi_i^{(m)}$ mod I in $M_p(k)$. Thus a point x of $\Omega(R)$ is just a class $[\mu_1, \dots, \mu_n]$ with $\mu_i \in M_p(k)$ for $i \in \{1, \dots, n\}$, such that $k\{\mu_1, \dots, \mu_n\} = M_p(k)$ and such that for each $f = f(\xi_1^{(m)}, \dots, \xi_n^{(m)}) \in I$ the specialization $\xi_i^{(m)} \to \mu_i$ annihilates f. The action of $\sigma \in Gal(k/k_o)$ on $[\mu_1, \dots, \mu_n]$ may be defined as follows : to $[\mu_1, \dots, \mu_n]$ there corresponds $M \in \Omega(R)$, to $\sigma_X(M)$ we have associated $[\nu_1, \dots, \nu_n]$, so we put $\sigma[\mu_1, \dots, \mu_n] = [\nu_1, \dots, \nu_n]$. On the other hand σ acts on $M_p(k)$ by conjugation on each entry.

Claim : for each $\sigma \in \mathrm{Gal}(k/k_o)$, $\sigma[\mu_1,\dots,\mu_n] = [\sigma\mu_1,\dots,\sigma\mu_n]$. Since the action of $\mathrm{Gal}(k/k_o)$ on $\Omega(R)$ extends to $\mathbb{A}_k^{m,n}$ in the obvious way (where $\Omega(R) \to \mathbb{A}_k^{m,n}$ is the affine embedding corresponding to $R \cong k\{\xi_1^{(m)},\dots,\xi_n^{(m)}\}/I$), it suffices to establish the claim in case $R = k\{\xi_1^{(m)},\dots,\xi_n^{(m)}\}$.

Consider the following diagram :

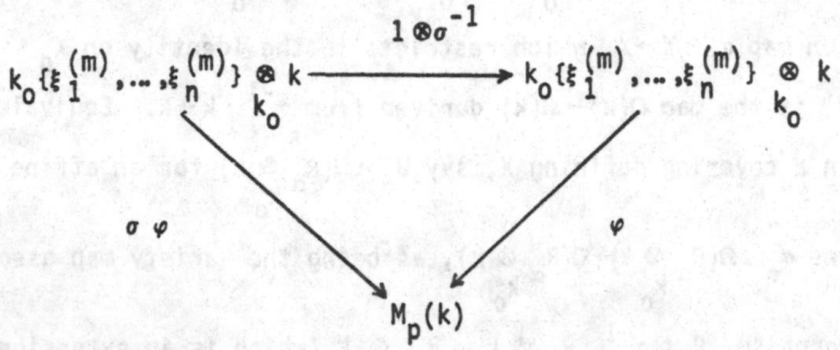

The map σ_X associates to the class of any φ, the class of $\sigma\varphi = \varphi \circ (1 \otimes \sigma^{-1})$. Using multi-indices for notational convenience, we have :

$$\sigma\varphi\left(\sum_{\underline{\alpha}} \lambda_{\underline{\alpha}} \xi_{\underline{\alpha}}^{(m)}\right) = \varphi \circ (1 \otimes \sigma^{-1})\left(\sum_{\underline{\alpha}} \lambda_{\underline{\alpha}} \xi_{\underline{\alpha}}^{(m)}\right)$$

$$= \varphi\left(\sum_{\alpha} \lambda_{\underline{\alpha}}^{\sigma^{-1}} \xi_{\underline{\alpha}}^{(m)}\right) = \sum_{\underline{\alpha}} \lambda_{\underline{\alpha}}^{\sigma^{-1}} \mu_{\underline{\alpha}} = \sigma^{-1}\left(\sum_{\underline{\alpha}} \lambda_{\underline{\alpha}} (\sigma\mu)_{\underline{\alpha}}\right)$$

Now if $\sum_{\underline{\alpha}} \lambda_{\underline{\alpha}} \xi_{\underline{\alpha}}^{(m)} \in \mathrm{Ker}(\sigma\varphi)$ then $\sum_{\underline{\alpha}} \lambda_{\underline{\alpha}} (\sigma\mu)_{\underline{\sigma}} = 0$ and conversely. This shows that $(\sigma\mu_1,\dots,\sigma\mu_n)$ does represent $\sigma_X(x)$ if φ represents x (or equivalently: if x corresponds to $[\mu_1,\dots,\mu_n]$ then $[\sigma\mu_1,\dots,\sigma\mu_n]$ corresponds to $\sigma(x)$).

<u>XI.1.3. Proposition.</u> If X_o is an algebraic k_o-(quasi) variety then for each point $x \in X_o$, the center $Z(\mathbb{k}_{X_o}(x))$ of the function ring at x is algebraic over k_o.

<u>Proof.</u> We may assume that X_o is affine i.e. $X_o = \Omega(R_o)$ for some affine k_o-algebra R_o. The structural inclusion $i : k \to R_o$ is a finitely generated extension of P.I. algebras and from Section II.4. it follows that for any maximal ideal M_o of R_o,

R_o/M_o is finite dimensional over $k_o = Z(k_o/k_o \cap M_o)$, (actually the result used is V.3.4 of [156]) hence so is $Z(R_o/M_o) = Z(\mathbb{k}_R(M_o))$. ∎

XI.1.4. Proposition. Let $X = X_o \times_{k_o} k$ be defined over k_o and let $p : X \to X_o$ be the canonical projection morphism.

1. If $X_o = \Omega(R_o)$ then p is just the variety map associated to the canonical inclusion $R_o \to R_o \otimes_{k_o} k$.

2. The map p is surjective.

Proof. 1. Obvious.

2. Reduce to the affine case, i.e. put $X_o = \Omega(R_o)$. If $M \in \Omega(R)$ and if $M \otimes_{k_o} k$ were non-maximal in $R \otimes_{k_o} k$ then it is included in some maximal ideal N of $R \otimes_{k_o} k$. Since $M = (M \otimes_{k_o} k) \cap R$ it follows that $N \cap R = M$ i.e. $p(N) = M$. ∎

XI.1.5. Proposition. Let R_o an affine k_o-algebra and $Z(R_o) = C_o$. Then $M \in \Omega_{k_o}(R_o)$ yields that $M \cap C \in \Omega_{k_o}(C)$.

Proof. Since R_o is a Jacobson ring and a finitely generated P.I. algebra over C_o we find that maximal ideals of R_o restrict to maximal ideals of C_o. The assumption $M_o \in \Omega_{k_o}(R_o)$ yields that $k_o = Z(R_o/M_o)$, therefore the monomorphic extensions $k_o \to C_o \to R_o$ induce extensions $k_o \to C/M_o \cap C_o \to R_o/M_o$. Consequently $k_o = C_o/M_o \cap C_o$ and the assertion follows.

XI.1.6. Corollary. Any M_o of p.i. degree equal to the p.i. degree of R_o satisfies $M_o \in \Omega_{k_o}(R_o)$ if and only if $M_o \cap C_o \in \Omega_{k_o}(C_o)$. As we will see below, this statement is false if the p.i. degree of M_o is not maximal.

XI.1.7. Proposition. If $y \in X_o$ has p.i. degree n then there is a bijective correspondence between points $x \in p^{-1}(y)$ and the set of equivalence classes (mod $PGL_n(k)$) of injective central k_o-extensions $\mathbb{k}_{X_o}(y) \to M_n(k)$.

<u>Proof.</u> It suffices to establish the affine case. Note that any $x\in p^{-1}(y)$ also has p.i. degree n. Indeed, $p:\Omega(R_o \otimes_{k_o} k)\to\Omega(R_o)$ is associated to the canonical $R_o\to R_o\otimes_{k_o} k$. Then, if $y=M_o$, we have that $N\in p^{-1}(y)$ if and only if $R_o\cap N=M_o$. Since $R_o\to R_o\otimes_{k_o} k$ is a central extension : p.i. $\deg(R_o\otimes_{k_o} k/N)$ = p.i. $\deg(R_o/M_o)$. Now, since k is algebraically closed, it follows that $(R_o\otimes_{k_o} k)/N\cong M_n(k)$. Define $i_N: R_o/N\cap R_o = \mathbb{k}_{X_o}(M_o)\to \mathbb{k}_X(N) = (R_o\otimes_{k_o} k)/N$ in the obvious way; then i_N is an injective central extension. We have the following commutative diagram of extensions :

$$\begin{array}{ccccc} & & k_o & \longrightarrow & \mathbb{k}_{X_o}(M_o) = \mathbb{k}(y) \\ & & \downarrow & & \downarrow i_N \\ k & \longrightarrow & M_n(k) & \xrightarrow{\cong} & \mathbb{k}_X(x) = \mathbb{k}_X(N) \end{array} .$$

This yields an injective central extension $\mathbb{k}(y)\to M_n(k)$ and determines an equivalence class modulo action of the k-automorphism group of $M_n(k)$. Conversely, such a class is represented by an injective central k_o-extension :

$$\begin{array}{ccc} \mathbb{k}(y) & \longrightarrow & M_n(k) \\ & \searrow \quad \swarrow & \\ & k_o & \end{array}$$

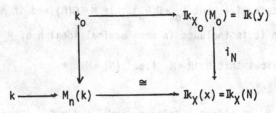

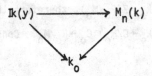

Therefore we have a central extension $\alpha: R_o\to R_o/M_o = \mathbb{k}(y)\to M_n(k)$ hence a morphism $\psi:\Omega(M_n(k))\to\Omega(R_o)$. If e is the unique point of $\Omega(M_n(k))$ then $\mathrm{Im}\,\psi=\psi(e)=M_o$ and the corresponding morphism $Q_{R_o-M_o}(R_o)\to M_n(k)$ is just the composition $Q_{R_o-M_o}(R_o)\to R_o/M_o=\mathbb{k}(y)\to M_n(k)$. The central extension α obviously yields a central extension $R_o\otimes_{k_o} k\to M_n(k)$ and the latter defines a morphism $\varphi:\Omega(M_n(k))\to$ $\to\Omega(R_o)\otimes_{k_o} k$. Let $\mathrm{Im}\,\varphi = x$, then $x\in p^{-1}(y)$. It is easily checked that this construction provides an inverse to the correspondence $x \mapsto$ class of i_N.

<u>XI.1.8. Remark</u>. If n=1 then elements in $p^{-1}(y)$ correspond to k_o-monomorphisms $\mathbb{k}(y) \to M_1(k) = k$. If k_o is perfect then for any $x_1 \in X$ of p.i. degree 1 we have that $p(x_1)$ is rational over k_o if and only if x_1 is fixed under the action of $Gal(k/k_o)$. If $n \neq 1$ such a statement cannot be expected to be true since, fixing k_o in $\mathbb{k}(y) \hookrightarrow M_n(k)$ still allows the image of $\mathbb{k}(y)$ to vary over k_o (and not only by k-automorphisms of $M_n(k)$ (which does not change the class of the extension)).

Let us now investigate the following problem : what conditions on X_o ensure that $X = X_o \times_{k_o} k$ is an algebraic k-quasivariety. The problem clearly reduces to the affine case and then it is equivalent to check whether for an affine k_o-algebra R_o it is true that $R_o \otimes_{k_o} k$ is an affine k-algebra. The only thing we have to check here is whether $R_o \otimes_{k_o} k$ is prime. The following lemma is well-known (cf.[105]).

<u>XI.1.9. Lemma</u>. Let ℓ and k be field extensions of k_o and assume that k is algebraically closed, then :

1. $\ell \otimes_{k_o} k$ has no nilpotent elements if and only if ℓ is separable over k_o.
2. $\ell \otimes_{k_o} k/rad(\ell \otimes_{k_o} k)$ is a domain if and only if k_o is separably algebraically closed in ℓ.
3. $\ell \otimes_{k_o} k$ is a domain if and only if ℓ is separable over k_o and k_o is algebraically closed in ℓ.

<u>XI.1.10. Proposition</u>. Let R_o be an affine k_o-algebra with $Z(R_o) = C_o$, and let k be an algebraically closed field containing k_o. The following statements are equivalent :

1. $R_o \otimes_{k_o} k$ is a prime ring (hence an affine k-algebra)
2. $F_o = Q(C_o)$ is a separable extension of k_o and k_o is algebraically closed in $Q(C_o)$.

Proof. If $R_o \otimes_{k_o} k$ is prime then so is $C_o \otimes_{k_o} k$, because $C_o \otimes_{k_o} k \to R_o \otimes_{k_o} k$ is an extension. But then $F_o \otimes_{k_o} k$ is prime too and then it follows from XI.1.9.3. that 1. ⇒ 2.

2. ⇒ 1. By XI.1.9., we know that 2. implies that $F_o \otimes_{k_o} k$ is a domain. Since $Q(R_o)$ is a central simple F_o-algebra it may be split by a finite separable extension L_o of F_o (Chapter I). Hence, L_o is separable over k_o and $Q(R_o) \otimes_{F_o} L_o \cong M_n(L_o)$. By XI.1.9.1. it follows that $L_o \otimes_{k_o} k$ is a semiprime ring, hence so is $M_n(L_o \otimes_{k_o} k)$, where $n =$ p.i. deg R_o.

However : $M_n(L_o \otimes_{k_o} k) = M_n(L_o) \otimes_{L_o} (L_o \otimes_{k_o} k)$

$$= M_n(L_o) \otimes_{k_o} k \cong (L_o \otimes_{F_o} Q(R_o)) \otimes_{k_o} k .$$

The fact that $Q(R_o) \otimes_{k_o} k \to L_o \otimes_{F_o} Q(R_o) \otimes_{k_o} k$ is a (central) extension entails that the ring $Q(R_o) \otimes_{k_o} k$ is a semiprime ring with center $F \otimes_{k_o} k$, a domain.
In a semiprime P.I. ring, each nonzero ideal intersects the center non-trivially, therefore $Q(R_o) \otimes_{k_o} k$ must be a prime ring. The conclusion, $R_o \otimes_{k_o} k$ is prime now follows because $R_o \otimes_{k_o} k \to Q(R_o) \otimes_{k_o} k$ is a (central) extension. ∎

XI.1.11. Convention.

From hereon we shall say that X_o is an algebraic k_o- (quasi) variety for those varietal spaces over k_o such that :

1. The center of the function ring of X_o is separable over k_o.
2. The field k_o is algebraically closed in the center of the function ring of X_o.

Hence we reserve the term "algebraic k_o- (quasi) variety" for those varietal spaces X_o over k_o which have the property that $X = X_o \times_{k_o} k$ is an algebraic k-(quasi) variety (cf. XI.1.10.).

XI.2. Non-commutative Curves.

Throughout this section k is an algebraically closed field and k_o is a subfield of k. An affine curve over k_o is an algebraic k_o-quasivariety $\Omega(R)$ for some prime affine P.I. algebra R over k_o which has Krull dimension 1.

XI.2.1. Lemma (L. Small. [166]). Let $R = A\{x_1,\dots,x_n\}$ be a finitely generated extension of a commutative Noetherian ring A. If R is a semiprime P.I. ring of Krull dimension 1 then R is a finite module over its center, hence R may be embedded in a full matrix algebra over a commutative ring K.

A special type of curve, studied by M. Artin, W. Schelter in [21] arises as follows. Let D be a commutative Dedekind domain with field of fractions K and let R be a D-order in some $M_n(K)$. If D is affine over k_o then $\Omega(R)$ is an affine curve. If $k_o = k$ then any affine k-curve $\Omega(R)$ such that $C = Z(R)$ is integrally closed (a Dedekind domain) is of the form described above. Note that in case $k = k_o$, Tsen's theorem implies that Q(R), a central simple Q(C)-algebra, is of the form $M_n(K)$, $K = Q(C)$, because that theorem states that the Brauer group of a function field in one variable over an algebraically closed field reduces to {1}.

The geometric structure of k-curves is easily visualized in the case $k_o = k$. Indeed, put n = p.i. deg R, the $M \in \Omega(R)_n$ correspond to those $m \in \Omega(C)$, $C = Z(R)$, such that $R \otimes_k \mathbb{k}_C(m) = M_n(k)$. Other $p \in \Omega(C)$ are such that $R \otimes_k \mathbb{k}_C(p)$ may possibly be non-simple. However, in this case $R \otimes_k \mathbb{k}_C(p)$ is semilocal and the finite number of maximal ideals of it are in bijective correspondence with those $P \in \Omega(R)$ lying over $p \in \Omega(C)$, i.e. such that $p = P \cap C$. Clearly, the splitting of $p \in \Omega(C)$ may also be described by looking at the function rings $\mathbb{k}_R(P)$, cf. [21].

In case $k_o \subset k$ is not algebraically closed, it is still true that $\mathbb{k}_R(P)$ is a central simple algebra and a finitely generated extension of k_o. Now $P \in \Omega_n(R)$ if and only if $\mathbb{k}_R(P)$ has degree n, if and only if $P \cap C$ is non-split. If splitting does occur for $p \in \Omega(C)$ then the multiplicities of the maximal ideals P over p are accurately described by G. Bergman, L. Small's results, cf. [29].

An example, showing which difficulties may arise in handling central maximal ideals

$$\begin{pmatrix} \mathbb{R}[X,Y]/(f) & X\mathbb{R}[X,Y]/(f)^* \\ \mathbb{R}[X,Y]/(f) & \mathbb{R}[X,Y]/(f) \end{pmatrix},$$

where f is irreducible in $\mathbb{R}[X,Y]$ and $(f)^* = (f) \cap X\,\mathbb{R}[X,Y]$.

In this chapter an <u>algebraic function field K in one variable over k_o</u> is a field K containing k_o such that k_o is algebraically closed in K and K is a separable extension of k_o i.e. K is the function field of a commutative affine k_o-curve. A <u>function algebra in one variable over k_o</u> is a central simple K-algebra. By the dimension result $\cdot$ $\dim R = \text{tr.deg}_{k_o}(Q(C))$, cf.[136], it follows that for any curve $\Omega(R)$ the field of fractions $Q(Z(R))$ has transcendence degree 1 over k_o. Since R is an affine k_o-algebra it follows that $Q(Z(R))$ is finitely generated over k_o; combined with the foregoing this yields that $Q(C)$, $C = Z(R)$, is a function field in one variable over k_o. By Posner's theorem $Q(R)$ is a central simple $Q(C)$-algebra, so the following lemma is easily proved :

<u>XI.2.2. Lemma</u>. For any affine prime P.I. algebra R over k_o the following statements are equivalent :

1. $\Omega(R)$ is an affine k_o-curve.
2. $Q(R)$ is a function algebra.

Recall from Chapter X that a rational map between varietal k_o-spaces X and Y is an equivalence class of pairs (U,φ_U) where U is a nonempty open subset of X, and $\varphi_U : U \to Y$ a morphism of varietal spaces. A birational map is just an isomorphism in the category of dominant rational maps.
Note that the extension of the notion of (bi) rational maps introduced in Chapter X to varietal spaces (i.e. not just quasivarieties) presents no problems. A <u>birational invariant</u> is an integer $z(X)$ associated to a varietal space X such that $z(X) = z(Y)$ for every varietal space Y which is birationally equivalent to X. If z is a birational invariant and if R is an affine P.I. k_o-algebra with p.i. deg R=n,

then $z(\Omega_n(R)) = z(\Omega(R))$, where $\Omega_n(R)$ has the induced structure of a varietal space.

The fact that we have assumed throughout that our varietal spaces were to be irreducible implies that birationally equivalent varietal spaces have the same (isomorphic!) function ring. Therefore the genus, studied in XI.3, which is associated to the function algebra of a curve, will be a birational invariant.

In the non-commutative case the notion of "regular points" presents some difficulties (indeed the splitted points P of $\Omega(R)$ lying over the same p of $\Omega(C)$ look unregular "the commutative way" but in the non-commutative framework it seems that these points should be called regular when certain regularity conditions do hold.) The notion of regularity we propose here should be called "strong regularity" since we think it is actually too strong, however since we do not use any of the possible weaker notions here, we have agreed to say that $P \in \Omega(R)$, R an affine prime P.I. algebra over k_o with p.i. deg $R = n$ and $Z(R) = C$, is a regular point if $Q_{R-P}(R) \subset Q(R)$ is a maximal order in $Q(R)$ over a regular local ring D in $Z(Q(R)) = K = Q(C)$, such that D is an affine k_o-algebra such that $Q(D) = Q(C) = K$. Recall that such a ring D is a regular ring if the Krull dimension dim D of D equals $\dim_{k_o} (m_D/m_D^2)$ where m_D is the maximal ideal of D. The following Proposition is now easily verified.

XI.2.3. Proposition. If $\Omega(R)$ is an affine k_o-curve then $P \in \Omega(R)$ is a regular point if and only if $Q_{R-P}(R)$ is a maximal order in $Q(R)$ over a valuation ring of K/k_o.

XI.2.4. Remark. The valuation rings of K containing k_o are discrete valuation rings since K is a function field in one variable over k_o. Furthermore $Q_{R-P}(R)$ has center D, a discrete valuation ring, moreover $Q_{R-P}(R)$ is finitely generated as a D-module and it is also a Zariski central ring (being a classical maximal order in a c.s.a.) and a left and right principal ideal ring.

XI.2.5. Lemma. Let D be a commutative integrally closed domain such that principal ideals are projective. If Λ is an Azumaya algebra over D then Λ is a maximal D-order in $\Sigma = \Lambda \otimes_D K$, where $K = Q(D)$.

Proof. Since Λ is D-flat, Λ is contained in Σ and it is clearly a D-order of Σ. Let Γ be a D-order of Σ containing Λ. By our assumptions it follows that $\Gamma = \Lambda \otimes_D Z_\Gamma(\Lambda)$. Now $Z_\Gamma(\Lambda) \subset \Gamma \cap K$ and since D is integrally closed in K, $\Gamma \cap K = D$. Consequently $\Gamma = \Lambda$ follows i.e. Λ is a maximal D-order. ∎

XI.2.6. Proposition. Consider a curve $\Omega(R)$ over k_o, where p.i. degree $R = n$. Take $P \in \Omega_n(R)$ and put $p = P \cap C$, $C = Z(R)$, then the following statements are equivalent :

1. p is a regular point of $\Omega(C)$.
2. P is a regular point of $\Omega(R)$.

Proof. First note that $P \in \Omega_n(R)$ implies that $Q_{R-P}(R)$ is an Azumaya algebra over C_p.
1. ⇒ 2. Regularity of $p \in \Omega(C)$ yields that C_p is a discrete valuation ring of $Q(C)=K$. By Lemma XI.2.6. it follows that $Q_{R-P}(R)$ is a maximal C_p-order in $Q_{R-P}(R) \otimes_{C_p} K = (R \otimes_C C_p) \otimes_{C_p} K = Q(R)$.

2. ⇒ 1. Obvious since $Q_{R-P}(R)$ is an Azumaya algebra over C_p. ∎

XI.2.7. Corollary. The set of regular points of an affine k_o-curve is dense. This statement as well as XI.2.6. may be generalized to regular points of algebraic k_o-

XI.2.8. Remark. If $P \in \Omega(R)$ is a regular point of an affine k-curve then $P \in \Omega(R)_n$. Indeed $Q_{R-P}(R)$ is then a maximal order in $Q(R) = M_n(K)$ over the discrete valuation ring $C_p, p = P \cap C$. The completion Λ of $Q_{R-P}(R)$ with respect to the valuation of the center is a maximal order of $M_n(\hat{K})$ over $\hat{C}_p$. Results of Chapter I imply that $Q_{R-P}(R)$ is an Azumaya algebra, i.e. $P \in \Omega_n(R)$.

XI.2.9. Corollary. Let $\Omega(R)$ be an affine k-curve, then $\Omega(R)$ is a non-singular curve (i.e. each point of $\Omega(R)$ is regular) if and only if R is an Azumaya algebra over a Dedekind ring. (It is this consequence that indicates that our notion of regularity is perhaps too restrictive. A more satisfactory approach may be found but this would lead us to far so we do not go into this here).

Let us fix notations once and for all, from here.

Throughout R is an affine prime P.I. algebra over k_o with ring of fractions $A = Q(R)$. Put $C = Z(R)$, $K = Q(C) = Z(A)$ and we assume that K is a function field in one variable over k_o (i.e. with our conventions this includes the hypothesis that K is a separable extension of k_o, a hypothesis which is superfluous as for us the Riemann-Roch theorem is concerned, cf. [169], [171]). We view A as a full matrix ring $M_r(\Delta)$ over a skewfield Δ such that it is identified with $Hom_A(V,V)$ for some minimal left ideal V of A which is an r-dimensional Δ-vectorspace. If D is a Dedekind domain in K and Λ a D-order in Δ then $M = \Sigma_{i=1}^{r} e_i \Lambda$ is a free Λ-lattice in V, where $\{e_1,\ldots,e_r\}$ is a right Δ-basis for V. It follows that $Hom_\Lambda(M,M)$ and $M_r(\Lambda)$ may be identified within $M_r(\Delta)$. Obviously $M_r(\Lambda)$ is a maximal D-order in $M_r(\Delta)$. Let $C_{k_o}(K)$ be the set of all k_o-valuation rings O_v of K (these are discrete!). For every $O_v \in C_{k_o}(K)$ we choose (and fix) a maximal order Λ_v of Δ over O_v and we write $C_{k_o}(\Delta)$ for this set. Finally we let $C_{k_o}(A)$ be the set consisting of all $Q_v = M_r(\Lambda_v)$ where Λ_v varies through the set $C_{k_o}(\Delta)$. If $k_o = k$ then $C_k(A)$ is the set of $M_n(O_v)$ with $O_v \in C_{k_o}(K)$. On $C_k(A)$ we put the topology of finite complements and endow an open set U_A in $C_k(A)$ with a ring of functions as follows. If U_K is the open set in $C_k(K)$ corresponding (in the obvious way) to U_A we put :

$$O(U_K) = \cap \{O_v,\ O_v \in U_K\}$$

$$Q(U_A) = \cap \{Q_v,\ Q_v \in U_A\}\ .$$

Now $O(U_K)$ is a Dedekind ring and $Q(U_A)$ is integral over $O(U_K)$ because the minimal polynomial over K of an element in $Q(U_A)$ has coefficients in O_v for all $O_v \in U_K$ i.e. it has coefficients in $O(U_K)$. It follows that $Q(U_A)$ is a maximal $O(U_K)$-order in A. The correspondence $U_A \to Q(U_A)$ defines a sheaf $\underline{Q}$ of maximal $\underline{O}$-orders, where $\underline{O}$ is the sheaf constructed on $C_{k_o}(K)$ in the usual way. The ringed space $(\Omega_{k_o}(A), \underline{Q})$ may be viewed as a regular model for the curve $\Omega(R)$. Without much problems we obtain a structure of a varietal space on $C_{k_0}(A)$ which is birationally equivalent to $\Omega(R)$.

Let us indicate how these maximal orders may be used in generalizing the properties of the Riemann surface of a function field in one variable to the non-commutative case. Let S be an arbitrary ring. A couple (P,S') such that S' is a subring of S and P is a prime ideal of S' is called a <u>prime</u> of S if $xS'y \subset P$ with $x,y \in S$ yields $x \in P$ or $y \in P$. We say that P is the kernel of the prime and that S' is the domain. The set Prim(S) of all primes of S may be endowed with a topology such that the sets $D(F) = \{P \in \text{Prim}(S),\ P \cap F = \phi\}$ where F is a finite subset of S form a basis of open sets.

For a prime (P,S'), the idealizer $S^P = \{s \in S;\ sP \subset P,\ Ps \subset P\}$ is a maximal domain for P, i.e. (P,S^P) is again a prime. Domination of couples (P,S') is defined as follows : $(P_1,S_1) < (P_2,S_2)$ if $S_1 \subset S_2$ and $P_1 = P_2 \cap S_1$. A prime (P,S') is said to be <u>dominating</u> if it is maximal with respect to the domination relation in the set of couples (P_1,S_1'), S_1' a subring of S, $P_1 \in \text{Spec}(S_1')$. For example, if A is a K-central simple algebra and O_v a valuation ring of K, then for any maximal O_v-order Λ in A the couple $(J(\Lambda),\Lambda)$ is a dominating prime in A. So it is clear that the maximal orders considered above may be linked to certain primes of A.

Let S be a prime Zariski central ring. Then the maximal symmetric ring of quotients Q(S) is simple. <u>Fractional ideals of</u> S in Q(S) are just the twosided S-submodules I of Q(S) such that $cI \subset S$ for some central $c \in Z(S)$. If the fractional ideals of S commute then we call S an <u>arithmetical ring</u>. Let F be the set of fractional ideals of an arithmetical ring S and let Γ be a totally ordered semigroup. A <u>pseudovaluation</u> v on F (or on $Q = Q(S)$) is a function $v : F \to \Gamma \cup \{\infty\}$ which satisfies :

P.v.1. $v(IJ) \geq v(I) + v(J)$ for all $I,J \in F$.

P.v.2. $v(I+J) \geq \inf(v(I),\ v(J))$ for all $I,J \in F$.

P.v.3. $v(S) = 0$, $v(0) = \infty$.

P.v.4. If $I \subset J$ then $v(I) \geq v(J)$, where $I,J \in F$.

Note that P.v.2. and P.v.4. yield $S(I+J) = \min(v(I),v(J))$. We say that v is an <u>arithmetical pseudovaluation</u> (a.p.v.) if for all $I,J \in F$, $v(IJ) = v(I) + v(J)$. We put $v(q) = v(SqS)$ for any $q \in Q$. Then two a.p.v's, v_1 and v_2 are said to be <u>equivalent a.p.v.'s</u> if for each $q \in Q$, $v_1(q) > o$ if and only if $v_2(q) > o$.

<u>XI.2.10. Proposition.</u> Let v be an a.p.v. on Q, then $P = \{q \in Q, v(q) > o\}$ defines a prime (P,Q^P) of Q. Conversely, if (P,Q^P) is a prime of Q such that $S \subset Q^P$ then there is an a.p.v., v, which is unique up to equivalence, such that

$$P = \{q \in Q;\ v(q) > o\}.$$

<u>Proof.</u> Proposition II.3.2. of [171].

<u>Note.</u> If the value semigroup of an a.p.v. is a group then the corresponding prime (P,Q^P) is dominating.

<u>XI.2.11. Definition.</u> A prime (P,Q^P) of Q is <u>discrete</u> if :

1. Q^P contains an arithmetical ring S.
2. Q^P satisfies the a.c.c. on ideals.
3. P is the unique maximal ideal of Q^P.
4. $P = \pi Q^P$ for some invertible element π of Q.

If (P,Q^P) is a discrete prime of Q then the value semigroup of the associated pseudovaluation is infinite cyclic, moreover Q^P is itself an arithmetical ring in Q and the fractional Q^P-ideals form an abelian group.
Indeed $O = Z(Q^P)$ is a valuation ring of Z(Q) and O is obviously discrete. Since Q^P is a localization of the arithmetical ring S it contains it follows that Q^P is Zariski central (and an arithmetical ring itself). Now Q^P is prime because $Z(Q)Q^P = Q$ is simple. The ideals of Q^P are all of the form $\pi^e Q^P$ where the elements π^e, $e \in \mathbb{N}$, are invariant in Q^P (i.e. $\pi^e Q^P = Q^P \pi^e$). It easily follows that fractional ideals of Q^P commute and that these form an abelian group.

A set of discrete primes of Q is said to be <u>proper</u> if the associated a.p.v.'s are inequivalent and the restrictions of these yield inequivalent valuations of

$Z(Q)$. A proper set of discrete primes Q is said to be divisorial if for each $q \in Q$, we have $v(q) = o$ for almost all a.p.v. associated to elements of $\mathcal{P}$. Assume now that $\mathcal{P}$ is a proper set of discrete primes such that for almost all $(P, Q^P) \in P$ we have that Q^P contains some fixed arithmetical ring S. If for every central $z \in Z(Q)$ we have that $z \in Q^P$ for almost all $(P, Q^P) \in P$, then P is a divisorial set of discrete primes (this is easily verified). In this situation it follows that for every fractional ideal I of S, $v(I) = o$ for almost all a.p.v associated to elements of P.

With notations as before, the set $C_{k_o}(A)$ may be identified with the set of discrete primes $(J(Q_v), Q_v)$ which is clearly a divisorial set. Let us normalize the a.p.v. associated to elements of $C_{k_o}(A)$ as follows. A fractional ideal I with respect to a discrete prime (P, Q^P) is of the form $\pi^{e_I} Q^P$ with $e_I \in \mathbb{Z}$, for the selected invertible element $\pi \in Q$; now define v_P by sending I to $e_I \in \mathbb{Z}$.

The elements of a divisorial set P are called prime divisors. A divisor ∂ of Q, associated with a fixed divisorial set P, is a formal product $\partial = \prod_{v \in P} v^{\gamma_v}$ with $\gamma_v \in \mathbb{Z}$ and $\gamma_v = o$ for almost all $v \in P$, (we identify P with the set of associated pseudovaluations). The integer γ_v is said to be the order of γ in v, we write $\gamma_v = \text{ord}_v \partial$.

The divisor ∂ is integral if $\text{ord}_v \partial \geq o$ for all $v \in P$. We say that ∂_1 divides ∂_2 (notation : $\partial_1 | \partial_2$) if for all $v \in P$, $\text{ord}_v \partial_1 \leq \text{ord}_v \partial_2$. If ∂_1, ∂_2 are divisors then we may define divisors $\partial_1 + \partial_2$ and $\partial_1 . \partial_2$ by : $\text{ord}_v(\partial_1 + \partial_2) = \min(\text{ord}_v(\partial_1), \text{ord}_v(\partial_2))$ $\text{ord}_v(\partial_1 . \partial_2) = \text{ord}_v \partial_1 + \text{ord}_v \partial_2$, for all $v \in P$.

Next reconsider the case where P consists of discrete primes (P, Q^P) such that almost all Q^P contain a fixed arithmetical ring S in Q. To a fractional ideal I of S we may associate the ideal divisor of I defined by $\partial_I = \prod_{v \in P} v^{v(I)}$. The ideal divisor ∂_I is principal if $I = (q)$ for some $q \in Q$. Now for any divisorial set P and any $q \in Q$ we say that the divisors $\partial_q = \prod_{v \in P} v^{v(q)}$ are the principal divisors of P. To a principal divisor ∂_q we associate the divisor of zeros ∂_q^+ and the divisor of poles ∂_q^- of q as follows :

$\text{ord}_v\, \partial_q^+ = v(q)$ if $v(q) > o$ and $\text{ord}_v\, \partial_q^+ = o$ otherwise,

$\text{ord}_v\, \partial_q^- = -v(q)$ if $v(q) < o$ and $\text{ord}_v\, \partial_q^- = o$ otherwise.

Note that, if S is an arithmetical ring contained in all the domains Q^P of elements of P then principal divisors are just principal ideal divisors.

XI.2.11. The Approximation Property. Let P be a divisorial set on Q, let $\{v_1,\dots,v_t\}$ be a finite set of a.p.v. in P. For given $\{\gamma_1,\dots,\gamma_t\} \subset \mathbb{Z}$, $\{q_1,\dots,q_t\} \subset Q$, there exists a $q \in Q$ such that $v_i(q-q_i) \geq \gamma_i$ for $i \in \{1,\dots,t\}$ while $v(q) \geq o$ for all $v \in P - \{v_1,\dots,v_t\}$.

In [171] the approximation property is related to similar properties (A_1 and A_2 in loc. cit.) but since we do not need these properties here we do not go into full detail. Let us give an example where the approximation property (A.P.) holds.

XI.2.12. Example. Let S be an arithmetical ring for which the fractional ideals form an abelian group (e.g. $S = A[X,\varphi]$, A a central simple algebra, φ an automorphism of A such that φ^e is inner for some $e \in \mathbb{N}$). To a proper prime ideal P of S we may associate an a.p.v., v_P, which associates to a fractional ideal I the integer $v_P(I)$ defined by the product expansion $I = P^{v_P(I)}.J$ of I, where P does not occur as a factor in the fractional ideal J. It is not hard to verify that $S = \bigcap_P Q^P$, where Q^P is the domain of the prime associated to S_P, and that the A.P. holds.

Recall from [171] that the validity of A.P. implies that the correspondence $I \to \partial_I$ defines an isomorphism between the group of fractional ideals of the arithmetical ring S and the group of divisors of Q (still in the case that P is such that almost all Q^P contain a fixed arithmetical ring S). The correspondence $I \to \partial_I$ maps integral S-ideals to integral divisors and proper prime ideals of S correspond to prime divisors in P.

We return to the framework of k_o-curves in the following.

XI.2.13. Example. (With notations as set after XI.2.9.) Let $\{v_i,\ i \in J\}$ be a set of inequivalent discrete valuations of K, and let O_{v_i} be the k_o-valuation ring associated to v_i. The maximal orders Q_{v_i} in A are Dedekind prime P.I. rings and also Zariski central rings having a unique maximal ideal $J(Q_{v_i})$. Therefore Q_{v_i} is a discrete prime in A. Now choose the set $\{v_i, i \in J\}$ such that A.P. holds, then $O = \bigcap_i O_{v_i}$ is a Dedekind ring in K and if Λ is a maximal order of A over O then $\{\Lambda_{v_i}, \Lambda_{v_i}$ the localization of Λ at $v_i\}$ yields a divisorial set of discrete primes satisfying A.P. Explicitely, there is a finite subset $V_K \subset C_{k_o}(K)$ such that $C_{k_o}(K) - V_K$ is divisorial in K and satisfies A.P. (this is a well-known property of commutative curves!). Let V_A be the set corresponding to V_K in $C_{k_o}(A)$, then $C_{k_o}(A) - V_A$ is divisorial and satisfies A.P. We may find a set $C'_{k_o}(A)$ of ideal divisors consisting of divisors which are defined with respect to the divisorial set $C_{k_o}(A) - V_A$. The elements of $C'_{k_o}(A)$ may be viewed as ideals in some maximal order in A.

If $O_v \in C_{k_o}(K)$ has maximal ideal m_v, and $k_v = O_v/m_v$ as residue field, then the maximal order Q_v with maximal ideal $M_v = J(Q_v)$ has residue algebra $Q_v/M_v = A_v$ which is a k_o-algebra. It is well-known that k_v is a finite field extension of k_o and we call $f_v = [k_v : k_o]$ the <u>absolute residue class degree</u> of v. Obviously A_v is a central simple algebra containing k_v in its center. The integer $\psi_v = [A_v : k_v]$ is called the <u>relative residue class degree</u> of v. If v^e is the normalized ($v^e(\pi) = 1$ for some uniformizing parameter π of O_v) restriction of v to K then $f_{v^c} = [k_{v^c} : k_o]$ is the <u>residue class degree of v^e</u>. The <u>ramification index</u> e_v of v is defined by $m_v Q_v = M_v^{e_v}$. As usual, v^e is said to be <u>unramified</u> if $e_v = 1$. The following relation holds : $e_r \psi_v = N = [A : K]$. Moreover if π is a uniformizing parameter for O_v and if Π generates M_v, then $v(\pi) = v(\Pi^{e_v}) = e_v$. Note that in case k_o is algebraically closed no ramification occurs and all residue class degrees reduce to $N = [A : K] = n^2$.

Write D_A, (resp. D_K), for the group of divisors generated by $C_{k_o}(A)$, (resp. $C_{k_o}(K)$). A divisor $d = \Sigma (v^e)^{\gamma_v}$ of D_K extends to a divisor $\partial_d = \Sigma v^{e_v \gamma_v}$ in D_A which is called the <u>extension of d to A</u>. The <u>degree of a divisor</u> $\partial \in D_A$ is the integer $\deg \partial = \Sigma f_v \operatorname{ord}_v \partial$. If ∂_d is the extension of d to A then $\deg \partial_d = N.\deg d$.

The ring $\ell = \cap\{Q_v, Q_v \in C_{k_o}(A)\}$ is called the ring of (k_o-) constants in A with respect to $C_{k_o}(A)$. With these notations :

XI.2.14. Proposition. 1. The ring ℓ is algebraic over k_o.
2. ℓ is a central simple algebra and ℓ is finite dimensional over k_o.

Proof. 1. The minimal polynomial over K for any $\alpha \in \ell$ has coefficients in $k_o = \cap\{O_v, O_v \in C_{k_o}(K)\}$, hence ℓ consists of k_o-algebraic elements.
2. The choice of the Q_v implies that $\ell = M_r(\ell_\Delta)$, where $\ell_\Delta = \cap\{\Lambda_v, \Lambda_v \in C_{k_o}(\Delta)\}$. Therefore it suffices to check 2. for ℓ_Δ in the skewfield Δ. The reduced norm $N : \Delta \to K$ maps an $\alpha \in \ell_\Delta$ in O_v for each $O_v \in C_{k_o}(K)$. Since $N(\alpha) = o$ if and only if $\alpha = o$, it follows that $N(\alpha) \notin m_v$ for all m_v. Consequently $\alpha \notin J(\Lambda_v)$ for all $\Lambda_v \in C_{k_o}(\Delta)$ and therefore ℓ_Δ embeds into some $\Lambda_v/J(\Lambda_v)$ which is finite dimensional over k_o. Since $[\ell_\Delta : k_o] < \infty$ and ℓ_Δ is contained in a skewfield it follows that ℓ_Δ is a skew-field. ∎

XI.2.15. Corollary. 1. $\ell = \{a \in A, a = o \text{ or } v(a) = o \text{ for all } v \in C_{k_o}(A)\}$.
2. ℓ is contained in A for each $v \in C_{k_o}(A)$.

An abstract regular commutative curve is just an open subset U in $C_{k_o}(K)$ with the induced topology and sheaf. The term "nonsingular" will be reserved to abstract regular curves over an algebraically closed field. It is clear how morphisms between abstract regular curves should be defined etc.... Now in the non-commutative case let us consider the set $C_{k_o}(A)^*$ of all maximal orders over the $O_v \in C_{k_o}(K)$. If k_o is algebraically closed then $C_{k_o}(A)^*$ reduces to $C_{k_o}(A)$. It is clear that $C_{k_o}(A)^*$ may be endowed with a topology and a sheaf (similar to the commutative case, or similar to the structures defined earlier on $C_{k_o}(A)$). If $[A : K] = n^2$, then an abstract regular (non-commutative) curve of degree n is just an open subset $U \subset C_{k_o}(A)^*$ endowed with the induced topology and sheaf. If $k_o = k$ is algebraically closed then we speak of abstract nonsingular curves.

XI.2.16. Proposition. A nonsingular affine k-curve of degree n is isomorphic to an abstract nonsingular curve.

Proof. If $\Omega(R)$ is a nonsingular affine curve of degree n then XI.2.9. yields that R is an Azumaya algebra of degree n over its center C which is a Dedekind ring. Clearly for each $P \in \Omega(R)$, $p = P \cap C$, we have that $Q_{R-P}(R)$ is an Azumaya algebra over the discrete valuation ring C_p of $K = Q(C) = Z(Q(R))$. Therefore we obtain an injective map $\varphi_R : \Omega(R) \to C_{k_o}(A)^*$ where $A = Q(R)$, defined by $P \to Q_{R-P}(R)$. If $\eta : \Omega(R) \to \Omega(C)$ is given by $P \to P \cap C$, and $\xi : C_{k_o}(A)^* \to C_{k_o}(K)$ is given by mapping a maximal order to its center, then we obtain a commutative diagram :

$$\begin{array}{ccc} \Omega(R) & \xrightarrow{\varphi_R} & C_{k_o}(A)^* \\ \big\downarrow \eta & & \big\downarrow \xi \\ \Omega(C) & \xrightarrow{\varphi_C} & C_{k_o}(K) \end{array}$$

Since an Azumaya maximal order over a discrete valuation ring is the unique maximal order over that discrete valuation ring, it follows that ξ is one-to-one on Im φ_R. Now η is a homeomorphism and φ_C is open because it is the commutative embedding of a nonsingular quasiprojective curve into an abstract nonsingular commutative curve. Therefore, by continuity of ξ, it follows that $U = \text{Im } \varphi_R$ is open in $C_{k_o}(A)^*$. Moreover, for any open $X(I)$ of $\Omega(R)$ we have that $Q_I(R) = \bigcap_{P \in X(I)} Q_{R-P}(R)$ and this states that the ringed space $(\Omega(R), \underline{O}_R)$ is isomorphic to U with the induced topology and sheaf. Now the fact that R is an Azumaya algebra of degree n entails that $\underline{\Omega}(R) = (\Omega(R), \underline{O}_R, \underline{O}_R, \pi_R, \underline{K}_R)$, where $\underline{K}_R$ is the constant sheaf with stalks $M_n(k)$. Consequently, the isomorphism of ringed spaces obtained above, may be viewed as an isomorphism of varietal spaces i.e. $\Omega(R)$ is isomorphic to the abstract curve U.

If $k_o = k$ is algebraically closed, we may define a <u>nonsingular curve</u> to be an algebraic k-quasivariety X such that each point $x \in X$ has an open neighborhood U_x which is isomorphic to an affine k-curve. With these conventions :

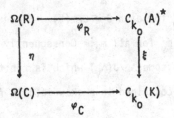

XI.2.17. Proposition. Let $k_o = k$ be algebraically closed. The abstract nonsingular curve $C_k(A)^*$ is isomorphic to a nonsingular k-curve.

Proof. By Tsen's theorem $A = M_n(K)$. A point of $C_k(A)^*$ is a maximal order Λ_v over a discrete k-valuation ring $O_v = Z(\Lambda_v)$ of K. The commutative theory yields the existence of a nonsingular affine curve $V = \Omega(C)$ and a point $p \in V$ such that $C_p = O_v$, the curve V may be identified with an open set of $C_k(K)$. Put $R = M_n(C)$ and $U = \Omega(M_n(C))$, $P \in U$ a point lying over $p \in V$. Since $A = M_n(K)$, Λ_v is to be identified with $Q_{R-P}(R)$. Therefore U may be identified with an open subset of $C_k(A)^*$. So we have shown that each point of $C_k(A)^*$ possesses an open neighborhood U which is isomorphic to an affine curve (modulo some straightforward verification of the fact that the identifications of topological spaces actually exist on the sheaf level and define identifications of varietal spaces). ■

XI.2.18. Remark. If two (irreducible! as always) Azumaya schemes X,Y have points $P \in X$, $Q \in Y$ having isomorphic local rings, then these points have isomorphic neighborhoods too; consequently the Azumaya schemes are birationally equivalent. The generalization of this to general algebraic k-varieties is probably false, but we have not yet established a counter-example.

XI.3. Genus and the Riemann-Roch theorem.

Let $C_{k_o}(A)'$ be any open set of $C_{k_o}(A)$ i.e. the complement of any finite subset in $C_{k_o}(A)$. Put $\Theta = \cap\{Q_v, Q_v \in C_{k_o}(A)'\}$. Fix a K-basis for A within Θ, say $\{u_1, \ldots, u_N\}$; note that $v(u_i) \geq o$ for any i, for almost all $v \in C_{k_o}(A)$. The group of divisors D_A generated by the elements of $C_{k_s}(A)'$ consists of the ideal divisors of Θ corresponding to the fractional Θ-ideals of Q.

Now that the stage is set we get into some technicalities, very similar to the preliminaries for the Riemann-Roch theorem in the commutative case. Aside from some expectable modifications, in particular whenever the ring of constants ℓ associated to $C_{k_o}(A)$ appears, we do not encounter new problems in this noncommutative version of the Riemann-Roch theorem.

To $v \in C_{k_o}(A)$ we associate the completion functor with respect to v i.e. $\hat{A}_v = \hat{K}_{v^c} \otimes_K A$, where $\hat{K}_{v^c}$ is the completion of K with respect to the induced central valuation v^c (normalized!). The product $\hat{V}_A = \prod_v \hat{A}_v$ is a k_o-algebra in the obvious way. The <u>algebra of valuation vectors</u> is by definition the k_o-subalgebra of $\hat{V}_A$ given by $V_A = \{\xi \in \hat{V}_A, v(\xi) \geq o$ for almost all $v \in C_{k_o}(A)\}$, where $V(\xi)$ is defined to be $v(\xi_v)$. An $\alpha \in V_A$ is called an <u>idèle</u> of A if and only if $v(\alpha) = o$ for almost all $v \in C_{k_o}(A)$. The idèles form a multiplicative subgroup I_A of V_A. To an idèle α there corresponds a divisor $\partial_\alpha = \prod_v v^{v(\alpha)}$ and it is easy enough to check that all divisors may be obtained this way. Since A may be embedded diagonally into I_A, we may view A as a k_o-subalgebra of V_A.

<u>XI.3.1. Definition</u>. The <u>∂-parallelotope</u> of V_A associated to the divisor ∂ is the k_o-subspace $\Pi_\partial = \{\xi \in V_A;\ v(\xi) \geq \text{ord}_v \partial$ for all $v \in C_{k_o}(A)\}$ of V_A. The parallelotop s Π_{∂_α}, resp. Π_{∂_a}, for $\alpha \in I_A$, resp. $a \in A$, will be called the <u>α-parallelotope,</u> resp. the <u>a-parallelotope</u> of V_A.

If V_K is the k_o-algebra of valuation vectors with respect to K then we may define a k_o-linear $\varphi : V_K^N \to V_A$, by $\varphi(\xi_i) \to \sum_{i=1}^N \xi_i u_i$. Here, for the first time, we make use of the fixed K-basis $\{u_1, \dots, u_N\}$ of A. Obviously $\varphi(K^N) = A$. Moreover, it is easily verified that for every divisor ∂ we have $\varphi(V_K^N) + \Pi_\partial = V_A$. An easy reduction to the commutative case yields that there is an idèle $\alpha \in V_A$ such that $\Pi_\alpha + A = V_A$, cf. [171] for more detail.

If S is a (finite) subset of V_A and ∂ a divisor of A then we may consider the k_o-space $\Gamma(\partial|S) = \{a \in A, v(a) \geq \text{ord}_v \partial$ for all $v \in S\}$. If $\partial_1 | \partial_2$ then $\Pi_{\partial_1} \supset \Pi_{\partial_2}$ and $\Gamma(\partial_1|S) \supset \Gamma(\partial_2|S)$. The relation between parallelotopes and spaces Γ is expressed in :

<u>XI.3.2. Lemma</u>. Consider divisors ∂_1 and ∂_2 of A such that $\partial_1 | \partial_2$. Put $S = \{v \in C_{k_o}(A),\ \text{ord}_v \partial_1 \neq o$ or $\text{ord}_v \partial_2 \neq o\}$. Then $\pi_{\partial_1} | \pi_{\partial_2}$ and $\Gamma(\partial_1|S)/\Gamma(\partial_2|S)$ are isomorphic k_o-vectorspaces.

Proof. For any $a \in \Gamma(\partial_1|S)$ define $\xi_a \in \Pi_{\partial_1}$ by : $(\xi_a)_v = a$ if $v \in S$ and $(\xi_a)_v = o$ if $v \notin S$. The image of a under the thus obtained k_o-linear map $\rho : \Gamma(\partial_1|S) \to \Pi_{\partial_1}$ is in Π_{∂_2} if and only if $a \in \Gamma(\partial_2|S)$ (because $\rho(a) \in \Pi_{\partial_2}$ yields that $v(a) \geq \mathrm{ord}_v\, \partial_2$ for all $v \in S$). Consequently ρ induces a monomorphism $i : \Gamma(\partial_1|S)/\Gamma(\partial_2|S) \to \Pi_{\partial_1}/\Pi_{\partial_2}$. Now we are in a situation where the approximation property of XI.2.11. holds (proof as in the commutative case, or cf. [FO]), therefore for any $\alpha \in \Pi_{\partial_1}$ we may find $a \in A$ such that $v(a-\alpha) \geq \mathrm{ord}_v\, \partial_2$ for all $v \in S$. Hence $\xi_a - \alpha \in \Pi_{\partial_2}$ and so the selected a which is in $\Gamma(\partial_1|S)$, because $v(a) \geq \min(\mathrm{ord}_v\, \partial_2, v(\alpha)) \geq \mathrm{ord}_v\, \partial_1$, defines an element $\bar{a}$ of $\Gamma(\partial_1|S)/\Gamma(\partial_2|S)$ which maps to α under i. ∎

XI.3.3. Lemma. Consider divisors ∂_1, ∂_2 of A such that $\partial_1 | \partial_2$. Then $\dim_{k_o}(\Gamma(\partial_1|S)/\Gamma(\partial_2|S)) = \deg \partial_2 - \deg \partial_1$.

Proof. Without loss of generality we may assume that $\mathrm{ord}_v\, \partial_2 = \mathrm{ord}_v\, \partial_1 + 1$ while $\mathrm{ord}_{v'}\, \partial_1 = \mathrm{ord}_{v'}\, \partial_2$ for all $v' \neq v$; hence $\deg \partial_2 - \deg \partial_1 = f_v$. Choose $u \in A$ such that $v(u) = \mathrm{ord}_v\, \partial_1$ and $v'(u) \geq \mathrm{ord}_{v'}\, \partial_1$ for all $v' \neq v$ in S. Since A.P. holds we may choose u to be regular and invariant for Q_v i.e. $uQ_v = Q_v u$. If $\{a_1,\dots,a_{1+f_v}\} \subset \Gamma(\partial_1|S)$ then $v(a_i u^{-1}) \geq o$ for all $i \in \{1,\dots,1+f_v\}$, i.e. $\{a_1 u^{-1},\dots,a_{1+f_v} u^{-1}\} \subset Q_v$. However, $[Q_v/M_v : k_o] = f_v$, therefore there exists a relation $\Sigma \lambda_i a_i u^{-1} \in M_v$ with $\lambda_i \in k_o$ not all equal to zero, and this yields that $\Sigma \lambda_i a_i \in \Gamma(\partial_2|S)$ and $\dim_{k_o}(\Gamma(\partial_1|S)/\Gamma(\partial_2|S)) \leq f_v$. Conversely if $\bar{b}_1,\dots,\bar{b}_{f_v}$ are k_o-independent in Q_v/M_v where b_i represents $\bar{b}_i$ in A choose b_i' in A such that $v(b_i' - b_i) > o$, $v'(b_i) > o$ for all $v' \neq v$ in S and $b_i' \equiv b_i \mod M_v$. Now it is clear that $b_i' u \in \Gamma(\partial_1|S)$ but then the existence of a non-trivial relation $\Sigma \lambda_i b_i' u \in \Gamma(\partial_2|S)$ would entail a non-trivial relation $\Sigma \lambda_i b_i' \in M_v$ i.e. $\Sigma \lambda_i \bar{b}_i = o$, a contradiction. Hence, $f_v = \dim_{k_o}(\Gamma(\partial_1|S)/\Gamma(\partial_2|S))$ follows. ∎

XI.3.4. Corollary. For any $\partial_1, \partial_2 \in D_A$ with $\partial_1 | \partial_2$ we have :

$$\dim_{k_o}(\Pi_{\partial_1}/\Pi_{\partial_2}) = [\Pi_{\partial_1} : \Pi_{\partial_2}] = \deg \partial_2 - \deg \partial_1.$$

If in particular $S = V_A$ then we write $L(\partial)$ for $\Gamma(\partial|S)$ and obviously $L(\partial) = A \cap \Pi_\partial$. We write $\ell(\partial) = \dim_{k_o} L(\partial)$.

XI.3.5. Proposition. $\ell(\partial)$ is finite and a bounded function of ∂.

Proof. If $\partial = \varepsilon$ is the unit divisor (i.e. $\mathrm{ord}_r\ \varepsilon = o$ for all $v \in C_{k_o}(A)$) then $L(\varepsilon) = 1$. and $\ell(\varepsilon) = [\ell : k] = n$. Next, if ∂_1/∂_2 then we have $[\Pi_{\partial_1} : \Pi_{\partial_2}] = [\Pi_{\partial_1} \cap A : \Pi_{\partial_2} \cap A] + [\Pi_{\partial_1} + A : \Pi_{\partial_2} + A]$. Hence $\deg \partial_2 - \deg \partial_1 = \ell(\partial_1) - \ell(\partial_2) + [\Pi_{\partial_1} + A : \Pi_{\partial_2} + A]$. Putting $m(\delta) = [V_A : \pi_g + A]$ and choosing ∂_o such that $M + A = V_A$, then we may select a divisor ∂_1 such that $\Pi_{\partial_1} = \Pi_{\partial_o} + \Pi_\partial$. Then $\partial_1|\partial$ and $\deg \partial - \deg \partial_1 = \ell(\partial_1) - \ell(\partial) + m(\partial)$. Consequently $m(\partial)$ is finite because $\deg \partial - \deg \partial_1$ is finite and also $\dim_{k_o}(L(\partial_1)/L(\partial_2)) \leq \dim_{k_o}(\Gamma(\partial_1|S)/\Gamma(\partial_2|S))$ where S is the set of v such that either $\mathrm{ord}_v\ \partial_1 \neq o$ or $\mathrm{ord}_v\ \partial_2 \neq o$.
So $\ell(\partial)$ is finite because $\ell(\varepsilon)$ is finite. Rewrite the above equality as $\deg \partial_2 - \deg \partial_1 = \ell(\partial_1) - \ell(\partial_2) + m(\partial_2) - m(\partial_1)$, then $\deg \partial_1 + \ell(\partial_1) - m(\partial_1) = \deg \partial_2 + \ell(\partial_2) - m(\partial_2)$. For any pair of divisors ∂ and ∂' one can define a divisor γ such that $\gamma|\partial$ and $\gamma|\partial'$ by putting $\mathrm{ord}_v(\gamma) = \min(\mathrm{ord}_v\ \partial, \mathrm{ord}_v\ \partial')$, and we may compare γ to both ∂ and ∂'. This makes it clear that the expression $c = \deg \partial + \ell(\partial) - m(\partial)$ is independent of ∂ and finite. ∎

XI.3.6. Definition. The integer $g_A = 1 - c$ is called the genus of A. If X is an affine k_o-curve then its genus g_X is defined to be g_A where $A = Q(R)$, $X = \Omega(R)$.

The parallelotopes Π_∂ may be used as a fundamental system of neighborhoods for the topologies on V_A and on V_K. Both V_K and V_A are complete. If we give V_K^N the product topology then :

XI.3.7. Corollary. The map $\varphi : V_K^N \to V_A$ is a bicontinuous k_o-linear map i.e. V_K^N and V_A are isomorphic.

Let $V_A^\star$ be the dual k_o-space of V_A. A valuation form ω is an element of $V_A^\star$ which vanishes on $\Pi_\partial + A$ (i.e. it should be continuous with respect to the restricted product topology.

If $\partial_1|\partial_2$ then $\Pi_{\partial_2}+A\subset\Pi_{\partial_1}+A$, hence a valuation form vanishing on $\Pi_{\partial_1}+A$ vanishes on $\Pi_{\partial_2}+A$. Let us write $M(\partial)$ for the set of valuation forms vanishing on $\Pi_{\partial}{-1}+A$, i.e. $M(\partial)$ is nothing but the k_o-dual of $V_A/\Pi_{\partial^{-1}}+A$. Therefore, $\dim_{k_o} M(\partial) = \dim_{k_o} V_A/\Pi_{\partial^{-1}}+A = m(\partial^{-1})$.

Now recall some properties of valuation forms in the commutative case, where valuation forms turn out to be differentials. We refer to Artin [15] for details and proofs :

D.1. The valuation forms of a function field K form a one dimensional vectorspace over K.

D.2. For every valuation form ω of a function field K there exists an upper bound for the parallelotopes $\Pi_{\partial^{-1}}$ such that $\omega\in M(\partial)$.

D.3. The divisors describing the maximal parallelotopes in which valuation forms vanish (see D.2.) belong to some fixed divisor class : the canonical class.

D.4. The degree of any divisor in the canonical class is $2-2g_K$.

One of the main purposes of this section is to derive similar properties of valuation forms in the non-commutative case.

Let Tr be the reduced trace map for the c.s.a. A. We write Tr(u-) for the K-linear map $x\to Tr(u\cdot x)$. For any K-basis $\{u_1,\dots,u_N\}$ of A the forms $Tr(u_i-)$, $i=1,\dots,N$, form a basis for A^*. If $f\in A^*$ then $f=\sum_{i=1}^N a_i\,Tr(u_i-)=Tr(\sum_{i=1}^N a_iu_i-)$, consequently every K-linear $f:A\to K$ is of the form Tr(a-) for some $a\in A$ suitably chosen. Any K-linear $f:A\to K$ extends to a K-linear $f:V_A\to V_K$ as follows : if $f(u_i)=c_i$ then for any $\xi=\Sigma\,\eta_{K,i}\,u_i\in V_A$ we put $f(\xi)=\Sigma\,\eta_{K,i}\,c_i$ and it is obvious that $f(\eta_K\xi)=\eta_K f(\xi)$ holds for every $\eta_K\in V_K$ and $\xi\in V_A$.

Let the extension of Tr to V_A again be denoted by $Tr:V_A\to V_K$. The foregoing yields that every K-linear and V_K-homogeneous map $V_A\to V_K$ is of the form Tr(a-) for some $a\in A$.

Finally, if Ω is a valuation form on V_A and $\xi=\Sigma\,\xi_{K,i}\,u_i\in V_A$ then $\Omega(\xi)=\Sigma\,\Omega(\xi_{K,i}\,u_i)$

and the maps $\Omega(-u_i)$ define valuation forms on V_K. If ω is a fixed valuation form on K then by D.1 we have that $\Omega(-u_i)=\omega(c_i-)$ for c_i in K.
This entails that $\Omega(\xi)=\omega(\Sigma\, c_i\, \xi_{K,i})$. Since the map $V_A \to V_K$ given by : $\Sigma\, \xi_{K,i}\, u_i \to \Sigma\, c_i\, \xi_{K,i}$ is K-linear and V_K- homogeneous, it is actually equal to some Tr(a-) for $a\in A$, hence :

XI.3.8. Proposition. Every valuation form Ω on V_A is of the form $\omega(\mathrm{Tr}(a-))$ for some $a\in A$ and a fixed valuation form ω on V_K.

Fix a valuation form (= differential) ω on V_K.

XI.3.9. Proposition. There is an upperbound for the parallelotopes $\Pi_{\partial^{-1}}$ such that $\omega(\mathrm{Tr}(-))\in M(\partial)$.

Proof. If $x\in L(\partial^{-1})$ we may define a valuation form $x\omega(\mathrm{Tr}(-))$ by putting $x\omega(\mathrm{Tr}(\xi))=\omega\mathrm{Tr}(x\xi)$ for every $\xi\in V_A$. This yields a k_o-linear map $X: L(\partial^{-1})\to M(\varepsilon)$, $x\to x\omega(\mathrm{Tr}(-))$. If $\omega(\mathrm{Tr}(x\xi))=o$ for all $\xi\in V_A$ then $\mathrm{Tr}(x\xi)=o$ for all ξ in V_A because Tr(x-) is either surjective or zero! Hence $x=o$ follows, proving that X is injective.
Now comparing k_o-dimensions, we obtain :

$$\ell(\partial^{-1})+\dim_{k_o} M(\varepsilon)=\ell(\varepsilon)+\deg(\varepsilon)+g_A-1=n+g_A-1 ,$$

and $\ell(\partial^{-1})+\deg(\partial^{-1})=1-f_A+\dim_{k_o} M(\partial)\geqslant 2-2g-n$. Hence if $\omega(\mathrm{Tr}(-))\in M(\partial)$ then the degree of ∂^{-1} is bounded below and therefore there must be an upper bound for the parallelotope chain :

$$\ldots \subset \Pi_{\partial_i^{-1}} \subset \Pi_{\partial_{i+1}^{-1}} \subset \ldots$$

such that $\omega(\mathrm{Tr}(-))\in M(\partial_i)$ for all i, because $\deg \partial_i^{-1} > \deg \partial_{i+1}^{-1}$. Finally, in order to obtain an upper bound for the parallelotopes in which $\omega(\mathrm{Tr}(-))$ vanishes it suffices to note that, if $\omega(\mathrm{Tr}(-))\in M(\partial_1)$ and $\omega(\mathrm{Tr}(-))\in M(\partial_2)$ then $\omega(\mathrm{Tr}(-))\in M(\gamma)$ where γ is the greatest common divisor of δ_1 and δ_2.

XI.3.10. Remark. In case A is a skewfield, the above is valid for every valuation form Ω on V_A. To establish this, write $\Omega = \omega(Tr(a-))$ for suitable $a \in A$ and use the fact that in this case a is a unit of A.

If $\beta \in D_A$ we may argue similar as in the proof of XI.3.9. and find that $X_\beta : L(\partial) \to M(\beta\partial)$, $x \to x\omega(Tr(-))$ is a k_o-monomorphism. If $\Pi_{\partial^{-1}}$ is chosen to be maximal such that $\omega(Tr(-)) \in M(\partial)$ then we claim that X_β is surjective. Indeed, pick $\Omega \in M(\beta\partial)$ then $\Omega = \omega(Tr(a-))$ for some suitable $a \in A$. If $a \notin L(\beta)$ then there is a v such that $v(a) < ord_v(\beta)$. Pick $\xi \in \Pi_{\beta^{-1}\partial^{-1}}$ then we have that $\omega(Tr(a\xi)) = o$ and therefore the smallest parallelotope containing a $\Pi_{\beta^{-1}\partial^{-1}}$ maps to zero (continuity of Ω). Since $\Pi_{\partial^{-1}}$ is supposed to be maximal with that property, $\Pi_{\beta^{-1}\partial^{-1}} \subset \Pi_{\partial^{-1}}$ follows. As a consequence of the approximation property we may choose $\xi \in \Pi_{\beta^{-1}\partial^{-1}}$ such that ξ_v is right invariant in Q_v and $v(\xi) = ord_v(\beta^{-1}\partial^{-1})$, i.e. $v(a\xi) < ord_v(\partial^{-1})$, hence a $\Pi_{\beta^{-1}\partial^{-1}} \subset \Pi_{\partial^{-1}}$ is a contradiction. Therefore $a \in L(\beta)$ and $L(\beta) \cong M(\beta\delta)$ follows for arbitrary $\beta \in D_A$. A comparison of dimensions yields: $m(\beta^{-1}\partial^{-1}) = \ell(\beta)$ or $m(\beta) = \ell(\beta^{-1}\partial^{-1})$. By definition $g_A = 1 - deg(\beta) + \ell(\beta) - m(\beta)$ and because of the foregoing calculation of $m(\beta)$ we finally obtain :

XI.3.11. Theorem (Riemann-Roch). If $\beta \in D_A$ is arbitrary and if ∂ is the divisor of XI.3.9. then :

$$deg(\beta) + \ell(\beta) = \ell(\beta^{-1}\partial^{-1}) + 1 - g_A .$$

XI.3.12. Remark. The divisor ∂^{-1} corresponds to the canonical class if A is commutative i.e. $A = K$. If $A = Q(R)$ for an Azumaya curve $\Omega(R)$ then the proof of XI.3.11. given here reduces directly to the center because divisors correspond to central divisors up to the suitable normalization.

XI.3.13. Corollary. Pick $\beta = \varepsilon$ the unit divisors then $deg(\varepsilon) + \ell(\varepsilon) = \ell(\partial^{-1}) + 1 - g_A$, hence we find :

1. $\ell(\partial^{-1}) = n - 1 + g_A$ where $n = [\ell : k_o]$.

On the other hand, choosing $\beta = \partial^{-1}$ we find that, $\deg(\partial^{-1}) + \ell(\partial^{-1}) = \ell(\varepsilon) + 1 - g_A$, and therefore :

2. $\deg(\partial^{-1}) = 2-2g_A$.

Still ω is a fixed valuation form of K and $\Omega = \omega(\mathrm{Tr}(-))$. The <u>local components</u> of ω and Ω are defined to be the valuation forms: $\omega_v(x) = \omega(x_v)$, $\Omega_v(\xi) = \Omega(\xi_v)$. Let $\Pi_{d^{-1}}$, resp. $\Pi_{\partial^{-1}}$, be maximal parallelotopes in V_K, resp. V_A, on which ω, resp. Ω vanishes. The v-components of the divisor d, resp. ∂, may be viewed as an $\hat{O}_v$-, resp. $\hat{Q}_v$- ideal, say $\hat{m}_v^{\gamma_v}$, resp. $\hat{M}_v^{\mu_v}$. The <u>inverse local different</u> of $\hat{Q}_v$ is defined to be :

$$\mathcal{D}_v^{-1} = \{x \in A,\ \mathrm{Tr}(x\Theta) \subset \bar{O}_v\} .$$

The ideal $\mathcal{D}_v = (\mathcal{D}_v^{-1})^{-1}$ is the <u>local different</u> of $\hat{Q}_v$. The local different $\mathcal{D}_v$ is the unit ideal if and only if $e_v = 1$ i.e. if v^c is unramified in A. The <u>different</u> of D_A is defined to be the divisor : $\mathcal{D} = \prod_{v \in C_{k_0}} v^{e_v - 1}$. It is clear that the local component of $\mathcal{D}$ at v is the local different $\mathcal{D}_v$.

<u>XI.3.14. Corollary</u>. If ∂ and d are as above and if ∂_d is the extension of d to D_A, then $\partial = \partial_d \cdot \mathcal{D}$.

<u>Proof</u>. $\Omega_v(\Pi_{\partial^{-1}}) = o$ if and only if $\omega_v(\mathrm{Tr}(M_v^{-\mu_v})) = o$ if and only if $\mathrm{Tr}(M_v^{-\mu_v}) \subset m_v^{-\gamma_v}$ i.e. $\mathrm{Tr}(M_v^{-\mu_v} m_v^{\gamma_v}) \subset \hat{O}_v$. Consequently $M_v^{-\mu_v} m_v^{\gamma_v}$ is the inverse local different of $\hat{Q}_v$. Expressing this in all local components yields immediately $\partial = \partial_d\ \mathcal{D}$. ∎

<u>XI.3.15. Remark</u>. The ramification e_v of v^c in A does not depend on the choice of the maximal O_v-order Q_v in A. The genus g_K is an invariant for K. The expressions $\deg(d^{-1}) = 2-2g_K$, $\deg(\partial^{-1}) = 2-2g_A$ together with the relation $\partial = \partial_d \mathcal{D}$ yield that g_A is not depending on the choice of the divisorial set $C_{k_0}(A)$ and the genus g_A is therefore an invariant of A and a birational invariant of the curve.

<u>XI.3.16. Corollary</u>. $g_A = N\ g_K - N + 1 + \frac{1}{2} \Sigma\ f_v(e_v - 1)$.

Proof. Direct calculation from $\partial = \partial_d . D$. ∎

XI.3.17. Example. Suppose that ℓ contains a K-basis for A, then $[A_v : k_v] = [A : K] = N$ and every v^c is unramified in A. Hence $g_A = N g_K - N + 1$. This may be applied to the case $A = M_n(K)$.

XI.3.18. Theorem. Let $X = \Omega(R)$ be an affine k-curve and let $Y = \Omega(Z(R))$ be the central curve underlying X. If g_X, resp. g_Y is the genus of X, resp. Y then $g_X = N g_Y - N + 1.$, where $N = [Q(R) : Q(C)] = (\text{p.i. deg } R)^2$.

Proof. Since k is algebraically closed, $Q(R) = A = M_n(K)$ by Tsen's theorem. ∎

XI.3.19. Proposition. If $A = M_r(\Delta)$ then $g_A = r^2\ g_\Delta - r^2 + 1$.

Proof. Ramification of v^c in Δ is the same as ramification of v^c in A. Hence $e_v = e_{v,A} = e_{v,\Lambda}$ and therefore $f_{v,A} = r^2\ f_{v,\Delta}$.
Put $N = r^2 m^2$ with $m^2 = [\Delta : K]$, then we obtain :

$$\begin{aligned} g_A &= Ng_K - N + 1 + \tfrac{1}{2}\ \Sigma\ f_{v,A}(e_v - 1) \\ &= r^2(m^2\ g_K - m^2 + \tfrac{1}{2}\ \Sigma\ f_{v,\Delta}(e_v - 1)) + 1 \\ &= r^2(m^2 g_K - m^2 + \tfrac{1}{2}\ \Sigma\ f_{r,\Delta}(e_v - 1) + 1) + 1 - r^2 \\ &= r^2\ g_\Delta - r^2 + 1. \quad ∎ \end{aligned}$$

XI.3.20. Remark. If $A = M_r(\Delta)$ then the reduced genus $\tilde{g}_A = \frac{1}{r^2}(-1 + r^2 + g_A)$ may be viewed as an invariant associated to the class of A in Br K.

XI.3.21. Remark. If K is a function field of genus 1 and A a c.s.a. over K such that there exists a valuation v^c of K which ramifies in A then $g_A > 2$. Indeed $e_v\ \psi_v = N = [A : K]$ is a square and ψ_v is also a square, $\psi_v \neq N$, hence $e_v \geq 4$. Therefore : $\frac{1}{2}\ f_v(e_v - 1) \geq \frac{3}{2}\ f_v > 1$, thus $g_A = N - N + 1 + \frac{1}{2}\ \Sigma\ f_v(e_v - 1) > 2$.

XI.3.22. Example. Consider $A = B(X,\varphi)$ where B is a c.s.a. over the field k', φ an automorphism of B such that φ^e is inner (take $e \in \mathbb{N}$ minimal as such). Put $k_o =$

$= (k')^{\varphi}$. The center of A is $k_o(T)$ where $T = \lambda X^c$ for a certain unit λ of B inducing the inner automorphism φ^e of B. Since $K = k_o(T)$ we have $g_K = o$. For $C_{k_o}(A)$ we take $\{B[X,\varphi]_{(X)}, B[X^{-1},\varphi^{-1}]_{(X^{-1})}, B[X,\varphi]_p$ where p is a central irreducible element of $k_o[T]$.$\}$

It follows that :

1. The c.s.a. of $C_{k_o}(A)$-constants in A is B.

2. Let v_X be the a.p.v. associated to $B[X,\varphi]_{(X)}$.

Then $k_{v_X} \cong B$ and therefore $f_{v^e} = 1$, $f_{v_X} = \psi_{v_X} = [B : k'] = n$. The ramification index e_{v_X} equals the integer e and $ne = N = [A : K]$.

3. Let $v_{X^{-1}}$ be the a.p.v. associated to $B[X^{-1}, \varphi^{-1}]_{(X^{-1})}$. Then $k_{v_{X^{-1}}} \cong B$,

$f_{v^e_{X^{-1}}} = 1$, $f_{v_{X^{-1}}} = \psi_{v_{X^{-1}}} = n$, $e_{v_{X^{-1}}} = e$.

4. Since M_{v_p} is generated by p, every v_p^c is unramified in A.

Furthermore we have that $f_{v_p^c} = \deg_T p$.

Thus $\psi_{v_p} = N$, $e_{v_p} = 1$, and $f_{v_p} = N.\deg_T p$.

Calculating g_A we obtain : $g_A = N.g_K - N + 1 + \frac{1}{2}.2.n.e$

$$= -N + 1 + N - n,$$

Hence $g_A = 1-n$. In particular if $A = \mathbb{C}(X,-)$ then we obtain that $g_A = -1$.

<u>XI.3.23. Proposition.</u> If $A = \Delta$ is a skewfield then $\deg \partial_a = o$ for all $a \in A$. Consequently if β is a divisor in D_A such that $\deg(-\beta) > 2g_A - 2$ then $\ell(\beta) + \deg(\beta) = 1 - g_A$.

<u>Proof</u>. The first statement follows from [171] and the second is just straightforward calculation.

XI.3.24. Comments and References.

The first non-commutative version of the Riemann-Roch theorem was established by E. Witt in [202] over a perfect groundfield. In this paper the ring of "constants" does not intervene in an explicit way, but the duality with respect to the class of the different of the c.s.a. A over K plays the important part. J.P. Van Deuren, J. Van Geel and F. Van Oystaeyen used "valuation" rings in skewfields, cf. [169], and derived the Riemann-Roch theorem for skewfields over function fields without any restriction on the groundfield k_o. The divisors used in this set up i.e. the so-called geometrical divisors are linked to certain extensions of valuation rings in the center to non-commutative valuation rings of the skewfield. Whereas this paper contains more information on the ring of constants its disadvantage however is that the canonical divisor (different) cannot be recovered as a geometrical divisor. Meanwhile the non-commutative theory of primes in algebras, started off by F. Van Oystaeyen in [178], has been developed by J. Van Geel in his Ph.D. thesis [171] to a very useful tool. Combination of this powerful tool and the nice properties of arithmetical rings introduced by E. Nauwelaerts, F. Van Oystaeyen in [125], gives rise to the presentation of the Riemann-Roch theorem we have included here (following J. Van Geel, [171]). Since the study of curves did not require full generality everywhere, we have made no point about the fact that need not be perfect. Both the study of rational points and regular curves are very much open ended, in particular (as pointed out before) the study of orders over regular commutative rings should be carried a lot further if one wants to attack the surface case, aiming for a non-commutative substitute for many interesting results as in O. Zariski's [204] for example.

Let us also mention the following problem: if X is a curve (non-singular, if one wishes), is $g + n = \dim_{k_o} H_1(X, \underline{O}_X)$, where n is the k_o-dimension of the ring of constants in the function-algebra of X?

XII. WORK IN PROGRESS, [91].

In this section we give a very short survey of some recent developments, present some problems whether or not accompanied by more or less extensive hints of possible ways to attack them, point out some directions to take and indicate possible applications.

XII.1. Geometric Homomorphisms and Curves.

By now the reader is probably strongly convinced that one of the really basic facts in the development of the theory of algebraic k-varieties over P.I. rings is the use of extensions of P.I. rings for the morphisms of affine k-algebras. Let us now destroy this conviction by adopting another point of view for a while, namely M. Artin and W. Schelter's point of view, cf. [20]. Any ring homomorphism $\varphi : R \to S$ between affine prime P.I. k-algebras induces a nonempty correspondence $\Omega(S) \multimap \Omega(R)$. Indeed, for any $M \in \Omega(S)$, the ring $R/\varphi^{-1}(M)$ is nonzero and possesses a finite number of maximal ideals $\overline{N}_\alpha$ because $R/\varphi^{-1}(M)$ is a subring of some $M_n(k)$. The inverse images N_α, in R, of $\overline{N}_\alpha$, are the maximal ideals corresponding to M. Therefore φ induces a finite-valued correspondence $\Omega_n(S) \multimap \Omega_1(R) \cup \ldots \cup \Omega_n(R)$ and one may verify that this correspondence is actually algebraic with respect to the variety structure on $\Omega_n(S)$ and the $\Omega_i(R)$, $i = 1, \ldots, n$, cf. [18]. Following Artin [19] we call a ring homomorphism $\varphi : R \to S$ integral if every $s \in S$ satisfies a relation of the form $s^n = \sum_i m_i$ where each m_i is a word in elements of $\{s\} \cup \varphi(R)$ which has degree less than n in s. If one considers P.I. rings, the notion of integrality gains importance because it may be linked to certain finiteness conditions as follows.

A ring morphism, $\varphi : R \to S$, is said to be finite if for every prime P.I. quotient S' of S of Krull dimension 1, the center Z(S') is integral over R. Whereas integral morphisms are clearly finite, the converse fails in general.

For affine P.I. rings however, cf. [21], the notions of finite morphisms and integral morphisms coincide. Let us reserve the term "curve in $\Omega(R)$" for a central extension $R \to A$ where A is an order of $M_n(K)$ over an affine Dedekind ring D with field of fractions K. The reader may tie this up with the general theory of curves expounded in Chapter XI.

With these conventions one may prove that a morphism $\varphi : R \to S$ is finite if and only if the induced correspondence $\Omega(S) \circ\!\!\to \Omega(R)$ is proper. Here <u>proper</u> means that the valuative criterion, [52] of (commutative) Algebraic Geometry holds (translated in terms of curves as defined above). From this it follows that composition of integral morphisms between affine P.I. rings yields an integral morphism. It remains to be seen however, how these notions work in dealing with geometrical problems, because up to now only the topological structure of the varieties has been used.

A second possibility in trying to avoid the use of ring extensions is to introduce the so-called geometric morphisms. A ring morphism $\varphi : R \to S$ is said to be <u>geometric</u> if for each $M \in \Omega(S)$, $\varphi^{-1}(M) \in \Omega(R)$ and $R/\varphi^{-1}(M)$ has p.i. degree equal to p.i. deg(S/M). With the usual assumptions i.e. all k-algebras affine and k algebraically closed, this states simply that φ induces an isomorphism $R/\varphi^{-1}(M) \cong S/M$. Recall from [20] that φ is geometric if and only if for each $P \in \mathrm{Spec}_n(S)$ we have that $\varphi^{-1}(P) \in \mathrm{Spec}_n(R)$. Central extensions and morphisms of the form $R \to R\{q^{-1}\}$, where $q \in R$ is invertible in some overring Q are examples. Note that if $R \hookrightarrow S$ is an injective geometric morphism between prime P.I. rings, then R and S have the same p.i. degree. This kind of restrictive conditions which arise on P.I. rings if there exists a geometric morphism between them, is somewhat of a drawback for this theory. On the other hand, if the reader is willing to go through a lot of technicalities, he may check that the varietal space $\underline{\Omega}(R)$ behaves functorially (in RI) with respect to geometric morphisms, i.e. the sheaf theory established before also works for geometric morphisms (and compositions of extensions and geometric morphisms). An interesting project is to study the notion of properness introduced in [21], but express it in terms of the curve criterion in a more intrinsic way, cf. XII.3.

XII.2. Zariski's Main Theorem.

First reconsider some commutative Algebraic Geometry. If $\varphi : R \to S$ is a morphism of commutative affine rings, then Zariski's Main Theorem states the following : if R is integrally closed and $P \in \mathrm{Spec}(S)$ is a point which is a component of the fibre of $\mathrm{Spec}(\varphi) : \mathrm{Spec}(S) \to \mathrm{Spec}(R)$, then the local rings S_P and R_Q, with $Q = \mathrm{Spec}(\varphi)(P)$, are isomorphic. Utilizing the notions introduced in XII.1. we may state the following non-commutative version of this well-known theorem, cf. M. Artin, W. Schelter [20]

XII.2.1. Theorem (Non-commutative Version of Zariski's Main Theorem.)

Let $\varphi : R \to S$ be a morphism of affine k-algebras, which is a composition of central extensions and integral geometric morphisms. There is a subring $S' \subset S$ which is integral and geometric over R and such that the fibres of the map $\mathrm{Spec}(S') \to \mathrm{Spec}(R)$ either consists of one point, or contain no component of dimension zero.

Let us reproduce the example given in [20].

Put $A = k[x]$, $\bar{A} = k[x] / (x^2 - x) \cong k \oplus k$,

$$R = \begin{pmatrix} k & k \\ o & k \end{pmatrix} , \qquad S = \begin{pmatrix} \bar{A} & \bar{A} \\ o & A \end{pmatrix} .$$

Then the (non-central) element $\begin{pmatrix} x & o \\ o & o \end{pmatrix}$ generates the desired subring S' of S over R.

Let us point out, perhaps stating the obvious, that there is a considerable amount of work left to be done concerning regularity, normal varieties etc. Many preliminary results obtained are related to integral morphisms, finite morphisms and properness. Still, it is our conviction that this topic could benefit a lot from a proper theory of hereditary orders over regular (in the geometrical sense) rings, in any Krull dimension. We fear, or should we say hope, that this part of non-commutative geometry may completely change face in the near future, and that is why we have chosen not to include it in this monograph in full depth.

XII.3. Some remarks on Cohomology.

Again we go back to some results of M. Artin and W. Schelter, cf. [20]. Let X be a commutative scheme with structure sheaf $\underline{O}_X$ and let $\underline{a}$ be a quasi-coherent sheaf of associative $\underline{O}_X$-algebras. If $U \subset X$ is open affine then $\underline{a}(U)$ is an $\underline{O}(U)$-algebra and so we may define Spec($\underline{a}$) to be the union of the spaces Spec($\underline{a}(U)$) which happen to stick together well. We say that (Spec($\underline{a}$), $\underline{a}$) is a scheme of algebras over X (or just "scheme" if no ambiguity arises) and we will write $\underline{a}$ as $\underline{O}$ if we refer to this scheme as Y. Projective space over R is then just the scheme ($\mathbb{P}^n_R$, $R \otimes O_{\mathbb{P}^n}$), which has been considered in some detail in Section XIII. . A scheme Y over $\mathbb{P}^n_R$ is a morphism $Y \to \mathbb{P}^n_R$ of schemes over $\mathbb{P}^n_k$ and this is defined by a k-morphism $R \to \underline{O}_Y$, or a map of $O_{\mathbb{P}^n}$-algebras $R \otimes \underline{O}_{\mathbb{P}^n}$. The fact that the notion of integrality is local, cf. [11], implies that it makes sense to speak about "schemes integral over $\mathbb{P}^n_R$". If Y is a scheme integral over $\mathbb{P}^n_R$ then the morphisms $Y \to$ Spec(R) satisfies the following restricted curve criterion for properness; cf. XII.1.

If A is an order over a Dedekind ring D as in XII.1. then a curve in Y is a central morphism (this too is a local notion !) Spec(A) $\to$ Y lying over a map Spec(D) $\to$ X if Y lies over X. Put $D' = D[t^{-1}]$ for some $t \neq o$ in D and consider a D'-order A' in $M_n(K)$; note that D' is also a Dedekind domain, still affine and with field of fractions K. Let φ' : Spec(A') $\to$ Y be a curve in Y and let R [D] be the subring of A' generated by R and D, then, if $t^{-1} \notin R[D]$ there is a unique D-order A and a unique central morphism φ : Spec(A) $\to$ Y such that φ' is the localization of φ with respect to t. It follows that for any scheme Y integral over $\mathbb{P}^n_R$, the correspondence $Y \to$ Spec(R) takes closed sets to closed sets.

One of the reasons why this notion works well is that here we can develop a satisfactory cohomology theory.

If Y is a scheme of algebras over X then a left $\underline{O}_Y$-module is just a quasi coherent sheaf of $\underline{O}_Y$-modules $\underline{M}$ such that for each affine open set U of X the sections $\underline{M}(U)$ form a left $\underline{O}_Y(U)$-module. The cohomology groups of M will then be the cohomology groups over X i.e. we put $H^q(Y,\underline{M}) = H^q(X,\underline{M})$. These groups are left modules over the ring $H^o(Y,\underline{O}_Y)$. If Y is integral over $\mathbb{P}^n_R$ then every $s \in H^o(Y,\underline{O}_Y)$ is integral over R, cf. [20]. For example, we may consider a scheme $\mathbb{P}^1_R$ as the result of gluing together two affine spaces. An $\underline{O}_Y$-Module $\underline{M}$ is then given by a diagram :

$$\begin{array}{ccc} M_o & \xrightarrow{d_1} & \\ & & M_{o1} \\ M_1 & \xrightarrow{d_o} & \end{array}$$

where M_o, M_1 are corresponding to the affine pieces and M_{o1} corresponds to their intersection.

The cohomology of $\underline{M}$ is then given by :

$$o \to H^o(Y,\underline{M}) \to M_o \oplus M_1 \to M_{o1} \to H^1(Y,\underline{M}) \to o$$

Obviously, all of this works because M. Artin and W. Schelter's constructions are rather close to the commutative case. The quasicoherent sheaves of modules encountered in Chapter VII are more general and therefore they cannot be dealt with in the same way. One of the main difficulties resides in the fact that, in the non-commutative case, the universal module of differentials satisfies different (from the commutative case) universal properties. Here the ideas of M. Artin and W. Schelter could still be valuable ones. Indeed, as the Riemann-Roch formula for curves shows, there do exist relations between the birational invariants of non-commutative varieties and those of the corresponding central variety (see for example the relation between g_K and g_A in the curve case, XI.3.). Therefore it is not unlikely that "central-methods" might carry cohomology theory a long way.

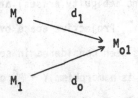

XII.4. Cohn's Scheme.

The field spectrum X of a general ring R consists of the prime matrix ideals i.e. the singular kernels of R-fields, cf. [41], with the following topology. For each square matrix a over R define the support of a to be the set $D(a) = \{x \in X, a \notin P_x\}$ where P_x is the prime matrix ideal corresponding to $x \in X$, the sets D(a) form a basis for a topology on X. If Σ is a set of square matrices over R then the universal Σ-inverting ring R_Σ is obtained from R, by adjoining for each nxn matrix A in Σ, n symbols a_{ij} such that $AA' = A'A = I$, $A' = (a'_{ij})$. Associating to D(a) the universal (X-D(a))-inverting ring defines a presheaf of rings over X; let $\tilde{R}$ denote the associated sheaf. The stalk of $\tilde{R}$ at $x \in X$ is a local ring which is the Σ-inverting ring where Σ is the complement of P_x in the set of all square matrices over R. If R is commutative then this construction reduces to the usual Spec(R), because taking determinants provides an isomorphism of ringed spaces $\tilde{R} \cong$ Spec(R). Functoriality of the construction with respect to ring morphisms may be checked but on the other hand R cannot be recovered as the ring of global sections of $\tilde{R}$, it makes it hard to consider $\tilde{R}$ as a geometrical object associated to R.

Still, if one could link $\tilde{R}$ and Spec(R) in the noncommutative case by using Dieudonné-determinants in a certain way (or determine a class of rings for which such link exists), then one could consider the situation (Spec(R), $\tilde{R}$, det), much in the way the k-algebraic varieties in our set up arose from consideration of both $\underline{O}_R$ and $\underline{O}_R^{bi}$.

For more detail on the field spectrum cf. [42].

XII.5. Brauer Groups of Projective Curves.

Let E be a projective curve with homogeneous coordinate ring $\Gamma_h(E)$ and let us restrict our attention to nonsingular plane curves here. There is a morphism from the "graded" Brauer group $Br^g\,\Gamma_h(E)$, cf. III.2.12, to the (Grothendieck) Brauer group of $\underline{E}$ which may in this case also be viewed as the Brauer group of the ringed space associated to E in B. Auslander's sense, cf. [12]. We conjecture that in fact $Br^g\,\Gamma_h(E) \cong Br\,E$. (*)
If we invert the image x_i of X_i in $\Gamma_h(E)$ (under the canonical $k[X_1, X_2, X_3] \to \Gamma_h(E)$) then $Br^g\,\Gamma_h(E)[x_i^{-1}]$ equals the Brauer group of the part of degree o of $\Gamma_h(E)[x_i^{-1}]$ which is a Dedekind ring. Instead of studying the behaviour of the Brauer group under "blowing up" of the top-singularity on the affine cone associated to E, one may study the "graded" Brauer group of $\Gamma_h(E)$ and obtain an equally interesting invariant of the curve E but with the property that in the stalks of $Proj(\Gamma_h(E))$ (i.e. points of the curve) the local Brauer groups embed into the function field of the curve.
Moreover, from the non-commutative Riemann-Roch theorem we obtain birational invariants for E, e.g. the genera of the skewfields over the function field, which deserve closer investigation. For example it is an intringuing problem to detect in which way the genera of the possible quaternion algebras (skewfields or not) over the function field of an elliptic curve over Q do relate to the elliptic invariant j(E) or to the isomorphism class of E, e.g. do these genera distinguish between the cases where E has complex multiplication or not, and many other related questions which seem to deal more with the commutative geometry of E than with the geometry of the non-commutative curves over E corresponding to elements of Br (E).
Some references : [186], [38], [47], [95].

(*) This has now been proved by the authors for arbitrary normally projective curves in general one finds that $Br\,E = Br^g N(\Gamma_k(E))$, where $N(\Gamma_k(E))$ is the graded localization of $\Gamma_k(E)$ at the graded ideal $\Gamma_k(E)_{>0}$, cf. III.1.

REFERENCES

[1] A.A. Albert, Structure of Algebras,A.M.S. Colloq. Publ., New York, 1939.

[2] A.A. Albert, New Results on Associative Division Algebras, J. of Algebra 5, 1967, 110-132.

[3] S.A. Amitsur, Some Results on Central Simple Algebras, Ann. Math. 63, 1956, 285-293.

[4] S.A. Amitsur, A Generalization of Hilbert's Nullstellensatz, Proc. A.M.S. 8, 1957, 643-656.

[5] S.A. Amitsur, Finite Dimensional Central Simple Algebras, Proc. A.M.S. 11, 1960, 28-31.

[6] S.A. Amitsur, Prime Rings having Polynomial Identities with Arbitrary Coefficients, Proc. A.M.S. 17, 1967, 470-486.

[7] S.A. Amitsur, Embeddings in Matrix Rings, Pac. J. Math. 36, 1971, 21-27.

[8] S.A. Amitsur, On Rings of Quotients, Symp. Math. vol. 8, Academic Press, London, 1972, 149-164.

[9] S.A. Amitsur, C. Procesi, Jacobson Rings and Hilbert Algebras With Polynomial Identities, Ann. Mat. Pura Applicata 71, 1960, 61-72.

[10] S.A. Amitsur, D. Saltman, Generic Abelian Crossed Products and p-Algebras, J. of Algebra 51, 1978, 76-87.

[11] S.A. Amitsur, L. Small, Prime P.I. Rings, Bull. A.M.S. 83, 1977, 249-251.

[12] S.A. Amitsur, L. Small, Prime Ideals in P.I. Rings, J. of Algebra, 62, 2, 1980, 358-383.

[13] E. Armendariz, On Semiprime P.I. Algebras over Commutative Regular Rings, Pac. J. Math. 66, 1976, 23-28.

[14] E. Artin, Theory of Algebraic Numbers, Göttingen 1963.

[15] E. Artin, Algebraic Numbers and Algebraic Functions, Gordon and Beach, London 1967.

[16] M. Artin, Grothendieck Topologies, Seminar Notes, Harvard University, 1962.

[17] M. Artin, On Azumaya Algebras and Finite Dimensional Representations of Rings, J. of Algebra II, 1969, 532-563.

[18] M. Artin, Specialization of Representations of Rings, INTl. Symp. on Algebraic Geometry, Kyoto 1977, Tokyo 1978, 237-247.

[19] M. Artin, Integral Homomorphisms of P.I. Rings, Proc. Durham 1979, 70-86.

[20] M. Artin, W. Schelter, A Version of Zariski's Main Theorem for P.I. Rings, Ann. J. Math. 101, 1979, 301-330.

[21] M. Artin, W. Schelter, Integral Ring Homomorphisms, to appear.

[22] B. Auslander, The Brauer Group of a Ringed Space, J. of Algebra 4, 1966, 220-273.

[23] M. Auslander, O. Goldman, Maximal Orders, Trans. Amer. Math. Soc. 97, 1960, 1-24.

[24] M. Auslander, O. Goldman, The Brauer Group of a Commutative Ring, Trans. Amer. Math. Soc. 97, 1960, 367-409.

[25] G. Azumaya, On Maximally Central Algebras, Nagoya Math. J. 2, 1951, 119-150.

[26] H. Bass, Algebraic K-Theory, Math. L.N., Benjamin, New York 1968.

[27] E. Becker, Partial Orders on a Field and Valuation Rings, Comm. in Algebra 7, 1979, 1933-1976.

[28] G. Bergman, Zero-divisors in Tensor Products, Kent State, L.N.M. 545, Springer Verlag, Berlin 1977.

[29] G. Bergman, L. Small, P.I. Degrees and Prime Ideals, J. of Algebra 33, 1975, 435-462.

[30] A. Borel, Linear Algebraic Groups, Benjamin, New York 1969.

[31] N. Bourbaki, Commutative Algebra, Hermann, Paris 1972.

[32] N. Bourbaki, Algèbre, chap. I, II, III, Hermann, Paris 1973.

[33] G. Cauchon, Les T-anneaux et les anneaux à identités polynomiales Noethériens Thèse, Univ. Paris XI, Centre d'Orsay, 1977.

[34] A.W. Chatters, Localization of P.I. Rings, J. London Math. Soc. 2, 1970, 763-768.

[35] A.W. Chatters, W.M. Ginn, Localization in Hereditary Rings, J. of Algebra 22, 1972, 82-88.

[36] A.W. Chatters, A.G. Heinicke, Localization at a Torsion Theory in Hereditary Noetherian Rings, Proc. London Math. Soc. 27, 1973, 193-204.

[37] C. Chevalley, Introduction to the Theory of Algebraic Functions of one Variable, A.M.S. Math. Surv. VI, 1951.

[38] L. Childs, The Brauer Group of Some Normal Local Rings, Evanston 1975, L.N.M. 549, Springer Verlag, Berlin 1976, 1-15.

[39] P.M. Cohn, Algebra II, J. Wiley and Sons, London 1977.

[40] P.M. Cohn, Skewfield constructions, L.M.S. Lect. Notes 27, Cambridge Univ. Press, London 1977.

[41] P.M. Cohn, The Affine Scheme of a General Ring, L.N.M. 753, Springer Verlag, Berlin 1979, 197-211.

[42] I. Connell, A Natural Transform of the Spec Functor, J. of Algebra 10, 1968, 69-91.

[43] J. Cozzens, C. Faith, Simple Noetherian Rings, Cambridge Univ. Press, Cambridge, 1975.

[44] J. Dauns, K. Hofmann, The Representation of Biregular Rings by Sheaves, Math. Zeitschr.91, 1971, 103-123.

[45] J.P. Delale, Sur le spectre d'un anneau noncommutatif, Sem. d'Alg. noncomm. 44, Publ. Math. d'Orsay n°4, 1973.

[46] J.P. Delale, Sur le spectre d'un anneau noncommutatif, Thèse Univ. de Paris-Sud, Centre d'Orsay 1974.

[47] F. De Meyer, The Brauer Group of Affine Curves, Evanston 1975, L.N.M. 549, Springer Verlag, Berlin 1976, 16-24.

[48] F. De Meyer, E. Ingraham, Separable Algebras over Commutative Rings, L.N.M. 181, Springer Verlag, Berlin 1970.

[49] M. Deuring, Algebren, Springer Verlag, Berlin 1968.

[50] M. Deuring, Lectures on the Theory of Algebraic Functions in One Variable, L.N.M. 394, Springer Verlag, Berlin 1973.

[51] J. Dieudonné, Cours de Géométrie Algébrique, P.U.F. Paris 1976.

[52] J. Dieudonné, A. Grothendieck, Elements de Géométrie Algébrique, I, II, III, Publ. Math. I.H.E.S., 1960, 1961,1963.

[53] D. Eisenbud, J.C. Robson, Modules over Dedekind Prime Rings, J. of Algebra 16, 1970, 67-85.

[54] D. Eisenbud, J.C. Robson, Hereditary Noetherian Prime Rings, J. of Algebra 16, 1970, 86-104.

[55] C. Faith, Algebra.Rings, Modules and Categories, I, Springer Verlag, Berlin, 1973.

[56] C. Faith, Algebra II, Ring Theory, Springer Verlag, Berlin 1976.

[57] E. Formanek, Central Polynomials for Matrix Rings, J. of Algebra 23, 1972, 129-133.

[58] E. Formanek, Noetherian P.I. Rings, Comm. Algebra 1, 1974, 79-86.

[59] E. Formanek, The Center of the Ring of 3x3 Generic Matrices, Inst. Adv. Study, Hebrew Univ., report 10/78.

[60] W. Fulton, Algebraic Curves, Benjamin, Reading Mass. 1969.

[61] P. Gabriel, Des Catégories Abéliennes, Bull. Soc. Math. France 90, 1962, 323-448.

[62] R. Godement, Théorie des Faisceaux, Hermann, Paris 1958.

[63] R. Godement, Les Fonctions ζ des Algèbres Simples, I, II, Séminaire Bourbaki 1958, 171-01-171-23 et 176-01-176-20.

[64] J. Golan, Localization of noncommutative Rings, Marcel Dekker, New York, 1975.

[65] J. Golan, J. Raynaud, F. Van Oystaeyen, Sheaves over the Spectra of Certain noncommutative Rings, Comm. Alg. 4, 1976, 491-502.

[66] A. Goldie, Localization in Non-commutative Noetherian Rings, J. of Algebra 5, 1967, 89-105.

[67] A. Goldie, Some Aspects of Ring Theory, Bull London Math. Soc. 1, 1969, 129-154.

[68] A. Goldie, The Structure of Noetherian Rings, Lectures on Rings and Modules, Tulane Univ., L.N.M. 246, Springer Verlag, Berlin 1972.

[69] O. Goldman, Hilbert Rings and the Hilbert Nullstellensatz, Math. Zeit 54, 1951, 136-140.

[70] O. Goldman, Rings and Modules of Quotients, J. of Algebra 13, 1969, 10-47.

[71] O. Goldman, Elements of Noncommutative Arithmetic I, J. of Algebra 35, 1975, 308-341.

[72] K. Goodearl, Ring Theory, Marcel Dekker, New York, 1978.

[73] A. Grothendieck, Sur Quelques Points d'Algèbre Homologique, Tohoku Math. J., 1958, 119-221.

[74] A. Grothendieck, Le Groupe de Brauer, I, Dix Exposés sur la Cohomologie des Schémas, North Holland, Amsterdam 1968, 46-65.

[75] W. Gwynne, J.C. Robson, Completions of Non-commutative Dedekind Prime Rings, J. London Math. Soc. 4, 1971, 346-352.

[76] C.R. Hajarnavis, T. Lenagan, Localization in Asano Orders, J. of Algebra 21, 1972, 441-449.

[77] M. Hall, Jr., The Theory of Groups, MacMillan Co., New York, 1959.

[78] D.K. Harrison, Finite and Infinite Primes for Rings and Fields, Mem. Ann. Math. Soc. vol. 68, 1974.

[79] R. Hartshorne, Algebraic Geometry, Springer Verlag, Berlin 1977.

[80] A. Heinicke, On the Ring of Quotients at a Prime Ideal of a Right Noetherian Ring, Canad. J. Math. 24, 1972, 703-712.

[81] I. Herstein, Noncommutative Rings, Carus Monograph 15, Math. Assoc. of Am. 1968.

[82] I. Herstein, Topics in Ring Theory, Univ. of Chicago Press, 1969.

[83] M. Hochster, Prime Ideal Structure in Commutative Rings, Trans. Amer. Math. Soc. 142, 1969, 43-60.

[84] K. Hofmann, Representations of Algebras by Continuous Sections, Bull. Amer. Math. Soc. 78, 1972, 291-373.

[85] H.S. Huang, Rings of Quotients in Ring Extensions with Central Basis, Chinese J. of Math. 4, n°1, 1976, p. 1.

[86] B. Iversen, Noetherian Graded Modules, I, Aarhus preprint 29, 1972.

[87] N. Jacobson, The Theory of Rings, Am. Math. Soc. Surveys II, Amer. Math. Soc., Providence 1943.

[88] N. Jacobson, P.I. Algebras, An Introduction, L.N.M. 441, Springer Verlag, Berlin 1975.

[89] A.V. Jategaonkar, Jacobson's Conjuncture and Modules over Fully Bounded Noetherian Rings, J. of Algebra 30, 1974, 103-121.

[90] A.V. Jategaonkar, Principal Ideal Theorem for Noetherian P.I. Rings, J. of Algebra 35, 1975, 17-22.

[91] J. Joyce, Finnegans Wake, Faber and Faber, London 1939.

[92] I. Kaplansky, Rings with Polynomial Identity, Bull. Amer. Math. Soc. 54, 1948, 575-580.

[93] I. Kaplansky, Fields and Rings, Univ. of Chicago Press, 1963.

[94] E. Kirkman, J. Kuzmanovich, Orders over Hereditary Rings, J. of Algebra 55, 1978, 1-27.

[95] M. Knus, M. Ojanguren, Théorie de la Descente et Algèbres d'Azumaya, L.N.M. 389, Springer Verlag, Berlin 1974.

[96] G. Krause, On Fully Bounded Left Noetherian Rings, J. of Algebra 23, 1972, 88-99.

[97] W. Krull, Jacobsonsche Ringe, Hilbertscher Nullstellensatz, Dimension Theorie, Math. Zeit. 54, 1951, 354-387.

[98] J. Kuzmanovich, Localization of Dedekind Prime Rings, J. of Algebra 21, 1972, 378-393.

[99] J. Lambek, Lectures on Rings and Modules, Waltham, Toronto 1966.

[100] J. Lambek, Torsion Theories, Additive Semantics and Rings of Quotients, L.N.M. 197, Springer Verlag, Berlin 1971.

[101] J. Lambek, Noncommutative Localization, Bull. Am. Math. Soc. 79, 1973, 857-872.

[102] J. Lambek, G. Michler, The Torsion Theory at a Prime Ideal of a Right Noetherian Ring, J. of Algebra 25, 1973, 364-389.

[103] J. Lambek, B. Rattray, Localization and Duality in Additive Categories, Houston J. of Math. 1, 1975, 87-100.

[104] T. Lenagan, Bounded Asano Orders are Hereditary, Bull. London Math. Soc. 3, 1971, 67-69.

[105] T. Lenagan, Bounded Hereditary Noetherian Prime Rings, J. London Math. Soc. 6, 1973, 241-246.

[106] P. Lestman, Simple Going Down in P.I.Rings, Proc. Amer. Math. Soc. 63, 1977, 41-45.

[107] P. Lestman, Going Down and the Spec Map in P.I. Rings, Comm.Alg. 6, 1978, 1567-1692

[108] J. Levitzki, A Theorem on Polynomial Identities, Proc. Amer. Math. Soc. 1, 1950, 334-341.

[109] S. MacLane, Category Theory for the Working Mathematician, G.T.M. 5, Springer Verlag, Berlin 1971.

[110] A. Magid, Pierce's Representation and Separable Algebras, Ill. J. of Math. 15, 1971, 114-121.

[111] A. Magid, The Separable Galois Theory of Commutative Rings, Marcel Dekker, New York 1974.

[112] W. Martindale, Prime Rings Satisfying a Generalized Polynomial Identity, J. of Algebra 12, 1969, 576-584.

[113] H. Matsumura, Commutative Algebra, Benjamin 1970.

[114] G. Michler, Asano Orders, Proc. London Math. Soc. 19, 1969, 421-443.

[115] B. Mueller, Localization in Fully Bounded Noetherian Rings, Mc. Master Univ. report 78, 1975.

[116] B. Mueller, Localization of Non-commutative Noetherian Rings at Semiprime Ideals, Lect. Notes, Mc. Master Univ. 1974.

[117] B. Mueller, Ideal Invariance and Localization, Comm. Algebra 6, 1979, 415-441.

[118] D. Mumford, Introduction to Algebraic Geometry, Harvard Lect. Notes.

[119] D. Mumford, Algebraic Geometry I, Complex Projective Varieties, Springer Verlag, New York, 1973.

[120] D. Murdoch, F. Van Oystaeyen, Noncommutative Localization and Sheaves, J. of Algebra 35, 1975, 500-515.

[121] C. Năstăsescu, F. Van Oystaeyen, Graded and Filtered Rings and Modules, L.N.M. 758, Springer Verlag, Berlin 1979.

[122] E. Nauwelaerts, Localization of P.I. Algebras, Comm. Algebra to appear soon ; U.I.A. preprint 1977.

[123] E. Nauwelaerts, Symmetric Localization of Asano Orders, Ring Theory at U.I.A. 1977, Lect. Notes vol. 40, Marcel Dekker, New York 1978, 65-80.

[124] E. Nauwelaerts, Zariski Extensions of Rings, Ph. D. Thesis, Univ. of Antwerp U.I.A., 1979.

[125] E. Nauwelaerts, F. Van Oystaeyen, Birational Hereditary Prime Rings, Comm. Algebra 8, 1980, 309-338.

[126] E. Nauwelaerts, F. Van Oystaeyen, Zariski Extensions and Biregular Rings, to appear in Israel J. of Math.

[127] B. Osofsky, On Twisted Polynomial Rings, J. of Algebra 18, 1971, 597-607.

[128] N. Popescu, Abelian Categories with Applications to Rings and Modules, Academic Press 1973.

[129] E. Posner, Prime Rings Satisfying a Polynomial Identity, Proc. Am. Math. Soc. 11, 1960, 180-182.

[130] C. Procesi, Noncommutative Affine Rings, Atti. Acc. Naz. Lincei, 1.VIII, 5.VIII, 6, 1967, 239-255.

[131] C. Procesi, Noncommutative Jacobson Rings, Annale Sci. Norm. Sup. Pisa, 5.XXI, II, 1967, 381-390.

[132] C. Procesi, Sugli anelli commutativi zero dimensionali con identitá polinomiale, Rend. Circolo Mat. Palermo, s.II, T.XVII, 1968, 5-11.

[133] C. Procesi, Sulle identita delle algebre semplici, Rend. Circolo Mat. Palermo, s.II, T.XVII, 1968, 13-18.

[134] C. Procesi, On a Theorem of M. Artin, J. of Algebra 22, 1972, 309-315.

[135] C. Procesi, Sulla rappresentazioni degli anelli e loro invariante, Ist. Naz. Alta. Mat., Symposia Matematica XI, 1973, 143-159.

[136] C. Procesi, Rings with Polynomial Identities, Marcel Dekker, New York, 1973.

[137] C. Procesi, The Invariant Theory of nxn Matrices, Advances in Math. 19, 3, 1976, 306-381.

[138] C. Procesi, L. Small, Endomorphism Rings of Modules over P.I. Algebras Math. Zeit. 106, 1968, 178-180.

[139] J. Raynaud, Localisations et Spectres d'Anneaux, Thèse, Univ. Claude-Bernard, Lyon I, France.

[140] Yu. P. Razmyslov, On a Problem of Kaplansky, Math.U.S.S.R.Izv. vol. 7, 1973, 479-496.

[141] Yu. P. Razmyslov, Trace Identities of Full Matrix Algebras over a Field of Characteristic Zero, Math. U.S.S.R. Izv. 8, 1974, n° 4.

[142] Yu. P. Razmyslov, The Jacobson Radical in P.I. Algebras, Algebra and Logic 13, 1974, 337-360 (in Russian), transl. Algebra and Logic, 1975, 394-399.

[143] A. Regev, Existence of Identities in A⊗B, Israel J. of Math. 11, 1972, 131-152.

[144] I. Reiner, Maximal Orders, Academic Press, New York, 1975.

[145] P. Ribenboim, Rings and Modules, Tracts in Math. 24, Interscience, New York, 1969.

[146] L. Risman, Non-cyclic Division Algebras, J. Pure Applied Algebra 11, 1977, 199-215.

[147] J.C. Robson, Non-commutative Dedekind Rings, J. of Algebra 9, 1968, 249-265.

[148] J.C. Robson, Idealizers and Hereditary Noetherian Prime Rings, J. of Algebra, 22, 1972, 45-81.

[149] J.C. Robson, L. Small, Hereditary Prime P.I. Rings are Classical Hereditary Orders, J. London Math. Soc. 8, 1974, 499-503.

[150] J.C. Robson, L. Small, Idempotent Ideals in P.I. Rings, J. London Math. Soc. 14, 1976, 120-122.

[151] L. Rowen, Some Results on the Center of a Ring with Polynomial Identity, Bull. Amer. Math. Soc. 79, 1973, 219-223.

[152] L. Rowen, On Rings with Central Polynomials, J. of Algebra, 31, 1974, 393-426.

[153] L. Rowen, Classes of Rings Torsion-free over their Centers, Pac. J. of Math. 69, 1977, 527-534.

[154] W. Schelter, Integral Extensions of Rings Satisfying a Polynomial Identity, J. of Algebra 40, 1976, 245-257.

[155] W. Schelter, Affine P.I. Rings are Catenary, Bull. Amer. Math. Soc. 83, 1977, 1309-1310.

[156] W. Schelter, Affine P.I. Rings are Catenary, J. of Algebra 51, 1978, 12-18.

[157] O. Schilling, The Theory of Valuations, A.M.S. Math. Survey IV, 1950.

[158] F.K. Schmidt, Zur Arithmetischen Theorie der Algebraischen Funktionen I, Math. Zeitschr. 41, 1936.

[159] J.P. Serre, Faisceaux Algébriques Cohérents, Ann. of Math. 61, 1955, 197-278.

[160] J.P. Serre, Géométrie Algébrique et Géométrie Analytique, Ann. Inst. Fourier 6, 1956, 1-42.

[161] Sibirskii, Algebraic Invariants of a System of Matrices, Sib. Mat. Z. 9, 1968, 152-164.

[162] L. Silver, Noncommutative Localization and Applications, J. of Algebra 7, 1967, 44-76.

[163] S.K. Sim, Prime Ideals and Symmetric Idempotent Kernel Functors, Nanta Math. 9, 1976, 121-124.

[164] L. Small, An Example in P.I. Rings, J. of Algebra, 17, 1971, 434-436.

[165] L. Small, Localization of P.I. Rings, J. of Algebra 18, 1971, 269-270.

[166] L. Small, Lecture, Proceedings of the Durham Symposium, 1979.

[167] B. Stenström, Rings of Quotients, An Introduction to Methods of Ring Theory, Springer Verlag, Berlin 1975.

[168] B.L. Van Der Waerden, Algebra, Springer Verlag, Berlin 1966.

[169] J.P. Van Deuren, J. Van Geel, F. Van Oystaeyen, Genus and the Riemann-Roch Theorem for Noncommutative Function Fields in One Variable, Sem. Dubreil, 1979 to appear, Springer Verlag, Berlin 1980.

[170] J. Van Geel, Primes in Algebras and the Arithmetic of Central Simple Algebras, Comm. Algebra, to appear.

[171] J. Van Geel, Primes and Value Functions, Ph. D. Thesis, University of Antwerp, U.I.A., 1980.

[172] F. Van Oystaeyen, Generic Division Algebras, Bull. Soc. Math. Belg. XXV, 1973, 259-285.

[173] F. Van Oystaeyen, Extensions of Ideals under Symmetric Localization, J. Pure and Applied Alg. 6, 1975, 275-283.

[174] F. Van Oystaeyen, Prime Spectra in Non-commutative Algebra, L.N.M. 444, Springer Verlag, Berlin 1975.

[175] F. Van Oystaeyen, Pointwise Localization in Presheaf Categories, Indag. Math. 80, 1976, 114-121.

[176] F. Van Oystaeyen, Compatibility of Kernel Functors and Localization Functors, Bull. Soc. Math. Belg. XVIII, 1976, 131-137.

[177] F. Van Oystaeyen, Localization of Fully Left Bounded Rings, Comm. Algebra 4, 1976, 271-284.

[178] F. Van Oystaeyen, Primes in Algebras over Fields, J. Pure Applied Alg. 5, 1977, 239-252.

[179] F. Van Oystaeyen, Stalks of Sheaves over the Spectra of Certain Noncommutative Rings, Comm. Algebra 5, 1977, 899-901.

[180] F. Van Oystaeyen, Zariski Central Rings, Comm. Algebra 6, 1978, 799-821.

[181] F. Van Oystaeyen, On Graded Rings and Modules of Quotients, Comm. Algebra 6, 1978, 1923-1959.

[182] F. Van Oystaeyen, Graded and Non-graded Birational Extensions, Ring Theory 1977, Lct. Notes 40, Marcel Dekker, New York 1978, 155-180.

[183] F. Van Oystaeyen, Birational Extensions of Rings, Ring Theory 1978, Lect. Notes 51, Marcel Dekker, New York 1979, 287-328.

[184] F. Van Oystaeyen, Graded Prime Ideals and the Left Ore Conditions, Comm. Algebra 8, 1980, 861-868.

[185] F. Van Oystaeyen, Arithmetically Graded Rings and Generalized Rees Rings, J. of Algebra to appear soon.

[186] F. Van Oystaeyen, On Brauer Groups of Graded Rings, in Ring Theory 1980, L.N.M. 825 Springer Verlag, Berlin, 1980

[187] F. Van Oystaeyen, A. Verschoren, Localization of Presheaves of Modules, Indag. Math. 79, 1976, 335-348.

[188] F. Van Oystaeyen, A. Verschoren, Reflectors and Localization. Application to Sheaf Theory. Lect. Notes 41, Marcel Dekker, New York 1978.

[189] F. Van Oystaeyen, A. Verschoren, Relative Localizations, Bimodules and Semiprime P.I. Rings, Comm. Algebra 7, 1979, 955-988.

[190] F. Van Oystaeyen, A. Verschoren, Fully Bounded Grothendieck Categories. Part II. Graded Modules, J. Pure Applied Alg. 21, 1981, 189-203

[191] A. Verschoren, A note on Strictly Local Kernel Functors, Bull. Soc. Math. Belg. XVIII, 1976, 133-146.

[192] A. Verschoren, Localization and the Gabriel-Popescu Embedding, Comm. Algebra 6, 1978, 1563-1587.

[193] A. Verschoren, Perfect Localizations and Torsion-free Extensions, Comm. Algebra 8, 1980, 839-860.

[194] A. Verschoren, Some Ideas in Noncommutative Algebraic Geometry, Ph. D. Thesis, Univ. of Antwerp U.I.A. 1979.

[195] A. Verschoren, Birationality of P.I. Rings and Noncommutative Algebraic Varieties, Ring Theory 1980, L.N.M. 825 Springer Verlag, Berlin 1980

[196] A. Verschoren, Pour une Géométrive Algébrique non-commutative, Sem. Dubreil 1979, L.N.M. Springer Verlag Berlin, to appear.

[197] A. Verschoren, Localization of Bimodules and P.I. Rings, Ring Theory 1978, Lect. Notes 51, Marcel Dekker, New York 1979, 329-351.

[198] A. Verschoren, Les Extensions et les Schémas Non Commutatifs, Publ. Math. Debrecen, to appear soon.

[199] A. Verschoren, When is a Noncommutative Scheme Affine ? Bull. Soc. Math. Belg., 32, 1980, 47-56

[200] A. Verschoren, F. Van Oystaeyen,Localization of Sheaves of Modules, Indag. Math. 79, 1976, 470-481.

[201] D. Webber, Ideals and Moduls of Simple Noetherian Hereditary Rings, J. of Algebra 16, 1970, 239-242.

[202] E. Witt, Riemann-Rochser Satz und z-Funktionen in Hyperkomplexen, Math. Ann. 110, 1934, 12-28.

[203] O. Zariski, Algebraic Sheaf Theory, A.M.S. 2nd Summer Inst., 1956, 117-141.

[204] O. Zariski, An Introduction to the Theory of Algebraic Surfaces, L.N.M. 83, Springer Verlag, Berlin 1969.

[205] O. Zariski, P. Samuel, Commutative Algebra, I, II, Van Nostrand Publ., London 1960.

INDEX

Vol. 728: Non-Commutative Harmonic Analysis. Proceedings, 1978. Edited by J. Carmona and M. Vergne. V, 244 pages. 1979.

Vol. 729: Ergodic Theory. Proceedings, 1978. Edited by M. Denker and K. Jacobs. XII, 209 pages. 1979.

Vol. 730: Functional Differential Equations and Approximation of Fixed Points. Proceedings, 1978. Edited by H.-O. Peitgen and H.-O. Walther. XV, 503 pages. 1979.

Vol. 731: Y. Nakagami and M. Takesaki, Duality for Crossed Products of von Neumann Algebras. IX, 139 pages. 1979.

Vol. 732: Algebraic Geometry. Proceedings, 1978. Edited by K. Lønsted. IV, 658 pages. 1979.

Vol. 733: F. Bloom, Modern Differential Geometric Techniques in the Theory of Continuous Distributions of Dislocations. XII, 206 pages. 1979.

Vol. 734: Ring Theory, Waterloo, 1978. Proceedings, 1978. Edited by D. Handelman and J. Lawrence. XI, 352 pages. 1979.

Vol. 735: B. Aupetit, Propriétés Spectrales des Algèbres de Banach. XII, 192 pages. 1979.

Vol. 736: E. Behrends, M-Structure and the Banach-Stone Theorem. X, 217 pages. 1979.

Vol. 737: Volterra Equations. Proceedings 1978. Edited by S.-O. Londen and O. J. Staffans. VIII, 314 pages. 1979.

Vol. 738: P. E. Conner, Differentiable Periodic Maps. 2nd edition, IV, 181 pages. 1979.

Vol. 739: Analyse Harmonique sur les Groupes de Lie II. Proceedings, 1976-78. Edited by P. Eymard et al. VI, 646 pages. 1979.

Vol. 740: Séminaire d'Algèbre Paul Dubreil. Proceedings, 1977-78. Edited by M.-P. Malliavin. V, 456 pages. 1979.

Vol. 741: Algebraic Topology, Waterloo 1978. Proceedings. Edited by P. Hoffman and V. Snaith. XI, 655 pages. 1979.

Vol. 742: K. Clancey, Seminormal Operators. VII, 125 pages. 1979.

Vol. 743: Romanian-Finnish Seminar on Complex Analysis. Proceedings, 1976. Edited by C. Andreian Cazacu et al. XVI, 713 pages. 1979.

Vol. 744: I. Reiner and K. W. Roggenkamp, Integral Representations. VIII, 275 pages. 1979.

Vol. 745: D. K. Haley, Equational Compactness in Rings. III, 167 pages. 1979.

Vol. 746: P. Hoffman, τ-Rings and Wreath Product Representations. V, 148 pages. 1979.

Vol. 747: Complex Analysis, Joensuu 1978. Proceedings, 1978. Edited by I. Laine, O. Lehto and T. Sorvali. XV, 450 pages. 1979.

Vol. 748: Combinatorial Mathematics VI. Proceedings, 1978. Edited by A. F. Horadam and W. D. Wallis. IX, 206 pages. 1979.

Vol. 749: V. Girault and P.-A. Raviart, Finite Element Approximation of the Navier-Stokes Equations. VII, 200 pages. 1979.

Vol. 750: J. C. Jantzen, Moduln mit einem höchsten Gewicht. III, 195 Seiten. 1979.

Vol. 751: Number Theory, Carbondale 1979. Proceedings. Edited by M. B. Nathanson. V, 342 pages. 1979.

Vol. 752: M. Barr, *-Autonomous Categories. VI, 140 pages. 1979.

Vol. 753: Applications of Sheaves. Proceedings, 1977. Edited by M. Fourman, C. Mulvey and D. Scott. XIV, 779 pages. 1979.

Vol. 754: O. A. Laudal, Formal Moduli of Algebraic Structures. III, 161 pages. 1979.

Vol. 755: Global Analysis. Proceedings, 1978. Edited by M. Grmela and J. E. Marsden. VII, 377 pages. 1979.

Vol. 756: H. O. Cordes, Elliptic Pseudo-Differential Operators – An Abstract Theory. IX, 331 pages. 1979.

Vol. 757: Smoothing Techniques for Curve Estimation. Proceedings, 1979. Edited by Th. Gasser and M. Rosenblatt. V, 245 pages. 1979.

Vol. 758: C. Năstăsescu and F. Van Oystaeyen; Graded and Filtered Rings and Modules. X, 148 pages. 1979.

Vol. 759: R. L. Epstein, Degrees of Unsolvability: Structure and Theory. XIV, 216 pages. 1979.

Vol. 760: H.-O. Georgii, Canonical Gibbs Measures. VIII, 190 pages. 1979.

Vol. 761: K. Johannson, Homotopy Equivalences of 3-Manifolds with Boundaries. 2, 303 pages. 1979.

Vol. 762: D. H. Sattinger, Group Theoretic Methods in Bifurcation Theory. V, 241 pages. 1979.

Vol. 763: Algebraic Topology, Aarhus 1978. Proceedings, 1978. Edited by J. L. Dupont and H. Madsen. VI, 695 pages. 1979.

Vol. 764: B. Srinivasan, Representations of Finite Chevalley Groups. XI, 177 pages. 1979.

Vol. 765: Padé Approximation and its Applications. Proceedings, 1979. Edited by L. Wuytack. VI, 392 pages. 1979.

Vol. 766: T. tom Dieck, Transformation Groups and Representation Theory. VIII, 309 pages. 1979.

Vol. 767: M. Namba, Families of Meromorphic Functions on Compact Riemann Surfaces. XII, 284 pages. 1979.

Vol. 768: R. S. Doran and J. Wichmann, Approximate Identities and Factorization in Banach Modules. X, 305 pages. 1979.

Vol. 769: J. Flum, M. Ziegler, Topological Model Theory. X, 151 pages. 1980.

Vol. 770: Séminaire Bourbaki vol. 1978/79 Exposés 525-542. IV, 341 pages. 1980.

Vol. 771: Approximation Methods for Navier-Stokes Problems. Proceedings, 1979. Edited by R. Rautmann. XVI, 581 pages. 1980.

Vol. 772: J. P. Levine, Algebraic Structure of Knot Modules. XI, 104 pages. 1980.

Vol. 773: Numerical Analysis. Proceedings, 1979. Edited by G. A. Watson. X, 184 pages. 1980.

Vol. 774: R. Azencott, Y. Guivarc'h, R. F. Gundy, Ecole d'Eté de Probabilités de Saint-Flour VIII-1978. Edited by P. L. Hennequin. XIII, 334 pages. 1980.

Vol. 775: Geometric Methods in Mathematical Physics. Proceedings, 1979. Edited by G. Kaiser and J. E. Marsden. VII, 257 pages. 1980.

Vol. 776: B. Gross, Arithmetic on Elliptic Curves with Complex Multiplication. V, 95 pages. 1980.

Vol. 777: Séminaire sur les Singularités des Surfaces. Proceedings, 1976-1977. Edited by M. Demazure, H. Pinkham and B. Teissier. IX, 339 pages. 1980.

Vol. 778: SK_1 von Schiefkörpern. Proceedings, 1976. Edited by P. Draxl and M. Kneser. II, 124 pages. 1980.

Vol. 779: Euclidean Harmonic Analysis. Proceedings, 1979. Edited by J. J. Benedetto. III, 177 pages. 1980.

Vol. 780: L. Schwartz, Semi-Martingales sur des Variétés, et Martingales Conformes sur des Variétés Analytiques Complexes. XV, 132 pages. 1980.

Vol. 781: Harmonic Analysis Iraklion 1978. Proceedings 1978. Edited by N. Petridis, S. K. Pichorides and N. Varopoulos. V, 213 pages. 1980.

Vol. 782: Bifurcation and Nonlinear Eigenvalue Problems. Proceedings, 1978. Edited by C. Bardos, J. M. Lasry and M. Schatzman. VIII, 296 pages. 1980.

Vol. 783: A. Dinghas, Wertverteilung meromorpher Funktionen in ein- und mehrfach zusammenhängenden Gebieten. Edited by R. Nevanlinna and C. Andreian Cazacu. XIII, 145 pages. 1980.

Vol. 784: Séminaire de Probabilités XIV. Proceedings, 1978/79. Edited by J. Azéma and M. Yor. VIII, 546 pages. 1980.

Vol. 785: W. M. Schmidt, Diophantine Approximation. X, 299 pages. 1980.

Vol. 786: I. J. Maddox, Infinite Matrices of Operators. V, 122 pages. 1980.

Vol. 787: Potential Theory, Copenhagen 1979. Proceedings, 1979. Edited by C. Berg, G. Forst and B. Fuglede. VIII, 319 pages. 1980.

Vol. 788: Topology Symposium, Siegen 1979. Proceedings, 1979. Edited by U. Koschorke and W. D. Neumann. VIII, 495 pages. 1980.

Vol. 789: J. E. Humphreys, Arithmetic Groups. VII, 158 pages. 1980.

Vol. 790: W. Dicks, Groups, Trees and Projective Modules. IX, 127 pages. 1980.

Vol. 791: K. W. Bauer and S. Ruscheweyh, Differential Operators for Partial Differential Equations and Function Theoretic Applications. V, 258 pages. 1980.

Vol. 792: Geometry and Differential Geometry. Proceedings, 1979. Edited by R. Artzy and I. Vaisman. VI, 443 pages. 1980.

Vol. 793: J. Renault, A Groupoid Approach to C*-Algebras. III, 160 pages. 1980.

Vol. 794: Measure Theory, Oberwolfach 1979. Proceedings 1979. Edited by D. Kölzow. XV, 573 pages. 1980.

Vol. 795: Séminaire d'Algèbre Paul Dubreil et Marie-Paule Malliavin. Proceedings 1979. Edited by M. P. Malliavin. V, 433 pages. 1980.

Vol. 796: C. Constantinescu, Duality in Measure Theory. IV, 197 pages. 1980.

Vol. 797: S. Mäki, The Determination of Units in Real Cyclic Sextic Fields. III, 198 pages. 1980.

Vol. 798: Analytic Functions, Kozubnik 1979. Proceedings. Edited by J. Ławrynowicz. X, 476 pages. 1980.

Vol. 799: Functional Differential Equations and Bifurcation. Proceedings 1979. Edited by A. F. Izé. XXII, 409 pages. 1980.

Vol. 800: M.-F. Vignéras, Arithmétique des Algèbres de Quaternions. VII, 169 pages. 1980.

Vol. 801: K. Floret, Weakly Compact Sets. VII, 123 pages. 1980.

Vol. 802: J. Bair, R. Fourneau, Etude Géometrique des Espaces Vectoriels II. VII, 283 pages. 1980.

Vol. 803: F.-Y. Maeda, Dirichlet Integrals on Harmonic Spaces. X, 180 pages. 1980.

Vol. 804: M. Matsuda, First Order Algebraic Differential Equations. VII, 111 pages. 1980.

Vol. 805: O. Kowalski, Generalized Symmetric Spaces. XII, 187 pages. 1980.

Vol. 806: Burnside Groups. Proceedings, 1977. Edited by J. L. Mennicke. V, 274 pages. 1980.

Vol. 807: Fonctions de Plusieurs Variables Complexes IV. Proceedings, 1979. Edited by F. Norguet. IX, 198 pages. 1980.

Vol. 808: G. Maury et J. Raynaud, Ordres Maximaux au Sens de K. Asano. VIII, 192 pages. 1980.

Vol. 809: I. Gumowski and Ch. Mira, Recurences and Discrete Dynamic Systems. VI, 272 pages. 1980.

Vol. 810: Geometrical Approaches to Differential Equations. Proceedings 1979. Edited by R. Martini. VII, 339 pages. 1980.

Vol. 811: D. Normann, Recursion on the Countable Functionals. VIII, 191 pages. 1980.

Vol. 812: Y. Namikawa, Toroidal Compactification of Siegel Spaces. VIII, 162 pages. 1980.

Vol. 813: A. Campillo, Algebroid Curves in Positive Characteristic. V, 168 pages. 1980.

Vol. 814: Séminaire de Théorie du Potentiel, Paris, No. 5. Proceedings. Edited by F. Hirsch et G. Mokobodzki. IV, 239 pages. 1980.

Vol. 815: P. J. Slodowy, Simple Singularities and Simple Algebraic Groups. XI, 175 pages. 1980.

Vol. 816: L. Stoica, Local Operators and Markov Processes. VIII, 104 pages. 1980.

Vol. 817: L. Gerritzen, M. van der Put, Schottky Groups and Mumford Curves. VIII, 317 pages. 1980.

Vol. 818: S. Montgomery, Fixed Rings of Finite Automorphism Groups of Associative Rings. VII, 126 pages. 1980.

Vol. 819: Global Theory of Dynamical Systems. Proceedings, 1979. Edited by Z. Nitecki and C. Robinson. IX, 499 pages. 1980.

Vol. 820: W. Abikoff, The Real Analytic Theory of Teichmüller Space. VII, 144 pages. 1980.

Vol. 821: Statistique non Paramétrique Asymptotique. Proceedings, 1979. Edited by J.-P. Raoult. VII, 175 pages. 1980.

Vol. 822: Séminaire Pierre Lelong-Henri Skoda, (Analyse) Années 1978/79. Proceedings. Edited by P. Lelong et H. Skoda. VIII, 356 pages, 1980.

Vol. 823: J. Král, Integral Operators in Potential Theory. III, 171 pages. 1980.

Vol. 824: D. Frank Hsu, Cyclic Neofields and Combinatorial Designs. VI, 230 pages. 1980.

Vol. 825: Ring Theory, Antwerp 1980. Proceedings. Edited by F. van Oystaeyen. VII, 209 pages. 1980.

Vol. 826: Ph. G. Ciarlet et P. Rabier, Les Equations de von Kármán. VI, 181 pages. 1980.

Vol. 827: Ordinary and Partial Differential Equations. Proceedings, 1978. Edited by W. N. Everitt. XVI, 271 pages. 1980.

Vol. 828: Probability Theory on Vector Spaces II. Proceedings, 1979. Edited by A. Weron. XIII, 324 pages. 1980.

Vol. 829: Combinatorial Mathematics VII. Proceedings, 1979. Edited by R. W. Robinson et al.. X, 256 pages. 1980.

Vol. 830: J. A. Green, Polynomial Representations of GL_n. VI, 118 pages. 1980.

Vol. 831: Representation Theory I. Proceedings, 1979. Edited by V. Dlab and P. Gabriel. XIV, 373 pages. 1980.

Vol. 832: Representation Theory II. Proceedings, 1979. Edited by V. Dlab and P. Gabriel. XIV, 673 pages. 1980.

Vol. 833: Th. Jeulin, Semi-Martingales et Grossissement d'une Filtration. IX, 142 Seiten. 1980.

Vol. 834: Model Theory of Algebra and Arithmetic. Proceedings, 1979. Edited by L. Pacholski, J. Wierzejewski, and A. J. Wilkie. VI, 410 pages. 1980.

Vol. 835: H. Zieschang, E. Vogt and H.-D. Coldewey, Surfaces and Planar Discontinuous Groups. X, 334 pages. 1980.

Vol. 836: Differential Geometrical Methods in Mathematical Physics. Proceedings, 1979. Edited by P. L. García, A. Pérez-Rendón, and J. M. Souriau. XII, 538 pages. 1980.

Vol. 837: J. Meixner, F. W. Schäfke and G. Wolf, Mathieu Functions and Spheroidal Functions and their Mathematical Foundations Further Studies. VII, 126 pages. 1980.

Vol. 838: Global Differential Geometry and Global Analysis. Proceedings 1979. Edited by D. Ferus et al. XI, 299 pages. 1981.

Vol. 839: Cabal Seminar 77 – 79. Proceedings. Edited by A. S. Kechris, D. A. Martin and Y. N. Moschovakis. V, 274 pages. 1981.

Vol. 840: D. Henry, Geometric Theory of Semilinear Parabolic Equations. IV, 348 pages. 1981.

Vol. 841: A. Haraux, Nonlinear Evolution Equations- Global Behaviour of Solutions. XII, 313 pages. 1981.

Vol. 842: Séminaire Bourbaki vol. 1979/80. Exposés 543–560. IV, 317 pages. 1981.

Vol. 843: Functional Analysis, Holomorphy, and Approximation Theory. Proceedings. Edited by S. Machado. VI, 636 pages. 1981.

Vol. 844: Groupe de Brauer. Proceedings. Edited by M. Kervaire and M. Ojanguren. VII, 274 pages. 1981.

Vol. 845: A. Tannenbaum, Invariance and System Theory: Algebraic and Geometric Aspects. X, 161 pages. 1981.

Vol. 846: Ordinary and Partial Differential Equations, Proceedings. Edited by W. N. Everitt and B. D. Sleeman. XIV, 384 pages. 1981.

Vol. 847: U. Koschorke, Vector Fields and Other Vector Bundle Morphisms – A Singularity Approach. IV, 304 pages. 1981.

Vol. 848: Algebra, Carbondale 1980. Proceedings. Ed. by R. K. Amayo. VI, 298 pages. 1981.

Vol. 849: P. Major, Multiple Wiener-Itô Integrals. VII, 127 pages. 1981.

Vol. 850: Séminaire de Probabilités XV. 1979/80. Avec table générale des exposés de 1966/67 à 1978/79. Edited by J. Azéma and M. Yor. IV, 704 pages. 1981.

Vol. 851: Stochastic Integrals. Proceedings, 1980. Edited by D. Williams. IX, 540 pages. 1981.

Vol. 852: L. Schwartz, Geometry and Probability in Banach Spaces. X, 101 pages. 1981.

Vol. 853: N. Boboc, G. Bucur, A. Cornea, Order and Convexity in Potential Theory: H-Cones. IV, 286 pages. 1981.

Vol. 854: Algebraic K-Theory. Evanston 1980. Proceedings. Edited by E. M. Friedlander and M. R. Stein. V, 517 pages. 1981.

Vol. 855: Semigroups. Proceedings 1978. Edited by H. Jürgensen, M. Petrich and H. J. Weinert. V, 221 pages. 1981.

Vol. 856: R. Lascar, Propagation des Singularités des Solutions d'Equations Pseudo-Différentielles à Caractéristiques de Multiplicités Variables. VIII, 237 pages. 1981.

Vol. 857: M. Miyanishi. Non-complete Algebraic Surfaces. XVIII, 244 pages. 1981.

Vol. 858: E. A. Coddington, H. S. V. de Snoo: Regular Boundary Value Problems Associated with Pairs of Ordinary Differential Expressions. V, 225 pages. 1981.

Vol. 859: Logic Year 1979–80. Proceedings. Edited by M. Lerman, J. Schmerl and R. Soare. VIII, 326 pages. 1981.

Vol. 860: Probability in Banach Spaces III. Proceedings, 1980. Edited by A. Beck. VI, 329 pages. 1981.

Vol. 861: Analytical Methods in Probability Theory. Proceedings 1980. Edited by D. Dugué, E. Lukacs, V. K. Rohatgi. X, 183 pages. 1981.

Vol. 862: Algebraic Geometry. Proceedings 1980. Edited by A. Libgober and P. Wagreich. V, 281 pages. 1981.

Vol. 863: Processus Aléatoires à Deux Indices. Proceedings, 1980. Edited by H. Korezlioglu, G. Mazziotto and J. Szpirglas. V, 274 pages. 1981.

Vol. 864: Complex Analysis and Spectral Theory. Proceedings, 1979/80. Edited by V. P. Havin and N. K. Nikol'skii, VI, 480 pages. 1981.

Vol. 865: R. W. Bruggeman, Fourier Coefficients of Automorphic Forms. III, 201 pages. 1981.

Vol. 866: J.-M. Bismut, Mécanique Aléatoire. XVI, 563 pages. 1981.

Vol. 867: Séminaire d'Algèbre Paul Dubreil et Marie-Paule Malliavin. Proceedings, 1980. Edited by M.-P. Malliavin. V, 476 pages. 1981.

Vol. 868: Surfaces Algébriques. Proceedings 1976–78. Edited by J. Giraud, L. Illusie et M. Raynaud. V, 314 pages. 1981.

Vol. 869: A. V. Zelevinsky, Representations of Finite Classical Groups. IV, 184 pages. 1981.

Vol. 870: Shape Theory and Geometric Topology. Proceedings, 1981. Edited by S. Mardešić and J. Segal. V, 265 pages. 1981.

Vol. 871: Continuous Lattices. Proceedings, 1979. Edited by B. Banaschewski and R.-E. Hoffmann. X, 413 pages. 1981.

Vol. 872: Set Theory and Model Theory. Proceedings, 1979. Edited by R. B. Jensen and A. Prestel. V, 174 pages. 1981.

Vol. 873: Constructive Mathematics, Proceedings, 1980. Edited by F. Richman. VII, 347 pages. 1981.

Vol. 874: Abelian Group Theory. Proceedings, 1981. Edited by R. Göbel and E. Walker. XXI, 447 pages. 1981.

Vol. 875: H. Zieschang, Finite Groups of Mapping Classes of Surfaces. VIII, 340 pages. 1981.

Vol. 876: J. P. Bickel, N. El Karoui and M. Yor. Ecole d'Eté de Probabilités de Saint-Flour IX – 1979. Edited by P. L. Hennequin. XI, 280 pages. 1981.

Vol. 877: J. Erven, B.-J. Falkowski, Low Order Cohomology and Applications. VI, 126 pages. 1981.

Vol. 878: Numerical Solution of Nonlinear Equations. Proceedings, 1980. Edited by E. L. Allgower, K. Glashoff, and H.-O. Peitgen. XIV, 440 pages. 1981.

Vol. 879: V. V. Sazonov, Normal Approximation – Some Recent Advances. VII, 105 pages. 1981.

Vol. 880: Non Commutative Harmonic Analysis and Lie Groups. Proceedings, 1980. Edited by J. Carmona and M. Vergne. IV, 553 pages. 1981.

Vol. 881: R. Lutz, M. Goze, Nonstandard Analysis. XIV, 261 pages. 1981.

Vol. 882: Integral Representations and Applications. Proceedings, 1980. Edited by K. Roggenkamp. XII, 479 pages. 1981.

Vol. 883: Cylindric Set Algebras. By L. Henkin, J. D. Monk, A. Tarski, H. Andréka, and I. Németi. VII, 323 pages. 1981.

Vol. 884: Combinatorial Mathematics VIII. Proceedings, 1980. Edited by K. L. McAvaney. XIII, 359 pages. 1981.

Vol. 885: Combinatorics and Graph Theory. Edited by S. B. Rao. Proceedings, 1980. VII, 500 pages. 1981.

Vol. 886: Fixed Point Theory. Proceedings, 1980. Edited by E. Fadell and G. Fournier. XII, 511 pages. 1981.

Vol. 887: F. M. J. van Oystaeyen, A. H. M. J. Verschoren, Non-commutative Algebraic Geometry. VI, 404 pages. 1981.